Sleep and Psychosomatic Medicine

Second Edition

Sleep and Psychosomatic Medicine

Second Edition

S.R. Pandi-Perumal
President and CEO
Somnogen Inc.
New York, USA

Meera Narasimhan, MD
University of South Carolina School of Medicine
Columbia, USA

Milton Kramer
Emeritus Professor of Psychiatry
University of Cincinnati, USA

 CRC Press
Taylor & Francis Group
Boca Raton London New York

CRC Press is an imprint of the
Taylor & Francis Group, an **informa** business

CRC Press
Taylor & Francis Group
6000 Broken Sound Parkway NW, Suite 300
Boca Raton, FL 33487-2742

First issued in paperback 2020

© 2016 by Taylor & Francis Group, LLC
CRC Press is an imprint of Taylor & Francis Group, an Informa business

No claim to original U.S. Government works

ISBN 13: 978-0-367-57479-6 (pbk)
ISBN 13: 978-1-4987-3728-9 (hbk)

Visit the Taylor & Francis Web site at
http://www.taylorandfrancis.com

and the CRC Press Web site at
http://www.crcpress.com

To our families
who continue to support us selflessly and
unreservedly in this and
all our personal and professional endeavors

Contents

Preface to the first edition

The topic of sleep medicine has emerged during the last few decades as an area of intense medical and scientific interest. It also reflects the philosophy of the editors that the aspects of psychosomatics should be well understood so as to produce skilled sleep practitioners. Sleep and psychosomatic medicine is an intensely personal, inherently interesting and fascinating field of medical science. Its subject matter touches all facets of our health and well-being, particularly in the sleep field; it overlaps with neuropharmacology, pharmacology, psychiatry, and basic and clinical medical disciplines. We have striven to present valuable chapters in order to make the reading enjoyable as well as meaningful.

The chapters have been written by experts in the field in order to provide physicians of widely ranging interests and abilities with a highly readable exposition of the principal results, which include numerous well-articulated examples and a rich discussion of applications. In particular, we have tried to indicate how psychosomatics can be applied to sleep medicine, and vice versa. This book is intended primarily for sleep researchers, general and neuropharmacologists, psychiatrists and physicians who evaluate and treat sleep disorders. In addition, the volume will be extremely useful for pharmacologists, pharmacists, medical students, and clinicians of various disciplines who want to get an overall grasp of both sleep and psychosomatic medicine. Nineteen chapters have been written for this purpose by authors who are experts in their fields. The chapters have been written in a readable and easily understood manner. The book presents topics of current interest, ranging from attachment disorders to personality disorders and stress and the psychosomatic factors implicated in various medical illnesses. This book explores many of these new and exciting developments in the field of sleep and psychosomatics. Unfortunately, it is impossible in a volume such as this to include all recent advances—but that is what makes this field unique and such an exciting field to study and read and write about. New concepts are discussed in this book. The reader may feel confident that the information presented is based on the most recent literature on sleep and psychosomatic medicine. Furthermore, the importance of the neuroimaging of sleep is stressed. It is our hope that we have succeeded in accomplishing this goal. As usual, we welcome communications from our readers concerning this book, especially any errors or deficiencies that may remain.

Preface to the second edition

Sleep and Psychosomatic Medicine was originally conceived and written to address the needs of students, trainees, and clinicians as a resource to integrate sleep medicine and psychiatry into the routine clinical practice of sleep medicine. We found the response to the first edition extremely rewarding—the text was critically acclaimed and sustained broad national and international distribution. It enjoyed readership by many working in different capacities in the occupational and environmental field, not merely clinicians.

The first edition of *Sleep and Psychosomatic Medicine* was extremely successful and well received by the worldwide sleep medicine community. In the second edition, we have kept a similar format for those who are familiar with our earlier edition. Yet we have also added new content and features, and included color figures based on the valuable feedback from readers.

This edition has been revised in an effort to further broaden the scope of the second edition with the inclusion of several new chapters such as "Sleep and dermatology," "Fatigue in chronic medical conditions," "Occupational sleep medicine," "Restless legs syndrome and neuropsychiatric disorders" and "Sleep dysfunction after traumatic brain injury," to name just a few.

We have strived to maintain a balance of including the most important information for clinical practitioners while also limiting the size of the volume so that it did not become a formal textbook. Although this volume offers treatment plans and suggestions, it remains the professional responsibility of the clinical practitioner to rely on experience given the signs and symptoms of the patient to determine individually tailored treatment interventions for a given condition.

When we published the first edition in 2007, we thought that a second edition of *Sleep and Psychosomatic Medicine* would follow within the next 5 years or so due to the continued advancement in the field. Additionally, the first edition did not include all of the topics that we wanted to cover. Our initial assumption was supported and encouraged by fellow readers who contacted us in person and via email to discuss what they wanted to see in future editions. We thank the readership for their encouragement and feedback, and have made every effort to take into account their suggestions and critiques to improve the second edition.

As editors and authors, we have tried to keep the information current whenever possible, with some chapters having been included from the first edition. Change in medical science is a dynamic state of progress and advance, and we want to achieve perfection in the forthcoming edition by giving readers what they want.

It is against this background that the editors of the first edition have drawn upon the prodigious efforts of two colleagues for this second edition. As with the first edition, we have worked closely with our distinguished contributors to provide a consistent format throughout the book, and we are extremely grateful for the spirit in which our contributors responded to this objective. We have also undertaken some changes: some chapters have been deleted, many more added, and all others significantly updated.

We are grateful to all of the contributing authors, including multiple new ones, for their hard work and dedication. We owe them a debt of gratitude for their excellent chapters and their promptness that made the task of editing this volume much easier. The support, encouragement, organizational skills and experience of the editorial and production staff at CRC Press have led this second edition from vision to volume. On this note, we would like to thank Lance Wobus and Amy Blalock, who made this project possible and a smooth ride. Above all, we owe special thanks to our families for their patience during our work on this new edition.

In this book, the authors have sought to bridge the gap between mental and physical health by introducing a holistic approach, which will unite physicians, psychologists, family therapists, social workers, nurses, psychological counselors and psychotherapists of all theoretical orientations in working with both individuals and families across a wide range of professional settings. We also believe that this volume will serve as an excellent resource for students, trainees, and physicians pursuing consultation liaison psychiatry. We believe

Sleep and Psychosomatic Medicine will continue to be an invaluable resource for students, resident physicians, fellows, and other practitioners of psychiatry or sleep medicine, particularly those who practice within the domains of consultation and liaison psychiatry.

We are grateful for the extraordinary privilege of sharing this second edition of *Sleep and Psychosomatic Medicine*.

Seithikurippu R. Pandi-Perumal
Toronto

Meera Narasimhan
South Carolina

Milton Kramer
New York

Contributors

Ridhwan Y. Baba
Division of Pulmonary, Critical Care and Sleep
Medicine
Department of Medicine
MetroHealth Medical Center
Case-Western Reserve University
Cleveland, Ohio

Karl Heinz Brisch
Leiter der Abteilung Pädiatrische Psychosomatik
und Psychotherapie
LMU—Klinikum der Universität München
Dr. von Haunersches Kinderspital, Kinderklinik
und Poliklinik
Pettenkoferstr, München (Postanschrift),
Pettenkoferstr
München (Hausanschrift), Germany

Elizabeth Brush
Division of Pulmonary, Critical Care and Sleep
Medicine
Icahn School of Medicine at Mount Sinai
New York, New York

Blynn G. Bunney
Department of Psychiatry and Human Behavior
University of California
Irvine, California

Kenneth Butto
Pickering, Whitby and Bowmanville Centre for
Sleep Disorders
Pickering, Ontario, Canada

Chen-Lin Chen
Department of Medicine
Hualien Tzu Chi Hospital and Tzu Chi University
Hualien, Taiwan

C. Robert Cloninger
Department of Psychiatry
Washington University School of Medicine
St. Louis, Missouri

Thien Thanh Dang-Vu
Center for Studies in Behavioral Neurobiology
and
Department of Exercise Science
Concordia University
and
Institut Universitaire de Gériatrie de Montréal
Université de Montréal
Montréal, Quebec, Canada

Achim Elfering
Department of Work and Organizational
Psychology
Institute of Psychology
Universität Bern
Bern, Switzerland

and

National Centre of Competence in Research,
Affective Sciences
University of Geneva
CISA, Geneva, Switzerland

Jason Ellis
Health and Life Sciences
Northumbria University
Newcastle upon Tyne, United Kingdom

Christin Gerhardt
Department of Work and Organizational
Psychology
Universität Bern
Bern, Switzerland

Deepak Goel
Department of Neurology and Sleep Clinic
Himalayan Institute of Medical Sciences
Uttarakhand, India

Meeta Goswami
Narcolepsy Institute and the Department
of Neurology
Montefiore Medical Center
Bronx, New York

Zoe Marie Gotts
Institute of Health & Society
Newcastle University Institute for Ageing
Newcastle University
Newcastle upon Tyne, United Kingdom

Raymond Gottschalk
Division of Respirology
Department of Medicine
McMaster University
Hamilton, Ontario, Canada

Aditya K. Gupta
Division of Dermatology
Department of Medicine
University of Toronto
London, Ontario, Canada

Madhulika A. Gupta
Department of Psychiatry
Schulich School of Medicine
and Dentistry
University of Western Ontario
London, Ontario, Canada

Ravi Gupta
Department of Psychiatry
and Sleep Clinic
Himalayan Institute of Medical Sciences
Uttarakhand, India

Benjamin Hatch
Center for Studies in Behavioral
Neurobiology
Concordia University
Montréal, Quebec, Canada

Ripu D. Jindal
Birmingham VA Medical Center
University of Alabama at Birmingham
Birmingham, Alabama

Matcheri S. Keshavan
Stanley Cobb Professor of Psychiatry
Department of Psychiatry
Beth Israel Deaconess Medical Center
Boston, Massachusetts

Katie Knapp
Department of Psychiatry
Schulich School of Medicine and
Dentistry
University of Western Ontario
London, Ontario, Canada

Maria U. Kottwitz
Department of Work and Organizational
Psychology
University of Marburg
Marburg, Germany

Milton Kramer
Department of Psychiatry
University of Cincinnati
Cincinnati, Ohio

Tatyana Mollayeva
Acquired Brain Injury Laboratory
Rehabilitation Science Institute
University of Toronto
Toronto, Ontario, Canada

Meera Narasimhan
Health Sciences
and
Department of Neuropsychiatry and
Behavioral Science
University of South Carolina
Columbia, South Carolina

Julia Newton
Clinical Academic Office
Newcastle University
Newcastle upon Tyne, United Kingdom

Zvjezdan Nuhic
Psychiatric Center
Chula Vista, California

William C. Orr
Lynn Health Science Institute
University of Oklahoma Health Sciences
Center
Oklahoma City, Oklahoma

James F. Pagel
Department of Family Practice
University of Colorado
Boulder, Colorado

and

Sleep Disorders Center of Southern Colorado
Parkview Neurological Institute Pueblo
and
Rocky Mountain Sleep Laboratory
Pueblo, Colorado

Seithikurippu R. Pandi-Perumal
Somnogen Canada Inc.
Toronto, Ontario, Canada

Donald B. Penzien
Department of Anesthesiology
Wake Forest School of Medicine
Medical Center Boulevard
Winston-Salem, North Carolina

Diana Pereira
Department of Work and Organizational
Psychology
Universität Bern
Bern, Switzerland

Samara Perez
Department of Psychology
McGill University
Montreal, Quebec, Canada

Lampros Perogamvros
Department of Psychiatry
University Hospitals of Geneva
Geneva, Switzerland

J. Steven Poceta
Sleep Center and Division of Neurology
Scripps Clinic
La Jolla, California

Jeanetta C. Rains
Center for Sleep Evaluation
Elliot Hospital—River's Edge
Manchester, New Hampshire

Zeev Rosberger
Louise Granofsky Psychosocial Oncology Program
Institute of Community and Family Psychiatry
Jewish General Hospital
Montreal, Quebec, Canada

Josée Savard
École de psychologie
Pavillon Félix-Antoine-Savard
Université Laval
Quebec City, Quebec, Canada

Neomi Shah
Hispanic Community Health Study/Study of
Latinos (Bronx)
Division of Pulmonary, Critical Care and Sleep
Icahn School of Medicine at Mount Sinai
New York, New York

Gilla K. Shapiro
Department of Psychology
McGill University
and
Lady Davis Institute for Medical
Research
Jewish General Hospital
Montreal, Quebec, Canada

Colin Michael Shapiro
Department of Psychiatry and Ophthalmology
University of Toronto
Toronto, Ontario, Canada

Arthur J. Spielman (Deceased)
Cognitive Neuroscience Program
The City College of New York
and
Center for Sleep Medicine
Weill Cornell Medical School
Cornell University
New York, New York

Michael J. Thorpy
Sleep–Wake Disorders Center
Montefiore Medical Center
Bronx, New York

Daniel J. Wallace
Cedars-Sinai Medical Center
David Geffen School of Medicine at UCLA
University of California
Los Angeles, California

Joseph C. Wu
Department of Psychiatry and Human
Behavior
University of California
Irvine, California

Chien-Ming Yang
Department of Psychology
The Research Center for Mind, Brain,
and Learning
National Chengchi University
Taipei, Taiwan

Dora M. Zalai
Department of Psychology
Ryerson University
Toronto, Ontario, Canada

Classification of sleep disorders

MICHAEL J. THORPY

Sleep disorders have had a formal classification since 1979, when the Association of Sleep Disorder Centers published the Classification of Sleep and Arousal Disorders in the journal *Sleep*.[1] In 2013, the revised version of the *Diagnostic and Statistical Manual of Mental Disorders, Fifth Edition* (DSM-V) included a section entitled "Sleep Wake Disorders," an update of the DSM-IV section (Table 1.1).[2] This produced a classification for mental health and general medical clinicians who are not experts in sleep medicine. The classification differs from that of the *International Classification of Sleep Disorders* (ICSD-3) that was produced by the American Academy of Sleep Medicine and updated in 2014 (Table 1.2).[3] The presence of two current competing classifications is not ideal, as it can cause confusion with health insurance companies and for epidemiological research. The *International Classification of Diseases* modified version—the ICD-10-CM (Table 1.3)—will be adopted in the United States in 2016 and contains a classification that more closely conforms to the ICSD-3.[4]

DSM-V

Insomnia disorder, the first entry in the DSM-V (Table 1.1), is a diagnostic entry that requires the presence of at least one sleep complaint such as difficulty initiating sleep that must be present at least 3 nights per week for at least 3 months. The diagnosis is coded along with other mental, medical, and sleep disorders. The diagnosis can be specified as being episodic if it occurs for at least 1 month; however, acute and short-term insomnia, which have symptoms of less than 3 months, should be coded as "other specified insomnia disorder."

Hypersomnolence disorder requires a 3-month history of excessive sleepiness in the presence of significant distress or other impairment. Objective documentation by electrophysiological tests, such as the multiple sleep latency test (MSLT), is not required. This diagnosis is coded along with any other concurrent mental, medical, and sleep disorder. Narcolepsy is defined as recurrent episodes of sleep that occur for at least 3 months along with one of three additional features, such as cataplexy, hypocretin deficiency, or polysomnographic features, either a sleep-onset rapid eye movement period (SOREMP) on a nighttime polysomnogram (PSG) or an MSLT that shows a mean sleep latency of 8 minutes or less and two or more SOREMPs. So narcolepsy can be diagnosed in DSM-V if just sleepiness occurs for 3 months and there is a SOREMP on the nocturnal PSG. This has the potential for leading to errors in diagnosis, as other disorders, including obstructive sleep apnea

Table 1.1 DSM-V

Sleep–wake disorders

- Insomnia disorder
- Hypersomnolence disorder
- Narcolepsy
  - Subtypes:
    - Without cataplexy but with hypocretin deficiency
    - With cataplexy but without hypocretin deficiency
    - Autosomal dominant cerebellar ataxia, deafness and narcolepsy (ADCADN)
    - Autosomal dominant narcolepsy, obesity and type 2 diabetes (ADNOD)
    - Secondary to another medical condition
- Obstructive sleep apnea syndrome
- Central sleep apnea
- Sleep-related hypoventilation
  - Subtypes:
    - Idiopathic hyperventilation
    - Congenital central alveolar hypoventilation
    - Comorbid sleep-related hypoventilation
- Circadian rhythm sleep disorders
  - Delayed sleep phase type
  - Advance sleep phase type
  - Irregular sleep–wake type
  - Non-24-hour sleep–wake type
  - Shift work disorder
  - Unspecified type
- Parasomnias
  - Non-rapid eye movement sleep arousal disorder
    - Subtypes:
      - Sleepwalking type
      - Sleep terror type
    - Nightmare disorder
    - Rapid eye movement sleep behavior disorder
- Restless legs syndrome
- Substance/medication-induced sleep disorder
- Other specified insomnia disorder
- Other specified hypersomnolence disorder
- Unspecified sleep–wake disorder
- Unspecified insomnia disorder
- Unspecified hypersomnolence disorder
- Unspecified sleep–wake disorder

Source: Adapted from American Psychiatric Association. *Diagnostic and Statistical Manual of Mental Disorders, 5th Edition (DSM-V).* Washington, DC: American Psychiatric Association, 2013.

syndrome (OSA), can produce similar features. Five subtypes of narcolepsy are specified according to: without cataplexy but with hypocretin deficiency; with cataplexy but without hypocretin deficiency; autosomal dominant cerebellar ataxia, deafness and narcolepsy; autosomal dominant narcolepsy, obesity and type 2 diabetes; or secondary to another medical condition.

OSA consists of an apnea hypopnea index (AHI) of at least 5 per hour, along with typical nocturnal respiratory symptoms, or daytime excessive sleepiness or fatigue. Alternatively, the diagnosis requires an AHI of at least 15, regardless of accompanying symptoms. Mild OSA is regarded as having an AHI of less than 15, moderate 15–30, and severe greater than 30. Central sleep apnea (CSA) requires the presence of five or more central apneas per hour of sleep. Sleep-related hypoventilation has PSG evidence of decreased ventilation with either elevated CO_2 levels or persistent oxygen desaturation unassociated with apneic/hypopneic events. Subtypes of sleep-related hypoventilation include idiopathic hypoventilation, congenital central alveolar hypoventilation, and comorbid sleep-related hypoventilation.

Circadian rhythm sleep–wake disorders (CRSWDs) with five subtypes is caused by a persistent recurrent pattern of sleep disruption due to an alteration or misalignment of the endogenous circadian rhythm and the individual's required sleep–wake schedule, along with symptoms of either insomnia or excessive sleepiness or both. Subtypes are delayed sleep phase type, advanced sleep phase type, irregular sleep–wake type, non-24-hour sleep–wake type, and shift work type, none of which have specific diagnostic criteria.

The parasomnias are subdivided into five disorders: non-rapid eye movement (REM) sleep arousal disorder, nightmare disorder, REM sleep behavior disorder (RBD), restless legs syndrome, and substance/medication-induced sleep disorder. Non-REM sleep arousal disorder consists of two types, with the typical features of either sleepwalking or sleep terrors. The sleepwalking type can be subtyped into sleep-related eating or sleep-related sexual behavior (sexsomnia). Nightmare disorder consists of repeated occurrences of extended, dysphoric, and well-remembered dreams that threaten the individual. Rapid orientation and alertness follows the episode, but it causes significant distress. RBD consists of recurrent episodes of arousal with vocalization

Table 1.2 ICSD-3

	ICD-9-CM code:	ICD-10-CM code:
Insomnia disorders:		
Chronic insomnia disorder	342	F51.01
Short-term insomnia disorder	307.41	F51.02
Other insomnia disorder	307.49	F51.09
Isolated symptoms and normal variants:		
Excessive time in bed		
Short sleeper		
Sleep-related breathing disorders:		
Obstructive sleep apnea disorders:		
Obstructive sleep apnea, adult	327.23	G47.33
Obstructive sleep apnea, pediatric	327.23	G47.33
Central sleep apnea syndromes:		
Central sleep apnea with Cheyne–Stokes breathing	786.04	R06.3
Central apnea due to a medical disorder without Cheyne-Stokes breathing	327.27	G47.37
Central sleep apnea due to high altitude periodic breathing	327.22	G47.32
Central sleep apnea due to a medication or substance	327.29	G47.39
Primary central sleep apnea	327.21	G47.31
Primary central sleep apnea of infancy	770.81	P28.3
Primary central sleep apnea of prematurity	770.82	P28.4
Treatment-emergent central sleep apnea	327.29	G47.39
Sleep-related hypoventilation disorders:		
Obesity hypoventilation syndrome	278.03	E66.2
Congenital central alveolar hypoventilation syndrome	327.25	G47.35
Late-onset central hypoventilation with hypothalamic dysfunction	327.26	G47.36
Idiopathic central alveolar hypoventilation	327.24	G47.34
Sleep-related hypoventilation due to a medication or substance	327.26	G47.36
Sleep-related hypoventilation due to a medical disorder	327.26	G47.36
Sleep-related hypoxemia disorder:		
Sleep-related hypoxemia	327.26	G47.36
Isolated symptoms and normal variants:		
Snoring		
Catathrenia		
Central disorders of hypersomnolence:		
Narcolepsy type 1	347.01	G47.411
Narcolepsy type 2	347.00	G47.419
Idiopathic hypersomnia	327.11	G47.11
Kleine–Levin syndrome	327.13	G47.13
Hypersomnia due to a medical disorder	327.14	G47.14
Hypersomnia due to a medication or substance	292.85 (drug-induced) 291.82 (alcohol-induced)	F11-F19

(Continued)

Table 1.2 (*Continued*) ICSD-3

	ICD-9-CM code:	ICD-10-CM code:
Hypersomnia associated with a psychiatric disorder	327.15	F51.13
Insufficient sleep syndrome	307.44	F51.12
Isolated symptoms and normal variants:		
Long sleeper		
Circadian rhythm sleep–wake disorders:		
Delayed sleep–wake phase disorder	327.31	G47.21
Advanced sleep–wake phase disorder	327.32	G47.22
Irregular sleep–wake rhythm disorder	327.33	G47.23
Non-24-hour sleep–wake rhythm disorder	327.34	G47.24
Shift work disorder	327.36	G47.26
Jet lag disorder	327.35	G47.25
Circadian sleep–wake disorder not otherwise specified (NOS)	327.30	G47.20
Parasomnias:		
NREM-related parasomnias:		
Disorders of arousal (from NREM sleep):		
Confusional arousals	327.41	G47.51
Sleepwalking	307.46	F51.3
Sleep terrors	307.46	F51.4
Sleep-related eating disorder	327.40	G47.59
REM-related parasomnias:		
REM sleep behavior disorder	327.42	G47.52
Recurrent isolated sleep paralysis	327.43	G47.51
Nightmare disorder	307.47	F51.5
Other parasomnias:		
Exploding head syndrome	327.49	G47.59
Sleep-related hallucinations	368.16	H53.16
Sleep enuresis	788.36	N39.44
Parasomnia due to a medical disorder	327.44	G47.54
Parasomnia due to a medication or substance	292.85 (drug-induced) 291.82 (alcohol-induced)	F11-F19
Parasomnia, unspecified	327.40	G47.50
Isolated symptoms and normal variants:		
Sleep talking		
Sleep-related movement disorders:		
Restless legs syndrome	333.94	G25.81
Periodic limb movement disorder	327.51	G47.61
Sleep-related leg cramps	327.52	G47.62
Sleep-related bruxism	327.53	G47.63
Sleep-related rhythmic movement disorder	327.59	G47.69
Benign sleep myoclonus of infancy	327.59	G47.69
Propriospinal myoclonus at sleep onset	327.59	G47.69
Sleep-related movement disorder due to a medical disorder	327.59	G47.69

(*Continued*)

Table 1.2 (*Continued*) ICSD-3

	ICD-9-CM code:	ICD-10-CM code:
Sleep-related movement disorder due to a medication or substance	292.85 (drug-induced) 291.82 (alcohol-induced)	F11-F19
Sleep-related movement disorder, unspecified	327.59	G47.69
Isolated symptoms and normal variants: Excessive fragmentary myoclonus Hypnagogic foot tremor and alternating leg muscle activation Sleep starts (hypnic jerks)		
Other sleep disorder	327.8	G47.8
Appendix A:		
Fatal familial insomnia	046.8	A81.83
Sleep-related epilepsy	345	G40.5
Sleep-related headaches	784.0	R51
Sleep-related laryngospasm	787.2	J38.5
Sleep-related gastroesophageal reflux	530.1	K21.9
Sleep-related myocardial ischemia	411.8	I25.6
Appendix B:		
ICD-10-CM coding for substance-induced sleep disorders		F10-F19

Source: Adapted from American Academy of Sleep Medicine. *International Classification of Sleep Disorders*, 3rd edition. Darien, IL: American Academy of Sleep Medicine, 2014.

and/or complex movements from REM sleep that is documented by either PSG or a history suggesting a synucleinopathy. Restless legs syndrome—an urge to move the legs—is accompanied by uncomfortable sensation in the legs, with typical features occurring at least 3 times per week for at least 3 months. Sleep/medication-induced sleep disorder is a sleep disturbance either during or soon after substance intoxication or after withdrawal, or the substance is known to cause sleep disturbance.

Other specified insomnia disorder is diagnosed when the insomnia does not meet the criteria for insomnia disorder, and other hypersomnolence disorder is diagnosed when the excessive sleepiness does not meet the criteria for hypersomnolence disorder. Similarly, unspecified sleep–wake disorder is diagnosed when the sleep–wake disorder does not meet the full criteria for the specified sleep–wake disorders. Unspecified forms consist of insomnia disorder, hypersomnolence disorder, and sleep–wake disorder.

ICSD-3

The ICSD-3 was published in March of 2014 (Table 1.2). The main change was the simplification of the insomnia disorders and an expansion of the sleep-related breathing disorders.

Insomnia disorders

The insomnia disorders mainly consist of one major disorder termed chronic insomnia disorder. The clinical features of insomnia are believed to be the result of a primary or secondary process, but the consequences are similar, no matter the etiology.[5] The diagnosis rests upon a sleep symptom of difficulty initiating sleep, difficulty maintaining sleep, early-morning awakening and, mainly for pediatric age groups, resistance to going to bed and difficulty in sleeping without a caregiver intervention. The symptoms occur 3 times per week for at least 3 months and have daytime consequences such as fatigue/malaise, cognitive impairment, mood lability, daytime sleepiness or social, family, academic or occupational impairment. Psychophysiological insomnia and other insomnia disorders of the ICSD-2 are mentioned in the text description of the disorder. Objective testing by polysomnography is not required for diagnosis, but should be considered if features suggest a concurrent sleep-related

Table 1.3 ICD-10-CM sleep disorders

F51 Sleep disorders not due to a substance or known physiological condition
    F51.01 Primary insomnia
    F51.02 Adjustment insomnia
    F51.03 Paradoxical insomnia
    F51.04 Psychophysiologic insomnia
    F51.05 Insomnia due to other mental disorder
    F51.09 Other insomnia not due to a substance or known physiological condition
  F51.1 Hypersomnia not due to a substance or known physiological condition
    F51.11 Primary hypersomnia
    F51.12 Insufficient sleep syndrome
    F51.13 Hypersomnia due to other mental disorder
    F51.19 Other hypersomnia not due to a substance or known physiological condition
  F51.3 Sleepwalking (somnambulism)
  F51.4 Sleep terrors (night terrors)
  F51.5 Nightmare disorder
  F51.8 Other sleep disorders not due to a substance or known physiological condition
  F51.9 Sleep disorder not due to a substance or known physiological condition, unspecified

G47 Organic sleep disorders
  G47.0 Insomnia, unspecified
    G47.01 Insomnia due to medical condition
    G47.09 Other insomnia
  G47.1 Hypersomnia, unspecified
    G47.11 Idiopathic hypersomnia with long sleep time
    G47.12 Idiopathic hypersomnia without long sleep time
    G47.13 Recurrent hypersomnia
    G47.14 Hypersomnia due to medical condition
    G47.19 Other hypersomnia
  G47.20 Circadian rhythm sleep disorder, unspecified type
    G47.21 Circadian rhythm sleep disorder, delayed sleep phase type
    G47.22 Circadian rhythm sleep disorder, advanced sleep phase type
    G47.23 Circadian rhythm sleep disorder, irregular sleep wake type
    G47.24 Circadian rhythm sleep disorder, free running type
    G47.25 Circadian rhythm sleep disorder, jet lag type
    G47.26 Circadian rhythm sleep disorder, shift work type
    G47.27 Circadian rhythm sleep disorder in conditions classified elsewhere
    G47.29 Other circadian rhythm sleep disorder
  G47.30 Sleep apnea, unspecified
    G47.31 Primary central sleep apnea
    G47.32 High-altitude periodic breathing
    G47.33 Obstructive sleep apnea (adult) (pediatric)
    G47.34 Idiopathic sleep-related non-obstructive alveolar hypoventilation
    G47.35 Congenital central alveolar hypoventilation syndrome
    G47.36 Sleep-related hypoventilation in conditions classified elsewhere
    G47.37 Central sleep apnea in conditions classified elsewhere
    G47.39 Other sleep apnea

(Continued)

Table 1.3 (*Continued*) ICD-10-CM sleep disorders

G47.4 Narcolepsy and cataplexy
  G47.41 Narcolepsy
  G47.411 Narcolepsy with cataplexy
  G47.419 Narcolepsy without cataplexy
  G47.42 Narcolepsy in conditions classified elsewhere
  G47.421 Narcolepsy in conditions classified elsewhere with cataplexy
  G47.429 Narcolepsy in conditions classified elsewhere without cataplexy
G47.50 Parasomnia, unspecified
  G47.51 Confusional arousals
  G47.52 Rapid eye movement sleep behavior disorder
  G47.53 Recurrent isolated sleep paralysis
  G47.54 Parasomnia in conditions classified elsewhere
  G47.59 Other parasomnia
G47.6 Sleep-related movement disorders
  G47.61 Periodic limb movement disorder
  G47.62 Sleep-related leg cramps
  G47.63 Sleep-related bruxism
  G47.69 Other sleep-related movement disorders
G47.8 Other sleep disorders
G47.9 Sleep disorder, unspecified

Z72.82 Problems related to sleep
  Z72.820 Sleep deprivation
  Z72.821 Inadequate sleep hygiene
Z73.8 Other problems related to life management difficulty
  Z73.810 Behavioral insomnia of childhood, sleep-onset association type
  Z73.811 Behavioral insomnia of childhood, limit setting type
  Z73.812 Behavioral insomnia of childhood, combined type
  Z73.819 Behavioral insomnia of childhood, unspecified type

Source: Adapted from *International Classification of Diseases, Tenth Revision, Clinical Modification (ICD-10-CM)*, National Center for Health Statistics. Hyattsville, MD: Centers for Disease Control and Prevention (CDC), 2010.

breathing disorder. There is the inclusion of short-term insomnia disorder with similar diagnostic criteria, as chronic insomnia disorder applies to insomnia that is less than 3 months in duration. Excessive time in bed and short sleeper are included as isolated symptoms and normal variants, not as specific disorders.

Sleep-related breathing disorders

The sleep-related breathing disorders section is organized into four main categories: OSA disorders, CSA syndromes, sleep-related hypoventilation disorders, and sleep-related hypoxemia disorder. The CSA syndromes are divided into eight types: two related to Cheyne–Stokes breathing (CSB), high altitude, substance, three primary CSA disorders (of which one is infancy and the other prematurity), and a new entity entitled treatment-emergent CSA. The latter category applies to central apnea that follows continuous positive airway pressure (CPAP) administration.

OSA maintains the criterion of five or more predominantly obstructive respiratory events per hour of sleep (including respiratory effort-related arousals) when studied in a sleep center or by out-of-center sleep studies, so long as typical symptoms are present; otherwise, 15 or more are sufficient to make the diagnosis. The OSA disorders are divided into adult and pediatric types. In the pediatric criteria, for those who are less than 18 years of age, one or more obstructive event(s) are required per hour of sleep so long as respiratory symptoms or sleepiness are present; alternatively, obstructive hypoventilation (with a $PaCO_2$ >50 mmHg) along with symptoms of snoring, labored breathing or sleepiness and other cognitive effects are required.

CSA with CSB consists of five or more central apneas or hypopneas per hour of sleep with a pattern that meets the criteria for CSB. This diagnosis can occur along with a diagnosis of OSA. CSA without CSB is diagnosed as CSA due to a medical disorder without CSB, which occurs as a consequence of a medical or neurological disorder, but not due to a medication or substance. CSA due to high-altitude periodic breathing is central apnea attributable to high altitude of at least 1500 meters, but usually above 2500 meters, with associated symptoms of sleepiness, difficulty in sleeping, awakening with shortness of breath or headache or witnessed apneas. Polysomnography is not required. CSA is due to a medicine or substance—most typically due to an opioid or respiratory depressant—but not associated with CSB. Polysomnography is required. Primary CSA consists of five or more central apneas or central hypopneas per hour of sleep in the absence of CSB and is of unknown etiology. Typical symptoms of sleepiness or abnormal breathing events are present, but may not be present in children. Primary CSA of infancy occurs in an infant at greater than 37 weeks of conceptional age with recurrent, prolonged (>20 seconds in duration) central apneas and periodic breathing for more than 5% of total sleep time during sleep. Apnea or cyanosis is seen by an observer or desaturation detected by monitoring. Primary CSA of prematurity occurs in an infant of less than 37 weeks of conceptional age with similar features and respiratory events.

Treatment-emergent CSA is a disorder that has become recognized since the application of positive airway pressure (PAP), despite resolution of obstructive respiratory events. It is diagnosed when there are five or more obstructive events during a PSG with PAP that shows resolution of obstructive events and presence of central apneas or hypopneas.[6]

Sleep-related hypoventilation disorders consist of seven disorders that meet diagnostic criteria for sleep-related hypoventilation with or without oxygen desaturation. Obesity hypoventilation is hypoventilation during wakefulness ($PaCO_2$ >45 mmHg) in the presence of obesity (BMI >30 kg/m^2). Congenital central alveolar hypoventilation syndrome is diagnosed when the sleep-related hypoventilation is associated with the *PHOX2B* gene. Late-onset central hypoventilation with hypothalamic dysfunction consists of sleep-related hypoventilation without symptoms in the first few years of life and the *PHOX2B* gene is not present. Idiopathic central alveolar hypoventilation consists of sleep-related hypoventilation with the presence of lung or airway disease or any other known cause. Sleep-related hypoventilation due to a medication or substance consist of sleep-related hypoventilation when a medication or substance is known to be the primary cause. Sleep-related hypoventilation due to a medical disorder consists of sleep-related hypoventilation due to a lung or airway disease, or other medical cause. Sleep-related hypoxemia disorder is arterial oxygen saturation of ≤88% in adults or ≤90% in children for ≥5 minutes. Polysomnography is not required. Snoring and catathrenia (prolonged expiratory expiration in REM sleep) are regarded as isolated symptoms or normal variants.

Central disorders of hypersomnolence

The central disorders of hypersomnolence comprise eight disorders. Narcolepsy has undergone a major revision with elimination of the disorder name terms, with or without cataplexy. Type 1 narcolepsy is presumed to be due to hypocretin loss with either measured reductions in cerebrospinal fluid (CSF) hypocretin or cataplexy with associated electrophysiological findings. Type 2 narcolepsy is confirmed by electrophysiological studies in the absence of cataplexy or with a normal CSF hypocretin level. A major change in the narcolepsy criteria is the addition of including a SOREMP on the nocturnal PSG as one of the two events that are required to meet the MSLT criteria of two SOREMPs for diagnosis. This finding is mainly based upon a study that indicates that the positive predictive value of a SOREMP on the nocturnal PSG for narcolepsy is 92%.[7] Approximately 50% of patients with narcolepsy will have a SOREMP of less than 15 minutes on the nocturnal PSG.

Idiopathic hypersomnia is now a single entity with elimination of the two ICSD-2 hypersomnia disorders that had specific sleep duration criteria. Idiopathic hypersomnia disorder requires sleepiness for at least 3 months and either an MSLT mean sleep latency of 8 minutes or less or a nocturnal sleep duration of at least 660 minutes. The ICSD-2 category of recurrent hypersomnia has been reduced to a single entry of Kleine–Levin syndrome, with a subtype of menstrual-related Kleine–Levin syndrome.[8] The sleepiness must persist for 2 days to 5 weeks, and at least once every 18 months. There can be only one symptom with sleepiness consisting of cognitive dysfunction, altered perception, eating disorder, or disinhibited behavior. Normal alertness and cognitive function must be present between episodes. Hypersomnia due to a medical disorder requires an association of sleepiness with any underlying medical or neurological disorder. Seven subtypes are mentioned: hypersomnia secondary to Parkinson's disease; posttraumatic hypersomnia; genetic disorders associated with primary central nervous system somnolence; hypersomnia secondary to brain tumors, infections, or other central nervous system lesions; hypersomnia secondary to endocrine disorder; hypersomnia secondary to metabolic encephalopathy; and residual hypersomnia in patients with adequately treated OSA.

Hypersomnia due to a medication or substance is sleepiness that occurs as a consequence of a current medication or substance or withdrawal from a wake-promoting medication or substance. Hypersomnia associated with a psychiatric disorder is sleepiness in association with a current psychiatric disorder. Insufficient sleep syndrome is the new term for the previous, more cumbersome term of behaviorally induced insufficient sleep syndrome. The reduced sleep must be present most days for at least 3 months. Extension of sleep time must result in resolution of symptoms. Long sleeper is no longer regarded as a disorder, but as a normal variant. There are no diagnostic criteria, but a total sleep time of 10 or more hours is suggested as being usually accepted.

Circadian rhythm sleep–wake disorders

The CRSWDs comprise six specific disorders, including delayed sleep–wake phase disorder, advanced sleep–wake phase disorder, irregular sleep–wake rhythm disorder, non-24-hour sleep–wake rhythm disorder, shift work disorder, and jet lag disorder. These disorders arise when there is a substantial misalignment between the internal circadian rhythm and the desired sleep–wake schedule. Specific general diagnostic criteria are given for CRSWD. A 3-month duration of symptoms is a requirement for diagnosing all of these disorders, except for jet lag disorder, which has a requirement of jet travel across at least two time zones.

Delayed sleep–wake phase disorder occurs when there is a significant delay in the phase of the major sleep episode in relation to the desired sleep and wake-up time. When allowed to sleep as desired, sleep duration and quality are age appropriate. Advanced sleep–wake phase disorder occurs when there is a significant advance in the phase of the major sleep episode in relation to the desired sleep and wake-up time. When allowed to sleep as desired, sleep duration and quality are age appropriate. Irregular sleep–wake rhythm disorder is a recurrent or chronic pattern of irregular sleep and wake episodes throughout the 24-hour period, with symptoms of insomnia and daytime sleepiness. Non-24-hour sleep–wake rhythm disorder is due to a progressively delayed sleep–wake pattern. Shift work disorder is insomnia or excessive sleepiness associated with a recurring work schedule that overlaps with the usual time for sleep. Sleep logs and actigraphy monitoring for at least 7 days is recommended for all of the above CRSWDs. Jet lag disorder is insomnia or sleepiness accompanied by transmeridian jet travel across at least two time zones.

A CRSWD not otherwise specified (NOS) is listed for patients who have a CRSWD and who meet all of the criteria for CRSWD, but not the specific types.

Parasomnias

The parasomnias are divided into three groups: the non-REM (NREM)-related parasomnias, the REM-related parasomnias, and an "other parasomnias" category. They are defined as undesirable physical events or experiences that occur during entry into sleep, within sleep, or during arousal from sleep.

The NREM-related parasomnias comprise general diagnostic criteria for the group heading of disorders of arousal (DA; from NREM sleep). Specific general diagnostic criteria are given for each DA and the detailed text applies to all of the DAs, as no text is presented for each of the specific DAs, except for diagnostic criteria. Confusional arousals are characterized by mental confusion or confused behavior that occurs while the patient is in bed. Sleep-related abnormal sexual behaviors is listed as a subtype to be classified under confusional arousals. Sleepwalking consists of arousals associated with ambulation and other complex behaviors out of bed. Sleep terrors are episodes of abrupt terror, typically beginning with alarming vocalizations, such as a frightening scream. The final NREM-related parasomnia is sleep-related eating disorder (SRED), which requires an arousal from the main sleep period to distinguish it from night eating syndrome disorder, which consists of excessive eating between dinner and bedtime, and SRED requires an adverse health consequence from the disorder.[9] The behavior consists of the consumption of peculiar combinations of food or inedible or toxic substances, injurious behaviors while in the pursuit of food, or adverse effects from recurrent nocturnal eating. There is partial or complete loss of conscious awareness during the episode, with subsequent impaired recall.

The REM-related parasomnias include RBD, recurrent isolated sleep paralysis (RISP), and nightmare disorder. RBD, which consists of repeated episodes of vocalizations and/or complex motor behaviors, requires the polysomnographic evidence of REM sleep without atonia.[10] RISP is the recurrent inability to move the trunk and all of the limbs at sleep onset or upon awakening from sleep that causes distress or fear of sleep. Nightmare disorder consists of repeated occurrences of extended, extremely dysphoric and well-remembered dreams that usually involve threats to survival, security, or physical integrity that are associated with significant distress or psychosocial, occupational, or other areas of impaired functioning.

The other parasomnias section includes three specific disorders: exploding head syndrome (EHS), sleep-related hallucinations, and sleep enuresis. EHS is a complaint of a sudden noise or sense of explosion in the head either at the wake–sleep transition or upon awakening during the night associated with abrupt arousal. Sleep-related hallucinations are predominantly visual hallucinations that are experienced just prior to sleep onset or upon awakening during the night

or in the morning. Sleep enuresis consists of involuntary voiding during sleep at least twice a week in people older than 5 years of age. Parasomnias associated with medical disorders, and medication or substance and unspecific parasomnia, comprise the other entries in this category. Sleep talking is a normal variant that can occur in both NREM and REM sleep and can be associated with parasomnias such as RBD or DAs.

The sleep-related movement disorders (SRMD) comprise seven specific disorders: restless legs syndrome; periodic limb movement disorder (PLMD); sleep-related leg cramps; sleep bruxism; sleep-related rhythmic disorder (RMD); benign sleep myoclonus of infancy (BSMI); and propriospinal myoclonus at sleep onset (PSM). SRMDs are relatively simple, usually stereotyped movements that disturb sleep or its onset.

Restless legs syndrome (also known as Willis–Eckbom disease) is an urge to move the legs, usually accompanied by or thought to be caused by uncomfortable and unpleasant sensations in the legs. The ICSD-3 criteria do not include any frequency or duration criteria as are contained in the DSM-V criteria.

PLMD is defined by the polysomnographic demonstration of periodic limb movements of >5/hour in children and >15/hour in adults that cause significant sleep disturbance or impairment of functioning. Sleep-related leg cramps are painful sensations that occur in the leg or foot with sudden, involuntary muscle hardness or tightness. Sleep-related bruxism is tooth grinding during sleep that is associated with tooth wear or morning jaw muscle pain or fatigue. RMD consists of repetitive, stereotyped, and rhythmic motor behaviors involving large muscle groups that are sleep related. BSMI consists of repetitive myoclonic jerks that involve the limbs, trunk, or whole body that occur from birth to 6 months of age during sleep. As PSM mainly occurs during relaxed wakefulness and drowsiness as the patient attempts to sleep, the term "at sleep onset" has been added to the propriospinal myoclonus name. The three final categories are related to a medical disorder, medication of substance, and an unspecified parasomnia.

Isolated symptoms and normal variants include excessive fragmentary myoclonus (EFM), hypnagogic foot tremor (HFT), alternating muscle activation (ALMA), and sleep starts (hypnic jerks). EFM is now regarded as a normal variant found on polysomnographic EMG recordings that is characterized by small movements of the corners of the mouth, fingers, or toes or without visible movement. HFT consists of rhythmic movement of the feet or toes that occurs in the transition between wake and sleep or in light NREM sleep. ALMA consists of brief activation of the anterior tibialis in one leg with alternation in the other leg. Sleep starts (hypnic jerks) are brief, simultaneous contractions of the body or one or more body segments occurring at sleep onset.

The final category in the ICSD-3 is a general other sleep disorder category for disorders that cannot be classified elsewhere.

CONCLUSION

The new ICSD-3 is a major advance over previous versions, but it is unfortunate that some of the diagnostic criteria differ from those of the DSM-V; for example, the criteria for narcolepsy. However, the DSM-V serves as an entry-level classification, mainly for psychiatrists, and it is to be hoped that in the future, the two classifications will be merged into one that will cause less confusion not only for clinicians, but also for agencies that reimburse for healthcare and provide for treatment options.

Appendix A lists several disorders that are coded in other sections of ICD-10 other than the sleep sections and include: fatal familial insomnia, sleep-related epilepsy, sleep-related headaches, sleep-related laryngospasm, sleep-related gastroesophageal reflux, and sleep-related myocardial ischemia. Appendix B lists the ICD-10 sleep-related substance-induced sleep disorders.

REFERENCES

1. Association of Sleep Disorder Centers. Classification of sleep and arousal disorders. *Sleep* 1979; 2(1): 1–154.
2. American Psychiatric Association. *Diagnostic and Statistical Manual of Mental Disorders 5th Edition (DSM-V)*. Washington, DC: American Psychiatric Association, 2013.
3. American Academy of Sleep Medicine. *International Classification of Sleep Disorders*, 3rd edition. Darien, IL: American Academy of Sleep Medicine, 2014.
4. *International Classification of Diseases, Tenth Revision, Clinical Modification (ICD-10-CM)*. National Center for Health Statistics. Hyattsville, MD: Centers for Disease Control and Prevention (CDC), 2010.
5. Edinger JD, Wyatt JK, Stepanski EJ et al. Testing the reliability and validity of DSM-IV-TR and ICSD-2 insomnia diagnoses: Results of a multi-method/multi-trait analysis. *Arch Gen Psychiatry* 2011; 68: 992–1002.
6. Westhoff M, Arzt M, Litterst P. Prevalence and treatment of central sleep apnoea emerging after initiation of continuous positive airway pressure in patients with obstructive sleep apnoea without evidence of heart failure. *Sleep Breath* 2012; 16: 71–78.
7. Andlauer O, Moore H, Jouhier L et al. Nocturnal REM sleep latency for identifying patients with narcolepsy/hypocretin deficiency. *JAMA Neurol* 2013; 6: 1–12.
8. Arnulf I, Lin L, Gadoth N et al. Kleine–Levin syndrome: A systematic study of 108 patients. *Ann Neurol* 2008; 63: 482–493.
9. Brion A, Flamand M, Oudiette D, Voillery D, Golmard JL, Arnulf I. Sleep related eating disorder versus sleepwalking: A controlled study. *Sleep Med* 2012; 3: 1094–1101.
10. Schenck CH, Howell MJ. Spectrum of RBD (overlap between RBD and other parasomnias). *Sleep Biol Rhythms* 2013; 11(Suppl 1): 27–34.

2

Sleep and gastrointestinal functioning

CHEN-LIN CHEN AND WILLIAM C. ORR

INTRODUCTION

Interest in sleep and gastrointestinal (GI) functioning has increased substantially over the past several years. These discoveries have led to a remarkable broadening of the focus and importance of the applications of basic sleep physiology to numerous areas of clinical medicine. In this chapter, the manifestation and/or pathogenesis of GI functioning and their relationship with GI disorders during sleep will be reviewed.

GASTROESOPHAGEAL REFLUX DISEASE

Nocturnal symptoms

The manifestation of GI symptoms during sleep is quite familiar to the practicing gastroenterologist. Perhaps the most obvious and common example is the occurrence of epigastric pain, characteristically awakening the patient from sleep in the early morning hours. This pattern of awakening from sleep is quite predictable by the patient and can help significantly in establishing a diagnosis of duodenal ulcer disease. Patients may also have awakenings from sleep with symptoms that ostensibly are not related to GI disorders. For example, individuals may complain of sleep disruption secondary to awakening from sleep with chest pain, heartburn, or regurgitation into the throat. Asthmatics may awaken from sleep by the exacerbation of bronchial asthma that can be secondary to gastroesophageal reflux (GER). Numerous studies are accumulating to suggest that respiratory complications secondary to GER are common, and these symptoms are primarily noted secondary to sleep-related GER.[1]

Other symptoms encountered by the practicing gastroenterologist that may occur during the day but whose occurrence during sleep adds a disconcerting dimension to the symptom include nocturnal diarrhea, fecal incontinence, chest pain, or the respiratory disorders noted above. Although a denial of symptoms that are thought to be related to GI problems such as GER does not necessarily preclude the occurrence of the sleep-related abnormalities, a positive symptom history would enhance the probability of the existence of a nocturnal GI disorder, as may be the case in patients with functional bowel disorders such as irritable bowel syndrome (IBS) or functional dyspepsia.

In a recent work by Orr et al.,[2] the hypothesis of whether the complaint of nighttime heartburn (NHB) as opposed to daytime heartburn (DHB) is a reliable indicator of actual sleep-related reflux events was investigated with combined pH monitoring and polysomnography (PSG). Those subjects with NHB are characterized by significantly greater sleep-related reflux events, more total and sleep-related acid contact time, and decreased measures of subjective sleep quality when compared with the DHB and control (no heartburn) groups. These data suggest that frequent NHB is a reliable symptom of a potentially more complicated form of gastroesophageal reflux disease (GERD) and may warrant more aggressive initial treatment in order to avoid the possible serious complications of sleep-related GER.

Nocturnal acid secretion

Patients with duodenal ulcer disease maintain a circadian pattern of gastric acid secretion, and studies have shown that the levels of secretion are enhanced.[3] This particular study shows that the peak of basal acid secretion occurs at approximately midnight, with minimal acid secretion occurring during the day in the absence of food ingestion. In addition, there does not appear to be any relation between the stages of sleep and gastric acid secretion. However, a study by Orr et al.[4] demonstrated a failure to inhibit acid during the first 2 h of sleep in patients with duodenal ulcer disease. Multicenter trials with bedtime administration of histamine (H_2) receptor antagonists have documented the efficacy of healing duodenal ulcers through nocturnal acid suppression.[5,6] These studies uniformly documented that duodenal ulcer-healing rates were at least as good with a once-a-day, bedtime dose of these potent acid-suppressing compounds as with the more conventional multiple daily dosing regimens. Howden et al.[7] reviewed the published data on nocturnal dosing of H_2 receptor antagonists in more than 12,000 patients with duodenal ulcer disease. They concluded that nocturnal dosing showed a clear advantage over multiple daily doses. These data strongly support the notion that nocturnal acid suppression alone is sufficient to heal a duodenal ulcer.

Other studies in patients with refractory duodenal ulcer suggest that nocturnal acid suppression is not only sufficient, but also necessary for duodenal ulcer healing. In a study by Gledhill et al.,[8] it was demonstrated that a reduction in nocturnal acid secretion through parietal cell vagotomy produced an enhanced healing rate in patients who were unresponsive to conventional cimetidine treatment. In a similar study by Galmiche et al.,[9] 20 patients with duodenal ulcer who were resistant to conventional cimetidine treatment received 150 mg ranitidine twice daily for 6 weeks. They demonstrated that in 8 patients the ulcer was healed, whereas in 12 patients, the ulcer remained unhealed. Patients whose ulcers healed had a substantial suppression of nocturnal acid secretion, whereas patients whose ulcers failed to heal maintained a nocturnal peak in gastric acid secretion. A subsequent study found that in persons who have had a parietal cell vagotomy, nocturnal acid secretion was significantly greater in those who experienced ulcer recurrence than in those who did not.[10] Further support for the important role of nocturnal acid secretion in the pathogenesis of duodenal ulcer disease comes from data showing that the maintenance of a modest degree of nocturnal acid suppression will effectively prevent the recurrence of duodenal ulcer disease.[11,12] These studies compared the use of 150 mg ranitidine at bedtime with 400 mg cimetidine at bedtime and found ranitidine to be superior in the prevention of ulcer recurrence. This finding is most likely due to the increased potency of ranitidine and its enhanced effectiveness in producing nocturnal acid suppression. A study actually documenting effective nocturnal acid suppression with 150 mg ranitidine at bedtime was reported by Santana et al.,[13] who concluded that "it may be relevant to the pathogenesis of duodenal ulceration that the short lived decrease in nocturnal acidity observed in this study is sufficient to prevent relapse of ulceration in most patients."

GER during sleep

GER, particularly with its familiar symptom of heartburn, is recognized as a common phenomenon. Most normal people will experience occasional bouts of heartburn. About 7% of the normal population experiences heartburn nearly every day.[14] Furthermore, the majority of patients with frequent heartburn complain of nighttime GER symptoms, and a substantial proportion (>50%) of these patients report that their symptoms disrupt their sleep and affect their daytime functioning.[15,16] Many patients who present to a physician with a complaint of heartburn can be readily treated with simple alterations in lifestyle, such as the avoidance of certain provocative foods and the use of antacid therapy, although there are no data specifically on the utility of these measures in treating sleep-related symptoms.[17] The familiarity of this symptom and its rapid response, in most instances, to relatively simple therapeutic measures belie the severity and potential complications of this disease process. As will be reviewed, the complications of GER appear to be the result of recurrent episodes of sleep-related GER.

GER events do occur during sleep, but appear to occur most commonly during brief arousals from sleep. Two studies have been published that were remarkably similar in their results in that reflux events occurred much more frequently in patients with diagnosed GERD, and relatively infrequent in normal volunteers.[18,19] The majority of reflux events in both studies did occur in association with polygraphically determined brief arousal responses.

Attention has been focused on the importance of different patterns of GER associated with waking and sleeping.[20] These patterns were documented in studies involving 24-h monitoring of the distal esophageal pH. GER is identified when the pH falls below 4 for a period of more than 30 s, and the reflux episode is arbitrarily terminated when the pH reaches 4 or 5. In this landmark study, Johnson and DeMeester[20] described two different patterns of reflux. Reflux in the upright position occurs most often post-prandially and usually consists of two or three reflux episodes that are rapidly cleared (2–3 min). Reflux in the supine position is usually associated with sleep and with more prolonged clearance time.

These studies documented highly significant increases in acid–mucosa contact time in patients with esophagitis, and these differences were most impressive when the supine position or sleep interval was considered; that is, there was a greater difference between patients and control subjects in the supine position as opposed to in the upright position. These investigators have also asserted that even though acid–mucosa contact time may be equivalent in the upright and supine positions, the prolonged acid clearance times associated with sleep appeared to result in greater damage to the esophageal mucosa.[21] In another study, the same investigators attempted to correlate the relation between the patterns of GER, as determined by 24-h esophageal pH monitoring, and the endoscopic evaluation of the esophageal mucosa.[22] They identified patients as primarily *upright* (waking) *refluxers, supine* (sleep) *refluxers,* and *combined refluxers,* who have both types of reflux throughout 24 h of monitoring. The severity of endoscopic change according to three grades of esophagitis was determined. Grade 1 esophagitis was defined simply as distal erythema and friability; grade 2 esophagitis was defined when mucosal erosions were noted; and grade 3 esophagitis involved more severe ulcerations and strictures. Their data indicated that an increasing incidence of nocturnal acid exposure was associated with more severe esophageal mucosal damage. An additional study compared the results of 24-h esophageal pH monitoring in patients who had either normal findings on endoscopy or erosive esophagitis.[23] The results of this study showed that total acid exposure time and the number of reflux episodes requiring longer than 5 min to clear acid were found to most reliably discriminate these two groups of patients. Furthermore, these authors found that 50% of the patients with reflux symptoms had normal 24-h pH monitoring and that 29% of the patients with erosive esophagitis also had normal pH studies. It is of interest that the most effective variable in distinguishing the two groups of patients was the number of episodes requiring longer than 5 min to clear acid (to pH 4).

An extension of these findings comes from a study by Orr et al.,[24] in which 24-h pH monitoring was conducted in a group of symptomatic patients with heartburn and normal endoscopic results and a group of patients with severe complications of GER, including erosive esophagitis, stricture, and Barrett's esophagus. The results showed that the best discriminator between the two groups was the number of episodes of prolonged acid clearance (longer than 5 min) in the supine position. These episodes appear to be more likely to occur during sleep, and this finding certainly confirms the notion that prolonged acid clearance is an important determinant in the development of esophagitis. Other investigators have not been as enthusiastic in their support of nocturnal GER as an important factor in the pathogenesis of reflux esophagitis. De Caestecker et al.[25] found that postprandial acid exposure was the best predictor of the severity of esophagitis, and their results led them to conclude that nocturnal reflux was substantially less important in the production of esophagitis.

The other consideration is whether the pattern of esophageal acid exposure may be more indicative of the state of consciousness rather than posture.[26–28] A study performed by Dickman et al.,[29] utilizing a diary during pH testing, attempted to determine if the principal acid reflux characteristics of recumbent–awake are closer to recumbent–asleep or upright (waking). The study enrolled a total of 64 GERD patients based on the results of the upper endoscopy and/or 24-h pH monitoring. The patients were instructed to carefully document their upright, recumbent–awake and recumbent–asleep periods. It was found that the reflux parameters such as the mean number of reflux events per hour, mean percentage total time pH < 4, and number of sensed reflux events were similar between upright and recumbent–awake periods. In contrast, these parameters were significantly higher in both upright and recumbent–awake periods compared to recumbent–asleep. It was concluded that the principal characteristics of the acid reflux events in the recumbent–awake period are closer to the upright period than to the recumbent–asleep period. The implication for that observation is that the state of consciousness (awake vs. asleep) is physiologically more important for 24-h esophageal monitoring analysis than body posture (upright vs. recumbent).[29] The awareness of GER and its occurrence across the spectrum of states of consciousness will be helpful for optimizing treatment of GERD.

A recent study addressed the interrelationship between quality of sleep and GER.[30] This work elegantly assessed the correlation between the severity of GERD symptoms and esophageal acid contact time and subjects' perceived quality of sleep. The authors determined the correlation between reported quality of sleep of the night prior and severity of GERD symptoms and esophageal acid contact time the following day, as well as the correlation between acid reflux events and sleep architecture. Subjects with typical GERD symptoms ≥3 times a week underwent upper endoscopy and pH monitoring. These subjects subsequently completed the GERD Symptom Assessment Score (GSAS) and the Sleep Heart Health Study Sleep Habits (SHHS) questionnaire to assess baseline sleep symptoms and GERD symptoms, including an index of GERD symptom severity. Before and after the pH test, the patients completed a different instrument, the Sleep Quality Questionnaire,

utilized specifically to assess the quality of each subject's sleep before and after pH testing. A PSG study during the pH test was also conducted in all participants.

Using data from the GSAS and SHHS questionnaires, disorders of initiating and maintaining sleep were found to be positively associated with greater severity of the GERD symptom index ($r = 0.33$, $p < 0.05$). More frequent awakenings were also correlated with a higher GERD symptom index ($r = 0.4$, $p < 0.01$). Correlations between the Sleep Quality Questionnaire on the night before sleep testing and pH monitoring data showed that subjects with poorer sleep quality had longer acid reflux events ($r = -0.34$, $p < 0.05$). More perceived awakenings were correlated with the number of supine acid reflux events of >5 min ($r = 0.31$, $p < 0.05$) and the duration of the longest supine acid reflux event ($r = 0.28$, $p = 0.05$). There were inverse correlations between overall sleep quality on the pH testing night and a higher percentage of time spent with pH < 4 supine ($r = -0.432$, $p < 0.002$), as well as the duration of the longest supine acid reflux event ($r = -0.37$, $p < 0.02$). It was concluded that subjects with worse GERD symptoms reported poorer subject sleep quality; while poor sleep quality on the night prior to pH testing was associated with greater acid exposure on the following day. Greater acid exposure at night was related to a worse perception of sleep quality on the next day. This work highlights the important interactions between GERD and sleep quality.

Acid clearance during sleep

As previously noted, acid clearance during sleep seems to be an important contributing factor in the development of reflux esophagitis. The process of acid clearance has been well studied by Helm et al.,[31] who described a two-factor theory of acid clearance. They proposed that acid clearance takes place in two phases: an initial phase, termed *volume clearance*, and a second phase, termed *acid neutralization*. Their data indicate that the vast majority of the volume of refluxed material is cleared from the esophagus quite rapidly by the first two or three swallows. There remains a coating of acid on the esophageal mucosa, which keeps the esophageal pH well below 4. Subsequent swallows serve to deliver saliva to the distal esophagus, and with its potent buffering capacity, the distal esophageal pH is returned to its normal level (5.5–6.5). A subsequent study confirmed these findings in that acid clearance was found to be independent of the swallowing rate, but significantly altered by an anticholinergic drug that inhibits salivation.[32]

Both swallowing frequency and salivation have been shown to be markedly depressed during sleep; as a result, one would hypothesize a prolongation of acid clearance during sleep.[33,34] Lichter and Muir[33] have shown that swallowing occurs sporadically during sleep, and there are long periods (longer than 30 min) without swallowing. Overall, the rate of swallowing during sleep is approximately 6 swallows per hour, and the swallows usually occur in association with a movement arousal. The highest frequency of these events is in stages 1 and 2 and rapid eye movement (REM) sleep.[33] Similarly, a study by Orr et al.[27] showed a marked reduction in swallows associated with esophageal acid infusion during sleep. Studies have focused specifically on the issue of the parameters of esophageal acid clearance during sleep. A model that incorporates the clearance of infused acid (15.0 mL 0.1 N HCl) during sleep was used in these studies. As opposed to simply analyzing spontaneous GER, this allowed the infusion of acid into the distal esophagus during specific periods of documented sleep (REM versus non-REM [NREM]), and it allowed for the precise timing of infusions such that the amount of sleep before each infusion could be relatively well controlled. This model also permitted a precise comparison of acid clearance during waking and during sleep under well-controlled conditions.

The initial study in this series involved a comparison of acid clearance during sleep in normal volunteers and patients with mild-to-moderate esophagitis.[29] The results revealed that sleep infusions in both groups were associated with a statistically significant prolongation of acid clearance time. In minutes, the absolute clearance time was nearly doubled in both groups. However, there was no significant difference between the clearance times in the patients and in the control subjects. The latter finding is believed to be a somewhat academic point because, as noted previously, normal persons rarely have reflux during sleep, whereas it is somewhat more common in patients with esophagitis. In addition, it was clear from the PSG observations that clearance was invariably associated with an arousal from sleep, and if this did not occur, there was a marked prolongation in acid clearance time. To more precisely evaluate this notion, clearance intervals that were associated with greater or less than 50% of waking during the clearance interval were compared. The clearance trials that involved more than 50% of waking had significantly faster clearance times. These data led to the conclusion that both arousal responses and waking are important elements in the response to an acidic distal esophagus.

To evaluate the motor functioning of the esophagus during sleep and the associated arousals from sleep, a subsequent study was performed using a specially designed esophageal probe to monitor not only distal esophageal pH, but also esophageal peristalsis.[28] This study also confirmed the importance of arousal responses in the efficient clearance of acid from the distal esophagus. The test was performed on normal volunteers who had a negative acid perfusion test; that is, they did not show any sensitivity to acid dripped in the distal esophagus and could not distinguish acid from water in the esophagus. However, the determination of arousal responses to these two different substances infused during sleep revealed that the acid infusions produced a significantly greater number of arousal responses. In addition, an exponential relation was described between the percentage of waking during the acid clearance interval and the acid clearance time; that is, the greater the amount of waking during the acid clearance interval, the faster the clearance time. Again, this finding substantiates those from the previous study of patients with esophagitis. This study did not document any difference between peristaltic parameters during sleep and during waking.

In a recent study with high-resolution manometry (HRM) and 24-h multichannel impedance pH monitoring,[35] total bolus clearing time in the upright and supine positions and acid exposure time were evaluated in 40 GERD patients without hiatal hernia. It was shown that patients with a pathological number of large breaks in esophageal transit on HRM had significantly impaired supine bolus clearance time and higher acid exposure time. GERD patients with esophageal erosions were characterized by significantly higher supine bolus clearance time and longer acid exposure time. It was concluded that GERD patients with a pathological number of large breaks are characterized by a significantly prolonged supine reflux clearance and subsequently higher acid exposure time.

To more definitively test the hypothesis that complications of GER are associated with prolonged acid clearance, a group of 13 patients with Barrett's esophagus was studied.[36] Barrett's esophagus is a condition believed to be related to chronic, severe GER, which results in the replacement of normal esophageal squamous epithelium with gastric columnar epithelium. The results of this study proved to be quite surprising, in that the patients with Barrett's esophagus were shown to have significantly faster acid clearance during sleep and waking compared with the control subjects. These data were, however, quite compatible with previous results in documenting the importance of arousal responses in the clearance process. The patients with Barrett's esophagus showed both a higher frequency of arousal responses and a shorter latency to the first swallow than control subjects.

Further illustrating the importance of these parameters in differentiating the patients with Barrett's esophagus from the control subjects is the fact that they could not be distinguished on the basis of any parameters associated with esophageal motor functioning, such as the amplitude of the peristaltic contraction or the esophageal transit time. It is especially notable in this study that there was a remarkable number of episodes of spontaneous GER in the group with Barrett's esophagus compared with the control subjects. These data led to the conclusion that the severe esophagitis in the patients with Barrett's esophagus is acquired through repeated episodes of spontaneous GER during sleep, which are associated with a prolongation of the acid clearance time. Although this was demonstrated to be faster than in normal control subjects, the acid clearance still is substantially longer than that occurring during waking.

Therapeutic considerations

In a review of a variety of studies related to NHB and sleep-related GER, Orr put for the notion that NHB and sleep-related GER constitute a distinct clinical entity worthy of recognition by clinicians.[37] It is suggested in this review that early recognition of the presence of NHB in GERD patients suggests the presence of erosive esophagitis, and more aggressive treatment to suppress sleep-related GER will optimize GERD treatment. GER appears to be affected by sleeping position. In a study by Khoury and colleagues,[38] it was determined that sleeping in the left lateral position significantly reduced the incidence of GER. The results of previously cited studies suggest that the pattern of GER during waking and sleep is important in that sleep-related reflux produces a prolongation of acid clearance.[23,24] Additional documentation of the importance of this pattern of the prolongation of acid clearance comes from studies that have shown that the back diffusion of hydrogen ions in the esophagus is directly related to the duration of acid–mucosa contact time.[39] Further evidence for the importance of nocturnal GER comes from a clinical study that documented that individuals with symptoms of nocturnal heartburn, as well as dysphagia and chest pain, were much more likely to have demonstrable esophageal disease.[40]

These results, as well as those described in the cited studies, tend to substantiate the time-honored clinical approach to persistent reflux, which is the suggestion that the patient sleep with the head of the bed elevated. This clinical axiom has survived for decades with little in the way of objective documentation that it actually is an efficacious approach to GER. Johnson and DeMeester[41] specifically tested this clinical axiom. Using intraesophageal pH monitoring during sleep, they demonstrated that sleeping with the head of the bed elevated produced a 67% improvement in the acid clearance time; however, the frequency of reflux episodes remained unchanged. The use of a cholinergic drug (bethanechol) that produces an elevation in lower esophageal sphincter pressure and increases esophageal peristaltic efficiency resulted in a decrease in both reflux frequency (30%) and acid clearance time (53%). The authors concluded that nocturnal reflux was most responsive to these therapeutic modalities. Another clinical axiom—avoidance of late evening meals—has also been directly tested in a study by Orr and Harnish.[42] They noted that reflux events during monitored sleep were not increased by a late evening provocative meal in patients with symptomatic GER disease. However, they did note that an OTC dose (75 mg) of ranitidine at bedtime was effective in significantly reducing reflux events during sleep. Along these lines, it has been demonstrated that the time interval between the evening meal and sleep can influence the occurrence of sleep-related GER. Piesman and colleagues demonstrated in a prospective, crossover study that less time between the evening meal and sleep onset results in significantly fewer sleep-related reflux events compared to a similar meal given with a longer interval between the evening meal and sleep onset.[43]

The use of H_2 receptor antagonists to suppress gastric acid secretion has been shown to be effective in the relief of heartburn.[44,45] One study using 24-h ambulatory esophageal pH assessments in patients with symptomatic heartburn and documented GER demonstrated that increasing doses (40 mg at bedtime, 20 mg twice daily and 40 mg twice daily) of an H_2 receptor antagonist (famotidine) produced increasing reductions in daytime and total acid–mucosa contact.[46] The three dosing regimens were equally effective in reducing nocturnal acid contact time. Thus, in contrast to duodenal ulcer disease, it does not appear that only bedtime dosing is adequate to treat GER. However, these data suggest that GER can be adequately controlled through effective gastric acid suppression. A study by Cohen and colleagues reported on the effect of a Nissen fundoplication on sleep-related GER and heartburn symptoms, as well as objective and subjective sleep measures.[47] They noted a significant decrease in both sleep-related reflux events and symptoms after surgery. Subjective and objective sleep measures were improved after surgery. More specifically, there was a very small but significant increase (approximately 5%) in NREM sleep, while there were more robust improvements in difficulty falling asleep, sleep quality, and daytime sleepiness.

An interesting finding was reported by Kerr et al.,[48] who showed that the administration of nasal continuous positive airway pressure (CPAP) to patients who were being treated for obstructive sleep apnea had the additional therapeutic benefit of reducing nocturnal GER and consequent esophageal acid contact time. Similar improvement was reported in patients with GER without apnea.[49] Since these studies examined only a single night of CPAP, it is important to note that this effect was replicated in a protocol that utilized 1 week of continuous CPAP treatment in patients with documented moderate/severe obstructive sleep apnea and abnormal acid contact time documented by 24-h esophageal pH monitoring.[50] In this study, patients who were studied at the end of 1 week of CPAP treatment were shown to have a significant decrease in sleep-related acid contact time. Also of interest with regard to sleep apnea syndrome

is a study by Graf et al.[51] that found patients with sleep apnea to have a high incidence of GER. The authors further determined that there is no relation between severity of sleep apnea and GER, nor is there any relation between apneic events and reflux events. A similar conclusion was reached from a study reported by Tardif et al.[52] Similar results have been reported by Ing and associates,[53] but they also reported an overall increase in reflux events and acid clearance time in spite of the fact that there was no clear relationship between obstructive apneic events and reflux events.

The question remains as to what is the pathogenesis of increased sleep-related GER in patients with obstructive sleep apnea (OSA). They clearly share some risk factors, such as obesity, alcohol consumption, and perhaps hiatal hernia, but based on some clinical and physiologic data, obesity alone would not appear to be an adequate explanation for this relationship. Suganuma and colleagues described an association between obesity and symptoms of OSA, but they found no relationship between obesity and reflux symptoms.[54] On the other hand, in another questionnaire study, Green et al.[55] showed a significant reduction in NHB symptoms after CPAP treatment, and those with higher CPAP levels had a greater reduction in symptoms of NHB. In another study that evaluated reflux symptoms (i.e., heartburn and acid regurgitation) in patients with a diagnosis of OSA, Valipour and associates found that there was no difference in reflux symptoms between those with a diagnosis of OSA and those designated as "simple snorers."[56] Furthermore, they found no relationship between the severity of OSA and reflux symptoms. It should be noted that, in contrast to the Green et al. study, this investigation did not specifically address nighttime GER symptoms. In addition, Valipour et al. noted in their discussion that the incidence of reflux symptoms did not appear to be greater than that noted in some general population surveys.

In a broader and somewhat more definitive study, Shepherd and colleagues compared the prevalence of nighttime symptoms and sleep-associated risk factors in untreated OSA patients, patients treated with CPAP, and the general population.[57] They reported that frequent (>2/week) NHB symptoms were more frequent in untreated OSA patients compared to the general population. In addition, they noted that the prevalence of frequent NHB was greater in those in the general population who were at high risk of OSA compared to those identified as at low risk.

An interesting physiologic study addressed the issue of the relationship between GER and OSA by identifying individuals with significant OSA and GER and treating them with acid suppression with a proton pump inhibitor (PPI).[58] In this study, Senior and colleagues noted that after treatment with the PPI for 1 month, there was a significant reduction in complete obstructive events, although the overall respiratory disturbance index (RDI) only showed a strong trend towards a reduction after treatment (p < 0.06). On the basis of these data, it would appear that the incidence of NHB is elevated in patients with OSA. In the Green et al. study, this was noted to be 62%, which is similar to the rate noted above in patients with frequent heartburn complaints.[16,55] Furthermore, the frequency shows a very significant decline with CPAP treatment. This coincides nicely with other physiologic studies that have clearly shown a significant decline in esophageal acid contact time with CPAP treatment.[48,49] The mechanism by which CPAP reduces esophageal acid contact time remains controversial, but an interesting study from this same group has suggested that some residual lower esophageal sphincter (LES) pressure (>10 mmHg) may be necessary for nasal CPAP to affect nocturnal reflux.[58]

A new PPI that might be considered for NHB treatment is dexlansoprazole, which is a dual delayed-release formulation PPI with an initial release 1–2 h after administration and another 4–5 h after ingestion.[59] In addition, the drug's absorption and activation are not influenced by a meal and can thus be administered prior to bedtime without concern that optimal efficacy will be compromised.[60] Dexlansoprazole has also been shown to be efficacious in reducing symptoms of both DHB and NHB.[61] It is theoretically suggested that a high index of suspicion of nighttime GER would need a more aggressive approach to treatment and the selection of a treatment regimen that would provide maximal acid suppression during the sleeping interval in an effort to reduce or eliminate sleep-related GERD more effectively.

Of considerable interest is the fact that studies have documented that commonly consumed sedating drugs, such as hypnotic drugs and alcohol, have been shown to prolong acid clearance during sleep.[62,63] A study by Vitale et al.[63] showed that alcohol consumption approximately 3 h before

sleep resulted in marked prolongation of the clearance of spontaneous episodes of GER. In another investigation of the effect of the administration of commonly used hypnotic drugs, a decrease was shown in the arousal latency and a prolongation was shown in the acid clearance time with triazolam.[64] In a more recent study using more upto-date technology, as well as the most commonly prescribed hypnotic (i.e., zolpidem), Gagliardi and colleagues noted that acid clearance was prolonged and arousal responses to reflux events were suppressed.[65] These data suggest that caution should be exercised in prescribing any central nervous system depressant medication to GERD patients.

Baclofen, a $GABA_b$ agonist, has been shown to reduce episodes of GER.[66] Orr et al. determined the effect of baclofen on GER during sleep, as well as objective and subjective measures of sleep, by the application of PSG and simultaneous esophageal pH monitoring.[67] It was found that baclofen significantly reduced the number of reflux events compared with placebo during sleep. Sleep disturbance was alleviated by baclofen due to a significant improvement in PSG variables, such as total sleep time, sleep efficiency, wake after sleep onset, and proportion of stage 1 sleep. Therefore, it was concluded that baclofen could be a useful adjunct therapy in patients with NHB and sleep disturbance who continue to have heartburn and/or sleep complaints despite PPI therapy.

INTESTINAL MOTILITY AND IBS

Due to the technological and practical difficulties in monitoring intestinal activity, particularly in patients, relatively little has been done in terms of acquiring data on intestinal motility during sleeping and waking in patient populations of interest. However, some data have been gradually appearing in the medical literature. One study that used 24-h ambulatory monitoring of small intestinal motility in patients with IBS was accomplished by Kellow et al.[68] Although nocturnal motor patterns did not differentiate the patient group from the control subjects, the notable lack of motility activity during sleep led these investigators to suggest that the changes in motor functioning that were noted were primarily the result of reactions to various "stressful" events occurring in the waking state. However, in a subsequent study, Kumar and colleagues noted a marked increase in REM sleep in patients with IBS.[69] These data were interpreted as lending additional support to their speculation that this syndrome has a central nervous system pathogenesis.

The observation by Kumar et al.[69] that REM sleep was enhanced in patients with IBS has prompted a number of investigations into sleep and functional bowel disorders to include IBS and functional dyspepsia. Using subjective reports of sleep quality, Goldsmith and Levin[70] documented that the exacerbation of IBS symptoms and poor sleep show a strong correlation. The obvious problems of a subjective study without physiological measurement of sleep are noted by Wingate,[71] one of the authors of the original study on IBS and sleep. Wingate points out that the occurrence of waking symptomatology may unduly influence the perception of the previous night's sleep, and without PSG documentation of sleep, this influence cannot be discounted. However, a more recent study included both objective and subjective measures of sleep, along with daily measures of pain.[72] In this study, subjective reports of poor sleep did predict subsequent daytime pain; however, daytime pain was not associated with nighttime sleep quality. Objective sleep measures (i.e., actigraphy) also predicted anxiety and fatigue, but not pain. In a study in patients with non-ulcer dyspepsia (characterized by epigastric postprandial bloating, nausea, or early satiety), two-thirds of 65 patients with non-ulcer dyspepsia complained of general sleep disturbances suggestive of non-restorative sleep (i.e., numerous awakenings after sleep onset and morning awakenings without feeling rested).[73] These were significantly more common than noted in control subjects, and 65% of those complaining of sleep disturbance attributed their sleep problem to their abdominal symptoms. Ten patients were also studied for a 24-h period with intestinal manometric monitoring, and a change in the rhythmicity and frequency of intestinal contractions was found in the functional dyspeptics during the nocturnal recording interval. In a subsequent study, Fass and colleagues documented a marked increase in sleep complaints in patients with functional bowel disorders (i.e., IBS and functional dyspepsia) compared to normal controls.[74] In addition, the dyspeptic patients had more complaints of sleep disturbance, perhaps due to more intense abdominal pain.

Increased sleep complaints have also been documented in studies by Elsenbruch et al., Heitkemper and colleagues and Rotem et al.[75–77] In these studies, full PSG was conducted. In the study by Elsenbruch and colleagues, no difference was found in any of the sleep measures reported between IBS subjects and normals.[75] However, the Heitkemper et al. study did show a significant prolongation of REM sleep.[76] However, the study by Rotem and colleagues revealed sharply contrasting PSG data.[77] In this study, they noted a number of significant PSG differences in the IBS subjects compared to control groups, including decreased slow wave sleep, increased arousal responses, and increased waking after sleep onset. Further research with larger sample sizes and stratification by IBS diagnostic subtypes may help elucidate and resolve these discrepancies. The original study by Kumar et al.[69] documenting the enhancement of REM sleep in patients with IBS reported on only six individuals who had a single night of sleep subsequent to small bowel intubation for monitoring of intestinal motility. With a small number of subjects and no attempt to adapt individuals to the laboratory setting (even though a control group was used), there is a high probability that these results were spurious. Orr et al.[78] attempted to replicate this study while at the same time noninvasively monitoring a GI measure so that more natural sleep could be obtained. Nine patients with IBS and nine control subjects were studied with full PSG monitoring, and gastric electrical activity was monitored by surface electrogastrogram. In this study, a statistically significant increase in REM sleep was documented in the patients with IBS, but the absolute level of REM sleep was not nearly in the range reported by Kumar et al.[69] In addition, specific electrogastrographic changes were found to be associated with sleep in normal subjects that were not noted in patients with IBS. Normal volunteers showed a significant decrease in the spectral amplitude of the EGG three-cycle-per-minute rhythm during NREM sleep compared with the waking state. Of interest is the fact that during REM sleep, the amplitude was significantly increased to levels approaching those in the waking state. The patients with IBS failed to significantly modulate the amplitude of the dominant frequency of the gastric electrical rhythm during any of these states of consciousness. The lack of modulation of the dominant frequency of the electrogastrographic amplitude during sleep in patients with IBS raises the possibility that other autonomic abnormalities may be unmasked by further study of physiological functioning during sleep.

In a subsequent study by Thompson and colleagues,[79] it was noted that IBS patients who had dyspeptic symptoms in addition to classic symptoms of abdominal pain and cramping did not exhibit the increase in sympathetic dominance in REM sleep that was noted only in the patients with abdominal cramping alone as part of their IBS symptomatology. In a series of subsequent studies by Orr and colleagues, none replicated the finding of an increase in REM sleep in IBS patients.[77,79,80]

A previous study investigated whether sleep dysfunction would influence anorectal motility in IBS patients.[81] It was found that IBS patients were characterized by lower rectal thresholds for defecation as compared with age-matched controls. Of importance is that IBS patients with sleep dysfunction had a significantly lower threshold than those without sleep dysfunction. Additionally, poor sleep quality was associated with a decrease in rectal sensation. It was concluded that IBS patients with sleep dysfunction are characterized by lower thresholds for rectal perception. This work implies that sleep disturbance might be associated with anorectal dysfunction and that it appears to create some degree of rectal hyperalgesia in patients with IBS.

Collectively, these results from various sleep investigations in patients with functional bowel disorders suggest not only that there are sleep disturbances noted in this patient population, but also that the sleep disturbances may contribute to altered GI functioning. Certainly, these studies confirm the notion that there are central nervous system alterations in patients with functional bowel disorders, and that these alterations are perhaps uniquely identified during sleep. Future studies on sleep in patients with functional bowel disorders will undoubtedly provide additional understanding of the pathophysiology of the brain–gut axis and its alterations during sleep.

COLONIC MOTILITY

Bassotti et al.[82] monitored colonic motility for 24 h in normal volunteers and in patients with chronic constipation. Although they documented a decrease in the number and duration of mass movements in the patients with chronic constipation, as well as a circadian pattern of decrease in

mass movements during the night, no significant difference was noted between patients and control subjects with regard to the circadian pattern itself. In a similar study, Ferrara et al.[83] monitored the motor activity of the distal colon, rectum, and anal canal over a 24-h interval in patients with slow-transit constipation. These patients were compared with ten healthy control subjects. The patients with slow-transit constipation were noted to have impaired responses to feeding on awakening from sleep in the morning.

Another interesting observation regarding alterations in anorectal functioning during sleep concerns a study by Orkin et al.,[84] in which the authors monitored rectal motor activity during sleep in patients who had undergone ileal–anal anastomosis. They noted that decreases in anal resting pressure coupled with marked minute-to-minute variations in pressure during sleep occurred in control subjects and in patients, and when particularly profound, these led to nocturnal fecal incontinence in some patients.

SLEEP AND INFLAMMATORY BOWEL DISEASE

Several studies have shown that disruption of the normal sleep cycle is associated with an increased risk of impairing homoeostasis of GI function,[85] including GERD,[86] peptic ulcer disease,[87] and IBS.[88] Disturbed sleep can alter immune function and thus impact on disease course in GI inflammatory disorders such as Crohn's disease and ulcerative colitis. Inflammatory cytokines such as TNF-α and IL-1 are known to disrupt sleep,[89] and therefore may provoke a vicious cycle and positive feedback loop that maintain and perpetuate inflammation. Thus, sleep alterations have been demonstrated in patients with inflammatory bowel disease (IBD),[90] and the concept of an exacerbation of mucosal damage via altered sleep patterns is supported by evidence from animal studies. For example, sleep restriction has been associated with increased clinical markers of IBD activity, such as TNF-α, IL-6, and C-reactive protein.[91] In another animal study assessing the effects of acute and chronic intermittent sleep deprivation on the dextran sodium sulfate-induced colitis model, it was noted that both acute and chronic intermittent sleep deprivation exacerbated colonic inflammation, as evidenced by marked increases in tissue myeloperoxidase activity. The effects of sleep deprivation on colonic inflammation were dose dependent, with more severe inflammation in the chronic intermittent sleep deprivation group.[92]

It is well established that disrupted sleep can cause fatigue in inflammatory diseases and may be one of the causes of fatigue in IBD patients.[93] Of importance is that fatigue is one of the most common complaints in patients with IBD, having a major negative impact on quality of life, regardless of the status of active inflammation.[93] Moreover, sleep disturbance has been shown to be associated with an increased risk of disease flares in Crohn's disease, but not ulcerative colitis.[94] Other studies have also shown a significant relationship between disease activity and sleep disturbance. In a large prospective cohort study known as the Manitoba IBD Cohort Study, patients with active IBD (77%) were found to have lower sleep quality compared with patients who had inactive disease (49%).[95] Interestingly, the sleep quality of patients with inactive disease was also lower than the estimated prevalence of poor sleep in the general adult population (32%). The study found a significant positive correlation of sleep quality and daytime dysfunction, as well as between perceived stress and psychological distress. There was also a strong negative correlation between sleep quality, number of sleep hours per night, and IBD-related quality of life.[95] These findings suggest that the evaluation and treatment of sleep disturbance in patients with Crohn's disease will optimize treatment and improve clinical outcome.

CONCLUSIONS

There appears to be an important relationship between sleep and the development of various acid peptic diseases, such as duodenal ulcer disease and GERD. The pathogenesis and treatment of these disorders rely on the control of acid secretion during sleep. The occurrence of nocturnal GER is an unquestionably important aspect of the development of the serious complications of this disorder. Continuous monitoring of the distal esophageal pH in order to document nocturnal GER is becoming an important and useful tool in the overall management of GERD patients. In addition, prolonged monitoring of small and large bowel motility appears to be a promising tool for further understanding of the pathogenesis of various GI diseases and how these diseases may be altered by sleep. The intervention and treatment of sleep disorders in IBD patients should be regarded as an important part of the clinical management of IBD. These sleep-related phenomena are becoming important factors in the practice of state-of-the-art gastroenterology, and future research will undoubtedly further substantiate the important role of sleep in the pathogenesis of GI disease.

REFERENCES

1. Cuttitta G, Cibella F, Visconti A, Scichilone N, Bellia V, Bonsignore G. Spontaneous gastroesophageal reflux and airway patency during the night in adult asthmatics. *Am J Respir Crit Care Med* 2000; 161: 177–181.
2. Orr WC, Goodrich S, Estep ME, Shepherd K. The relationship between complaints of night-time heartburn and sleep-related gastroesophageal reflux. *Dis Esophagus* 2014; 27: 303–310.
3. Feldman M, Richardson CT. Total 24-hour gastric acid secretion in patients with duodenal ulcer. Comparison with normal subjects and effects of cimetidine and parietal cell vagotomy. *Gastroenterology* 1986; 90: 540–544.
4. Orr WC, Hall WH, Stahl ML, Durkin MG, Whitsett TL. Sleep patterns and gastric acid secretion in duodenal ulcer disease. *Arch Intern Med* 1976; 136: 655–660.
5. Colin-Jones DG, Ireland A, Gear P et al. Reducing overnight secretion of acid to heal duodenal ulcers. Comparison of standard divided dose of ranitidine with a single dose administered at night. *Am J Med* 1984; 77: 116–122.
6. Kildebo S, Aronsen O, Bernersen B et al. Cimetidine, 800 mg at night, in the treatment of duodenal ulcers. *Scand J Gastroenterol* 1985; 20: 1147–1150.
7. Howden CW, Jones DB, Hunt RH. Nocturnal doses of H$_2$ receptor antagonists for duodenal ulcer. *Lancet* 1985; 1: 647–648.
8. Gledhill T, Buck M, Paul A, Hunt RH. Cimetidine or vagotomy? Comparison of the effects of proximal gastric vagotomy, cimetidine and placebo on nocturnal intragastric acidity and acid secretion in patients with cimetidine resistant duodenal ulcer. *Br J Surg* 1983; 70: 704–706.

9. Galmiche JP, Tranvouez JL, Denis P et al. Does nocturnal monitoring of gastric pH permit the prediction of therapeutic response in severe duodenal ulcer treated with ranitidine? *Gastroenterol Clin Biol* 1985; 9: 583–589.

10. Gotthard R, Strom M, Sjodahl R, Walan A. 24-h study of gastric acidity and bile acid concentration after parietal cell vagotomy. *Scand J Gastroenterol* 1986; 21: 503–508.

11. Gough KR, Korman MG, Bardhan KD et al. Ranitidine and cimetidine in prevention of duodenal ulcer relapse. A double-blind, randomised, multicentre, comparative trial. *Lancet* 1984; 2: 659–662.

12. Silvis SE. Final report on the United States Multicenter Trial comparing ranitidine to cimetidine as maintenance therapy following healing of duodenal ulcer. *J Clin Gastroenterol* 1985; 7: 482–487.

13. Santana IA, Sharma BK, Pounder RE, Wood EC, Masters S, Talbot M. 24 hour intragastric acidity during maintenance treatment with ranitidine. *Br Med J (Clin Res Ed)* 1984; 289: 1420.

14. Nebel OT, Fornes MF, Castell DO. Symptomatic gastroesophageal reflux: Incidence and precipitating factors. *Am J Dig Dis* 1976; 21: 953–956.

15. Farup C, Kleinman L, Sloan S et al. The impact of nocturnal symptoms associated with gastroesophageal reflux disease on health-related quality of life. *Arch Intern Med* 2001; 161: 45–52.

16. Shaker R, Castell DO, Schoenfeld PS, Spechler SJ. Nighttime heartburn is an underappreciated clinical problem that impacts sleep and daytime function: The results of a Gallup survey conducted on behalf of the American Gastroenterological Association. *Am J Gastroenterol* 2003; 98: 1487–1493.

17. Orr W. Lifestyle measures for the treatment of gastroesophageal reflux disease. In: Bayless TM, Diehl AM (eds), *Advanced Therapy in Gastroenterology and Liver Disease* 2005; 5th edition. Hamilton: B.C. Decker Inc.

18. Freidin N, Fisher MJ, Taylor W et al. Sleep and nocturnal acid reflux in normal subjects and patients with reflux oesophagitis. *Gut* 1991; 32: 1275–1279.

19. Penzel T, Becker HF, Brandenburg U, Labunski T, Pankow W, Peter JH. Arousal in patients with gastro-oesophageal reflux and sleep apnoea. *Eur Respir J* 1999; 14: 1266–1270.

20. Johnson LF, DeMeester TR. Twenty-four-hour pH monitoring of the distal esophagus. A quantitative measure of gastroesophageal reflux. *Am J Gastroenterol* 1974; 62: 325–332.

21. DeMeester TR, Johnson LF, Joseph GJ, Toscano MS, Hall AW, Skinner DB. Patterns of gastroesophageal reflux in health and disease. *Ann Surg* 1976; 184: 459–470.

22. Johnson LF, DeMeester TR, Haggitt RC. Esophageal epithelial response to gastroesophageal reflux. A quantitative study. *Am J Dig Dis* 1978; 23: 498–509.

23. Schlesinger PK, Donahue PE, Schmid B, Layden TJ. Limitations of 24-hour intraesophageal pH monitoring in the hospital setting. *Gastroenterology* 1985; 89: 797–804.

24. Orr WC, Allen ML, Robinson M. The pattern of nocturnal and diurnal esophageal acid exposure in the pathogenesis of erosive mucosal damage. *Am J Gastroenterol* 1994; 89: 509–512.

25. De Caestecker, Blackwell JH, Brown J et al. When is acid reflux most damaging to the esophagus? [abstract]. *Gastroenterology* 1985; 88: 1360.

26. Orr WC, Elsenbruch S, Harnish MJ, Johnson LF. Proximal migration of esophageal acid perfusions during waking and sleep. *Am J Gastroenterol* 2000; 95: 37–42.

27. Orr WC, Johnson LF, Robinson MG. Effect of sleep on swallowing, esophageal peristalsis, and acid clearance. *Gastroenterology* 1984; 86: 814–819.

28. Orr WC, Robinson MG, Johnson LF. Acid clearance during sleep in the pathogenesis of reflux esophagitis. *Dig Dis Sci* 1981; 26: 423–427.

29. Dickman R, Shapiro M, Malagon IB, Powers J, Fass R. Assessment of 24-h oesophageal pH monitoring should be divided to awake and asleep rather than upright and supine time periods. *Neurogastroenterol Motil* 2007; 19: 709–715.

30. Dickman R, Green C, Fass SS et al. Relationships between sleep quality and pH monitoring findings in persons with

gastroesophageal reflux disease. *J Clin Sleep Med* 2007; 3: 505–513.

31. Helm JF, Dodds WJ, Hogan WJ, Soergel KH, Egide MS, Wood CM. Acid neutralizing capacity of human saliva. *Gastroenterology* 1982; 83: 69–74.

32. Allen ML, Orr WC, Woodruff DM, Duke JC, Robinson MG. The effects of swallowing frequency and transdermal scopolamine on esophageal acid clearance. *Am J Gastroenterol* 1985; 80: 669–672.

33. Lichter I, Muir RC. The pattern of swallowing during sleep. *Electroencephalogr Clin Neurophysiol* 1975; 38: 427–432.

34. Schneyer LH, Pigman W, Hanahan L, Gilmore RW. Rate of flow of human parotid, sublingual, and submaxillary secretions during sleep. *J Dent Res* 1956; 35: 109–114.

35. Ribolsi M, Balestrieri P, Emerenziani S, Guarino MP, Cicala M. Weak peristalsis with large breaks is associated with higher acid exposure and delayed reflux clearance in the supine position in GERD patients. *Am J Gastroenterol* 2014; 109: 46–51.

36. Orr WC, Lackey C, Robinson MG et al. Acid clearance and reflux during sleep in Barrett's esophagus. *Gastroenterology* 1983; 84: 1265.

37. Orr WC. Review article: Sleep-related gastro-oesophageal reflux as a distinct clinical entity. *Aliment Pharmacol Ther* 2010; 31: 47–56.

38. Khoury RM, Camacho-Lobato L, Katz PO, Mohiuddin MA, Castell DO. Influence of spontaneous sleep positions on nighttime recumbent reflux in patients with gastroesophageal reflux disease. *Am J Gastroenterol* 1999; 94: 2069–2073.

39. Johnson LF, Harmon JW. Experimental esophagitis in a rabbit model. Clinical relevance. *J Clin Gastroenterol* 1986; 8(Suppl 1): 26–44.

40. Andersen LI, Madsen PV, Dalgaard P, Jensen G. Validity of clinical symptoms in benign esophageal disease, assessed by questionnaire. *Acta Med Scand* 1987; 221: 171–177.

41. Johnson LF, DeMeester TR. Evaluation of elevation of the head of the bed, bethanechol, and antacid form tablets on gastroesophageal reflux. *Dig Dis Sci* 1981; 26: 673–680.

42. Orr WC, Harnish MJ. Sleep-related gastro-oesophageal reflux: Provocation with a late evening meal and treatment with acid suppression. *Aliment Pharmacol Ther* 1998; 12: 1033–1038.

43. Piesman M, Hwang I, Maydonovitch C, Wong RK. Nocturnal reflux episodes following the administration of a standardized meal. Does timing matter? *Am J Gastroenterol* 2007; 102: 2128–2134.

44. Behar J, Brand DL, Brown FC et al. Cimetidine in the treatment of symptomatic gastroesophageal reflux: A double blind controlled trial. *Gastroenterology* 1978; 74: 441–448.

45. Sontag S, Robinson M, McCallum RW, Barwick KW, Nardi R. Ranitidine therapy for gastroesophageal reflux disease. Results of a large double-blind trial. *Arch Intern Med* 1987; 147: 1485–1491.

46. Orr WC, Robinson MG, Humphries TJ, Antonello J, Cagliola A. Dose–response effect of famotidine on patterns of gastro-oesophageal reflux. *Aliment Pharmacol Ther* 1988; 2: 229–235.

47. Cohen JA, Arain A, Harris PA et al. Surgical trial investigating nocturnal gastroesophageal reflux and sleep (STINGERS). *Surg Endosc* 2003; 17: 394–400.

48. Kerr P, Shoenut JP, Millar T, Buckle P, Kryger MH. Nasal CPAP reduces gastroesophageal reflux in obstructive sleep apnea syndrome. *Chest* 1992; 101: 1539–1544.

49. Kerr P, Shoenut JP, Steens RD, Millar T, Micflikier AB, Kryger MH. Nasal continuous positive airway pressure. A new treatment for nocturnal gastroesophageal reflux? *J Clin Gastroenterol* 1993; 17: 276–280.

50. Tawk MM, Kinasewitz GT, Hunter D, Orr WC. Effect of one week of CPAP on esophageal reflux in patients with obstructive sleep apnea. *Chest* 2003; 124: 74S–75S.

51. Graf KI, Karaus M, Heinemann S, Korber S, Dorow P, Hampel KE. Gastroesophageal reflux in patients with sleep apnea syndrome. *Z Gastroenterol* 1995; 33: 689–693.

52. Tardif C, Denis P, Verdure-Poussin A, Hidden F, Pasquis P, Samson-Dollfus D. Gastroesophageal reflux during sleep in obese patients. *Neurophysiol Clin* 1988; 18: 323–332.

53. Ing AJ, Ngu MC, Breslin AB. Obstructive sleep apnea and gastroesophageal reflux. *Am J Med* 2000; 108(Suppl 4a): 120S–125S.

54. Suganuma N, Shigedo Y, Adachi H et al. Association of gastroesophageal reflux disease with weight gain and apnea, and their disturbance on sleep. *Psychiatry Clin Neurosci* 2001; 55: 255–256.

55. Green BT, Broughton WA, O'Connor JB. Marked improvement in nocturnal gastroesophageal reflux in a large cohort of patients with obstructive sleep apnea treated with continuous positive airway pressure. *Arch Intern Med* 2003; 163: 41–45.

56. Valipour A, Makker HK, Hardy R, Emegbo S, Toma T, Spiro SG. Symptomatic gastroesophageal reflux in subjects with a breathing sleep disorder. *Chest* 2002; 121: 1748–1753.

57. Shepherd KL, James AL, Musk AW, Hunter ML, Hillman DR, Eastwood PR. Gastroesophageal reflux symptoms are related to the presence and severity of obstructive sleep apnoea. *J Sleep Res* 2011; 20: 241–249.

58. Senior BA, Khan M, Schwimmer C, Rosenthal L, Benninger M. Gastroesophageal reflux and obstructive sleep apnea. *Laryngoscope* 2001; 111: 2144–2146.

59. Vakily M, Zhang W, Wu J, Atkinson SN, Mulford D. Pharmacokinetics and pharmacodynamics of a known active PPI with a novel dual delayed release technology, dexlansoprazole MR: A combined analysis of randomized controlled clinical trials. *Curr Med Res Opin* 2009; 25: 627–638.

60. Lee RD, Vakily M, Mulford D, Wu J, Atkinson SN. Clinical trial: The effect and timing of food on the pharmacokinetics and pharmacodynamics of dexlansoprazole MR, a novel dual delayed release formulation of a proton pump inhibitor—Evidence for dosing flexibility. *Aliment Pharmacol Ther* 2009; 29: 824–833.

61. Metz DC, Howden CW, Perez MC, Larsen L, O'Neil J, Atkinson SN. Clinical trial: Dexlansoprazole MR, a proton pump inhibitor with dual delayed-release technology, effectively controls symptoms and prevents relapse in patients with healed erosive oesophagitis. *Aliment Pharmacol Ther* 2009; 29: 742–754.

62. Shoenut JP, Kerr P, Micflikier AB, Yamashiro Y, Kryger MH. The effect of nasal CPAP on nocturnal reflux in patients with aperistaltic esophagus. *Chest* 1994; 106: 738–741.

63. Vitale GC, Cheadle WG, Patel B, Sadek SA, Michel ME, Cuschieri A. The effect of alcohol on nocturnal gastroesophageal reflux. *JAMA* 1987; 258: 2077–2079.

64. Orr WC, Robinson MG, Rundell OH. The effect of hypnotic drugs on acid clearance during sleep. *Gastroenterology* 1985; 88: 1526.

65. Gagliardi GS, Shah AP, Goldstein M et al. Effect of zolpidem on the sleep arousal response to nocturnal esophageal acid exposure. *Clin Gastroenterol Hepatol* 2009; 7: 948–952.

66. Vela MF, Tutuian R, Katz PO, Castell DO. Baclofen decreases acid and non-acid post-prandial gastro-oesophageal reflux measured by combined multichannel intraluminal impedance and pH. *Aliment Pharmacol Ther* 2003; 17: 243–251.

67. Orr WC, Goodrich S, Wright S, Shepherd K, Mellow M. The effect of baclofen on nocturnal gastroesophageal reflux and measures of sleep quality: A randomized, cross-over trial. *Neurogastroenterol Motil* 2012; 24: 553–559, e253.

68. Kellow JE, Gill RC, Wingate DL. Prolonged ambulant recordings of small bowel motility demonstrate abnormalities in the irritable bowel syndrome. *Gastroenterology* 1990; 98: 1208–1218.

69. Kumar D, Thompson PD, Wingate DL, Vesselinova-Jenkins CK, Libby G. Abnormal REM sleep in the irritable bowel syndrome. *Gastroenterology* 1992; 103: 12–17.

70. Goldsmith G, Levin JS. Effect of sleep quality on symptoms of irritable bowel syndrome. *Dig Dis Sci* 1993; 38: 1809–1814.

71. Wingate D. An association between poor sleep quality and the severity of IBS symptoms. *Dig Dis Sci* 1994; 39: 2350–2351.

72. Buchanan DT, Cain K, Heitkemper M et al. Sleep measures predict next-day symptoms in women with irritable bowel syndrome. *J Clin Sleep Med* 2014; 10: 1003–1009.

73. David D, Mertz H, Fefer L et al. Sleep and duodenal motor activity in patients with severe non-ulcer dyspepsia. *Gut* 1994; 35: 916–925.

74. Fass R, Fullerton S, Tung S, Mayer EA. Sleep disturbances in clinic patients with functional bowel disorders. *Am J Gastroenterol* 2000; 95: 1195–2000.

75. Elsenbruch S, Harnish MJ, Orr WC. Subjective and objective sleep quality in irritable bowel syndrome. *Am J Gastroenterol* 1999; 94: 2447–2452.

76. Heitkemper M, Charman AB, Shaver J, Lentz MJ, Jarrett ME. Self-report and polysomnographic measures of sleep in women with irritable bowel syndrome. *Nurs Res* 1998; 47: 270–277.

77. Rotem AY, Sperber AD, Krugliak P, Freidman B, Tal A, Tarasiuk A. Polysomnographic and actigraphic evidence of sleep fragmentation in patients with irritable bowel syndrome. *Sleep* 2003; 26: 747–752.

78. Orr WC, Crowell MD, Lin B, Harnish MJ, Chen JD. Sleep and gastric function in irritable bowel syndrome: Derailing the brain–gut axis. *Gut* 1997; 41: 390–393.

79. Thompson JJ, Elsenbruch S, Harnish MJ, Orr WC. Autonomic functioning during REM sleep differentiates IBS symptom subgroups. *Am J Gastroenterol* 2002; 97: 3147–3153.

80. Elsenbruch S, Thompson JJ, Hamish MJ, Exton MS, Orr WC. Behavioral and physiological sleep characteristics in women with irritable bowel syndrome. *Am J Gastroenterol* 2002; 97: 2306–2314.

81. Chen CL, Liu TT, Yi CH, Orr WC. Evidence for altered anorectal function in irritable bowel syndrome patients with sleep disturbance. *Digestion* 2011; 84: 247–251.

82. Bassotti G, Gaburri M, Imbimbo BP et al. Colonic mass movements in idiopathic chronic constipation. *Gut* 1988; 29: 1173–1179.

83. Ferrara A, Pemberton JH, Hanson RB. Motor responses of the sigmoid, rectum and anal canal in health and in patients with slow transit constipation (STC). *Gastroenterology* 1991; 100: A441.

84. Orkin BA, Soper NJ, Kelly KA, Dent J. Influence of sleep on anal sphincteric pressure in health and after ileal pouch-anal anastomosis. *Dis Colon Rectum* 1992; 35: 137–144.

85. Hoogerwerf WA. Role of biological rhythms in gastrointestinal health and disease. *Rev Endocr Metab Disord* 2009; 10: 293–300.

86. Demeter P, Visy KV, Gyulai N et al. Severity of gastroesophageal reflux disease influences daytime somnolence: A clinical study of 134 patients underwent upper panendoscopy. *World J Gastroenterol* 2004; 10: 1798–1801.

87. Segawa K, Nakazawa S, Tsukamoto Y et al. Peptic ulcer is prevalent among shift workers. *Dig Dis Sci* 1987; 32: 449–453.

88. Nojkov B, Rubenstein JH, Chey WD, Hoogerwerf WA. The impact of rotating shift work on the prevalence of irritable bowel syndrome in nurses. *Am J Gastroenterol* 2010; 105: 842–847.

89. Shoham S, Davenne D, Cady AB, Dinarello CA, Krueger JM. Recombinant tumor necrosis factor and interleukin 1 enhance slow-wave sleep. *Am J Physiol* 1987; 253: R142–R149.

90. Keefer L, Stepanski EJ, Ranjbaran Z, Benson LM, Keshavarzian A. An initial report of sleep disturbance in inactive inflammatory bowel disease. *J Clin Sleep Med* 2006; 2: 409–416.

91. Vgontzas AN, Zoumakis E, Bixler EO et al. Adverse effects of modest sleep restriction on sleepiness, performance, and inflammatory cytokines. *J Clin Endocrinol Metab* 2004; 89: 2119–2126.

92. Tang Y, Preuss F, Turek FW, Jakate S, Keshavarzian A. Sleep deprivation worsens inflammation and delays recovery in a mouse model of colitis. *Sleep Med* 2009; 10: 597–603.

93. Marcus SB, Strople JA, Neighbors K et al. Fatigue and health-related quality of life in pediatric inflammatory bowel disease. *Clin Gastroenterol Hepatol* 2009; 7: 554–561.

94. Ananthakrishnan AN, Long MD, Martin CF, Sandler RS, Kappelman MD. Sleep disturbance and risk of active disease in patients with Crohn's disease and ulcerative colitis. *Clin Gastroenterol Hepatol* 2013; 11: 965–971.

95. Graff LA, Vincent N, Walker JR et al. A population-based study of fatigue and sleep difficulties in inflammatory bowel disease. *Inflamm Bowel Dis* 2011; 17: 1882–1889.

3

Sleep and attachment disorders in children

KARL HEINZ BRISCH

INTRODUCTION

An infant's quiet night of sleep is a source of happiness and empowerment for parents. In prenatal classes, many parents worry that their baby might develop a sleep disorder and that nighttime could become an intense scene of crying and responses. Indeed, quite a high percentage of infants and children develop sleep disorders, and nocturnal wakings and bed sharing are quite common during early childhood. During infancy, the frequency of nightwakings increases with maturation of locomotion.[1] Nocturnal wakings have been reported in 20%–30% of 1- to 3-year-olds.[2,3] These findings are despite the fact that methodological problems exist in assessing sleep problems in infants, and it is well documented that maternal reports do not objectively reflect the sleep patterns of their infants.[4] Although a sleep disorder does not necessarily lead to an attachment disorder, an infant's crying through the night can be the start of a disturbed parent–infant relationship that may conclude with this result. Conversely, attachment disorders in children are also associated with a range of psychosomatic problems, one of which is sleep problems. If a sleep disorder and an attachment disorder are a baby's predominant symptoms, the parent–infant and, later, parent–child relationship will be stressful and, in the worst case, can progress to a vicious circle of crying and physical abuse.[5] Parental personality, psychopathology, and related cognitions and emotions contribute to parental sleep-related behaviors and, ultimately, influence infant sleep. However, the links are bidirectional and dynamic, so that poor infant sleep may influence parental behaviors and poor infant sleep appears to be a family stressor and a risk factor for maternal depression. Therefore, it is necessary to understand more about the association of sleep and attachment and their disorders in children and parents, and to strategically plan prevention measures that can help parents and infants establish sleep patterns and regulate sleep rhythms from the beginning.[5–13]

ATTACHMENT THEORY AND DISORDERS

Attachment is a fundamental human motivation that helps the infant to survive. During the first year, an infant develops a specific, exclusive attachment relationship to an attachment figure that serves as a secure base for the infant and provides protection. Once the baby's attachment system starts to develop, which can be observed from 12 weeks onward, the infant reacts on separation with attachment behavior, such as crying to protest separation from the attachment figure, followed by seeking physical contact and reunion.[14] We can distinguish three different patterns of attachment quality: a securely attached infant will protest after separation from his or her attachment figure and will calm down quickly after reunion. An insecurely avoidant attached infant will appear not to be stressed by separation and will not actively seek physical contact with the attachment figure after reunion, whereas an insecurely anxious–ambivalent attached infant will react with extreme arousal and will take a long time to settle down after his or her attachment figure has returned. It is typical that the attachment system of the infant, once activated, can be preferentially calmed by physical contact with the attachment figure. Only if the primary attachment figure, such as the mother, is not present does the infant allow a secondary attachment figure, such as the father, to soothe him or her.[15–19]

Attachment disorders are caused by an infant's early experiences of repeated separation and multiple traumas. Such disorders commonly evolve from traumatic events such as physical, sexual, or emotional violence and severe deprivation, often perpetrated by attachment figures. In addition, if an attachment figure is sometimes a source of emotional availability and protection for the child and at other times a source of violence and anxiety, it will be difficult for the child to organize these disparate experiences into a coherent internal working model of attachment.[19,20]

On a behavioral level, attachment disorders may emerge as strange patterns. Two forms of attachment disorder are included in the *International Classification of Diseases* (ICD-10) and *Diagnostic and Statistical Manual of Mental Disorders* (DSM-IV/V). One pattern involves nonselective, undifferentiated attachment behavior. Children possessing this pattern exhibit promiscuous attachments, rapidly and seemingly randomly seeking physical contact with strangers. They are indiscriminately friendly toward strangers, who by definition can never be real attachment figures. Other children

display a type of disorder that is characterized by inhibited attachment behavior: These children, although anxious, do not show their attachment behavior, instead suppressing their attachment activities, which results in a continuous state of high arousal. Additional types of attachment disorder have been classified, including attachment disorders with psychosomatic symptoms (e.g., sleep problems).[19] Further types of attachment disorders (such as nonattachment behavior in attachment-relevant situation, aggressive behavior, role reversal, aggressive symptoms, and a hyperactivation of attachment behavior) also show pathological behavior patterns in attachment-relevant situations.[21]

Separation at night for sleep is one of the attachment-related situations leading to activation of the attachment system. Children with different types of attachment disorder may have disturbed sleep patterns or even sleep disorders. For example, some attachment-disordered children cannot calm down easily at night or wake up often and suffer from nightmares and night-walking. These disorders may manifest through hyperactivity of their attachment system, or the children may have difficulty separating before sleep. Other children may suffer from an inhibited attachment disorder and will anxiously lie in bed and not cry at night to seek the attachment figure. Caregivers of these latter children may thus think that the infants are easily cared for, whereas the babies are instead lying in bed in a state of hyperarousal. Their hyperarousal and inhibition of showing attachment may cause them to complain of stomachaches or headaches, vomit or develop an elevated temperature. If attachment figures do not understand these signals and prefer children who do not cry at night, children may develop chronic psychosomatic symptoms. Still other children may suffer from undifferentiated attachment disorders (as most foster infants do) and will be happy when anyone picks them up from bed. They might calm down for a short while, but again cry until another person comes along. No secure attachment representation results from this undifferentiated attachment behavior, so that while the children may receive physical contact from various people, there is no decrease in the level of arousal.[22] Higley and Dozier reported that mothers of securely attached infants at age 12 months had nighttime interactions that were generally more consistent, sensitive, and responsive than those of insecurely attached infants.

Specifically, in secure dyads, mothers generally picked up and soothed infants when they fussed or cried after an awakening.[22]

Infants or children with hyperactivation of their attachment systems normally cannot separate until they fall asleep in close physical contact with their parents in the children's or the parents' bed. It is important to note that many parents also have attachment problems and have difficulty separating, and sometimes it is not clear who is clinging to whom. Some parents, especially those with prior trauma experiences, also have their own sleep problems. Attachment anxiety was associated with self-reported sleep difficulties in men and women; even if depressed affect was included as a control variable, the effect of attachment anxiety remained significant.[23] If a mother has an attachment disorder with role reversal, she may carry her infant into her bed and take the infant as a secure base to help herself fall asleep. Mothers with panic disorders, when describing parenting behaviors concerning infant sleep, reported less sensitivity toward their infants, who showed more ambivalent/resistant attachment, higher salivary cortisol levels, and more sleep problems.[24] Mothers with high symptoms of depression and anxiety were more likely to have ambivalent attached infants and used high levels of active physical comforting, and their infants developed high initial levels of sleep problems that continued as infant sleep disturbances over time.[25] Benoit et al.[26] have shown that a mother's own insecure status of attachment is highly correlated with attachment and sleep disturbance in her infant: every insecurely attached mother in their study had a child or children with sleep disturbances. Therefore, at the start of treatment, it is vital that the therapist learns something about the parents' own histories of attachment and their experiences of unresolved loss and separation so that treatment can also address their needs, or else the therapy of the sleep-disordered child will not be successful. The importance of focusing on the parents' status of attachment when treating their infant's sleep problem cannot be overstated.

Finally, sleep disturbances and sleep disorders of infants caused by traumatizing experiences with insensitive care by attachment figures can lead to attachment disorders, but if a child is securely attached during the day, inconsistent caregiving or unresponsiveness to attachment signals at night will not necessarily lead to a complete attachment disorder, but perhaps only to subtle irritations in the attachment system. It may be that infants with insensitive nighttime care become more clingy or ambivalent in their daytime attachments, which makes separation for sleep more difficult and may result in long-lasting behavioral problems.[27–29]

The presence of parents when an infant separates for sleep and sleeps during the night may support him or her in developing a secure attachment representation. Children from kibbutzim who were home sleepers with their parents developed a secure attachment relationship with their parents, while infants who slept in the group setting without their parents available at night developed attachment relationships with their metapelet (caregiver in the kibbutzim).[30–33]

ATTACHMENT, SEPARATION, AND SLEEP

Looking at attachment behavior from an evolutionary point of view, most infants around the world have slept and continue to sleep in close physical contact with their parents for the first year of life and possibly longer, so these infants do not experience separation at sleeping hours.[34] Thus, a crying baby at night is not a question in most countries. Only in Western countries and especially in Europe and North America do parents expect an infant to separate at night and sleep in his or her own bed or own room. This form of separation between infants and attachment figures during the night is not consistent with evolutionary development. In former times, when human beings were nomads, survival required that an infant remained in close contact with the attachment figure, usually the biological mother, during the daytime and even more so during nighttime. Since an infant is dependent on the attachment figure for all of his or her physical, social, and emotional needs, close physical contact was a great advantage for survival. It is likely that the attachment system in humans developed within the context of evolution, as those infants who showed attachment behavior when separated from the attachment figure and when experiencing anxiety had a higher survival rate than those who did not. This might explain why many children in Western countries do not stay in their beds at night, especially when they experience anxiety and initiate co-sleeping in the parents' bedroom once they can walk.[2] Through the lens of attachment, it is not surprising that once arrived and snuggling up to their parents, the children can fall asleep within seconds.

Considered in the context of evolution, then, it is quite natural that an infant reacts to nightly separation from his or her attachment figure with alarm, crying, and signaling a desire to be picked up. If the attachment figure does not arrive to soothe the infant, the attachment arousal can escalate to hyperarousal in the autonomous nervous system, leading to an increase in bowel movements, as with colicky infants, or to vomiting when the gastrointestinal tract reacts. Therefore, nighttime crying, seeking physical contact with the attachment figure, and protesting against separation from the attachment figure are correct evolutionary-based behavior patterns.[35–37]

Nonetheless, an infant can learn to sleep through the night without his or her attachment figure. If Western cultural standards indicate that it is proper for parents and children to sleep apart, parents must train children to tolerate this type of separation, even though it is contrary to evolution. Parents must listen for sounds from the baby after separating and leaving the room and be ready to provide the child with a positive, attachment-oriented experience. Whenever the infant starts crying energetically and increasingly loudly, the parent should return to the room and try to console the infant. The child will sometimes need physical contact to calm down, especially if he or she has become hyperaroused. Returning rapidly to the room when the child starts to cry intensely is key to not having the child's arousal escalate to hyperarousal. Parents may have to enter the room repeatedly during the first nights, but this frequency will decrease. If parents respond promptly to an infant's crying at night, the baby will cry less during the next few weeks. In contrast, if parents delay in answering the cry signal and consoling the child by physical contact, perhaps because of their philosophy of not spoiling the baby, the child will cry for longer periods in the future.[37] It has been found that each time the parents come in and respond, the infant learns that he or she is not lost, separate, and alone, but that the attachment figure is available and sensitive to his or her signaling. When parents consistently and reliably respond in this way, an infant will make an important discovery: even while separated at night when it is dark and anxiety can become intense,

attachment figures are present and emotionally and physically available. This comes to signify an important attachment representation within the context of sleep and nighttime separation, implying security and safety despite being separated from the parents.[38]

THERAPY OF ATTACHMENT-RELATED SLEEP DISORDERS

Sleep problems in babies can be subtle indicators of difficulties in parent–infant relationships. If a baby cries for several hours day after day, it is important to seek help with a specially trained psychotherapist who can quickly treat the dyad with an eye toward assessing the attachment and trauma experiences of the mother and father, in addition to the interactional irritability of the infant.[18,39,40] The aim of attachment-related therapy for sleep disorders in infants is to enable these children to separate from their attachment figures in the evening, fall asleep and remain in their own bed overnight without nightmares, anxiety, or panic attacks.

As mentioned earlier, attachment and separation concerns are present for parents as well as infants and children, and thus treatment must involve both parties. As in any attachment-related therapy, the therapist must become a therapeutic bonding figure (i.e., he or she must become a safe place for the parents, as well as for the infant or child). In the same way that parents' "sensitive behavior" is required for the positive development of a baby's secure attachments,[41] a therapist must become a secure base for parents—a framework for trust and a springboard for change.[18,19] Highly interactive therapeutic sensitivity, in which the therapist comes to recognize family signals (especially those of the parents), interprets these signals correctly, and reacts conscientiously and promptly, will lead to the development of such a therapeutic bond, which will become a model for the parent–child relationship. The therapist fosters the development of a secure therapeutic bonding with the parents, and as a result, parents can become a safe haven for their infants.[42]

The therapist can then help the parents to understand the nighttime needs of their infants, be sensitive about a child's anxiety, and react appropriately by going into the infant's bedroom and trying to soothe him or her. If the baby is in an elevated state of arousal, the parents should take the child out of bed and provide physical contact. Most hyperaroused children will quickly relax with physical contact. Securely attached infants will need more and longer periods of physical contact to calm down than insecurely avoidant-attached infants, but securely attached infants will have longer sleep durations than avoidant-attached infants.[35] Some parents may allow a child to briefly sleep with them to calm them down, after which the child can be placed back in his or her bed.

CASE STUDY

A mother, T., was referred by her pediatrician and telephoned that she urgently needed help to deal with the nighttime needs of her 6-week-old infant. Every night, Baby S. had awakened for a feeding session. After being fed and put back to bed, the infant started to whine and cry, whereupon the mother would go into S.'s bedroom, lift her out of bed, cuddle and soothe her, rock her, and lay her back down in bed. Despite these ministrations, the baby continued to cry. This interaction occurred several times each night, with the mother walking around and rocking S. for hours until the two fell asleep on the sofa during the morning hours. The whole family, especially the mother, was exhausted and did not know "how to survive." The partnership was in danger, as the husband had threatened to leave the family. The couple's first child, now 6, had also cried at night for 2 years, but the parents had decided to have another child despite their first "catastrophic" experience. For these parents, the first years of having a child were equated with regular nightmares and sleepless nights. As a result, the whole family was in an acute alarm state, and the children were at risk of harm from their parents. This is the moment when parents might start shaking their babies. Things were worst at night, but similar difficult sleeping interactions took place during Baby S.'s morning and afternoon sleeps. Several pediatric examinations had established a normal developmental pattern for her, with no indications of somatic disease to explain the symptom of sleep disturbance. Therefore, the sleep problem seemed to be a psychosomatic sleep disturbance.

A video diagnostic session of the mother changing diapers and playing with the infant as she would have done at home revealed an interesting interactional pattern. At first, the mother interacted sensitively with eye contact, fine vocal attunement and touch, responding to cues from the infant, and engaging in a very nice dialogue of rhythmic interaction. However, in between were switches in behavior and affect attunement: suddenly, the mother would stop, avert her gaze, and anxiously and sadly examine the child's feet. Her affect became simultaneously shut down, depressed, and highly aroused. This lasted about 20 seconds, after which she again attended to the infant, interacting vocally and visually, then switching back and examining the child's feet, saying that the feet were too cold. In 2 minutes of videotaping, there were several switches back and forth between mother and child. When the mother shut down eye contact with S. and became preoccupied with the infant's feet, the child's gaze also shifted.

When we watched the video recording with T. and tried to understand what we saw and how to interpret this, she told us that she was not aware of these switches, but remarked that she was checking the child's feet for signs of disability. T. related that, because of her age, she had undergone amniocentesis to check for possible fetal abnormalities. The first result of amniocentesis indicated an abnormal set of chromosomes and a handicapped child. T. and her husband were deeply shocked, and the gynecologist took another blood sample that revealed a normal set of chromosomes and a normal child. Of course, this double diagnosis of contrary results led to extreme arousal and stress for the parents. The mother was highly ambivalent about attaching prenatally to the child or holding back in case the baby was born disabled. After birth, externally and physically S. appeared normal, so the mother declined a third postnatal chromosome test. Nevertheless, she began constantly to check the child for signs of abnormality, such as the special foot or hand folds found in children with Down's syndrome, which she had learned about on the Internet and in books about disabled children. Although she did not find any such signs, the absence of abnormalities did not calm her, and she compulsively checked her child over and over. She had also read that disabled children sometimes exhibit a particular type of crying and wondered whether S.'s crying at night was the special kind of whimpering and crying called the cri-du-chat syndrome. On top of the erroneous prenatal diagnosis, S.'s crying was a trigger for anxiety and bonding ambivalence on the part of

T., alarming the mother and leading her to worry that the symptom was part of a disability that was as-yet undiagnosed.

During the process of diagnostics, we routinely perform an Adult Attachment Interview (AAI)[43] or an Adult Attachment Projective[44] test, as well as a Caregiving Interview,[45] for any mother presenting an infant with early interactional problems. These three interviews give us a lot of information about parents' own attachment representations and perhaps unresolved trauma experiences. During the AAI, T. was asked when she was first separated from her own parents. She remembered quite vividly that at the age of 3, she was admitted to a hospital for a tonsillectomy. Her mother sent both T. and her 8-year-old brother for tonsillectomies with the idea that the brother might calm her down when feelings of being lost and separated at nighttime would come up. T. felt very lonely at night in her unfamiliar bed in the hospital and experienced a tremendous, sick feeling in her stomach, which she did not interpret as anxiety and arousal. At this point in the AAI, I realized that the mother had previously told me that she felt sick to her stomach when Baby S. cried at night and she took the baby out of her bed and started walking about the apartment. T. had also experienced a second separation shortly after her discharge from the hospital when her mother gave birth to another child and all of the children left home to stay with a grandmother. Again, she felt lonely and separated from her mother and had the same gastrointestinal symptoms. From that point onward, she could never tolerate separation and stay elsewhere overnight. Any attempt at an overnight separation such as in kindergarten or during school excursions failed because she became sick and her parents had to pick her up during the night.

Attachment dynamics of the sleeping disorder

Within the context of attachment-oriented psychodynamic theory, the mother's history and Baby S.'s sleep problem become more understandable. When T. and her husband were confronted with the possibility of expecting a disabled child, triggers of anxiety and preoccupation emerged. T. was highly ambivalent about bonding with her infant and became preoccupied with searching for signs of disability after birth. Thus, the mother was in

a permanent state of arousal, which did not help to bring the child into a relaxed state and help her fall asleep. Baby S. might have sensed T.'s ambivalence—clinging to her infant on the one hand and being preoccupied and emotionally distant on the other—which might have led S. to cry louder and search for physical contact with the mother, as the child experienced emotional separation and detachment from T. Furthermore, the AAI revealed that the mother retained her own separation problems from childhood and had a high psychosomatic arousal and trigger when she had to separate from her infant: the 3-year-old within the mother's own representational world needed an attachment figure. Because of her own experiences, T. could not be a secure attachment base for her own infant. S.'s crying at night had triggered T.'s own separation experience from the past and brought the mother into a helpless state. Parents who become triggered by their infant's nighttime crying and whose own traumatic experiences are reactivated have a high probability of acting out at night or becoming hyperaroused and needing their own attachment figure, thus not being emotionally available to their infants.

Treatment

Using an attachment-oriented approach, the following treatment procedure was arranged. During the daytime just before putting Baby S. to sleep, T. telephoned me and we talked about her feelings of anxiety and feeling lost. This therapeutic phone contact helped her to feel reassured and secure and to separate more easily from the child. During the nighttime, there was still a great sleep disturbance, so we explained the attachment problem to the husband and asked him to get up at night with his wife. This led to the following situation: when T. had nursed the infant at night and put her to bed, the infant was still awake, whining a bit but not crying. The husband took T.'s hand and helped her separate from the infant, providing a secure base and becoming an attachment figure for her. Once the mother became calmer, Baby S. was already sound asleep.

Baby S.'s sleep problem disappeared rapidly, and it became quite clear that the infant's sleep problem was an entangled reenactment of acute insecurity because of the prenatal diagnosis and the early unresolved trauma of the mother. After the acute situation with Baby S. eased, T. came for further therapeutic sessions in order to work on her unresolved trauma. The result was quite remarkable, and the mother made an astonishing recovery. For the first time in her life, she could drive away for holidays and sleep in an unfamiliar bed. Furthermore, she was able, without hyperarousal and anxiety, to cross bridges and drive through tunnels, locations that she had previously avoided. After termination of the treatment, she phoned me only once, after her son's first day of school. The morning after the first day, her son told her he wanted to go to school with his friends and without her, and he separated quite easily with a quick goodbye. Standing at the window and watching him walk along the road, she experienced the same sick feeling and remembered it was related to her experience of early separation. At that moment, she decided to phone me, and we talked about how the situation came about and how it was triggered. She was aware of it and did not need to reenact that situation by holding back or accompanying her son, which would hinder his autonomy and individuation.

DISCUSSION

Sleep problems of infants and even older children can be strongly related to attachment problems. Children and adults with attachment disorders may have problems at night regarding falling asleep, staying in their bed in darkness, or going back to sleep after waking due to anxiety or nightmares. Depending on the attachment disorder, they long for physical contact or, in contrast, may not want physical contact and instead stay in bed in a hyperaroused state, suppressing their attachment needs and developing psychosomatic symptoms.*

Children who have experienced early trauma such as deprivation or violence are likely to develop attachment disorders. Typically, those children do not have an inner representation of security, and if they have to separate and sleep in the dark

---

* In addition to gastrointestinal symptoms, respiratory symptoms (e.g., asthmatic symptoms with coughing and wheezing) are quite common and should be considered outside diagnoses of allergies. A convincing study showed how asthma attacks and separation problems are associated.[49]

separate from any person, anxiety arises and activates the attachment system. Depending on the type of attachment disorder, they may start crying, shouting, fighting, or entering dissociative states and not showing signs of attachment behavior.

Since a baby cannot crawl or walk to search for the attachment figure, the only way to signal an attachment need is to cry through the night. If an infant is to form a secure attachment during the night, the parents must help the child to calm down by walking into the room and soothing the child, going away to help him or her to tolerate a short period of separation and returning if the child is aroused again. This helps the baby to learn a form of separation training in which the attachment figure is available and will consistently arrive to soothe him or her when anxiety becomes intolerable and the crying escalates to a panic state. This training requires more time, emotional and physical availability and sensitivity in a consistent and reliable way than leaving the child to cry through the night and get used to sleeping on his or her own.

Ultimately, if the child cannot calm down, a temporary period of having the child sleep with the parents may be wise, provided that there is no contraindication for co-sleeping such as drug addiction, alcoholism, smoking, elevated temperature in the parents' room, or a very soft mattress. Most children who bed in between the mother and father fall asleep fairly soon at night or after waking from nightmares, as the space between their attachment figures seems to provide the most security and reassurance.

Prevention

Many parents in Western countries themselves did not have the stressful experience of initially sleeping apart from their own parents, and so started co-sleeping with their infants, as most parents and children throughout the world still do.[46] Insufficient research is currently available that examines how parental status of attachment, correlated with co-sleeping and bedding-in, influences the emotional development of infants. Studies on sleep patterns in earlier days, which did not include attachment concerns in the research, showed that co-sleeping mothers and infants had the same sleep patterns in terms of depth and alertness. When the child became uneasy and irritable, the mother awoke and fed it, and both fell asleep again. Co-sleeping mothers were in tune with their babies and did not feel irritated during the night. In contrast, if the infant slept in a bed next to that of the mother, their sleep rhythms were not as well-tuned together, and if the child slept in a different room, the sleep rhythms of mother and infant were completely uncoordinated. Those mothers were the most exhausted in the morning.[46–48]*

Children who can reestablish close physical contact with their parents at bedtime or even sleep together in the same room may form more secure relationships than those who are separated from their parents at night.† If parents do not want to co-sleep or room-in with their infant, they must consider attachment theory and attachment needs and realize that they are subjecting the child to a behavior that is contrary to evolution. If parents want children to sleep on their own, the separation in the evening hours and calming down at night have to be done delicately and with awareness that the evening and night separations are the most sensitive phases for attachment needs. Parents have to reassure children again and again that they are physically and emotionally available and help make the separation tolerable. Here, significant teaching and training are necessary for parents.

---

* Coincidentally, bedding-in during the weeks after delivery seems to protect against postpartum depression, as the incidence of postpartum depression is much lower in Asian countries, where bedding-in is the traditional form of caring. Some researchers recommend that bedding-in after delivery should be practiced everywhere as a preventive method against maternal postpartum depression.[47,48] In addition, we hypothesize that if mothers and children do not co-sleep or bed-in, perhaps mothers become depressed because they cannot see their infants and worry about whether the children are still breathing and alive. Co-sleeping promotes breast feeding and might, consciously or unconsciously, reassure a mother during the night that her baby is breathing and is side by side with her in physical contact, and so she might relax and sleep more quietly. In addition, the child would feel secure about the mother's closeness.[49]

† Parenthetically, this is one reason why admittance of parents with their infants in children's hospitals should be the norm.

In our parent groups, one of the biggest fears is that if the child is brought to the parents' bed as a co-sleeper, he or she might stay there for 25 years. Of course, this will not happen, and most parents find places and times for sexual activity outside of the parental bed at night so that, among other things, co-sleeping need not be an obstruction to parental sexuality.

All of these subjects are part of our new prevention program SAFE®, standing for "Secure Attachment Formation for Educators" and for "Secure Attachment Family Education." Parents participate in this preventive program of four prenatal and six postnatal full-day workshops from the twentieth week of gestation until the end of the child's first year. In addition to receiving many instructions and having personal experiences, all parents are given the AAI. Parents with unresolved traumas receive supportive psychotherapy before birth and trauma-centered therapy after birth. The goals of this prevention program are to uncover parental unresolved traumas that could be risk factors leading to a reenactment with the infant and to treat these problems before and after birth so that harm to the infant is prevented.

CONCLUSION

Sleep problems in infants and children can be difficult psychological problems that always need early attention and treatment. Diagnosis should focus on the whole family (i.e., psychosocial and partnership problems, individual attachment problems, and traumatic experiences of the mother and father) and the infant's own experiences of attachment or trauma. Children with attachment disorders are high-risk candidates for sleep problems because separation and sleeping at night are important markers of the attachment relationship with the attachment figure. On the other hand, sleeping through the night does not mean that no attachment problems exist. The infant may have learned that no one is available at night, no matter how loud he or she might cry.

In addition to markedly helping individual families, education about attachment theory, attachment figures, and attachment relationships holds the potential to effect dramatic social change. Such information can be obviously and directly useful to parents of infants with sleep disorders, as we have seen in this chapter. Moreover, many powerful societal benefits could also accrue if knowledge about the concrete ramifications of attachment theory were disseminated more widely to adult clients, clinics, schools, and society at large.

ACKNOWLEDGMENTS

I am most grateful to the parents who allowed me to learn about their attachment problems and to increasingly understand the psychodynamics within families with infants who cry at night. Through these case histories and treatment experiences, I learned about the attachment-related problems of sleep disturbances in children with normal family backgrounds and those with attachment disorders and trauma-related experiences. Without these experiences, this paper would not have been possible.

REFERENCES

1. Scher A, Cohen D. Locomotion and night-waking. *Child Care Health Dev* 2005; 31(6): 685–691.

2. Jenni OG, Fuhrer HZ, Iglowstein I, Molinari L, Largo R. A longitudinal study of bed sharing and sleep problems among Swiss children in the first 10 years of life. *Pediatrics* 2005; 115(1): 223–240.

3. Thunström M. Severe sleep problems among infants in a normal population in Sweden: Prevalence, severity and correlates. *Acta Paediatr* 1999; 88: 1356–1363.

4. Sazonov E, Sazonova N, Schuckers S, Neuman M, CHIME Study Group. Activity-based sleep–wake identification in infants. *Physiol Meas* 2004; 25: 1291–1304.

5. Sadeh A, Tikotzky L, Scher A. Parenting and infant sleep. *Sleep Med Rev* 2010; 14(2): 89–96.

6. Anders TF. Infant sleep, nighttime relationships, and attachment. *Psychiatry* 1993; 57(1): 11–21.

7. Anders TF, Keener M, Bowe TR, Shoaff BA. A longitudinal study of nighttime sleep–wake patterns in infants from birth to one year. In: Call JD, Galenson E, Tyson PI (Eds). *Frontiers of Infant Psychiatry*. New York: Basic Books, 1983, pp. 150–170.

8. Minde K. Sleep problems in toddlers: Effects of treatment on their daytime behavior. *J Am Acad Child Adolesc Psychiatry* 1994; 33(8): 1114–1121.

9. Moore SM. Disturbed attachment in children: A factor in sleep disturbance, altered dream production and immune dysfunction. *J Child Psychother* 1989; 15(1): 99–111.

10. Sadeh A, Anders TF. Infant sleep problems: Origins, assessment, interventions. *Inf Mental Health J* 1993; 14(1): 17–34.

11. von Hofacker N, Papouseck M. Disorders of excessive crying, feeding, and sleeping: The Munich Interdisciplinary Research and Intervention Program. *Inf Mental Health J* 1998; 19(2): 180–201.

12. Wolke D, Meyer R, Ohrt B, Riegel K. Co-morbidity of crying and feeding problems with sleep problems in infancy: Concurrent and predictive associations. *Early Dev Parent* 1995; 4(4): 191–207.

13. Zucherman B, Stevenson J, Bailey V. Sleep problems in early childhood: Continuities, predictive factors and behavioral correlates. *Pediatrics* 1987: 80: 664–671.

14. Ainsworth MDS, Blehar M, Waters E, Wall S. *Patterns of Attachment: A Psychological Study of the Strange Situation*. Hillsdale, NJ: Lawrence Erlbaum, 1978.

15. Bowlby J. Affectional bonds: Their nature and origin. In: Freeman H. (Ed.). *Progress in Mental Health*. London: Churchill, 1969, pp. 319–327.

16. Bowlby J. *Attachment and Loss. Vol. II: Separation, Anxiety and Anger*. New York: Basic Books, 1973.

17. Bowlby J. *Attachment and Loss. Vol. III: Loss: Sadness and Depression*. New York: Basic Books, 1980.

18. Bowlby J. *A Secure Base: Clinical Implications of Attachment Theory*. London: Routledge, 1988.

19. Brisch KH. *Treating Attachment Disorders. From Theory to Therapy* (2nd ed.). New York, London: Guilford Press, 2002.

20. Bretherton I. In pursuit of the internal working model construct and its relevance to attachment relationships. In: Grossmann KE, Grossmann K, Waters E (Eds). *Attachment from Infancy to Adulthood. The Major Longitudinal Studies*. New York: Guilford, 2005, pp. 13–47.

21. Zeanah CH, Boris NW. Disturbances and disorders of attachment in early childhood. In: Zeanah CH (Ed.). *Handbook of Infant Mental Health* (2nd ed.). New York: Guilford, 2000, pp. 353–368.
22. Higley E, Dozier M. Nighttime maternal responsiveness and infant attachment at one year. *Attach Hum Dev* 2009; 11(4): 347–364.
23. Carmichael CL, Reis HT. Attachment, sleep quality, and depressed affect. *Health Psychol* 2006; 24(5): 526–531.
24. Warren SL, Gunnar MR, Kagan J et al. Maternal panic disorders: Infant temperament, neurophysiology, and parenting behavior. *J Acad Child Adolesc Psychiatry* 2003; 42(7): 814–825.
25. Morrell J, Steele H. The role of attachment security, temperament, maternal perception, and care-giving behavior in persistent infant sleeping problems. *Inf Mental Health J* 2003; 24(5): 447–468.
26. Benoit D, Zeanah CH, Boucher C, Minde K. Sleep disorders in early childhood: Association with insecure maternal attachment. *J Am Acad Child Adolesc Psychiatry* 1992; 31(1): 86–93.
27. Scher A, Asher R. Is attachment security related to sleep–wake regulation? Mothers' reports and objective sleep recordings. *Inf Behav Dev* 2004; 27: 288–302.
28. Scher A, Zuckerman S, Epstein R. Persistent night waking and settling difficulties across the first year: Early precursors of later behavioural problems? *J Reprod Infant Psychol* 2005; 23(1): 77–88.
29. Thunström M. Severe sleep problems in infancy associated with subsequent development of attention-deficit/hyperactivity disorder at 5.5 years of age. *Acta Paediatr* 2002; 91: 584–592.
30. Sagi-Schwartz A, Aviezer O. Correlates of attachment to multiple caregivers in Kibbutz children from birth to emerging adulthood. The Haifa longitudinal study. In: Grossmann KE, Grossmann K, Waters E (Eds). *Attachment from Infancy to Adulthood. The Major Longitudinal Studies.* New York: Guilford, 2005, pp. 165–197.
31. Sagi A, Lamb ME, Lewkowicz KS, Shoham R, Dvir R, Estes D. Security of infant–mother, –father, and –metapelet attachments among kibbutz-reared Israeli children. In: Bretherton I, Waters E (Hrsg). *Growing Points of Attachment Theory and Research: Monogr Soc Res Child Dev* 1985; 50(1–2): S. 257–275.
32. Sagi A, van IJzendoorn MH. Multiple caregiving environments: The kibbutz experience. In: Harel S, Shonkoff JP (Eds). *Early Childhood Intervention and Family Support Programs: Accomplishments and Challenges.* Jerusalem: JDC-Brookdale Institute, 1996, pp. 143–162.
33. Sagi A, van Ijzendoorn MH, Aviezer O, Donnell F, Mayseless O. Sleeping out of home in a kibbutz communal arrangement: It makes a difference for infant–mother attachment. *Child Dev* 1994; 65: 992–1004.
34. Latz S, Wolf AW, Lozoff B. Cosleeping in context: Sleep practices and problems in young children in Japan and the United States. *Arch Pediatr Adolesc Med* 1999; 153: 339–346.
35. Scher A. Attachment and sleep: A study of night waking in 12-month-old infants. *Dev Psychobiol* 2001; 38(4): 274–285.
36. Scher A. Mother–child interaction and sleep regulation in one-year-olds. *Inf Mental Health J* 2001; 22(5): 515–528.
37. St. James-Roberts I, Alvarez M, Csipke E, Abramsky T, Goodwin J, Sorgenfrei E. Infant crying and sleeping in London, Copenhagen and when parents adopt a "proximal" form of care. *Pediatrics* 2006; 117: 1146–1155.
38. Hayes MJ, Roberts SM, Stowe R. Early childhood co-sleeping: Parent–child and parent–infant nighttime interactions. *Inf Mental Health J* 1998; 17(4): 348–357.
39. Daws D. *Through the Night. Helping Parents and Sleepless Infants.* London: Free Association Books, 1993.
40. Daws D. Family relationships and infant feeding problems. *Health Visitor* 1994; 67(5): 162–164.
41. Grossmann K, Grossmann KE, Spangler G, Suess G, Unzner L. Maternal sensitivity and newborns' orientation responses as related to quality of attachment in Northern Germany. In: Bretherton I, Waters E (Eds). *Growing Points of Attachment Theory and Research.* Chicago, IL: University of Chicago Press, 1985, pp. 231–256.

42. Orlinsky DE, Grawe K, Parks BK. Process and outcome in psychotherapy: Noch einmal. In: Bergin AE, Garfield SL. (Eds). *Handbook of Psychotherapy and Behavior Change* (4th ed.). New York: Wiley, 1994, pp. 270–376.

43. George C, Kaplan N, Main M. *Adult Attachment Interview*. Berkeley, CA: University of California, 1985.

44. George C, West M, Pettem O. The Adult Attachment Projective: Disorganization of adult attachment at the level of representation. In: Solomon J, George C (Eds). *Attachment Disorganization*. New York, London: Guilford, 1999, pp. 318–346.

45. George C, Solomon J. Representational models of relationships: Links between caregiving and representation. *Inf Mental Health J* 1996; 17(3): 198–216.

46. Jenni OG, O'Connor BB. Children's sleep: An interplay between culture and biology. *Pediatrics* 2005, 115(1 Pt 2): 204–216.

47. Louis J, Cannard C, Bastuji H, Challamel MJ. Sleep ontogenesis revisited: A longitudinal 24 hour home polygraphic study on 15 normal infants during the first two years of life. *Sleep* 1997; 20: 323–333.

48. Minard KL, Freudigman K, Thoman EB. Sleep rhythmicity in infants: Index of stress or maturation. *Behav Processes* 1999; 47(3): 189–203.

49. Sandberg S, Paton JY, Ahola S et al. The role of acute and chronic stress in asthma attacks in children. *Lancet* 2000; 356: 982–987.

4

Fatigue in chronic medical conditions: A psychosomatic perspective

DORA M. ZALAI, RAYMOND GOTTSCHALK, AND COLIN MICHAEL SHAPIRO

INTRODUCTION

Fatigue is a dispiriting word for clinicians by virtue of the myriad of clinical possibilities that is spawned by this presenting complaint. The non-specificity of this symptom, when presented to a physician, induces a cascade of chemico-pathologic testing that is often disproportionate to the clinical history that is obtained. As medicine has evolved, the availability of diagnostic testing has allowed the clinician rapid access to a cornucopia of hematological analytic results that can greatly increase our diagnostic capability, but it seems that these tests are frequently to be as a fishing net cast into a diagnostic abyss.

Organic pathology is generally the first consideration vis-à-vis etiology in such a presentation. Accordingly, the approach has been to look for and attempt to find the errant chemical or hematological culprit. Differential diagnosis, even since the time of William Osler, has represented a unifying concept in medical education. Most medical education in North America, however, is directed by a specialist or subspecialists. This results in many students graduating from medical schools and residency programs having never experienced or been challenged with one of the top 10 differential diagnoses in primary care.

Fatigue is a presentation in amongst the top 10 diagnoses in primary care.[1] In this group of 10 are cough, fatigue, low back pain, fever, dyspnea, generalized abdominal pain, headache, vertigo, chest pain, and edema. The estimates are that only 15% of patients in primary care settings have an organic cause that can be identified to explain the fatigue. These possible causes include infections, anemia, endocrinopathies, medication side effects, sleep disturbances, and a rare presentation of malignancies. The diagnostic and clinical outcomes of these investigations often demonstrate a disquietingly high percentage labeled "not yet diagnosed" (NYD) or "other" as the outcome of investigating patients with fatigue. In individuals over the age of 45, fatigue NYD was present in 36% of those investigated, and the category "other" accounted for 33%.[1]

In specialist practice, the evaluation and education in fatigue and sleep medicine has been dishearteningly sparse and poor. In medical school curricula, education in sleep disorders is minimal, and even those who graduated from MBChB programs of 6 years of medical school often receive no lectures on sleep disorders. Matters are not much better in current programs, with fewer than 10 hours given to sleep disorders throughout the entire medical training curriculum in most programs.

In the iconic textbook of medicine, *Cecil* (18th edition, 1988), there is one index reference to sleepiness, resulting from melatonin.[2] The references indexed to fatigue relate only to medications as causes of sleepiness or organic pathology. The improvements in the last 25 years have been notable, but the majority of graduates who are active in medicine today will have had no primary background in the ubiquitous complaint of fatigue.

The goal of this chapter is to provide a comprehensive overview of fatigue in chronic medical conditions from a psychosomatic perspective. The chapter opens with the definition and conceptualization of fatigue. Next, it describes the assessment and the most common perpetuators of fatigue (i.e., sleep disorders, mood disorders, and cognitive–behavioral factors) in chronic medical conditions.

The chapter concludes with a summary of fatigue management methods.

DEFINITION OF FATIGUE

The categorization of fatigue as either a physiological or a psychological phenomenon is currently the prevailing approach in science (Table 4.1).

The concepts of physiological and psychological fatigue are then further segmented by the disciplines of science that provide distinct perspectives and conceptual models for understanding fatigue (Figure 4.1). The physiological and psychological conceptualizations of fatigue are elaborated below.

Physiological fatigue

Physiological fatigue in the basic sciences has been defined as a temporary loss of voluntary force-generating capacity during exercise.[3] The declining force in the muscles during contractions can originate from impairment of the muscles or the neuro-muscular junction (peripheral fatigue) or from decreased central input (central fatigue or central activation failure). The conceptualization of fatigue in neurology is similar to the above model. As in physiology, peripheral fatigue is attributed to impaired neuro-muscular transmission, neuro-metabolic disorders, or disorders in the contractile mechanisms, but a broader, neurological conceptualization also includes diseases of the peripheral, afferent nerves and of the lower motor neuron.[4] From a neurological perspective, central fatigue is defined as muscle fatigability due to decreased input from the nervous system at or above the level of the upper motor neuron.

An even broader conceptualization framework describes peripheral (physical) fatigue as

Table 4.1 The subtypes and assessment of physiological and psychological fatigue

Categories	Subtypes	Assessment
Physiological	Peripheral	Force transducer, surface electromyography
	Central	Magnetic stimulation, EEG, performance in neuropsychological tests, functional neuroimaging
Psychological (central, subjective, and cognitive)	Motivational Cognitive Physical Emotional	Questionnaire, clinical interview, qualitative research methods

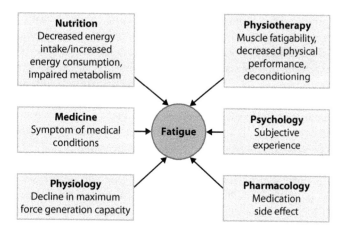

Figure 4.1 Conceptual framework for understanding fatigue.

a decreased ability to maintain force because of impairment of the muscles or the cardiovascular or nervous system without impairment of sustained mental tasks. Central fatigue, on the other hand, is defined as a difficulty with initiating or maintaining cognitive or physical activities that require self-motivation.[5]

Psychological fatigue

Psychological (experienced, subjective, or central) fatigue is the lived experience of fatigue that is constructed by our perceptions of the internal and external environment and our mood, life experiences, knowledge, and belief systems. The definitions of psychological fatigue typically refer to a multidimensional construct, a subjective sense of energy or effort imbalance and functional impairment as, for example "an overwhelming sense of tiredness, lack of energy and a feeling of exhaustion, associated with impaired physical and/or cognitive functioning."[6]

The above definitions of physiological and psychological fatigue imply that fatigue is a pathological condition associated with "overwhelming" sensations and anatomical or functional impairment. Alternatively, fatigue can be viewed as both a normative experience and a pathological condition.

Normal and pathological fatigue

A simple distinction between normal and pathological fatigue is based on characteristics of duration, associated conditions, functional impairment, and alleviation by usual adaptive behaviors (Table 4.2).[6]

The shortcoming of this approach is that some fatigue conditions do not fit neatly into these categories. For example, fatigue associated with acute viral infections could be considered normal, since it is temporary and is alleviated by rest. However, it could also be categorized as pathological in that it is associated with an illness and is functionally impairing.

From a neurophysiological perspective, fatigue has been depicted as an essentially normal adaptation mechanism.[4] According to this model, perceived exertion has an optimal set point that is maintained by the balance between the amount of voluntary effort and work output. This balance is the function of the interaction between control factors in the internal and external environment (e.g., hormone levels, humidity, and temperature) and the effortless information flow among sensory systems, the efferent motor system, and the brain areas regulating motivation and voluntary motor control. Pathological fatigue is an enhanced sense of normal

Table 4.2 Normal and pathological fatigue

Normal fatigue	Pathological fatigue
Normal, adaptive function	
Temporary	Persistent
Relieved by rest	Not relieved by rest
Can be pleasant	Distressing
Does not impact on normal functioning	Interferes with normal functioning
No impact on quality of life	Negative impact on quality of life

perceived exhaustion produced by a homeostatic imbalance due to abnormal changes in the system at any level. For example, decreased motivational input leads to perceived fatigue in depression; diseases of the motor cortex, neurons, or muscles result in fatigue in neurological disorders; changes in hormone levels cause fatigue in endocrine diseases; or working in a hot, humid environment is associated with amplified levels of perceived exhaustion.

Finally, as has been alluded to above, whether fatigue is regarded as a normal experience or a pathological condition is influenced by socio-cultural traditions and beliefs. These beliefs and attributions influence whether individuals seek medical help when they are feeling fatigued and whether they receive medical care for their fatigue (and if so, which specialist they are sent to and what will be the focus of treatment).

THE ASSESSMENT OF FATIGUE

The conceptualization of fatigue guides the choices of measurement methods used to assess it (Table 4.1).

The measurement of physiological fatigue

Physiological fatigue is measured by electrophysiological methods. A force transducer, for example, measures peripheral fatigue as a reduction of actual force produced by the muscle in response to electrical stimulation following exercise. An alternative, noninvasive method is using surface electromyography, which allows amplitude, frequency and muscle fiber conduction velocity to be measured online and simultaneously in several muscles. Central physiological fatigue can be measured as a decreased output of the motor cortex after magnetic stimulation following fatiguing mental activity.[7] A more simple measure of central fatigue is readiness potential: a movement-related negative cortical electroencephalograph (EEG) potential appearing approximately 1 second before a voluntary motor act. The amplitude of the readiness potential is related to the voluntary input and perceived effort and has been shown to change when individuals are engaged in fatigue-inducing physical tasks.[8]

The objective measurement of cognitive (central or mental) fatigue, as defined by decline of mental performance over the course of mentally demanding cognitive tasks, has traditionally been measured in neuropsychological experiments. In these situations, fatigue is typically induced in the following ways: (a) prolonged mental effort (e.g., participation in prolonged neuropsychological testing; typically 3–5 hours in length); (b) sustained mental effort (e.g., sustaining maximal or another determined level of mental performance during neuropsychological testing); (c) performing challenging mental tasks; or (d) performing challenging physical tasks. Alternatively, cognitive performance can be measured after prolonged wakefulness; for example, after partial or total sleep deprivation.

An intriguing pattern emerging from these experiments is that fatigue induction leads to cognitive fatigue, but there is no consistent difference in the cognitive performance between groups with medical conditions in which fatigue is a characteristic feature and healthy controls.[9] In other words, patients reporting moderate to severe fatigue before or over the course of testing show similar performance to healthy individuals who do not report increased fatigue. These findings raise the possibility that physiological and psychological fatigue are separate phenomena, or that experiential fatigue is related to, but is a more complex phenomenon than physiological fatigue.

Functional neuroimaging methods have also revealed that individuals after traumatic brain injury or with chronic fatigue syndrome or multiple sclerosis (MS; i.e., conditions in which the majority of patients suffer from fatigue) had more dispersed cerebral activation than healthy controls during cognitive tasks. Based on these findings, it has been speculated that persons with the above pathological conditions may need to recruit additional cerebral resources to compensate for brain damage in order to achieve the same performance as their healthy counterparts, and this increased effort is perceived as mental fatigue.

The assessment of psychological (experienced, subjective, and central) fatigue

QUESTIONNAIRES

The most widespread approach for the assessment of experienced fatigue is using self-report scales and questionnaires. Since there may be different types of fatigue and there is not a single "gold standard" definition of experienced fatigue, the fatigue questionnaires assess distinct fatigue constructs. A summary of the most frequently used questionnaires is found in Appendix A. The questionnaires are printed in the STOP, THAT and One Hundred Other Sleep Scales.[10]

Some tools (e.g., the Fatigue Assessment Scale) are based on the assumption that fatigue is a unidimensional construct; others are designed for the assessment fatigue as a multifaceted phenomenon.[11-14] The multidimensional fatigue scales distinguish between, for example, physical, cognitive, and motivational fatigue, as well as assessing the impact of fatigue on everyday functioning. There is variation in the items belonging to the above-mentioned dimensions of fatigue, sometimes with similar items falling into different categories. The Chalder Fatigue Scale, for example, has "physical symptoms" and "mental symptoms" subscales, and it lists both the general feeling of tiredness and motivational fatigue under "physical symptoms."[15] In contrast, the Multidimensional Fatigue Inventory separates the core feeling of tiredness into a "general fatigue" category, and links motivational fatigue to the "reduced motivation" subscale.[14]

An important issue related to construct validity is that some questionnaires do not differentiate between sleepiness and fatigue, despite the fundamental differences between the two constructs. Sleepiness correlates with a homeostatic drive to sleep and traditionally is measured as the degree of sleep propensity under standard circumstances in the sleep laboratory. The differentiation between sleepiness and fatigue is crucial in clinical assessment, since they have different pathophysiology, causes, consequences, and clinical implications. For example, certain sleep disorders (e.g., narcolepsy) are predominantly associated with excessive daytime sleepiness, whereas others (e.g., insomnia) are more related to daytime fatigue. When both are present, the alleviation of excessive daytime sleepiness should be the treatment priority, since it increases the risk for motor vehicle and occupational accidents.[16] Fatigue questionnaires that include items about sleepiness (e.g., the Chalder Fatigue Scale and the Fatigue Impact Scale) or other scales that were validated using established sleepiness scales (e.g., the Visual Analogue Scale for Fatigue)

neglect this important distinction.[13,15,17] One can argue that even the questionnaires using the word "tiredness," which in lay language means both sleepiness and fatigue, may fail to differentiate between the two phenomena.

An abundance of fatigue scales is available for use in clinical populations. For example, the Brief Fatigue Inventory, the FACT-F subscale, the Rhoten Fatigue Scale, the Fatigue Symptom Inventory, the Swartz Cancer Fatigue Scale, and the Piper Fatigue Scale have specifically been developed to assess cancer-related fatigue.[13,18-22] The Fatigue Severity Scale, the Fatigue Assessment Instrument, the Fatigue Impact Scale, and the Fatigue Questionnaire have been designed and validated for use in multiple medical conditions.[11,13,15,23]

Questionnaires provide a cost- and time-effective way of assessing fatigue in clinical practice and research. The choice of questionnaire should be based on the psychometric properties of the particular tests and on understanding of the characteristics of the fatigue constructs that they assess. It is important to note that responses are influenced by the respondents' interpretation of the items, their mood, and their recall bias. Unfortunately, most fatigue questionnaires fail to offer a definition of fatigue before prompting respondents to mark their answers. In order to circumvent recall bias, questionnaires or electronic devices (e.g., actigraphs or cellphones) can be used to detect the momentary experience of fatigue several times a day.[24,25] This Ecological Momentary Assessment Method can reveal individual fatigue profiles that retrospective methods may not illuminate.

QUALITATIVE METHODS

Qualitative methods are ideal for understanding the subjective experience and functional impact of fatigue. The advantage of these methods is that individuals are free to use their own words to describe what fatigue means to them without being constrained by a limited choice of the questionnaire items. Inquiry about the individual's fatigue validates this experience and can also reveal the processes that link experiential fatigue to changes in function and quality of life. As will be described below, this is especially important in medical settings, where fatigue is among the most common patient-reported factors contributing to distress and disability. Indeed, patient focus groups have expressed a preference for describing their fatigue experience, including the tandem life changes and coping strategies, with the healthcare team, rather than merely assigning numbers to fatigue symptoms on scales that do not fully capture the experiential aspects of their fatigue.[26] The value of qualitative research on fatigue has been demonstrated in several medical conditions. Qualitative studies have shown, for example, that fatigue is the most frequently reported concern of individuals with chronic hepatitis C infection.[27] Sleep restriction therapy may seem to be a low-risk intervention for insomnia, but a qualitative study showed that patients may experience severe fatigue in the first week of the treatment.[28] Patients described their fatigue management behaviors and beliefs about fatigue in qualitative studies on chronic fatigue syndrome, rheumatoid arthritis, and cancer.[29-33] These studies provided invaluable information for cognitive–behavioral interventions. The assumption underlying these qualitative assessment methods is that individuals are reliable reporters of their internal states. One can argue, however, that these approaches may be biased, since individuals have limited insight into their internal realities, or may be so much influenced by it that their self-report will be distorted.

FATIGUE IN CHRONIC GENERAL MEDICAL CONDITIONS

Fatigue is among the leading, most distressing and debilitating symptoms of many medical conditions. When fatigue is present in conjunction with a parallel disorder, it is usually considered to be the epiphenomenon of the primary medical condition. However, the relationship between the illness and fatigue appears to be more complex, in that it often does not dissipate with recovery from the illness, or it appears to be unrelated to the severity of the disorder.

The complexity of fatigue in the medically ill prompted the development of two-stage fatigue models. These models posit that the medical illness brings about biological changes that give rise to primary fatigue, which eventually becomes perpetuated by additional factors; for instance, sleep disorders, depression, or nonadaptive cognitions and behaviors.[34] The relationships between sleep disorders, depression, cognitive–behavioral factors, and fatigue are described below.

Sleep disorders and fatigue in chronic medical conditions

There are many sleep disorders, some of which result in a change in the quality of sleep, others reduce the duration of sleep, and some lead to excessive movement or behaviors or sensations that occur during sleep. When faced with a patient presenting with a complaint related to fatigue and/or sleepiness, it is imperative to gather information about the sleeping patterns and behaviors of that patient.

Excessive daytime sleepiness is a complaint that is noted in up to 20% of the general population.[35] Normal sleepiness is appreciated prior to a typical bedtime in all of us. Excessive sleepiness, however, occurs when one enters sleep inappropriately, or if there are episodes of unintentional sleep. Attempts at classifying the degree of sleepiness have been made with various sleepiness scales. The most popular scale used is the Epworth Sleepiness Scale.[36] This scale outlines the assessment of sleepiness based on eight situations and the likelihood of falling asleep in those situations (Figure 4.2). The term used to identify sleepiness in that situation is "chance of dozing." Mild sleepiness may be seen when individuals fall asleep while reading or in a rather unstimulated environment. Severe sleepiness, however, may be associated with irresistible sleepiness, such as while on the phone, having a discussion, or while driving. The evaluation of these patients would always include a question on the potential for sleep debt, some form of dyssomnia or medical or psychiatric causes. The top 10 causes of excessive daytime sleepiness are listed in Table 4.3.[37]

Many patients with excessive daytime sleepiness have coexistent complaints of decreased energy and fatigue. However, it is important to note that fatigue (without subjective sleepiness) can be the only symptom of sleep problems. For example, in a sleep clinic sample, 64% of patients reported high fatigue without excessive daytime sleepiness, whereas only 4% reported sleepiness without high fatigue.[38]

The most common sleep disorders perpetuating fatigue in chronic medical conditions are chronic insomnia and sleep apnea.

INSOMNIA AS A CAUSE OF FATIGUE SEPARATE FROM SLEEPINESS

The definition of insomnia is that of unsatisfactory sleep despite having adequate opportunity for sleep. It may include difficulties with sleep initiation or

EPWORTH SLEEPINESS SCALE

**HOW LIKELY ARE YOU TO DOZE OFF OR FALL ASLEEP
IN THE FOLLOWING SITUATIONS, IN CONTRAST TO FEELING JUST TIRED?**
*This refers to your usual way of life in recent times. Even if you have not done some of these things recently,
try to work out how they would have affected you. Use the following scale to choose the most
appropriate number for each situation:*

0 = Would **never** doze   2 = **Moderate** chance of dozing
1 = **Slight** chance of dozing   3 = **High** chance of dozing

Situation	Chance of Dozing			
Sitting and reading	0	1	2	3
Watching TV	0	1	2	3
Sitting, inactive in a public place (e.g., theatre or a meeting)	0	1	2	3
As a passenger in a car for an hour without a break	0	1	2	3
Lying down to rest in the afternoon when circumstances permit	0	1	2	3
Sitting and talking to someone	0	1	2	3
Sitting quietly after a lunch without alcohol	0	1	2	3
In a car, while stopped for a few minutes in the traffic	0	1	2	3

Note: The summary score of 10 or above indicate excessive subjective daytime sleepiness.

Figure 4.2 Epworth Sleepiness Scale.

maintenance, early-morning wakening, and/or apparently normal continuity of sleep but waking up unrefreshed.[39] Insomnia is a clinical disorder if the sleep difficulty is associated with daytime symptoms; for example, fatigue, mood problems, or impairment of functioning. Approximately a quarter of patients with chronic medical conditions suffer from comorbid insomnia.[40] It is notable that in studies looking at individuals with insomnia, when objective testing is used, they frequently do not exhibit daytime somnolence.[41]

Fatigue is the leading daytime symptom of insomnia. One can speculate that insomnia leads to fatigue via multiple pathways, including the disruption of sleep and/or the circadian rhythm because of the irregular sleep pattern. Additionally, nonadaptive coping behaviors that people with insomnia engage in can maintain insomnia and fatigue. For example, anxiety about insomnia and fatigue and the daytime safety behaviors fueled by this anxiety (e.g., monitoring one's appearance for signs of fatigue, cancelling programs and spending more time at home) aggravate both insomnia and its daytime consequences, including fatigue.[42]

Additionally, those taking hypnotic medications may describe mental tiredness or sleepiness, because many medications that are used to treat insomnia have a duration of action that leads to an accumulation of this medication, with persistent daytime sleepiness and cognitive impairment related to the hypnotic. Furthermore, there is apparent downregulation of the GABA system from hypnotic use. The GABA system is the main

Table 4.3 Ten differential diagnosis considerations for sleepiness

1. Sleep restriction and sleep debt
2. Sleep apnea and other sleep-related breathing disorders
3. Narcolepsy
4. Fragmented sleep of unknown origin
5. Periodic leg movement disorder
6. Drugs and alcohol
7. Medical disorders (e.g., acute infections, brain injury, stroke, and acute organ failures)
8. Psychiatric disorders (e.g., depression and seasonal affective disorder)
9. Circadian sleep disorders
10. Idiopathic hypersomnia

Source: Adapted from Shapiro CM et al. *Fighting Sleepiness and Fatigue.* Toronto, ON: Joli Joco Publications, Inc., 2005.

inhibitory neural circuitry system accounting for about a third of the neural representation of the central nervous system. As this is the main off switch to the natural orientation that is alertness, the downregulation of the sleep initiation system with hypnotics results in a further degradation of sleep quality. The downregulation of any receptor system in response to a medication (hypnotic or GABA agonist) can now be anticipated to occur as the human body attempts to maintain allostasis (the process of achieving stability or homeostasis through physiological or behavioral change; this can be carried out by means of alteration of hypothalamic–pituitary–adrenal [HPA] axis hormones, the autonomic nervous system, cytokines or a number of other systems, and is generally adaptive in the short term). As we push in one direction, the response is to reduce that directional change. Hence, we often see that a state of tolerance may quite rapidly occur to the effects of that medication.

It is also key to recognize that only with the advent of the *Diagnostic and Statistical Manual of Mental Disorders* (DSM)-5 has insomnia been recognized as a condition in its own right that is maintained by insomnia-specific psychopathological factors.[39] Perhaps the next decade will provide the evidence necessary to make the same statement about fatigue.

SLEEP BREATHING DISORDERS AS CAUSES OF FATIGUE

Obstructive sleep apnea (OSA) is the most common form of sleep-related breathing disorder. It is characterized by periodic (at least 10-second) reductions or complete cessations of airflow due to repeated upper airway obstructions during sleep. It is present in anywhere up to 24% of middle-aged men and 9% of middle-aged women.[43] The prevalence of OSA increases with age: at least 25% of older adults have OSA.[44,45] Notably, approximately 80% of individuals with OSA never receive a diagnosis.[46]

Physiologically, the sleep-related breathing events are terminated by arousals from sleep. The repeated arousals lead to sleep fragmentation and consequently to non-restorative sleep, daytime sleepiness, and fatigue. The breathing cessations during sleep are also associated with hypoxia, abrupt and chronic hemodynamic changes, endothelial damage, increased cytokine levels, and inflammation.[47,48] These physiological changes may be the shared pathways between untreated OSA and the medical conditions that are commonly associated with it, including hypertension, myocardial infarction, congestive heart failure, transient ischemic attacks, stroke, insulin resistance, and diabetes.[49–54]

Despite the commonly held belief that daytime sleepiness is the leading symptom of OSA, in reality, approximately a third of those with OSA report daytime fatigue as a key complaint.[55] Critically, individuals with OSA and high fatigue report more severe impairment of functioning than those with high sleepiness only, despite the fact that the severity of OSA does not differ between the high-fatigue and the high-sleepiness groups.[55] It is therefore imperative to screen for and treat OSA (even mild OSA) in those who suffer from chronic medical conditions and high fatigue. The diagnosis and effective treatment of OSA yields a double benefit: it alleviates fatigue and enhances the outcomes of chronic medical conditions in which fatigue is common (e.g., stroke or heart failure).

Mood disorders maintain fatigue in chronic medical conditions

Two decades ago, Fuhrer and Wessely[56] showed that, in a sample of more than 3700 individuals attending their family doctors' office, persistent fatigue was present in over 40% of the sample, but less than a fifth had shared this complaint with their doctors. Of those patients who had complained about fatigue, women were more likely to be diagnosed with depression. Despite a strong relationship between depression and fatigue, fatigue was not a sufficiently specific or sensitive complaint for the diagnosis of depression. Subsequently, a Danish epidemiological study showed that two characteristics predicted fatigue across a range of fatigue types: low social class and depression.[57] It was also found that chronic medical illnesses had a heightened impact on mental fatigue across the sample, and these conditions were also associated with increased general and physical fatigue in the elderly. A more recent epidemiological study showed that chronic fatigue was independently related to both depression and generalized anxiety in community-dwelling older adults (above 65 years of age), and emphasized the strong influence of fatigue on quality of life in the elderly.[58]

The interplay between mood and fatigue is complex. It would appear that fatigue may lead to low mood (in the same way as other common complaints, such as pain or poor sleep, may do so). Low mood may perpetuate fatigue, and in this situation, one can presume that without dealing with both, one will not come to a resolution of either in many cases.

The problem facing clinicians is that many antidepressants (which may be tried in order to resolve the mood component) will lead to sleep disruption that, in itself, may amplify depression and also may increase fatigue directly. For this reason, in patients with prominent complaints of fatigue or sleep difficulties, a "sleep-friendly" (i.e., sedating) antidepressant may be desirable. Unfortunately, the range of options is limited, and as with all medications, they have specific side effects. In a study carefully assessing mood, sleepiness, and fatigue using a variety of measures, there was clearly more sleepiness and fatigue in depressed patients as compared to normal controls. Much of the fatigue and sleepiness resolved over an 8-week treatment program with mirtazapine.[59] It should be noted that residual insomnia and sleepiness after depression treatment is predictive of relapse.[60] Tricyclic antidepressants, especially clomipramine, may also be useful in those with depression, sleep problems, and fatigue.

Dysfunctional cognitions and behaviors perpetuate fatigue in chronic medical conditions

The recent realization that the experience of fatigue involves high-level mental processes represents a paradigm shift and opens a path to a cognitive–behavioral conceptualization of fatigue. From a psychological perspective, patients' understanding of their symptoms (perceptions of their illness, fatigue, and their consequences) and views about themselves guide the coping behaviors that they engage in to manage the illness and fatigue. These cognitions and coping behaviors are usually accurate and adaptive in acute illnesses and help the individual to recover from the illness. Conversely, the same cognitions and behaviors can perpetuate fatigue and lead to functional impairment in chronic medical conditions. For example, patients holding the belief that the only cause of their fatigue is the biology of their chronic illness may feel helpless, passive, and threatened by the experience of fatigue.[61-63] Focusing on the distressing sensation

of fatigue and adopting passive coping behaviors (e.g., excessive rest, avoidance of physical activities, and cancelling social events) in turn may amplify fatigue and lead to fatigue-related functional impairment. Patients' beliefs that their fatigue will have a chronic course in parallel with the course of their disease may also contribute to the experience of chronic fatigue. For example, expectations of the chronicity of fatigue symptoms, even in a context of acute virus infection, predicted the development of chronic fatigue within 6 months.[64] Coping with fatigue in a reactive (rather than in a proactive) way—in other words, engaging in passive rest when one feels fatigued and cramming all duties into the time slots when one feels more energetic (instead of creating and following through with an activity plan)—is also common and nonadaptive in high-fatigue conditions. These nonadaptive cognitions and behaviors are targeted in cognitive–behavioral therapy (CBT), as described below.

FATIGUE MANAGEMENT METHODS

Pharmaceutical treatment of fatigue

A comprehensive review of modafinil emphasizes that fatigue is not a U.S. Food and Drug Administration (FDA)-approved reason for using this drug, and that limited benefits were detected only in small trials in patients with chronic fatigue syndrome, post-polio fatigue, or MS, indicating that the role of modafinil in these conditions may be limited.[65] However, a study conducted subsequent to this review showed a clear benefit with modafinil for treating fatigue, in patients with HIV, with almost threefold improvement with active treatment compared to placebo.[66] The same group replicated this result in a smaller sample size using armodafinil (the isomer of modafinil).[67] Notably, sleep disorders were not assessed in this study; therefore, it is difficult to tell whether the effect was specific to HIV fatigue or if it was due to symptom relief in those with sleep disorders, particularly sleep apnea. Finally, a meta-analysis of the results of five studies on the efficacy of psychostimulants (methylphenidate and dexamphetamine) for the treatment of cancer fatigue concluded that only one study (using methylphenidate) showed a significant effect (compared to placebo), and the effect size was small.[68] Given that promising results are sporadic, more research involving larger samples is needed in order to evaluate the possible utility of psychostimulants for the treatment of fatigue in chronic medical conditions.

Cognitive–behavioral therapy

CBT is a well-established, evidence-based treatment of some of the most common conditions that maintain fatigue in the medically ill, including chronic insomnia and depression.[69,70] CBT for fatigue has only been evaluated in a few research trials in patients with cancer and MS and has not entered the mainstream of clinical practice.

CBT for cancer-related fatigue

Five randomized trials for CBT in cancer-related fatigue generally concluded that the interventions were more efficacious than usual care or no treatment.[71–75] Fatigue is a common experience during radiotherapy and chemotherapy, affecting most patients undergoing these treatments.[75] Two of these five cancer trials explored the efficacy of CBT for the management of fatigue during cancer therapy.[73,74]

In the first trial, women with breast cancer were provided with the combination of CBT and hypnosis.[74] The cognitive component of the CBT focused on recognizing and challenging catastrophic thoughts about radiotherapy and fatigue in order to reduce emotional and physiological distress; the behavioral fatigue-management strategies involved exercise and activity scheduling. In the hypnosis module, patients were given suggestions about reduced radiotherapy-related fatigue and distress. Overall, although fatigue had increased both in the intervention and in the control groups during radiotherapy, the rate of increase was significantly lower in the intervention groups, and the effect size was in the moderate to high range. One potential advantage of this intervention was that it required minimal client–therapist contact, since most of the practice sessions were done at home, between radiotherapy treatments.

The second randomized trial during cancer treatment compared the efficacy of CBT to a brief nursing intervention and to usual care in a sample of patients receiving chemotherapy for various cancer types.[73] The nursing intervention coached the patients to maintain and increase their level of physical activity in order to prevent physical deconditioning. The CBT, in addition to the module from the nursing intervention, encompassed cognitive elements focusing on fatigue and cancer. These included strategies of changing nonadaptive fatigue-related cognitions, developing adaptive coping to the effects of cancer, and overcoming interpersonal difficulties related to the cancer experience. Additionally, patients were instructed to maintain a fixed bedtime and to avoid daytime naps in order to improve their sleep. At least 2 months after the completion of the chemotherapy, participants receiving usual care were, on average, more fatigued than they had been before chemotherapy; those receiving CBT, on the other hand, were less fatigued. Overall, patients after CBT were significantly less tired than those receiving only usual care, whereas the nursing intervention did not result in significant gains compared to the usual care. Further analysis revealed that patients reporting more concentration and memory problems before the CBT intervention benefited the most from the treatment.[77] Monthly tracking of fatigue after treatment showed that the CBT group maintained the treatment gains until 7 months post-intervention, but thereafter the difference between the CBT and usual care groups disappeared.[77]

Approximately a quarter of disease-free cancer survivors experience fatigue years after they complete curative treatment. Two randomized trials have explored the efficacy of CBT for fatigue in this group of patients. One of these studies included patients within 3 months after treatment completion, irrespective of their level of fatigue, and found that CBT was more effective than usual care in managing fatigue, although CBT was not more effective than the other active control condition (physical training).[75] Notably, in this study, there was an overlap between the interventions.

In contrast to the above two post-treatment studies, another study selected cancer-free patients with severe fatigue who completed treatment at least 1 year prior to their assignment to the psychotherapy.[72] In this post-treatment study, the behavioral elements of CBT included general suggestions regarding sleep-related habits (to keep a regular bedtime and wake-up time and to avoid naps) and the establishment of a balanced activity schedule before gradually increasing the overall activity level. Additionally, patients were encouraged to openly express their cancer-related experiences and to voice their fears of disease recurrence. Dysfunctional thoughts about fatigue and unrealistic social expectations were also challenged. Patients in the CBT conditions showed significantly greater reductions of fatigue and functional impairment than those in the waiting list condition. Approximately two-thirds of those receiving CBT reported significant improvements in their well-being. Moreover, participants maintained these treatment gains over a mean of 2 years.[71]

CBT for multiple sclerosis-related fatigue

The efficacy of CBT in patients suffering from MS has been studied in two randomized trials. The first study compared the efficacy of telephone-based CBT to supportive emotion-focused therapy in reducing depression, but fatigue was also included as an outcome measure.[78] The CBT intervention was based on a standard CBT protocol for depression, but in the second half of the treatment, the management of fatigue was directly addressed. The fatigue management content was based on the *Clinical Practice Guidelines for Fatigue and Multiple Sclerosis*,[79] which suggest prioritizing activities and scheduling naps and rest during the day, similar to adaptive pacing in chronic fatigue syndrome (CFS). The emotion-focused therapy did not directly focus on fatigue. The authors found significant reductions of fatigue in both conditions, which were related to reductions in depression. Additionally, in the CBT condition, a significant portion of the decline in physical fatigue was independent of depression. Although it may be tempting to conclude that the fatigue component may have been responsible for this change, it is important to consider that behavioral activation of the depression treatment may have also contributed to this effect. It is also interesting to point out that the fatigue intervention in this trial was very similar to adaptive pacing—an intervention that did not appear to be effective in chronic fatigue syndrome.[80]

In the more recent MS fatigue trial by van Kessel et al.,[81] the CBT intervention was based on a MS fatigue model that the authors had previously developed.[36] Participants were encouraged to keep a balanced, moderate activity level, to have a regular bedtime and wake-up time and

to leave their bedroom if they were unable to sleep. The patients were expected to find alternative explanations for their somatic symptoms rather than the illness, and they were assisted to challenge negative thoughts. The importance of social support was also discussed. The control arm received training in various relaxation techniques. Both intervention groups achieved significant reductions of fatigue, with a large between-group effect size post-treatment and a moderate between-group effect size at the 6–month follow-up due to the significantly larger treatment effect in the CBT group. The fatigue level of participants in the relaxation group dropped to normal levels after the completion of treatment and 6 months thereafter, whereas patients in the CBT condition experienced less fatigue than did healthy controls. Depression, anxiety, stress, and fatigue-related impairments also decreased in both groups; the gains were more significant in the CBT group at post-treatment, but this difference disappeared by 6 months after treatment completion. Recently, an eight-session self-help, internet-based CBT program was piloted by the same group of researchers and showed a large effect size compared to standard care.[82]

Taken together, from this preliminary data, one can conclude that patients with cancer (or a cancer history) and individuals with MS benefit from cognitive–behavioral, fatigue-focused interventions. In other words, paying attention to fatigue in the medical settings and providing some form of fatigue management program would likely increase the well-being of many patients. It seems that CBT, exercise therapy, and relaxation may contribute to controlling fatigue in these populations, although their long-term benefits and niches are not yet fully established. In addition, due to the dearth of treatment studies in other general medical conditions, the usefulness of these interventions remains to be explored.

Activity-based interventions

Chronic fatigue is associated with significantly reduced activity in a subgroup of patients with chronic fatigue. A sedentary lifestyle may lead to physical deconditioning and increased activity-induced fatigue. Accordingly, regular, gradually increasing aerobic or tension physical activity regimes should lead to increased vigor and energy and decreased fatigue. Indeed, only one session of low-level physical activity increases vigor and is associated with changes in EEG activity in healthy individuals.[83] Interestingly, the decrease of fatigue was not consistent, which confirms the notion that alertness/vigor and fatigue are distinct phenomena. In addition, moderate physical activity did not exert positive effects.

At the other end of the fatigue spectrum (i.e., in individuals with chronic fatigue syndrome), graded exercise therapy was as efficacious as CBT in alleviating chronic fatigue.[80] Exercise therapy has also shown promising results in those with cancer, although activity-based interventions appear to be less efficacious than psychological treatments.[76] Notably, patients often do not maintain an increased activity schedule after treatment completion. Furthermore, in CBT trials, increased activity does not seem to directly contribute to the treatment effects; instead, changes in cognition appear to mediate the effects of CBT on fatigue.[73,84] It is possible that temporary increases of activity help patients to realize that they are able to function despite feeling fatigued. This not only lessens the perceived functional impact of fatigue, but also decreases patients' preoccupation with their fatigue.[85] In the future, dismantling studies will need to explore the efficacy of individual treatment components (e.g., gradual increase of activity and cognitive interventions) on fatigue.

CONCLUSIONS

Fatigue has multiple conceptualizations and myriad assessment tools. The most common assessment tools in chronic medical conditions are questionnaires that capture the subjective intensity and/or impact of fatigue. These tools measure different constructs and often combine questions about fatigue with items assessing subjective sleepiness. Therefore, it is imperative to be knowledgeable about the properties of these assessment tools and the fatigue constructions that these measure. The experience and the impact of subjective fatigue that patients report in chronic medical conditions arise from the combined biological effects of the medical disorders and common comorbidities, such as sleep and mood problems. These effects can be aggravated by patients' cognitive and emotional responses, as well as their behavioral adaptation

to the chronic illness and fatigue. Since effective management of the primary medical condition may not be sufficient to alleviate chronic fatigue, the assessment and treatment of sleep and mood comorbidities is recommended in those struggling with chronic fatigue. There are no evidence-based, fatigue-specific treatments, but CBT and graded exercise are promising methods. The authors hope that the view of subjective fatigue as a specific, multifactorial condition, rather than simply an epiphenomenon of chronic disorders, will inspire research on fatigue-focused assessment and interventions.

REFERENCES

1. Ponka DKM. *Top 10 Differential Diagnosis in Primary Care*. Ottawa, ON, Canada: Department of Family Medicine, University of Ottawa, 2006.
2. *Cecil Textbook of Medicine* (18th ed.). St Louis, MO: W.B. Saunders Company, 1988.
3. Vollestad NK. Measurement of human muscle fatigue. *J Neurosci Methods* 1997; 74(2): 219–227.
4. Chaudhuri A, Behan PO. Fatigue in neurological disorders. *Lancet* 2004; 363(9413): 978–988.
5. Chaudhuri, A, Behan PO. Fatigue and basal ganglia. *J Neurol Sci* 2000; 179(S 1–2): 34–42.
6. Shen J, Barbera J, Shapiro CM. Distinguishing sleepiness and fatigue: Focus on definition and measurement. *Sleep Med Rev* 2006; 10(1): 63–76.
7. Gandevia, SC. Spinal and supraspinal factors in human muscle fatigue. *Physiol Rev* 2001; 81(4): 1725–1789.
8. Schillings ML, Kalkman JS, Janssen HM, van Engelen, BG, Bleijenberg G, Zwarts MJ. Experienced and physiological fatigue in neuromuscular disorders. *Clin Neurophysiol* 2007; 118(2): 292–300.
9. Leavitt VM, DeLuca J. Central fatigue: Issues related to cognition, mood and behavior, and psychiatric diagnoses. *PM R* 2010; 2(5): 332–337.
10. Shahid A, Wilkinson K, Marcu S, Shapiro CM. *STOP, THAT and One Hundred Other Sleep Scales*. New York, NY: Springer, 2011.

11. Fisk JD, Ritvo PG, Ross L, Haase DA, Marrie TJ, Schlech WF. Measuring the functional impact of fatigue: Initial validation of the fatigue impact scale. *Clin Infect Dis* 1994; 18(Suppl 1): S79–S83.

12. Michielsen HJ, De Vries J, Van Heck GL. Psychometric qualities of a brief self-rated fatigue measure: The Fatigue Assessment Scale. *J Psychosom Res* 2003; 54(4): 345–352.

13. Schwartz JE, Jandorf L, Krupp LB. The measurement of fatigue: A new instrument. *J Psychosom Res* 1993; 37(7): 753–762.

14. Smets EM, Garssen B, Bonke B, De Haes JC. The Multidimensional Fatigue Inventory (MFI) psychometric qualities of an instrument to assess fatigue. *J Psychosom Res* 1995; 39(3): 315–325.

15. Chalder T, Berelowitz G, Pawlikowska T et al. Development of a fatigue scale. *J Psychosom Res* 1993; 37(2): 147–153.

16. Teran-Santos J, Jimenez-Gomez A, Cordero-Guevara J. The association between sleep apnea and the risk of traffic accidents. Cooperative Group Burgos-Santander. *N Engl J Med* 1999; 340(11): 847–851.

17. Lee KA, Hicks G, Nino-Murcia G. Validity and reliability of a scale to assess fatigue. *Psychiatry Res* 1991; 36(3): 291–298.

18. Hann DM, Jacobsen PB, Azzarello LM et al. Measurement of fatigue in cancer patients: Development and validation of the Fatigue Symptom Inventory. *Qual Life Res* 1998; 7(4): 301–310.

19. Mendoza TR, Wang XS, Cleeland CS et al. The rapid assessment of fatigue severity in cancer patients: Use of the Brief Fatigue Inventory. *Cancer* 1999; 85(5): 1186–1196.

20. Piper BF, Rieger PT, Brophy L et al. Recent advances in the management of biotherapy-related side effects: Fatigue. *Oncol Nurs Forum* 1989; 16(6 Suppl): 27–34.

21. Rhoten D. Fatigue and the postsurgical patient. In: Norris C (ed.). *Concept Clarification in Nursing*. Rockville, MD: Aspen Systems, 1982, pp. 277–300.

22. Yellen SB, Cella DF, Webster K, Blendowski C, Kaplan E. Measuring fatigue and other anemia-related symptoms with the Functional Assessment of Cancer Therapy (FACT) measurement system. *J Pain Symptom Manage* 1997; 13(2): 63–74.

23. Krupp LB, LaRocca NG, Muir-Nash J, Steinberg AD. The Fatigue Severity Scale. Application to patients with multiple sclerosis and systemic lupus erythematosus. *Arch Neurol* 1989; 46(10): 1121–1123.

24. Buysse DJ, Thompson W, Scott J et al. Daytime symptoms in primary insomnia: A prospective analysis using ecological momentary assessment. *Sleep Med* 2007; 8(3): 198–208.

25. Dimsdale JE, Ancoli-Israel S, Ayalon L, Elsmore TF, Gruen W. Taking fatigue seriously, II: Variability in fatigue levels in cancer patients. *Psychosomatics* 2007; 48(3): 247–252.

26. Yorkston KM, Johnson K, Boesflug E, Skala J, Amtmann D. Communicating about the experience of pain and fatigue in disability. *Qual Life Res* 2010; 19(2): 243–251.

27. Kleinman L, Mannix S, Yuan Y, Kummer S, L'Italien G, Revicki D. Review of patient-reported outcome measures in chronic hepatitis C. *Health Qual Life Outcomes* 2012; 10: 92.

28. Kyle SD, Morgan K, Spiegelhalder K, Espie CA. No pain, no gain: An exploratory within-subjects mixed-methods evaluation of the patient experience of sleep restriction therapy (SRT) for insomnia. *Sleep Med* 2011; 12(8): 735–747.

29. Anderson VR, Jason LA, Hlavaty LE, Porter N, Cudia J. A review and meta-synthesis of qualitative studies on myalgic encephalomyelitis/chronic fatigue syndrome. *Patient Educ Couns* 2012; 86(2): 147–155.

30. Larun L, Malterud K. Identity and coping experiences in chronic fatigue syndrome: A synthesis of qualitative studies. *Patient Educ Couns* 2007; 69(1–3): 20–28.

31. Repping-Wuts H, Uitterhoeve R, van Riel P, van Achterberg T. Fatigue as experienced by patients with rheumatoid arthritis (RA): A qualitative study. *Int J Nurs Stud* 2008; 45(7): 995–1002.

32. Siegel K, Lekas HM, Maheshwari D. Causal attributions for fatigue by older adults with advanced cancer. *J Pain Symptom Manage* 2012; 44(1): 52–63.

33. Spichiger E, Rieder E, Muller-Frohlich C, Kesselring A. Fatigue in patients undergoing chemotherapy, their self-care and the role of health professionals: A qualitative study. *Eur J Oncol Nurs* 2012; 16(2): 165–171.

34. van Kessel K, Moss-Morris R. Understanding multiple sclerosis fatigue: A synthesis of biological and psychological factors. *J Psychosom Res* 2006; 61(5): 583–585.

35. Ohayon MM. From wakefulness to excessive sleepiness: What we know and still need to know. *Sleep Med Rev* 2008; 12(2): 129–141.

36. Johns MW. A new method for measuring daytime sleepiness: The Epworth Sleepiness Scale. *Sleep* 1991; 14(6): 540–545.

37. Shapiro CM, Ohayon MM, Huterer N, Grunstein R. *Fighting Sleepiness and Fatigue.* Toronto, ON: Joli Joco Publications, Inc., 2005.

38. Hossain JL, Ahmad P, Reinish LW et al. Subjective fatigue and subjective sleepiness: Two independent consequences of sleep disorders? *J Sleep Res* 2005; 14(3): 245–253.

39. *Diagnostic and Statistical Manual of Mental Disorders (DSM-5).* Washington, DC: American Psychiatric Publishing, 2013.

40. Budhiraja R, Roth T, Hudgel DW, Budhiraja P, Drake CL. Prevalence and polysomnographic correlates of insomnia comorbid with medical disorders. *Sleep* 2011; 34(7): 859–867.

41. Riedel BW, Lichstein KL. Insomnia and daytime functioning. *Sleep Med Rev* 2000; 4(3): 277–298.

42. Harvey AG. A cognitive model of insomnia. *Behav Res Ther* 2002; 40(8): 869–893.

43. Young T, Palta M, Dempsey J, Skatrud J, Weber S, Badr S. The occurrence of sleep-disordered breathing among middle-aged adults. *N Engl J Med* 1993; 328(17): 1230–1235.

44. Ancoli-Israel S, Kripke DF, Klauber MR, Mason WJ, Fell R, Kaplan O. Sleep-disordered breathing in community-dwelling elderly. *Sleep* 1991; 14(6): 486–495.

45. Johansson P, Alehagen U, Svanborg E, Dahlstrom U, Brostrom A. Sleep disordered breathing in an elderly community-living population: Relationship to cardiac function, insomnia symptoms and daytime sleepiness. *Sleep Med* 2009; 10(9): 1005–1011.

46. Young T, Peppard PE, Gottlieb DJ. Epidemiology of obstructive sleep apnea: A population health perspective. *Am J Resp Crit Care Med* 2002; 165(9): 1217–1239.

47. Beebe DW, Gozal D. Obstructive sleep apnea and the prefrontal cortex: Towards a comprehensive model linking nocturnal upper airway obstruction to daytime cognitive and behavioral deficits. *J Sleep Res* 2002; 11(1): 1–16.

48. Lopez-Jimenez F, Sert Kuniyoshi FH, Gami A, Somers VK. Obstructive sleep apnea: Implications for cardiac and vascular disease. *Chest* 2008; 133(3):793–804.

49. Johansson P, Svensson E, Alehagen U, Dahlstrom U, Jaarsma T, Brostrom A. Sleep disordered breathing, hypoxia and inflammation: Associations with sickness behaviour in community dwelling elderly with and without cardiovascular disease. *Sleep Breath* 2015; 19(1): 263–271.

50. Mokhlesi B, Finn LA, Hagen EW et al. Obstructive sleep apnea during REM sleep and hypertension. Results of the Wisconsin Sleep Cohort. *Am J Resp Crit Care Med* 2014; 190(10): 1158–1167.

51. Montesi SB, Bajwa EK, Malhotra A. Biomarkers of sleep apnea. *Chest* 2012; 142(1): 239–245.

52. Park JG, Ramar K, Olson EJ. Updates on definition, consequences, and management of obstructive sleep apnea. *Mayo Clin Proc* 2011; 86(6): 549–554; quiz 554–545.

53. Young T, Palta M, Dempsey J, Peppard PE, Nieto FJ, Hla KM. Burden of sleep apnea: Rationale, design, and major findings of the Wisconsin Sleep Cohort study. *WMJ* 2009; 108(5): 246–249.

54. Young T, Skatrud J, Peppard PE. Risk factors for obstructive sleep apnea in adults. *JAMA* 2004; 291(16): 2013–2016.

55. Bailes S, Libman E, Baltzan M et al. Fatigue: The forgotten symptom of sleep apnea. *J Psychosom Res* 2011; 70(4): 346–354.

56. Fuhrer R, Wessely S. The epidemiology of fatigue and depression: A French primary-care study. *Psychol Med* 1995; 25(5): 895–905.

57. Watt T, Groenvold M, Bjorner JB, Noerholm V, Rasmussen NA, Bech P. Fatigue in the Danish general population. Influence of

sociodemographic factors and disease. *J Epidemiol Commun Health* 2000; 54(11): 827–833.

58. Chou KL. Chronic fatigue and affective disorders in older adults: Evidence from the 2007 British National Psychiatric Morbidity Survey. *J Affect Disord* 2013; 145(3): 331–335.

59. Shen J, Hossain N, Streiner DL et al. Excessive daytime sleepiness and fatigue in depressed patients and therapeutic response of a sedating antidepressant. *J Affect Disord* 2011; 134(1–3): 421–426.

60. Fava M. Daytime sleepiness and insomnia as correlates of depression. *J Clin Psychiatry* 2004; 65(Suppl 16): 27–32.

61. Bol Y, Duits AA, Hupperts RM, Vlaeyen JW, Verhey FR. The psychology of fatigue in patients with multiple sclerosis: A review. *J Psychosom Res* 2009; 66(1): 3–11.

62. Glacken M, Coates V, Kernohan G, Hegarty J. The experience of fatigue for people living with hepatitis C. *J Clin Nurs* 2003; 12(2): 244–252.

63. Servaes P, Verhagen S, Bleijenberg G. Determinants of chronic fatigue in disease-free breast cancer patients: A cross-sectional study. *Ann Oncol* 2002; 13(4): 589–598.

64. Moss-Morris R, Spence MJ, Hou R. The pathway from glandular fever to chronic fatigue syndrome: Can the cognitive behavioural model provide the map? *Psychol Med* 2011; 41(5): 1099–1107.

65. Kumar R. Approved and investigational uses of modafinil: An evidence-based review. [Review]. *Drugs* 2008; 68(13): 1803–1839.

66. Rabkin JG, McElhiney MC, Rabkin R, McGrath PJ. Modafinil treatment for fatigue in HIV/AIDS: A randomized placebo-controlled study. *J Clin Psychiatry* 2010; 71(6): 707–715.

67. Rabkin JG, McElhiney MC, Rabkin R. Modafinil and armodafinil treatment for fatigue for HIV-positive patients with and without chronic hepatitis C. *Int J STD AIDS* 2011; 22(2): 95–101.

68. Minton O, Richardson A, Sharpe M, Hotopf M, Stone PC. Psychostimulants for the management of cancer-related fatigue: A systematic review and meta-analysis. *J Pain Symptom Manage* 2011; 41(4): 761–767.

69. Butler AC, Chapman JE, Forman EM, Beck AT. The empirical status of cognitive–behavioral therapy: A review of meta-analyses. *Clin Psychol Rev* 2006; 26(1): 17–31.

70. Morin CM, Colecchi C, Stone J, Sood R, Brink D. Behavioral and pharmacological therapies for late-life insomnia: A randomized controlled trial. *JAMA* 1999; 281(11): 991–999.

71. Gielissen MF, Verhagen CA, Bleijenberg G. Cognitive behaviour therapy for fatigued cancer survivors: Long-term follow-up. *Br J Cancer* 2007; 97(5): 612–618.

72. Gielissen MF, Verhagen S, Witjes F, Bleijenberg G. Effects of cognitive behavior therapy in severely fatigued disease-free cancer patients compared with patients waiting for cognitive behavior therapy: A randomized controlled trial. *J Clin Oncol* 2006; 24(30): 4882–4887.

73. Goedendorp MM, Peters ME, Gielissen MF et al. Is increasing physical activity necessary to diminish fatigue during cancer treatment? Comparing cognitive behavior therapy and a brief nursing intervention with usual care in a multicenter randomized controlled trial. *Oncologist* 2010; 15(10): 1122–1132.

74. Montgomery GH, Kangas M, David D et al. Fatigue during breast cancer radiotherapy: An initial randomized study of cognitive–behavioral therapy plus hypnosis. *Health Psychol* 2009; 28(3): 317–322.

75. van Weert E, May AM, Korstjens I et al. Cancer-related fatigue and rehabilitation: A randomized controlled multicenter trial comparing physical training combined with cognitive–behavioral therapy with physical training only and with no intervention. *Phys Ther* 2010; 90(10): 1413–1425.

76. Jacobsen PB, Donovan KA, Vadaparampil ST, Small BJ. Systematic review and meta-analysis of psychological and activity-based interventions for cancer-related fatigue. *Health Psychol* 2007; 26(6): 660–667.

77. Goedendorp MM, Gielissen MF, Peters ME, Verhagen CA, Bleijenberg G. Moderators and long-term effectiveness of cognitive behaviour therapy for fatigue during cancer treatment. *Psychooncology* 2012; 21(8): 877–885.

78. Mohr DC, Hart SL, Goldberg A. Effects of treatment for depression on fatigue in multiple sclerosis. *Psychosom Med* 2003; 65(4): 542–547.

79. Multiple Sclerosis Council for Practice Guidelines. Fatigue and multiple sclerosis: Evidence-based management strategies for fatigue in multiple sclerosis. Washington, DC: Paralyzed Veterans of America, 1998.

80. White PD, Goldsmith KA, Johnson AL et al. Comparison of adaptive pacing therapy, cognitive behaviour therapy, graded exercise therapy, and specialist medical care for chronic fatigue syndrome (PACE): A randomised trial. *Lancet* 2011; 377(9768): 823–836.

81. van Kessel K, Moss-Morris R, Willoughby E, Chalder T, Johnson MH, Robinson E. A randomized controlled trial of cognitive behavior therapy for multiple sclerosis fatigue. *Psychosom Med* 2008; 70(2): 205–213.

82. Moss-Morris R, McCrone P, Yardley L, van Kessel K, Wills G, Dennison L. A pilot randomised controlled trial of an Internet-based cognitive behavioural therapy self-management programme (MS Invigor8) for multiple sclerosis fatigue. *Behav Res Ther* 2012; 50(6): 415–421.

83. Dishman RK, Thom NJ, Puetz TW, O'Connor PJ, Clementz BA. Effects of cycling exercise on vigor, fatigue, and electroencephalographic activity among young adults who report persistent fatigue. *Psychophysiology* 2010; 47(6): 1066–1074.

84. Wiborg JF, Knoop H, Stulemeijer M, Prins JB, Bleijenberg G. How does cognitive behaviour therapy reduce fatigue in patients with chronic fatigue syndrome? The role of physical activity. *Psychol Med* 2010; 40(8): 1281–1287.

85. Knoop H, Prins JB, Moss-Morris R, Bleijenberg G. The central role of cognitive processes in the perpetuation of chronic fatigue syndrome. *J Psychosom Res* 2010; 68(5): 489–494.

5

Occupational sleep medicine: Role of social stressors

DIANA PEREIRA, CHRISTIN GERHARDT, MARIA U. KOTTWITZ, AND ACHIM ELFERING

INTRODUCTION

If we look at what shapes our daily experience and behavior, then work is a crucial factor for many people. Work has significant positive psychosocial functions, and at least some of them are tied to its social meaning. Besides maintaining and advancing our skills or giving us structure, work facilitates social contacts and appreciation, and it is often an important part of our identity.[1,2]

Despite its positive psychosocial functions, work can be an important source of stress. The term *stress* refers to a subjectively unpleasant state of strain arising from the fear of being unable to cope with an aversive situation.[3] Moreover, this psychological response is linked to biophysiological reactions, leading to a cascade of stress-related somatic changes. In general, adverse work conditions are called *work stressors*, for they increase the likelihood of employees experiencing stress.[4] There is now a consensus regarding work stressors affecting physical as well as psychological health (for reviews, see De Lange et al.,[5] Kivimäki et al.,[6] Nixon et al.[7] and Semmer et al.[8]).

Within the scope of stress, the social context of work is very important. Because of the changing nature of work, service jobs have become the major employment sector in western countries. Therefore social interactions with coworkers, supervisors and non-organizational members (e.g., customers or clients) are part of the everyday experience of a large number of employees.[9] Social interactions at work may have both positive and negative effects on employees.[9]

SOCIAL STRESSORS AT WORK

Social stressors at work may take many forms, such as social animosities, conflicts with coworkers and supervisors, unfair behavior, and a negative group climate.[9] Behavior like attacks, giving rude and reckless feedback, or undermining the successes of others constitute a direct expression of disrespectful social behavior.[10] The behavior can be distinguished with respect to intent, duration, directedness, and intensity (Table 5.1).[11–28] Mobbing, bullying, harassment, workplace incivilities, and discrimination are intense social stressors that have aggression in common.[29] The definition/distinctiveness of so many different phenomena within social stressors and the need for so many constructs are under debate, and the discussion is beyond the scope of the current chapter.[29–32]

Social stressors at work occur frequently.[33,34] According to a study conducted by Keenan and Newton,[35] 74% of the stressful incidents that occurred at work were social in nature. Verbal abuse seems to be particularly prevalent. In 2010, an average of 11% of European workers reported that they had experienced verbal abuse during the previous month.[36]

Social stressors have been found to be related to impaired well-being and health.[37,38] Moreover, the experience of social stressors can exacerbate impairments such as depression.[9] A meta-analysis found that aggression from supervisors, coworkers, and non-organizational members (e.g., customers or clients) elicited adverse attitudinal, behavioral, and health-related outcomes.[39] Thus, social stressors are not only common, but are also the most troublesome stressors at work.[40] However, compared with task-related stressors such as time pressures or interruptions in workflow, there is less research investigating social stressors at work.[9,41,42] In the last two decades, interest in social stressors at work has increased, and meta-analyses on social stressors and health underline the risk from adverse social conditions at work.[39,43]

SOCIAL STRESSORS AT WORK AND SLEEP: EVIDENCE FROM QUESTIONNAIRE STUDIES

Occupation sleep medicine has a long tradition of addressing sleep quantity and sleep quality (e.g., as antecedent of work-related injury).[44] However,

studies that test a link between sleep and social stressors are rare. A recent meta-analysis on bullying and health-related outcomes (including sleep) by Nielsen and Einarsen[45] summarizes previous studies published until 2011. The authors expected a significant positive relationship between bullying and sleep problems. However, based on four studies, they found a nonsignificant effect size of −0.10. Recently, a meta-analysis on work conditions and sleep based on five studies[46] reported a significant association between sleep quality and workplace bullying (−0.23, k = 3), as well as between sleep quality and workplace violence (−0.20, k = 2). The number of studies reviewed is still small, but it has increased. Therefore, for this chapter, we conduct a new search of social stressors and sleep, resulting in a total of 14 studies (Table 5.2).[47–77]

Nine out of ten of these studies listed in Table 5.2 report significant correlation coefficients that could be aggregated to indicate a mean effect size, weighted for sample size and tested to be significant. Table 5.3 shows the effect sizes for each of the studies and an aggregated value. The weighted mean effect size is 0.21 (95% CI = 0.12–0.31). Thus, the relation between social stressors at work and impaired sleep is meaningful but small to moderate. Moreover, further analysis of the variation of effect sizes across studies showed that only a very small part of this variation was due to measurement error. Thus, the association between social stressors and sleep is not homogeneous, and moderators have to be taken into account.

This small meta-analysis shows that evidence for a meaningful association is increasing, but the underlying mechanisms remain unclear. Insights in real occupational life via continuous ambulatory monitoring at work and at home can shed some light on potential mechanisms.

MECHANISMS THAT LINK SOCIAL STRESSORS TO IMPAIRED SLEEP: AMBULATORY RESEARCH

Spending some time off work does not automatically imply that recovery will occur. Even when workers are not exposed to stressful demands after work, cognitive mechanisms can prolong or reactivate psycho-physiological activation, jeopardizing recovery and sleep.[78] Evidence is increasing that cognitive mechanisms like detachment and worry act as intervening variables.[79]

Table 5.1 Social stressor concepts

Social stressor concept	Definition	Author(s)
Abusive supervision	Subordinates' perceptions of the extent to which supervisors engage in the sustained displays of hostile verbal and nonverbal behaviors, excluding physical contact	Tepper[11]
Bullying	Harassing, offending, or socially excluding someone or negatively affecting someone's work tasks. In order for the label bullying (or mobbing) to be applied to a particular activity, interaction, or process, it has to occur constantly and repeatedly (e.g., weekly) and over a period of time (e.g., about 6 months)	Einarsen et al.[12]
Discrimination	Differential treatment of people based on actual or perceived membership in a particular group	Williams et al.[13]
Dysfunctional social support	Support that is given reluctantly, often combined with reproaches, expectations of infinite thankfulness, and indications that a person should have dealt with the problem him or herself	Semmer et al.[14]
Harassment	Repeated and persistent attempts by a person to torment, wear down, frustrate, or get a reaction from another person; it is treatment that persistently provokes, pressures, frightens, intimidates or otherwise causes discomfort in another person	Brodsky[15]
	Situations in which a person is exposed repeatedly and over time to negative actions on the part of one or more persons	Vartia[16]
	Repeated activities with the aim of bringing mental (but sometimes also physical) pain, and directed towards one or more individuals who, for one reason or another, are not able to defend themselves	Bjorkqvist et al.[17]
Hostility	Experience of anger and disgust	Barefoot[18]
Interpersonal unfairness	Refers to a lack of propriety, dignity, and respect from authority figures who are charged with implementing company procedures, making decisions, and distributing outcomes	Bies and Moag[19] Colquitt[20]
Mobbing	Involves hostile and unethical communication, which is directed in a systematic way by one or a few individuals mainly towards one individual who, due to mobbing, is pushed into a helpless and defenseless position, being held there by means of continuing mobbing activities	Leymann[21]
Role conflict	Incompatible demands placed upon an employee such that compliance with both would be difficult	Katz and Kahn[22]

(Continued)

Table 5.1 (*Continued*) Social stressor concepts

Social stressor concept	Definition	Author(s)
Social exclusion	The general perception of an individual of being excluded, rejected, or ignored by another individual at the workplace that hinders their ability to establish or maintain positive interpersonal relationships	Leary[23]
Supervisor interactional injustice	(Justice reversed) Focuses on the importance of the quality of the interpersonal treatment people receive when procedures are implemented	Bies and Moag[19]
Supervisor-rated unethical behavior	Ethical behavior is that which is both legal and morally acceptable to the larger community in which it occurs, whereas unethical behavior is either illegal or morally unacceptable	Jones[24]
Undermining	Involves the perception of being degraded, devalued, disliked, and discounted by others	Rook[25]
Workplace aggression	Any form of behavior by employees (which could be verbal or physical in nature) that is intended to harm current or previous coworkers or their organizations in general	Folger and Baron[26]
Workplace incivility	Low-intensity deviant behavior with ambiguous intent to harm the target, in violation of workplace norms for mutual respect	Andersson and Pearson[27]
Workplace violence	Includes physical assaults and threats of assault directed towards employees	Jenkins[28]

Many scientists have tried to explain how social stressors at work elicit harmful effects on health. Geurts and Sonnentag[80] argue that recovery including sleep may help explain how acute load reactions may develop into chronic load reactions and so impair health. According to Meijman and Mulder's[81] effort recovery theory, McEwen's[82,83] allostatic load theory and Geurts and Sonnentag's[80] conceptual approach of incomplete recovery, recovery might be the intervening variable in the hypothetical causal chain of stressful work characteristics and the development of health impairments in the long run.

According to Meijman and Mulder's[81] effort recovery theory, job demands require effort on the part of individuals in order to perform adequately. In principle, these efforts unavoidably involve adaptive physiological and psychological reactions, such as accelerated heart rate, elevated blood pressure, and fatigue. Under normal circumstances, the acute stress-related reactions are short lived and fully reversible. Thus, after spending a certain period of time in which the systems

concerned are not or are only slightly activated, the respective psycho-physiological systems will stabilize again at a specific baseline (pre-stressor) level and recovery starts. Therefore, normally recovery occurs after a short respite from work and is completed before the next morning starts. In this situation, health is not at risk.[80,84]

However, particularly stressful work conditions or continued exposure to job demands may lead to persistent physiological and psychological load reactions, and as a consequence, opportunities for recovery will be lacking or incomplete. In a case of incomplete recovery, the worker will have to begin work in a suboptimal condition, and will have to invest compensatory effort in order to perform adequately. This will lead to increased intensity of load reactions, higher demands on recovery processes and, finally, to an accumulative process resulting in less transitory or even permanent symptoms such as impaired sleep and psychosomatic complaints.[85]

Whereas the effort recovery model[81] does not describe in detail which psycho-physiological

Table 5.2 Studies that report an association between indicators of sleep quantity, sleep quality and social stressors at work

Source	Sample	Social stressor concept	Sleep measure	Association(s) reported	Association recoded: (low to high) social stressors and (low to high) sleep quality
Barling et al.[47]	591 employees of various occupations (67.5% female)	Study 1: • Frequency of sexual harassment • Items were included if they clearly described behaviors, actions, or heard comments	Sleeping habits[48]	Females main sample, r = 0.23** Females replication sample, r = 0.15 Males main sample, r = 0.21**, r = 0.12	r = −0.23** r = −0.15 r = −0.21** r = −0.12
	141 female undergraduate student volunteers	Study 2: • Frequency of sexual harassment • Items were included if they clearly described behaviors, actions, or heard comments			
Barnes et al.[49]	182 pairs of employees and supervisors (49% female)	Supervisor-rated unethical behavior (measure of unethical behavior at work[50])	Sleep quantity (how many hours of sleep do you usually get on a typical night?) Sleep quality (insomnia scale[51])	−0.20** −0.32**	−0.20** −0.32**
Giorgi[52]	715 Japanese workers (30.5% female)	Bullying (Negative Acts Questionnaire Revised [NAQ-R][53]) • Bullying at work • Work-related bullying • Personal bullying	Average hours of sleep (seven-point Likert scale from less than 5 hours to 10 or more hours)	−0.08* −0.11** −0.08*	−0.08* −0.11** −0.08*
Goldenhar et al.[54]	211 construction workers (100% female), United States	Sexual harassment and discrimination (questions from the Northwestern National Life Insurance Company[55]) Responsibility for safety of others	Insomnia (questions from the North-western National Life Insurance Company[55])	0.14* 0.003	−0.14 0.003

(Continued)

Table 5.2 (*Continued*) Studies that report an association between indicators of sleep quantity, sleep quality and social stressors at work

Source	Sample	Social stressor concept	Sleep measure	Association(s) reported	Association recoded: (low to high) social stressors and (low to high) sleep quality
Hansen et al.[56]	3382 employees (67% female) at baseline, 1671 at t_1	Bullying Have you been subjected to bullying at work within the past 6 months? • Not bullied • Occasionally bullied • Frequently bullied Have you witnessed bullying at work within the past 6 months? • Occasionally observed • Not observed • Frequently observed	Sleep problems (modified version of the Karolinska Sleep Questionnaire [KSQ][57]) with scales for disturbed sleep and awakening problems Sleep quality (how do you rate your overall quality of sleep?)	*Odds ratios* Subject to bullying: Disturbed sleep: 1 (sig.)/3.6 (sig.)/2.11 Unsatisfactory awakenings: 1 (sig.)/3.08 (sig.)/2.47 Poor quality of sleep: 1 (sig.)/5.22 (sig.)/3.56 (sig.) Observed bullying: Disturbed sleep: 1/1.71/1.68 Unsatisfactory awakenings: 1/1.31/1.29 Poor sleep quality: 1/1.38/1.21 → (all crude)	
Lallukka et al.[58]	n = 7332 (t_1) employees of City of Helsinki	Workplace bullying If they have been bullied: • No • Earlier in this or another workplace • Yes currently • I do not know If they have witnessed bullying at their working place: • No • Sometimes • Frequently • I do not know (Similar to Kivimäki et al.[59])	Sleep problems (insomnia scale[51])	*Odds ratios for women/men* 1/1 1.47/1.58 1.69/3.17 1.39/1.79 1/1 1.13/1.15 2/2.04 1.27/1.42 All adjusted for age	

(*Continued*)

Table 5.2 (*Continued*) Studies that report an association between indicators of sleep quantity, sleep quality and social stressors at work

Source	Sample	Social stressor concept	Sleep measure	Association(s) reported	Association recoded: (low to high) social stressors and (low to high) sleep quality
Niedhammer et al.[60]	7694 members of the working population in southeast France (59.3% female)	Workplace bullying (French version of the Leymann Inventory of Psychological Terror[61] by Niedhammer et al.[62])	Sleep disturbances (measured using two items evaluating difficulty initiating sleep and difficulty returning to sleep after experiencing a premature awakening)	*Odds ratios for men/woman* Exposed to bullying: No: 1***/1*** Yes: 4.4***/3.83*** Timeframe of exposure to bullying: None: 1***/1*** Past: 0.91***/2.63*** Current: 5.47***/4.35*** Frequency of exposure to bullying: None: 1***/1*** Weekly: 3.25***/3.38*** Daily or almost daily: 6.34***/4.28*** Duration of exposure to bullying: 0: 1***/1*** < 2: 4.25***/3.22*** ≥ 2 to <5: 4.2***/4.63*** 5+: 4.58***/3.91*** Observer of bullying: No: 1***/1*** Yes: 2.53***/2.2*** Was bullied or observed bullying: Neither: 1***/1*** Observed: 2.08***/1.7*** Was bullied: 5.33***/3.04***	

(*Continued*)

Table 5.2 (Continued) Studies that report an association between indicators of sleep quantity, sleep quality and social stressors at work

Source	Sample	Social stressor concept	Sleep measure	Association(s) reported	Association recoded: (low to high) social stressors and (low to high) sleep quality
				All adjusted for age, marital status, presence of children in the home, education level, occupation, number of hours working per week, working at night, and physical–chemical exposures	
Park et al.[63]	10,039 (42.1% female)	Organizational factors underlying work-related sleep problems (measured with single items) • Sexual harassment • No: n = 9976 • Yes: n = 63 • Sexual discrimination • No: n = 9894 • Yes: n = 145 • Age discrimination • No: n = 9696 • Yes: n = 343 • Violence at work • No: n = 9964 • Yes: n = 75 • Threat of violence • No: n = 9959 • Yes: n = 80	Sleep problems (do you currently suffer from work-related sleep problems?)	Odds ratios

1
6.25

1
3.02

1
3.21

1
6.09

1
5.27

All crude (without controlling for something else) | |

(Continued)

Table 5.2 (*Continued*) Studies that report an association between indicators of sleep quantity, sleep quality and social stressors at work

Source	Sample	Social stressor concept	Sleep measure	Association(s) reported	Association recoded: (low to high) social stressors and (low to high) sleep quality
Rafferty et al.[64]	175 Filipinos in a variety of occupations (43% female among subordinates)	Supervisor interactional injustice (measures of distributive and interactional justice[65] → reverse coded to index injustice) Abusive supervision[11]	Insomnia (partner rating scale[66])	0.18* 0.17*	−0.18* −0.17*
Rodríguez-Muñoz et al.[67]	2033 and 2035 (two samples) White- and blue-collar workers, other (44.3% female)	Workplace bullying (short version of the NAQ[68])	Sleep quality (Questionnaire on the Experience and Evaluation of Work [QEEW][69])	0.51** 0.51**	−0.51** −0.51**
Rogers and Kelloway[70]	194 bank tellers Canada	Direct violence Vicarious experience of violence	Physical well-being: sleep disturbances (subscale of the modified version of the Health Scale[48])	0.22 0.24	−0.22 −0.24
Slopen and Williams[71]	2983 from the Chicago Community Adult Health Study (CCAHS; 52.9% female)	Job harassment (adapted from Perceived Racism[72]) Unfair treatment on the job (adapted from Perceived Racism Scale[72])	Sleep duration (how many hours of sleep do you usually get at night?) Sleep difficulties[73]	$\beta = -0.14$*** $\beta = 0.04$**	

(*Continued*)

Table 5.2 (*Continued*) Studies that report an association between indicators of sleep quantity, sleep quality and social stressors at work

Source	Sample	Social stressor concept	Sleep measure	Association(s) reported	Association recoded: (low to high) social stressors and (low to high) sleep quality
Takaki et al.[74]	2634 workers	Workplace bullying (Japanese version of the NAQ[68])	Sleep disturbance (Pittsburgh Sleep Quality Index,[89] Japanese version[75])	Mediation effect of workplace bullying Men = 0.103 (sig.) Woman = 0.094 (sig.) (Only adjusted for age)	
Vartia[76]	949 in variety of occupations (85% female)	Bullying (revised version of the Leymann Inventory of Psychological Terrorization[77])	Use of sleep-inducing medicaments and sedatives were assessed with one question	Sleep-inducing medicaments: ($\chi^2 = **$) (in %) Bullied: no = 88; yes, occasionally, or regularly = 13 Observers: no = 90; yes, occasionally, or regularly = 10 Non-bullied: no = 95; yes, occasionally, or regularly = 5 Sedatives ($\chi^2 = ***$) (in %) Bullied: no = 84; yes, occasionally, or regularly = 16 Observers: no = 92; yes, occasionally, or regularly = 8 Non-bullied: no = 97; yes, occasionally, or regularly = 3	

Note: Significance level is indicated: * <0.05, ** <0.01, *** <0.001, two-tailed.

Table 5.3 Meta-analytic results on studies from Table 5.2 that report correlation coefficients between impaired sleep and social stressors

Study	n	Correlation	Significance	Lower	Higher
Barling et al.[47]	591	0.22	0.00	0.14	0.30
Barling et al.[47]	141	0.12	0.16	−0.05	0.28
Barnes et al.[49]	182	0.20	0.01	0.06	0.34
Giorgi[52]	715	0.11	0.00	0.04	0.18
Goldenhar et al.[54]	211	0.14	0.04	0.01	0.27
Rafferty et al.[64]	175	0.18	0.02	0.03	0.32
Rodríguez-Muñoz et al.[67]	2033	0.51	0.00	0.48	0.54
Rogers and Kelloway[70]	194	0.24	0.00	0.10	0.37
Slopen and Williams[71]	2983	0.14	0.00	0.10	0.18
Takaki et al.[74]	2634	0.10	0.00	0.06	0.14
Σ	9859	0.21	0.00	0.12	0.31

systems are involved, the allostatic load theory is more explicit in this regard.[80,82,83] According to McEwen,[82,83] occupational stressors require protective responses of the body in order to maintain homeostasis. Those responses include physiological responses of the autonomic nervous system, the hypothalamic–pituitary–adrenocortical (HPA) axis, and the cardiovascular, metabolic and immune systems. These systems lead to protection and adaptation of the organism from occupational stressors and are useful in order to cope with them adequately. According to McEwen,[82,83] allostatic load is an imbalance in these systems, being simply the result of too much repeated stress, and of adaptive systems that are out of balance and fail to shut off or fail to turn on adequately. In line with McEwen,[82,83] the failed shutdown (i.e., failed recovery) of the stress responses is particularly important, because it can cause diverse health impairments.

Based on the effort recovery theory[81] and the allostatic load theory,[82,83] Geurts and Sonnetag[80] have defined recovery as a process of psycho-physiological unwinding after effort expenditure, and have argued that the psycho-physiological activation may be sustained when stressful conditions are particularly high or frequent. This is a further development of the research of Frankenhaeuser,[86] who stated that occupational stressors may have an aftereffect on well-being by showing that, in periods of high occupational stress, adrenaline excretion rates remained elevated or even increased during leisure time. This shows that occupational stressors have an effect on individuals during leisure time related to their experiences during the working day. Moreover, sustained activation may manifest itself in subjective reports of incomplete recovery, such as impaired sleep quality. In a situation of incomplete recovery, workers have to mobilize compensatory effort, resulting in the accumulation of load effects, and over a longer period, allostatic load and chronic health impairments can develop.

Sleep and recovery

According to Cropley et al.,[87] sleep is one of the most important recovery mechanisms available to humans, and sleep impairments have been associated with a variety of negative consequences, including various health impairments. Most studies analyzing the effects of occupational stressors on sleep assessed self-reported sleep quality. Although self-reports are important, studies relying solely on self-report for both independent (e.g., stressors) and dependent (e.g., sleep) variables are open to problems of common method variance, which may lead to inflated or even spurious correlations and predictions.[88] There are a number of different ways in which associations between self-reports about stressful working conditions and self-reported well-being (e.g., sleep) may be spurious. Firstly, they may reflect response styles rather than substantive relations. Secondly, they may reflect respondents' hypotheses about their job stress and strain. Thirdly, responses may reflect personality characteristics that determine the way in which life in general is experienced. Thus, in order to avoid the problem of common method variance, it is suggested that alternative measures may yield more objective information about the phenomena under investigation.[88] Therefore, the question of whether social stressors at work affect objective indicators of sleep should be considered more carefully.

SOCIAL STRESSORS AT WORK AND SLEEP: EVIDENCE FROM AMBULATORY FIELD STUDIES

Although sleep quality is an accepted clinical construct, it still represents a complex phenomenon that is difficult to define and measure objectively.[89] Sleep quality includes quantitative aspects of sleep, such as sleep-onset latency, sleep efficiency, and sleep fragmentation, as well as more subjective aspects, such as "depth" or "restfulness" of sleep. The exact elements that compose sleep quality and their relative importance may vary between individuals.[89] Because of their relative importance for individuals, it seems of special interest to include both objective and subjective sleep quality parameters in research studies. There are many objective indicators of sleep such as the polysomnography (PSG) and electroencephalography (EEG) for describing brain waves, electro-oculography for describing eye movements and electromyography for describing muscle tension. Sleep evaluation in humans has been usually performed with PSG. This technique has come to be considered the gold standard for detecting sleep impairments.[90]

The merit of Åkerstedt and coworkers is that they combined laboratory and field research and, for the first time, used ambulatory EEG in order to assess sleep at home.[91] Although the research group did not specifically focus on social stressors, the studies showed for the first time that work stressors impair sleep quality in real working life, and showed that lack of a of social support predicted lower sleep quality, while cognitive anticipation of work issues interfered with sleep.[92–95]

The causes of low sleep quality are various. However, as with recovery, the evidence has increased that work stress may play an important role in the development of disturbed sleep quality.[96] According to Åkerstedt,[92] the increased physiological and psychological activation as a response to occupational stress should be incommensurate with the deactivation that is a main characteristic of sleep. Although this association seems plausible, there is relatively little empirical

evidence about stress and sleep, and the topic of sleep has not caught the interest of occupational health researchers, with the exception of shiftwork-related issues.[92] This is surprising, considering the importance of sleep for physiological balance, long-term health, and mental functioning.

Consecutive sleep actigraphy in working individuals

During the last decade, actigraphy (activity-based monitoring) has also become an essential tool in sleep research and sleep medicine.[97,98] The term "actigraphy" refers to methods using miniaturized computerized wristwatch-like devices to monitor and collect data generated by movements. Through algorithms, sleep quality can be derived. A major strength of actigraphy compared to PSG is the ability to monitor sleep–wake patterns continuously with minimal inconvenience over an extended period of time at home.[97] Actigraphy is a good way of providing low-cost, noninvasive, objective, and longitudinal data for the diagnosis of sleep disorders in an ambulatory setting.[81] The comparison of actigraphy with the "gold standard" of PSG has yielded agreement rates ranging from 78% to 95%.[99] Actigraphy is considered to be a valid way to determine sleep patterns in both normal, healthy populations and in patients suspected of certain sleep disorders.[100]

Even though the empirical evidence in terms of studies of causal connections is modest, insofar as evidence is available from cross-sectional studies, it indicates a strong link between occupational stress, sleep, and health impairments.[92] However, the question of whether psychosomatic complaints may be caused by insufficient or incomplete recovery can empirically be answered only by longitudinal studies.[81] Conclusive evidence from longitudinal research on work stress, recovery, and health that covers longer time periods is still lacking.[80] The first evidence came from recent actigraphy-based field research (see Box 5.1).[101]

Weekend sleep and recovery

During the working week, individuals have to deal with occupational stressors, resulting in depleted resources, more fatigue, and an increased need for recovery.[102] Weekends, however, as time during which work demands are absent or at least reduced, should provide time to rest, unwind, and recharge

BOX 5.1: A longitudinal ambulatory field study by Pereira and Elfering[101]

In order to answer whether insufficient recovery may result in psychosomatic health impairments, a two-wave longitudinal ambulatory field study investigated the role of social stressors at work and sleep fragmentation on psychosomatic health complaints. Sixty full-time employees working in a variety of Swiss industries (healthcare, finance, and management) participated in the study. Overall, hierarchical regression analysis revealed that social stressors at work were positively related to objectively assessed sleep fragmentation and to psychosomatic health complaints. Moreover, sleep fragmentation significantly mediated the effect of social stressors at work on psychosomatic health complaints. These results provided longitudinal evidence for the incomplete recovery approach,[80] showing that poor sleep quality is an important intervening variable that is linked both to stress at work and to the development of health impairments in the long run. In order to prevent the long-term negative effects on psychosomatic health, social stressors at work should be prevented, or at least minimized.

the depleted batteries.[102,103] With the exception of a few studies, the role of weekends in employees' recovery and other potential benefits (e.g., performance) has been largely ignored. However, as far as empirical evidence is provided, weekends seems to contribute significantly to individuals' recovery. Results obtained by Binnewies et al.[102] show that psychological detachment, relaxation, and mastery experiences during the weekend predicted a better recovery state after the weekend. Furthermore, being recovered after the weekend predicted task performance, personal initiative, organizational citizenship behavior, and low perceived effort. These results demonstrated the importance of recovery during weekends for both individuals and organizations. Moreover, in a longitudinal study, Fritz and Sonnentag[104] investigated the effects of specific weekend experiences on burnout and general well-being and found that nonwork hassles, absence of positive work reflection and low social activity during the weekend predicted burnout and poor well-being after the weekend. Even though empirical evidence shows that weekends are essential for recovery, some studies showed that occupational stressors at work may lead to incomplete recovery even during weekends. In a diary study, Van Hooff et al.[105] found that compared to low-effort workers, high-effort workers spent significantly more hours on overtime work during the weekend, experienced home activities as significantly more effortful, and reported higher levels of fatigue and less motivation to start the upcoming week. This suggests that occupational stressors may interfere with the recovery process and thus may lead to incomplete recovery, even during weekends. However, further empirical evidence for this assumption is needed.

To increase our understanding of the interference effect of occupational social stressors on weekend sleep, another field study was conducted (see Box 5.2).[106]

BOX 5.2: Field study by Pereira et al.[106]

Many employees in service work are required to work on Saturdays, recovering during work-free Sundays and work-again Mondays. This study examined the effects of social stressors at work on recovery status at Sunday noon and Monday noon, and investigated if sleep quality mediates the negative effects of social stressors at work on recovery. From Saturday until Monday morning, 41 participants wore actigraphs to measure sleep duration and sleep fragmentation. Social stressors at work were assessed by self-reported questionnaires administered on Saturday. Recovery status was reported on Sunday noon and Monday noon. Hierarchical regression analysis revealed that social stressors at work were negatively related to recovery status on Sunday and on Monday. More social stressors at work predicted higher sleep fragmentation in the night at Monday. A mediation effect of sleep quality, however, was not found. Future studies should test whether perseverative thoughts affect sleep quality from Sunday to Monday.

CONDITIONS AND MECHANISMS THAT MAY IMPEDE RECOVERY AND SLEEP: PERSEVERATIVE THOUGHTS

In the modern work context, in which individuals have to focus on getting their work done regardless of the location, being physically away from the workplace does not necessary imply leaving one's workplace behind in psychological terms.[78] At home, many individuals accomplish job-related tasks, continue to think about their jobs, ruminate about work-related problems, or reflect on future opportunities.[78] Therefore, spending some time off work does not automatically imply that recovery occurs. Cognitive mechanisms can prolong or reactivate psycho-physiological activation due to a stressor or their perception, thereby slowing down or hindering recovery. According to McEwen,[82,83] cognitive mechanisms are also likely to result in allostatic load. Moreover, Brosschot et al.[79] hypothesized that occupational stressors will only lead to prolonged activation when individuals cognitively perseverate about these stressors; so perseverative cognition should act as a mediator.[79] Perseverative cognition, such as the inability to detach psychologically, should be responsible for converting the immediate psychological and physiological concomitants of stressors into the prolonged physiological activation of the body's systems, which in turn may lead to incomplete or failed recovery and the development of health impairments in the long run.

Inability to detach psychologically

It is not only the amount of time that matters for recovery, but also the quality of the time off work.[78] In that sense, Sonnentag and Fritz[107] suggest that the ability to detach psychologically from work is crucial to any recovery process. Psychological detachment might best be described as an individual's sense of being away from the work situation[108] in both physical and psychological ways. The inability to detach psychologically not only implies that one is still occupied with work-related duties, such as receiving job-related calls, refraining from job-related activities, and thinking about work-related problems, but also means not being able to disengage mentally from work, nor to stop thinking or ruminating about work or work-related problems.[107] The inability to psychologically detach is often experienced as "failed switching off" when being away from one's workplace.[78] Thus, another field study tested lack of detachment as a mediator (see Box 5.3).[109]

DAILY RECOVERY

Drawing on existing recovery theories,[80–83] it is reasonable to assume that occupational stressors will have immediate effects on individuals' recovery by spilling over into the evening.[87] However, only a limited number of studies have examined the stress–recovery process on a day-to-day basis.[105] Thus, studies investigating the effects of work stressors on recovery have mostly focused on rather chronic work conditions over fairly long periods of time, as opposed to investigating the episodic

---

BOX 5.3: Field study by Pereira and Elfering[109]

The aim of this study was to test whether the inability to detach psychologically from work-related issues mediates the negative effect of social stressors at work on different sleep quality parameters on Sunday night. Sixty full-time employees participated in this study. In line with our hypothesis, social stressors at work were negatively related to sleep quality on Sunday night. Moreover, psychological detachment from work partly mediated the effects of social stressors at work on sleep-onset latency and sleep fragmentation. Thus, the study extends previous research by showing that social stressors at work are potential causes of disturbed sleep quality on Sunday night and by showing that the inability to detach psychologically from work on Sunday evening mediates this relationship. In line with Rook and Zijlstra,[110] the results provide further empirical support that the inability to detach psychologically converts the immediate psychological and physiological concomitants of stressors into prolonged physiological activation and decreasing sleep quality.

and immediate effects of work stressors or on cross-sectional relations between occupational stressors and recovery.[96] Daily studies, however, have shown that occupational stressors significantly account for after-work recovery. Cropley et al.[87] found that teachers reporting high occupational stressors took longer to unwind, ruminated more about work-related issues, and reported poorer sleep quality compared to teachers reporting lower occupational stressors. Even though occupational stressors and work rumination were significantly correlated with sleep quality, work rumination could not be found to mediate the relationship between occupational stressors and sleep quality. Therefore, in order to further shed light on the day-to-day relations between occupational (social) stressors, daily cognitive processes, and daily sleep quality, a study was conducted by Pereira et al. (see Box 5.4).[111]

DISCUSSION

The first aim of this chapter was to update evidence for social stressors at work impairing sleep. Overall, the evidence is increasing that social stressors at work have both short-term and long-term effects on sleep and psychosomatic health complaints. In a number of field studies, the Bernese research group found that the effects of social stressors at work on health were still significant when controlling for daily task-related stressors; this provides further evidence for the important rule of social stressors at work on the development of health impairments. Surprisingly, social stressors at work were predominantly related to sleep fragmentation, whereas a relationship between social stressors at work on sleep-onset latency, sleep efficiency, and self-reported sleep quality were found less consistently. According to Wesensten et al.,[112] sleep fragmentation systematically affects recuperation independently of total sleep time. Furthermore, the percentage of sleep spent in stage 1, which has little or no recuperative value, increased on fragmentation nights, even though total sleep time did not change.[112] This suggests that, compared with other sleep quality parameters, sleep fragmentation is a milder form of sleep impairment, being initially largely unperceived by individuals. Therefore, it is often not noticed and is not closely related to self-reported sleep quality. Furthermore, it is important to note that the samples in these Bernese field studies were rather healthy and exposed to relatively low levels of social stressors at work. Overall, this could have led to an underestimation of the examined effects.

The second aim of this chapter was to shed light on the mechanisms involved in the relationship between social stressors at work and recovery by examining the mediating effects of work-related psychological detachment and worries. Our results showed that the inability to detach psychologically from work partly mediated the effect of social stressors at work on sleep-onset latency and on sleep fragmentation. Our results provide further empirical evidence that being physically away from the workplace does not automatically imply that recovery will occur. Moreover, by prolonging or reactivating psycho-physiological activation due to a stressor involving HPA axis regulation, cognitive processes may play a crucial mediating role in the stressor recovery process.

BOX 5.4: Sleep quality study by Pereira et al.[111]

Pereira et al.[111] investigated the short-term effects of daily social stressors at work on various indicators of sleep quality and tested whether worries mediate this relationship. Ninety full-time workers, mostly employed in service jobs (e.g., as secretaries, nurses, or call-center agents) participated in a time-based diary study and ambulatory assessments of sleep quality. Multilevel regression analyses revealed that both daily workplace social exclusion and daily worries were positively related to fragmented sleep on the following night, and that worries were negatively related to sleep efficiency. Contrary to expectations, however, daily social exclusion at work and worries were not related to other sleep quality parameters, and neither could the author(s) find any support for the idea that the effect of daily workplace social exclusion on sleep quality was mediated by worries. Although this study failed to show the proposed mediation effect, it showed unique associations of both social exclusion at work and worries with sleep fragmentation.

SUGGESTIONS FOR FUTURE RESEARCH

Future studies should also test the reversed causation hypothesis that impaired sleep leads to increased social stressors at work. Indirect first evidence for reversed causation comes from studies that show a positive relation between sleep problems and subsequent hostility, as well as from studies that show impaired sleep in those who bully. Table 5.4 reviews the evidence for the reversed causation hypothesis.[113–133] It is noteworthy that Table 5.4 included three studies with nonwork samples. Evidence increases because the sleep–bullying relationship is already established in children and adolescents.[120,134] Research in adults should also investigate these early associations.

Studies analyzing the relationship between occupational stressors, recovery, and well-being have mostly relied on cross-sectional designs or on designs with rather limited time frames, such as diary studies. Although such designs are important in order to better understand the short-term effects of occupational stressors on recovery, health impairments, however, usually need longer periods of time to develop.[84] To analyze whether occupational stressors and incomplete recovery result in health impairments in the long run, we analyzed longitudinal effects. Our results supported the finding that exposure to occupational stressors (e.g., social stressors) and incomplete recovery result in health impairments in the long run. However, future studies including longer time frames should provide further evidence of this pattern.

Home-related social stressors, such as conflicts with family members, may cause strain themselves and, as with work-related stressors, result in decreased recovery and be related to sleep problems.[46] Although some studies have shown that home-related stressors may have unfavorable effects on recovery, future studies should include both work- and home-related social stressors in order to create a better understanding of the role of occupational- and home-related stressors in the development of incomplete recovery and health impairments.

The relation of task stressors and social stressors should be analyzed in depth. Semmer et al.[135] argued that—beyond social stressors—anything that signals a lack of appreciation and respect, and thus constitutes a threat to self-esteem, is especially upsetting and frustrating, and is likely to play a major role in the experience of stress (see also Lazarus[136]). An example can be found in the concept of illegitimate tasks, based on role theory, identity theory, and justice theory. Roles imply expectations; they specify what may appropriately be expected from a role occupant.[137,138] However, if roles specify what may be expected from someone, there also must be things that *cannot* be expected; this consideration constitutes the basis of the concept of illegitimate tasks. Tasks are considered legitimate to the extent that they conform to norms about what can appropriately be expected from a given person, and they are illegitimate to the extent that they violate such norms. Their (perceived) illegitimacy may derive from: (i) the perception that a task does not conform to an employee's professional role (*unreasonable* task), as when a company janitor is told to care for the private lawn of his or her boss or when experienced employees are assigned a novice's work; or (ii) the perception that a task is *unnecessary*, such as having to document information that no one will ever use, which many think cannot be expected from employees.[139] In a recent diary study by Pereira et al.,[140] daily illegitimate tasks predicted reduced sleep quality in the following night as assessed by actigraphy.

Overall, the reported evidence suggests that social stressors at work have negative effects on sleep (and vice versa in the reversed causation effect model). As with severe social stressors at work, like mobbing, it is in the organizations' best interests to take seriously the issue of milder social stressors at work such as conflicts with coworkers and supervisors, unfair behavior, and a negative group climate by developing effective policies and by fostering a positive organizational culture. The role of supervisors thereby seems to be the most important one.[39] According to Dormann and Zapf,[141] social support can buffer the effects of social stressors on psychological health. Thus, by providing support to their employees and by acting as role models, supervisors can help prevent the negative long-term effects of social stressors at work on recovery and health. The person-oriented approach to sleep in workers and employees so far includes sleep extension on weekends, especially for those who sleep less than 6 hours after work days,[142] and online sleep training intervention to increase sleep quality after work.[143]

Table 5.4 Studies that report sleep quality as an antecedent of hostility and social stressors

Source	Sample	Social stressor concept	Sleep measure	Association(s) reported	Association recoded: (low to high) sleep quality and (low to high) social stressors
Person-oriented					
Christian and Ellis[113]	171 nurses (82% female) 75 junior and senior business students (48% female)	Hostility (Positive and Negative Affect Schedule [PANAS-X][114])	Sleep deprivation (asked to indicate the number of whole hours slept the night before)	0.25** 0.41**	−0.25** −0.41**
Granö et al.[115]	5433 hospital employees (88.9% female)	Hostility (Finnish Twin Study Hostility [FTSH] scale[116])	Sleep duration (how many hours do you normally sleep during the day and night?) Sleep disturbances[52]	−0.02 0.13***	−0.02 −0.13***
Scott and Judge[117]	51 administrative employees (70.6% female)	Hostility (PANAS-X[114])	Insomnia[58,118]	0.17	−0.17
Situation-oriented					
Barnes et al.[50]	85 undergraduate students (62.4% female)	Supervisor-rated unethical behavior (measure of unethical behavior at work[51])	Sleep quantity (Pittsburgh Sleep Diary measure[119]) Sleep quality as control variable (insomnia scale[58] → reverse coded to index sleep)	Direct effect sleep quantity on unethical behavior β = 0.03 (n.s.) Indirect effect with mediator of cognitive self-control → Sobel test z = −1.99*	

(Continued)

Table 5.4 (Continued) Studies that report sleep quality as an antecedent of hostility and social stressors

Source	Sample	Social stressor concept	Sleep measure	Association(s) reported	Association recoded: (low to high) sleep quality and (low to high) social stressors
Fekkes et al.[120]	118 children aged 9–11 years	Dutch version of the Olweus Bully/Victim Questionnaire[121–123]	KIVPA, a Dutch instrument to measure psycho-social problems among children[124]	Incidence of bullying at end of school year (%/number of bullying cases/sample size/odds ratio [OR]) 11.2/83/741/ OR = 1 12.6/26/206/ OR = 1.1 (n.s.)	
Gini[125]	565 primary-school children	Involvement in bullying and victimization was measured through a self-report scale[126] based on the Bullying Behavior Scale[127] and the Peer Victimization Scale[128] Teachers rated the children's emotional and behavioral problems through the Goodman's 25-item strengths and difficulties questionnaire (SDQ)—Teacher version[129]	Sleeping problems (they were asked to report whether they had experienced this symptom in the last 3 months)	Association between reports of active bullying and psychosomatic problems Sleeping problems: Uninvolved children (n = 178): 28.1 Bullies (n = 63): 47.6 OR = 2.12*	
Hansen et al.[56]	3382 employees (67% female) at baseline, 1671 at t_1	• Bullying Have you been subjected to bullying at work within the past 6 months? Have you witnessed bullying at work within the past 6 months?	Sleep problems (modified version of the Karolinska Sleep Questionnaire [KSQ])[57] with scales for disturbed sleep and awakening problems	Disturbed/unsatisfactory awakenings/poor quality: $\beta = -0.34$***/−0.19***/−0.32*** $\beta = -0.78$***/−0.51***/−0.56*** $\beta = -0.24$***/−0.18***/−0.23*** $\beta = -0.58$***/−0.38***/−0.45***	

(Continued)

Table 5.4 (Continued) Studies that report sleep quality as an antecedent of hostility and social stressors

Source	Sample	Social stressor concept	Sleep measure	Association(s) reported	Association recoded: (low to high) sleep quality and (low to high) social stressors

		• Association between sleep difficulties and being subjected to bullying at baseline
		  • Occasionally bullied
		  • Frequently bullied
		• Association between sleep difficulties and witnessing bullying at baseline
		  • Observing occasionally
		  • Observing frequently
		• Association between sleep difficulties and witnessing bullying at follow-up
		  • Observing occasionally
		  • Observing frequently
		• Association between sleep difficulties and being subjected to bullying at follow-up
		  • Occasionally bullied
		  • Frequently bullied
			Sleep quality (how do you rate your overall quality of sleep?)	Disturbed/unsatisfactory awakenings/poor quality (ORs):
				1.44/1.37 (sig.)/0.9
				3.81/0.88/0.81
				1.69/1.68 (sig.)/1.13
				4.01/1.91/1.08	

(Continued)

Table 5.4 (Continued) Studies that report sleep quality as an antecedent of hostility and social stressors

Source	Sample	Social stressor concept	Sleep measure	Association(s) reported	Association recoded: (low to high) sleep quality and (low to high) social stressors
Kubiszewski et al.[130]	1422 students aged 10–18 years (mean = 14.3, SD = 2.7; 43% female)	Bullying with revised Bully–Victim Questionnaire (rBVQ)[131]	Sleep/wake patterns: four items from the School Sleep Habit Survey[132] Sleep schedule variability • Bedtime differences (differences between usual bedtime during the week and at the weekend) • Wake time differences (differences between usual wake-up time during the week and at the weekend) Subjective sleep disorders (insomnia symptoms and subjective nighttime and daytime distress) with Athens Insomnia Scale[133]	Persons who are bullies or victims and bullies (n = 155) Externalizing behaviors: Aggression Sleep duration: Sufficient: 13.45 (4.6) Insufficient: 16.40 (5.1) t = −3**, $P\eta^2 = 0.06$ Irregular bedtimes: No: 14.29 (4.7) Yes: 16.06 (5.3) t = −2.16*, $P\eta^2 = 0.03$ Irregular wake-up times: No: 13 (3.5) Yes: 15.53(5.2) t = −2.01*, $P\eta^2 = 0.02$ Antisocial behavior Sleep duration: Sufficient: 12.66 (4.7) Insufficient: 16.19 (5.6) t = −4.18***, $P\eta^2 = 0.1$ Irregular bedtimes: No: 13.62 (5.3) Yes: 15.84 (5.5) t = −2.6*, $P\eta^2 = 0.04$	

(Continued)

Table 5.4 (Continued) Studies that report sleep quality as an antecedent of hostility and social stressors

Source	Sample	Social stressor concept	Sleep measure	Association(s) reported	Association recoded: (low to high) sleep quality and (low to high) social stressors
				Irregular wake-up times: No: 12.26 (4.4) Yes: 15.13(5.6) $t = -2.03^*$, $P\eta^2 = 0.03$	
				Persons who are victims or victims and bullies (n = 257) Perceived social disintegration	
				Insomnia symptoms: No: 13.88 (5.7) Yes: 15.83 (6.7) $t = -2.27^*$, $P\eta^2 = 0.02$	
				Subjective nighttime and daytime stress: No: 13.85 (5.7) Yes: 15.86 (6.7) $t = -2.42^*$, $P\eta^2 = 0.02$	
				Psychological distress Insomnia symptoms: No: 29.36 (11.2) Yes: 37.43 (12.3) $t = -5.19^{***}$, $P\eta^2 = 0.1$	
				Subjective nighttime and daytime stress: No: 28.76 (10.7) Yes: 38.39 (12.2) $t = -6.47^{***}$, $P\eta^2 = 0.14$	

Note: Significance level is indicated: * <.05, ** <.01, *** <.001, two-tailed.

REFERENCES

1. Jahoda M. *Wie viel Arbeit braucht der Mensch? Arbeit und Arbeitslosigkeit im 20. Jahrhundert.* [How much work does man need? Employment and unemployment in the 20th century] Weinheim: Beltz, 1983.

2. Semmer NK, Meier LL. Bedeutung und Wirkung von Arbeit [Meaning and outcomes of work]. In: Schuler H, Moser K (Eds). *Lehrbuch Organisationspsychologie.* Bern: Huber, 2014, pp. 559–604.

3. Zapf D, Semmer NK. Stress und Gesundheit in Organisationen [Stress and health in organizations]. In: Schuler H (Ed.). *Enzyklopädie der Psychologie, Themenbereich D, Serie III.* Göttingen: Hogrefe, 2004, pp. 1007–1112.

4. Beehr TA, Franz TM. The current debate about the meaning of job stress. *J Organ Behav Manage* 1987; 8: 5–18.

5. De Lange AH, Taris TW, Kompier MAJ, Houtman ILD, Bongers PM. "The very best of the millennium": Longitudinal research and the demand–control–(support) model. *J Occup Health Psychol* 2003; 8: 282–305.

6. Kivimäki M, Nyberg ST, Batty GD et al. Job strain as a risk factor for coronary heart disease: A collaborative meta-analysis of individual participant data. *Lancet* 2012; 380: 1491–1497.

7. Nixon AE, Mazzola JJ, Bauer J, Krueger JR, Spector PE. Can work make you sick? A meta-analysis of the relationships between job stressors and physical symptoms. *Work Stress* 2011; 25: 1–22.

8. Semmer NK, McGrath JE, Beehr TA. Conceptual issues in research on stress and health. In: Cooper CL (Ed.). *Handbook of Stress Medicine and Health.* Boca Raton, FL: CRC Press, 2005, pp. 1–43.

9. Dormann C, Zapf D. Social stressors at work, irritation, and depressive symptoms: Accounting for unmeasured third variables in a multi-wave study. *J Occup Organ Psychol* 2002; 75: 33–58.

10. Semmer NK, Jacobshagen N, Meier LL, Elfering A, Kaelin W, Tschan F. Psychische Beanspruchung durch Illegitime Aufgaben [Psychological strain by illegitimate tasks]. In: Junghanns G, Morschhaeuser M. (Eds). *Immer schneller, immer mehr: Psychische Belastungen und Gestaltungsperspektiven bei Wissens- und Dienstleistungsarbeit.* Berlin: BAuA, 2013, pp. 97–112.

11. Tepper BJ. Consequences of abusive supervision. *Acad Manage J* 2000; 43: 178–190.

12. Einarsen S, Hoel H, Zapf D, Cooper CL. The concept of bullying at work: The European tradition. In: Einarsen S, Hoel H, Zapf D, Cooper CL (Eds). *Bullying and Emotional Abuse in the Workplace. International Perspectives in Research and Practice.* London: Taylor & Francis, 2003, pp. 3–30.

13. Williams DR, Lavizzo-Mourey R, Warren RC. The concept of race and health status in America. *Public Health Rep* 1994; 109: 26–41.

14. Semmer NK, Amstad F, Elfering A, Kälin W. Dysfunctional social support. Paper presented at the APA/NIOSH Conference on "Work, Stress, and Health," Miami, FL, March 2006.

15. Brodsky CM. *The Harassed Worker.* Toronto, Ontario, Canada: Lexington Books, DC Health.

16. Vartia M. Psychological harassment (bullying, mobbing) at work. In: Kauppinen K-T (Ed.), *OECD Panel Group on Women, Work, and Health.* Helsinki: Ministry of Social Affairs and Health, 1993, pp. 149–152.

17. Bjorkqvist K, Osterman K, Lagerspetz KM. Sex differences in covert aggression among adults. *Aggress Behav* 1994; 20: 27–33.

18. Barefoot JC. Developments in the measurement of hostility. In: Friedman HS (Ed.). *Hostility, Coping, and Health.* Washington, DC: American Psychological Association, 1992, pp. 13–31.

19. Bies RJ, Moag JF. Interactional justice: Communication criteria of fairness. In: Lewicki RJ, Sheppard BH, Bazerman MH (Eds). *Research on Negotiations in Organizations.* Greenwich, CT: JAI Press, 1986, pp. 43–55.

20. Colquitt J. On the dimensionality of organizational justice: A construct validation of a measure. *J Appl Psychol* 2001; 86: 386–400.

21. Leymann H. The content and development of mobbing at work. *Eur J Work Organ Psychol* 1996; 2: 165–184.

22. Katz D, Kahn RL. *The Social Psychology of Organizations*. New York, NY: Wiley, 1978.

23. Leary MR. Toward a conceptualization of interpersonal rejection. In: Leary MR (Ed.). *Interpersonal Rejection*. New York, NY: Oxford University Press, 2001, pp. 3–20.

24. Jones TM. Ethical decision-making by individuals in organizations—An issue-contingent model. *Acad Manage Rev* 1991; 16: 366–395.

25. Rook KS. The negative side of social interaction: Impact on psychological well-being. *J Pers Soc Psychol* 1984; 46: 1097.

26. Folger R, Baron RA. Violence and hostility at work: A model of reactions to perceived injustice. In: VandenBos GR, Bulatao EQ (Eds). *Violence on the Job: Identifying Risks and Developing Solutions*. Washington, DC: American Psychological Association, 1996, pp. 51–85.

27. Andersson LM, Pearson CM. Tit for tat? The spiraling effect of incivility in the workplace. *Acad Manage Rev* 1999; 24: 452–471.

28. Jenkins L. *Violence in the Workplace: Risk Factors and Prevention Strategies*. US Department of Health and Human Services, Centers for Disease Control and Prevention, National Institute for Occupational Safety and Health, Division of Safety Research, Washington, DC, 1996.

29. Hershcovis MS. Incivility, social undermining, bullying...Oh my! A call to reconcile constructs within workplace aggression research. *J Organ Behav* 2011; 32: 499–519.

30. Hershcovis MS, Reich TC. Integrating workplace aggression research: Relational, contextual, and method considerations. *J Organ Behav* 2013; S1: 26–42.

31. Spector PE. Introduction: Should distinctions be made among different forms of mistreatment at work? *J Organ Behav* 2011; 32: 485–486.

32. Tepper BJ, Henle CA. A case for recognizing distinctions among constructs that capture interpersonal mistreatment in work organizations. *J Organ Behav* 2010; 32: 487–498.

33. Bolger N, DeLongis A, Kessler RC, Schilling EA. Effects of daily stress on negative mood. *J Pers Soc Psychol* 1989; 57: 808–818.

34. Grebner S, Elfering A, Semmer NK, Kaiser-Probst C, Schlapbach ML. Stressful situations at work and in private life among young workers: An event sampling approach. *Soc Indic Res* 2004; 67: 11–49.

35. Keenan A, Newton TJ. Stressful events, stressors and psychological strains in young professional engineers. *J Organ Behav* 1985; 6: 151–156.

36. Eurofound. *Fifth European Working Conditions Survey*. Luxembourg: Publications Office of the European Union, 2012.

37. Elfering A, Semmer NK, Tschan F, Kälin W, Bucher A. First years in job: A three-wave analysis of work experiences. *J Vocat Behav* 2007; 70: 97–115.

38. Meier LL, Gross S, Spector PE, Semmer NK. Relationship and task conflict at work: Interactive short-term effects on angry mood and somatic complaints. *J Occup Health Psychol* 2013; 18: 144–156.

39. Hershcovis MS, Barling J. Towards a multi-foci approach to workplace aggression: A meta-analytic review of outcomes from different perpetrators. *J Organ Behav* 2010; 31: 24–44.

40. Smith CS, Sulsky L. An investigation of job-related coping strategies across multiple stressors and samples. In: Murphy LR, Hurrell JJ, Sauter SL, Keita GP (Eds). *Job Stress Interventions*. Washington, DC: American Psychological Association, 1995, pp. 109–123.

41. Sonnentag S, Frese M. Stress in organizations. In: Schmitt W, Highhouse S (Eds). *Handbook of Psychology*. Hoboken, NJ: John Wiley & Sons, 2012, pp. 560–592.

42. Spector PE, Jex SM. Development of four self-report measures of job stressors and strain: Interpersonal conflict at work scale, organizational constraints scale, quantitative workload inventory, and physical symptoms inventory. *J Occup Health Psychol* 1998; 3: 356–367.

43. Bowling NA, Beehr T. Workplace harassment from the victim's perspective: A theoretical model and meta-analysis. *J Appl Psychol* 2006; 91: 998–1012.

44. Uehli K, Mehta AJ, Miedinger D et al. Sleep problems and work injuries: A systematic review and meta-analysis. *Sleep Med Rev* 2014; 18: 61–73.

45. Nielsen MB, Einarsen S. Outcomes of exposure to workplace bullying: A meta-analytic review. *Work Stress* 2012; 26: 309–332.

46. Litwiller BJ. *The Relationship between Sleep and Work: A Meta-Analysis*. Unpublished Dissertation. Norman, OK: University of Oklahoma, 2014.

47. Barling J, Dekker I, Loughlin CA, Kelloway EK, Fullagar C, Johnson D. Prediction and replication of the organizational and personal consequences of workplace sexual harassment. *J Manag Psychol* 1996; 11: 4–25.

48. Spence JT, Helmreich RL, Pred RS. Impatience versus achievement strivings in the type A pattern: Differential effects on students' health and academic performance. *J Appl Psychol* 1987; 72: 522–528.

49. Barnes CM, Schaubroeck J, Huth M, Ghumman S. Lack of sleep and unethical conduct. *Organ Behav Hum Dec Process* 2011; 115: 169–180.

50. Akaah IP. Social inclusion as a marketing ethics correlate. *J Bus Ethics* 1992; 11: 599–608.

51. Jenkins DC, Stanton B-A, Niemcryk SJ, Rose RM. A scale for the estimation of sleep problems in clinical research. *J Clin Epidemiol* 1988; 41: 313–321.

52. Giorgi G. Workplace bullying partially mediates the climate-health relationship. *J Manag Psychol* 2010; 25: 727–740.

53. Giorgi G, Matthiesen SB, Einarsen S. Italian validation of the negative acts questionnaire revised. Poster presented at the *5th International Conference of Bullying and Harassment at Work*, Dublin, June 2006.

54. Goldenhar LM, Swanson NG, Hurrell JJ, Ruder A, Deddens J. Stressors and adverse outcomes for female construction workers. *J Occup Health Psychol* 1998; 3: 19–32.

55. Northwestern National Life Insurance Company. *Fear and Violence in the Workplace Survey Documenting the Experience of American Workers*. Minneapolis, MN: Northwestern National Life Insurance Co., 1993.

56. Hansen AM, Hogh A, Garde AH, Persson R. Workplace bullying and sleep difficulties: A 2-year follow-up study. *Int Arch Occup Environ Health* 2014; 87: 285–294.

57. Åkerstedt T, Hume K, Minors D, Waterhouse J. The meaning of good sleep: A longitudinal study of polysomnography and subjective sleep quality. *J Sleep Res* 1994; 3: 152–158.

58. Lallukka T, Rahkonen O, Lahelma E. Workplace bullying and subsequent sleep problems—The Helsinki Health Study. *Scand J Work Environ Health* 2011; 37: 204–212.

59. Kivimäki M, Virtanen M, Vartia M, Elovainio M, Vahtera J, Keltikangas-Järvinen L. Workplace bullying and the risk of cardiovascular disease and depression. *Occup Environ Med* 2003; 60: 779–783.

60. Niedhammer I, David S, Degioanni S, Drummond A, Philip P. Workplace bullying and sleep disturbances: Findings from a large scale cross-sectional survey in the French working population. *Sleep* 2009; 32: 1211–1219.

61. Leymann H. The content and development of mobbing at work. *Eur J Work Organ Psychol* 1996; 2: 165–184.

62. Niedhammer I, David S, Degioanni S. Et 143 médecins du travail. The French version of the Leymann's questionnaire on workplace bullying: The Leymann Inventory of Psychological Terror (LIPT). *Rev Epidemiol Sante* 2006; 54: 245–262.

63. Park JB, Nakata A, Swanson NG, Chun H. Organizational factors associated with work-related sleep problems in a nationally representative sample of Korean workers. *Int Arch Occup Environ Health* 2013; 86: 211–222.

64. Rafferty AE, Restubog SD, Jimmieson NL. Losing sleep: Examining the cascading effects of supervisors' experience of injustice on subordinates' psychological health. *Work Stress* 2010; 24: 36–55.

65. Niehoff BP, Moorman RH. Justice as a mediator of the relationship between methods of monitoring and organizational citizenship behavior. *Acad Manage J* 1993; 36: 527–556.

66. Greenberg J. Losing sleep over organizational injustice: Attenuating insomniac reactions to underpayment inequity with supervisory training in interactional injustice. *J Appl Psychol* 2006; 91: 58–69.

67. Rodríguez-Muñoz A, Notelaers G, Moreno-Jiménez B. Workplace bullying and sleep quality: The mediating role of worry and need for recovery. *Biol Psychol* 2011; 19: 453–468.

68. Einarsen S, Raknes BI. Harassment in the workplace and the victimization of men. *Violence Vict* 1997; 12: 247–263.

69. Veldhoven MJPM, Meijman TF. *Het Meten van Psychosociale Arbeidsbelasting met een Vragenlijst: De Vragenlijst Beleving en Beoordeling van de Arbeid (VBBA) [The Measurement of Psychosocial Strain at Work: The Questionnaire Experience and Evaluation of Work]*. Amsterdam: NIA, 1994.

70. Rogers K-A, Kelloway EK. Violence at work: Personal and organizational outcomes. *J Occup Health Psychol* 1997; 2: 63–71.

71. Slopen N, Williams DR. Discrimination, other psychosocial stressors, and self-reported sleep duration and difficulties. *Sleep* 2014; 37: 147–156.

72. McNeilly M, Robinson EL, Anderson NB et al. Effects of racist provocation and social support on cardiovascular reactivity in African-American women. *Int J Behav Med* 1995; 2: 321–338.

73. Strawbridge WJ, Shema SJ, Roberts RE. Impact of spouses' sleep problems on partners. *Sleep* 2004; 27: 527–531.

74. Takaki J, Taniguchi T, Fukuoka E et al. Workplace bullying could play important roles in the relationships between job strain and symptoms of depression and sleep disturbance. *J Occup Health* 2010; 52: 367–374.

75. Doi Y, Minowa M, Uchiyama M et al. Psychometric assessment of subjective sleep quality using the Japanese version of the Pittsburgh Sleep Quality Index (PSQI-J) in psychiatric disordered and control subjects. *Psychiatry Res* 2000; 97: 165–172.

76. Vartia MA. Consequences of workplace bullying with respect to the well-being of its targets and the observers of bullying. *Scand J Work Environ Health* 2001; 27: 63–69.

77. Leymann H. *Presentiation av LIPT-formuläret. Konstrution, Validering, Utfall [Presentation of the Leymann Inventory for Psyhological Terrorization]*. Stockholm: Violen inom Praktikertjänst, 1989.

78. Sonnentag S, Bayer U. Switching off mentally: Predictors and consequences of psychological detachment from work during off-job time. *J Occup Health Psychol* 2005; 10: 393–414.

79. Brosschot JF, Pieper S, Thayer JF. Expanding stress theory: Prolonged activation and perseverative cognition. *Psychoneuroendocrinology* 2005; 30: 1043–1049.

80. Geurts SAE, Sonnentag S. Recovery as an explanatory mechanism in the relation between acute stress reactions and chronic health impairment. *Scand J Work Environ Health* 2006; 32: 482–492.

81. Meijman TF, Mulder G. Psychological aspects of workload. In: Drenth PJD, Thierry H, Wolff CJD (Eds). *Handbook of Work and Organizational Psychology*. East Sussex: Psychology Press Ltd., 1998, pp. 5–33.

82. McEwen BS. Stress, adaptation, and disease: Allostasis and allostatic load. *Ann NY Acad Sci* 1998; 840: 33–44.

83. Juster RP, McEwen BS. Sleep and chronic stress: New directions for allostatic load research. *Sleep Med* 2015; 16(1): 7–8.

84. Demerouti E, Bakker AB, Geurts SAE, Taris TW. Daily recovery from work-related effort during non-work time. In: Sonnentag S, Perrewé PL, Ganster DC (Eds). *Current Perspectives on Job-Stress Recovery: Research in Occupational Stress and Well Being*. Bingley: JAI Press, 2009, pp. 85–123.

85. Sluiter JK, Frings-Dresen MHW, van der Beek AJ, Meijman TF. The relation between work-induced neuroendocrine reactivity and recovery, subjective need for recovery, and health status. *J Psychosom Res* 2001; 50: 29–37.

86. Frankenhaeuser M. Coping with stress at work. *Int J Health Serv* 1981; 11: 491–510.

87. Cropley M, Dijk DJ, Stanley N. Job strain, work rumination, and sleep in school teachers. *Eur J Work Organ Psychol* 2006; 15: 181–196.

88. Semmer NK, Grebner S, Elfering A. Beyond self-report: Using observational, physiological, and event-based measures in research on occupational stress. In: Perrewé PL, Ganster DC (Eds). *Emotional and Physiological Processes and Positive Intervention Strategies. Research in Occupational Stress and Well-being*. Amsterdam: JAI Press, 2004, pp. 205–263.

89. Buysse DJ, Reynolds CF, Monk TH, Berman SR, Kupfer DJ. The Pittsburgh Sleep Quality Index: A new instrument for psychiatric practice and research. *Psychiatry Res* 1989; 28: 193–213.

90. De Souza L, Benedito-Silva AA, Pires ML, Poyares D, Tufik S, Calil HM. Further validation of actigraphy for sleep studies. *Sleep* 2003; 26: 81–84.

91. Torsvall L, Åkerstedt T, Gillberg M. Age, sleep and irregular work-hours—A field-study with electroencephalographic recordings, catecholamine excretion and self-ratings. *Scand J Work Environ Health* 1981; 7: 196–203.

92. Åkerstedt T. Psychosocial stress and impaired sleep. *Scand J Work Environ Health* 2006; 32: 493–501.

93. Åkerstedt T, Kecklund G, Axelsson J. Impaired sleep after bedtime stress and worries. *Biol Psychol* 2007; 76: 170–173.

94. Kecklund G, Åkerstedt T. Apprehension of the subsequent working day is associated with a low amount of slow wave sleep. *Biol Psychol* 2004; 66: 169–176.

95. Petersen H, Kecklund G, Donofrio P, Akerstedt T. Sleep during days with work stress compared to weekend sleep—Home polysomnography in normal life settings. *J Sleep Res* 2012; 21: 145.

96. Åkerstedt T, Fredlund P, Gillberg M, Jansson B. Work load and work hours in relation to disturbed sleep and fatigue in a large representative sample. *J Psychosom Res* 2002; 53: 585–588.

97. Sadeh A, Acebo C. The role of actigraphy in sleep medicine. *Sleep Med Rev* 2002; 6: 113–124.

98. Wrzus C, Brandmaier AM, von Oertzen T, Müller V, Wagner GG, Riediger M. A new approach for assessing sleep duration and postures from ambulatory accelerometry. *PLoS One* 2012; 7: e48089.

99. Kushida CA, Chang A, Gadkary C, Guilleminault C, Carrillo O, Dement WC. Comparison of actigraphic, polysomnographic, and subjective assessment of sleep parameters in sleep-disordered patients. *Sleep Med* 2001; 2: 389–396.

100. Morgenthaler T, Alessi C, Friedman L et al. Practice parameters for the use of actigraphy in the assessment of sleep and sleep disorders: An update for 2007. *Sleep* 2007; 30: 519–529.

101. Pereira D, Elfering A. Social stressors at work, sleep fragmentation and psychosomatic complaints—A longitudinal ambulatory field study. *Stress Health* 2014; 29: 240–252.

102. Binnewies C, Sonnentag S, Mojza EJ. Recovery during the weekend and fluctuations in weekly job performance: A week-level study examining intra-individual relationships. *J Occup Organ Psychol* 2010; 83: 419–441.

103. Fritz C, Sonnentag S, Spector PE, McInroe JA. The weekend matters: Relationships between stress recovery and affective experiences. *J Organ Behav* 2010; 31: 1137–1162.

104. Fritz C, Sonnentag S. Recovery, health, and job performance: Effects of weekend experiences. *J Occup Health Psychol* 2005; 10: 187–199.

105. Van Hooff MLM, Geurts SAE, Kompier MAJ, Taris TW. Workdays, in-between workdays and the weekend: A diary study on effort and recovery. *Int Arch Occup Environ Health* 2007; 80: 599–613.

106. Pereira D, Gross S, Elfering A. Social stressors at work, sleep, and recovery. *Appl Psychophysiol Biofeedback* 2016; 41(1): 93–101. doi: 10.1007/s10484-015-9317-6.

107. Sonnentag S, Fritz C. The Recovery Experience Questionnaire: Development and validation of a measure assessing recuperation and unwinding at work. *J Occup Health Psychol* 2007; 12: 204–221.

108. Etzion D, Eden D, Lapidot Y. Relief from job stressors and burnout: Reserve service as a respite. *J Appl Psychol* 1998; 83: 577–585.

109. Pereira D, Elfering A. Social stressors at work and sleep quality on Sunday night—The mediating role of psychological detachment. *J Occup Health Psychol* 2014; 19: 85–95.

110. Rook J, Zijlstra FRH. The contribution of various types of activities to recovery. *Eur J Work Organ Psychol* 2006; 15: 218–240.

111. Pereira D, Meier LL, Elfering A. Short-term effects of social exclusion at work and worries on sleep. *Stress Health* 2013; 29: 240–252.

112. Wesensten NJ, Balkin TJ, Belenky G. Does sleep fragmentation impact recuperation? A review and reanalysis. *J Sleep Res* 1999; 8: 237–245.

113. Christian MS, Ellis AP. Examining the effects of sleep deprivation on workplace deviance: A self-regulatory perspective. *Acad Manage J* 2011; 54: 913–934.

114. Watson D, Clark LA. The PANAS-X: Manual for the positive and negative affect schedule—Expanded form. Unpublished manuscript. Iowa City, IA: University of Iowa, 1994.

115. Granö N, Vahtera J, Virtanen M, Keltikangas-Järvinen L, Kivimäki M. Association of hostility with sleep duration and sleep disturbances in an employee population. *Int J Behav Med* 2008; 15: 73–80.

116. Koskenvuo M, Kaprio J, Rose RJ et al. Hostility as a risk factor for mortality and ischemic heart disease in men. *Psychosom Med* 1988; 50: 330–340.

117. Scott BA, Judge TA. Insomnia, emotions, and job satisfaction: A multilevel study. *J Manage* 2006; 32: 622–645.

118. Jenkins DC, Jono RT, Stanton B-A. Predicting completeness of symptoms relief after major heart surgery. *Behav Med* 1996; 22: 45–57.

119. Monk TH, Reynolds CF, Kupfer DJ et al. The Pittsburgh Sleep Diary. *J Sleep Res* 1994; 3: 111–120.

120. Fekkes M, Pijpers FI, Fredriks AM, Vogels T, Verloove-Vanhorick SP. Do bullied children get ill, or do ill children get bullied? A prospective cohort study on the relationship between bullying and health-related symptoms. *Pediatrics* 2006; 117: 1568–1574.

121. Mooij T. *Pesten in het Onderwijs [Bullying in Education]*. Nijmegen: Instituut voor Toegepaste Sociale Wetenschappen, Katholieke Universiteit van Nijmegen, 1992.

122. Liebrand J, Van IJzerdoorn H, Van Lieshout C. *KRVL Klasgenoten Relatie Vragenlijst Junior*. Nijmegen: Vakgroep Ontwikkelingspsychologie, Katholieke Universiteit Nijmegen, 1991.

123. Olweus D. Bullying at school: Basic facts and effects of a school based intervention program. *J Child Psychol Psychiatry* 1994; 35: 1171–1190.

124. Reijneveld SA, Vogels AG, Brugman E, van Ede J, Verhulst FC, Verloove-Vanhorick SP. Early detection of psychosocial problems in adolescents: How useful is the Dutch Short Indicative Questionnaire (KIVPA)? *Eur J Public Health* 2003; 13: 152–159.

125. Gini G. Associations between bullying behaviour, psychosomatic complaints, emotional and behavioural problems. *J Paediatr Child Health* 2008; 44: 492–497.

126. Caravita S, Bartolomeo A. Bullying and the use of violent videogames. Presented at the *10th European Conference on Developmental Psychology*. Uppsala, August 27–31, 2001.

127. Austin S, Joseph S. Assessment of bully/victim problems in 8–11 year-olds. *Br J Educ Psychol* 1996; 66: 447–456.

128. Neary A, Joseph S. Peer victimization and its relationship to self-concept and depression among schoolgirls. *Pers Indiv Differ* 1994; 16: 183–186.

129. Goodman R. The strengths and difficulties questionnaire: A research note. *J Child Psychol Psychiatry* 1997; 38: 581–586.

130. Kubiszewski V, Fontaine R, Potard C, Gimenes G. Bullying, sleep/wake patterns and subjective sleep disorders: Findings from a cross-sectional survey. *Chronobiol Int* 2014; 31: 542–553.

131. Solberg ME, Olweus D. Prevalence estimation of school bullying with the Olweus Bully/Victim Questionnaire. *Aggress Behav* 2003; 29: 239–68.

132. Wolfson AR, Carskadon MA. Sleep schedules and daytime functioning in adolescents. *Child Dev* 1998; 69: 875–887.

133. Soldatos CR, Dikeos DG, Paparrigopoulos TJ. Athens Insomnia Scale: Validation of an instrument based on ICD-10 criteria. *J Psychosom Res* 2000; 48: 555–560.

134. Wolke D, Lereya ST. Bullying and parasomnias: A longitudinal cohort study. *Pediatrics* 2014; 134: E1040–E1048.

135. Semmer NK, Jacobshagen N, Meier LL, Elfering A. Occupational stress research: The "stress-as-offense-to-self" perspective. In: Houdmont J, McIntyre S (Eds). *Occupational Health Psychology: European Perspectives on Research, Education and Practice*. Castelo de Maia: ISMAI Publishing, 2007, pp. 43–60.

136. Lazarus RS. *Stress and Emotion: A New Synthesis*. London: Springer Publishing Company, 1999.

137. Burke PJ, Stets JE. *Identity Theory*. New York, NY: Cambridge University Press, 2009.

138. Katz D, Kahn RL. *The Social Psychology of Organizations*. New York, NY: Wiley, 1978.

139. Semmer NK, Jacobshagen N, Meier LL et al. Illegitimate tasks as a source of work stress. *Work & Stress* 2015; 29: 32–56.

140. Pereira D, Semmer NK, Elfering A. Illegitimate tasks and sleep quality: An ambulatory study. *Stress Health* 2014; 30: 209–221.

141. Dormann C, Zapf D. Social support, social stressors at work, and depressive symptoms: Testing for main and moderating effects with structural equations in a three-wave longitudinal study. *J Appl Psychol* 1999; 84: 874–884.

142. Kubo T, Takahashi M, Sato T, Sasaki T, Oka T, Iwasaki K. Weekend sleep intervention for workers with habitually short sleep periods. *Scand J Work Environ Health* 2011; 37: 418–426.

143. Thiart H, Lehr D, Ebert D, Sieland B, Berking M, Riper H. Log in and breathe out: Efficacy and cost-effectiveness of an online sleep training for teachers affected by work-related strain—Study protocol for a randomized controlled trial. *Trials* 2013; 14: 169–178.

6

Restless legs syndrome and neuropsychiatric disorders

RAVI GUPTA AND DEEPAK GOEL

Restless legs syndrome (RLS) is a common sensory–motor disorder with a population prevalence of around 2%–10% that varies across studies.[1-3] Women have a higher chance of suffering from RLS.[1] Although its prevalence increases with age, no age is spared, and RLS is also seen among children.[4] This disorder imparts a huge burden on the health system and also negatively influences the productivity of a person.[3,5] Not only does it impart a direct cost on the health system, but it also increases the indirect contact and utilization of the health system, since it is often associated with multiple comorbidities (e.g., attention deficit hyperactivity disorder [ADHD] in children, migraine, depression, poor sleep quality, fibromyalgia, somatic symptom disorders, and heart disease, to name a few).[6-10] These patients often have poor-quality sleep, and this can increase smoking at night, which directly and indirectly increases healthcare utilization.[11] Not only does RL increase financial burdens and worsen quality of life, but it can also worsen mortality rates.[12]

It is a well-known fact that RLS worsens sleep quality.[13] At the same time, it is also known that poor sleep quality or insomnia can have daytime symptoms that also fulfill the criteria for depression and anxiety. However, it is yet to be defined neurobiologically whether these neuropsychiatric disorders have any causal relationship with RLS, or whether they are just the daytime symptoms of RLS. One study reported that RLS patients have a higher incidence of depression, but on Beck's Depression Inventory, they reported low scores on cognitive items and high scores on somatic symptoms.[14] A number of psychiatric disorders (e.g., depression, anxiety, panic disorders, generalized anxiety disorder, ADHD, and behavioral disturbances) have been reported to be associated with RLS among children and adults.[15,16] In this chapter, we will discuss some of the neuropsychiatric disorders that have been reported alongside RLS.

RLS AND PERSONALITY PROFILE

In general terms, personality is the way in which a person perceives incoming stimuli, understands them and plans a behavioral output based upon them. Thus, personality characteristics color a person's behavior and are important for adjustment in the social structure. These factors also govern the way that a person deals with illnesses. Personality characteristics are not only governed by the genetic material that a person inherits from their parents, but also are influenced by learning. Personality profiles of RLS patients have been examined in a few studies by comparing the personality profiles of subjects with RLS with those of controls. It has been found that these patients have high scores on depressive, anxiety, hypochondriacal, and hysterical personality traits.[17] Using the NEO personality inventory, these patients were reported to have higher scores on neuroticism.[18] These factors translate into more dramatizing and help-seeking behaviors among these patients. They were found to have lower internal loci of control, which means that they are prone to being influenced by external stimuli.[19] Another study that evaluated personality traits based upon temperament and character showed higher scores on harm avoidance and lower scores on self-directedness in RLS patients.[20] All of these factors have been attributed as risks for the development of depression in these patients.[18,20] However, it must be remembered that these tests are subjective, and thus a response bias is always present. Moreover, it is difficult to estimate the influence of the RLS-related neurobiological changes that were present at the time of assessment on the results of these tests. Responses on the personality tests may be influenced by the presence of depression and anxiety disorders seen in patients with RLS. Thus, it would have been better if these factors had been analyzed after the resolution of symptoms.

RLS AND ANXIETY

Anxiety is a normal part of our emotions and can be an appropriate response in certain situations. However, when it is pervasive and longstanding, it may be considered pathological. One study analyzed state and trait anxiety among drivers and found that drivers with RLS had higher scores on both state and trait anxiety.[21] Patients with RLS have been found to have a higher prevalence of generalized anxiety disorder and panic disorder during the symptomatic periods.[16] In addition, these patients had higher lifetime prevalence of panic disorder. Considering these facts, this condition was given the name "anxietas tibiarum."[16]

RLS AND DEPRESSION

RLS has been found to be correlated with depression using various methodologies. One study found that nearly a third of RLS patients show clinical depression.[9] Duration of depression was not correlated with RLS, and nor were other factors (e.g., sleep disturbance and gender). Thus, it was concluded that depression was a comorbid condition with RLS.[9] Another study among pregnant women with RLS reported that women with pregnancy had higher chances of having antenatal and postnatal depression.[22] A longitudinal study involving healthy women reported higher chances of depression among women who were diagnosed with RLS.[23] Another epidemiological study showed that RLS increased the odds of depression by 1.8.[24] A similar relationship was seen in a cohort of patients with chronic kidney disease as well.[25] However, the prevalence of depression appears to be related to the severity of RLS, as it has been reported that subjects with mildly severe RLS did not have depression.[26] Thus, all RLS patients must be screened for the presence of depression and anxiety disorders. This issue becomes important as antidepressants may exacerbate RLS symptoms; however, this finding could not be confirmed in a retrospective chart review.[27,28] A more interesting finding is the fact that mild to moderate depression often responds to dopaminergic therapy for RLS, and thus in these cases, first the RLS should be treated, and if depression persists even after adequate resolution of the RLS symptoms, then this should be addressed.[29] Thus, by the time we have adequate reports, it can be said that all depressed patients should be screened for RLS, and antidepressants should be added after RLS has resolved.

RLS AND INSOMNIA

Insomnia is diagnosed when a person has difficulties with sleep onset and sleep maintenance, has early-morning awakenings, or has non-refreshing sleep, along with the daytime symptoms of insomnia.[30] It is a well-known fact that RLS patients often complain of poor-quality sleep, and patients with severe RLS report more sleep disturbance.[13,31] This insomnia could be related to the symptoms of RLS; however, in the long term, these patients may also present with dysfunctional beliefs related to sleep that could help in the maintenance of insomnia.[32]

In such cases, treatment of RLS may not improve insomnia completely, and it is important to address these dysfunctional beliefs through cognitive–behavioral therapy for insomnia. Moreover, many of these patients present with fatigue as a prominent daytime symptom, and adequate treatment of RLS can not only improve sleep, but also improve daytime fatigue.[33]

Insomnia in these patients could be related to hyperarousal, as RLS patients were found to have high-frequency electroencephalographic activity just before and after sleep onset.[34] In addition, these patients have been found to have higher levels of glutamine in the thalamus, which can lead to hyperarousal.[35] Thus, it can be said that insomnia in RLS has a biological basis, and hyperarousal should also be addressed in these patients. Moreover, nearly 11% of patients attending the sleep clinic were found to have RLS; thus, RLS should be screened in every patient who presents with some kind of sleep problem.[36]

RLS AND COGNITIVE DYSFUNCTION

RLS is associated with sleep loss, and this insomnia can impart cognitive dysfunction. However, it appears that brain cognitive dysfunction caused by insomnia is and that by RLS are different in terms of mechanism as well as magnitude. One study showed that RLS patients perform better than matched insomnia patients on tests of cognitive functions (i.e., letter fluency and category fluency). Thus, the authors concluded that RLS patients had a compensatory mechanism for the sleep-related cognitive impairment.[37] Moreover, like depression, cognitive dysfunction in RLS appears to be correlated with the severity of RLS, and cognitive dysfunction has not been reported in elderly subjects with mild RLS.[26] These patients have been found to have a reduced ability to decide in ambiguous situations, which increases the chances of taking risky decisions.[38,39] However, decision-making ability was found to improve after the adequate treatment of RLS, so much so that it became comparable to controls.[32]

RLS AND PAINFUL SYNDROMES

Clinical diagnosis and treatment plans for isolated RLS are straightforward, but become complicated once RLS is associated with various comorbidities.[27] RLS is commonly associated with other pain syndromes, leading to a more complex picture of disease. The comorbid pain syndrome with RLS needs to be addressed separately, as a combination of these disorders has a different therapeutic and prognostic implication. It is important for physicians to be aware of these comorbid pain syndromes while selecting therapies for RLS. The various painful syndromes associated with RLS can be grouped into four categories:

1. Headache syndromes
2. Fibromyalgia syndromes (FMSs)
3. Chronic multiple site pain syndrome (CMP)
4. Somatoform disorder, including persistent pain disorder

Headache syndromes and RLS

There are many studies addressing the issue of the coexistence of headache and RLS. Most studies have found a positive correlation between migraine headache and RLS. Chen et al.[40] in 2010 studied the associations between different functional headaches and RLS. Their results showed that a migraine headache was the most common type of headache associated with RLS, being present in 11.4% of patients. Tension-type headache and cluster headache were less frequent, as they were reported by 4.6% and 2% of RLS patients, respectively.[40] There is regional variation of the association between functional headache and RLS. In India, this association was found in more than 50% of patients. Gupta et al.[10] from India have reported that, among RLS subjects, coexistence of migraine was found in 51.5%, and other functional headache in 7.1%.[10]

If we consider the studies that had addressed the issue from other side, a third of patients suffering from migraine were found to have a coexistence of RLS.[41] Migraine without aura was found to be more commonly associated with RLS as compared to migraine with aura.[42,43] The migraine and RLS association is not only reported in adults, but also in pediatric patients.[6]

Not all research groups have confirmed the association between RLS and migraine. Gozubatik-Celik et al.[44] compared their results of migraine-like headache in RLS patients with a population-based study conducted in Turkey. The results showed no association of migraine in RLS, but severity of RLS was significantly higher in migraine patients. Moreover, family history of RLS was higher in migraine patients. According to the authors, the association between migraine and RLS needs further confirmation.[44]

One meta-analysis included 24 studies in order to determine the coexistence of migraine with RLS. The prevalence of RLS in migraine patients ranged from 8.7% to 39% without any differences in terms of gender or presence of aura. This meta-analysis suggests that migraine is a significant comorbidity of RLS, but the degree of the association is still unclear.[45]

Pathogenesis for explaining the coexistence of migraine and RLS is challenging in view of current evidence. The current literature has tried to explain the association between migraine and RLS, and these mechanisms can be grouped into four categories: first, many studies have tried to analyze the pathophysiological association and they have proposed many hypothesis, with the most fitting involving dopamine and melatonin.[46] Since the administration of dopaminergic drugs for RLS is found to have adverse effects on anti-migraine therapy, the dopaminergic correlation of both of these diseases is unlikely. Second, sets of studies have explored the role of iron metabolism and sleep disturbance in explaining the association between RLS and migraine. Comorbid illness in RLS patients with migraine-like insomnia, daytime sleepiness, depressive symptoms, headache-related disability, and increased serum phosphorus levels may provide a better understanding of RLS pathogenesis in migraine.[47] The third mechanism is based on genetic linkage between the two disorders. There are a few hypotheses to explain the association between both of these disorders, including the same genetic origin for migraine without aura and RLS in a single Italian family on chromosome 14q21; this gene encodes survival motor neuron-interacting protein 1 (SIP1), which can play a role in both diseases.[41] The last possible explanation of the coexistence of RLS and migraine is based on the alteration of cortical excitability in both migraine and RLS.[41] More recently, the role of basal ganglia in pain processing has been confirmed by functional imaging data in the caudate, putamen, and pallidum in migraine patients. A critical appraisal of all of these clinical and experimental data suggests that the extrapyramidal system is somehow related to migraine. Although the primary involvement of the extrapyramidal system

in the pathophysiology of migraine cannot as yet be proven, it may be suggested that a more general role in the processing of nociceptive information may be part of the complex behavioral adaptive response that characterizes migraine.[48]

FMS with RLS

FMS is a common, painful syndrome that is more common in females, with a prevalence rate of 2%–8% in population studies.[49] In clinical practice, diagnosis of FMS is based on American College of Rheumatology criteria.[50]

As both RLS and FMS are more common in females, about 2% of females in population studies have both RLS and FMS. Stehlik et al. (2009) evaluated 332 female patients aged 20–60 years with a primary diagnosis of fibromyalgia. In this cohort, 64% subjects had RLS symptoms. Sleep initiation and maintenance were worse in females having FMS with RLS as compared to females with only FMS. The females with a coexistence of FMS with RLS had more non-refreshing awakenings and daytime sleepiness as compared to females with FMS only.[51] Other studies have reported that quality of sleep and quality of life worsened when FMS was associated with RLS, and it is recommended that FMS should always be excluded in RLS patients, as it can change the treatment plan.[52,53]

CMP with RLS

Several studies have found that chronic multisite pain (CMP) is prevalent in 31%–64% of RLS patients.[54] Chronic multisite pain is a highly prevalent condition and common health problem today, remaining undiagnosed and poorly understood by general physicians. Many of these patients may have a diagnosis of depression. RLS patients with sleep disorders leading to painful syndromes can have poor quality of life and secondary depression or other psychiatric problems. It is important that clinicians are aware of the full spectrum of the painful syndromes that are associated with RLS for better management planning. In the general population, the prevalence of RLS increases with the severity of pain and spreading of pain to multiple sites. Both acute and chronic pain are associated with RLS and result in poor responses to pain-relieving treatment.[54]

Pain is a multidimensional response involving several levels of expression, ranging from the somatosensory to the emotional. The potential shared mechanisms between RLS and pain may involve sleep deprivation/fragmentation effects, inducing an increase in markers of inflammation and a reduction in pain thresholds. These are modulated by several different settings of neurotransmitters with a huge participation of monoaminergic dysfunctional circuits. A thorough comprehension of these mechanisms is of the utmost importance for establishing correct approaches and treatment choices.[55]

Somatic symptoms disorder

It has been reported that nearly half of the patients with somatoform pain disorder have symptoms of RLS.[8] Patients with both disorders had a higher prevalence of poor-quality sleep as compared to patients without RLS.[8] These patients often have longer durations of somatoform symptoms and continuous courses of somatoform disorder.[56] Thus, patients with somatoform disorder (especially those who are not remitting) should be screened for RLS. This is yet to be established whether the CMP described in the literature is actually the somatic symptoms disorder.

RLS AND ADHD

ADHD has been reported in 20%–25% of the children with RLS.[15,57] By contrast, another study found that nearly a third of ADHD children had RLS.[58] However, children with comorbid RLS–ADHD did not show any differences in severity of ADHD and behavioral disturbances as compared to ADHD children without RLS.[58] Children with RLS often have sleep disturbances from the age of 3 years, and these sleep disturbances were found to antedate RLS by 11 years, suggesting that children with sleep disturbances must be screened for RLS on each visit.[57] Not only children, but also adult RLS sufferers were found to have higher chances of having ADHD as compared to persons with insomnia and controls.[59] Not only patients with ADHD themselves, but their parents also showed a higher prevalence of RLS.[60] These conditions have been found to be comorbid disorders, having common neurobiological underpinnings of dopamine dysregulation, and some reports have emphasized that RLS can be misdiagnosed as ADHD not only during childhood, but also during adulthood.[59–61] However, the dopaminergic theory was challenged by one report that suggested that treatment with levodopa could improve RLS symptoms, but not ADHD symptoms.[62] Further evidence suggests that low serum ferritin may be the causal factor in the development of RLS among ADHD children.[58] Furthermore, the role of epileptic discharges during sleep has been proposed as causative.[63] These discharges responded to levetiracetam in a small study, and improvements in both RLS and ADHD were noticed.[63] Therefore, these studies have found that a subgroup of ADHD children has RLS, and these children have a higher prevalence of iron deficiency. Correction of iron can improve RLS and can bring about some improvement in ADHD symptoms as well.

RLS AND NOCTURNAL EATING

Sleep-related eating disorder (SRED) is characterized by episodes of eating from the onset of sleep to the offset of sleep, often during partial awakening, and can be akin to sleepwalking. Although patients with SRED often prefer high-calorie food, at times they are unaware of their eating on the morning after. On the other hand, night eating syndrome (NES) is characterized by eating at night, often of large portions, but this occurs during full wakefulness. Both of these disorders have been found to be associated with RLS.[64–66] These patients are often obese and are using sedative drugs; however, this is a controversial finding.[65,66] Another study examined the prevalence of RLS in SRED and found that nearly half of them suffered from RLS.[67] SRED appears to be linked to RLS rather than hypnotics, as one study reported that patients with RLS who were taking hypnotics had a higher prevalence of SRED as compared to insomnia patients taking hypnotics.[68] Moreover, nocturnal eating was not related to awakenings, as patients with insomnia had longer awakenings at night, yet less frequent eating episodes.[68] Thus, both SRED and NES appear to have relationships with RLS. One case report depicted the complete resolution of SRED by treatment with pramipexole.[69]

SCHIZOPHRENIA AND RLS

Schizophrenia is a chronic disorder that affects the thoughts and cognitions of the sufferer. It is hypothesized to be induced by a functional excess of dopamine in the mesolimbic tract of the brain.

Table 6.1 Candidate genes linking antipsychotic-induced RLS and schizophrenia

Author	Population	Study type	Gene	Results
Kang et al.[71]	190 schizophrenia patients	Cohort	Single-nucleotide polymorphisms of: 1. *DRD1* gene –48A/G 2. *DRD2* gene Taql A, 3. *DRD3* gene Ser9Gly 4. *DRD4* gene –521C/T	Nearly 50% had RLS; no differences in analyzed polymorphisms
Cho et al.[72]	190 schizophrenia patients	Cohort	Val81Met single-nucleotide polymorphism of tyrosine hydroxylase gene	Nearly half had RLS; this polymorphism was associated with RLS among women
Kang et al.[73]	190 schizophrenia patients	Cohort	1. Variable number of tandem repeat polymorphism of the *MAOA* gene 2. A644G polymorphism of the *MAOB* gene	No differences between allele frequencies
Kang et al.[74]	190 schizophrenia patients	Cohort	rs9357271 and rs3923809 polymorphisms of the *BTBD9* gene	rs9357271 was associated with RLS in a dominant and heterozygous model
Jung et al.[75]	190 schizophrenia patients	Cohort	*CLOCK* and *NPAS2* gene polymorphisms	*CLOCK* gene polymorphism (rs2412646) was associated with antipsychotic-induced RLS

Thus, antipsychotics, which act on dopamine receptors and antagonize their action, have been found to be effective in the treatment of schizophrenia. However, antipsychotics may produce some adverse effects that can mimic RLS. One of these is akathisia, which appears early during the course of treatment. Akathisia is characterized by inner motor restlessness, making it difficult for the person to sit in one place. This is more commonly seen during phases of inactivity and thus may be more severe at night. However, unlike RLS, these symptoms are widespread in the body, and patients do not complain about the sensory aspects of RLS. Nevertheless, it is a known fact that antipsychotics can induce RLS. A number of candidate genes have been identified as being associated with this condition, and these are depicted in Table 6.1.

Not only antipsychotics, but also schizophrenia itself have been found to increase the chances of having RLS.[70] These patients had more severe psychotic symptoms and higher scores on insomnia severity scales.[70]

CONCLUSION

RLS has a close relationship with several neuropsychiatric disorders, although it is difficult to establish causality between them. The available literature suggests that considering its high prevalence along with psychiatric disorders, all such patients should be screened for the RLS.

REFERENCES

1. Erer S, Karli N, Zarifoglu M, Ozcakir A, Yildiz D. The prevalence and clinical features of restless legs syndrome: A door to door population study in Orhangazi, Bursa in Turkey. *Neurol India* 2009; 57: 729–733.

2. Ulfberg J, Nyström B, Carter N, Edling C. Prevalence of restless legs syndrome among men aged 18 to 64 years: An association with somatic disease and neuropsychiatric symptoms. *Mov Disord* 2001; 16: 1159–1163.

3. Rangarajan S, Rangarajan S, D'Souza GA. Restless legs syndrome in an Indian urban population. *Sleep Med* 2007; 9: 88–93.

4. Picchietti DL, Underwood DJ, Farris WA et al. Further studies on periodic limb movement disorder and restless legs syndrome in children with attention-deficit hyperactivity disorder. *Mov Disord* 1999; 14: 1000–1007.

5. Allen RP, Bharmal M, Calloway M. Prevalence and disease burden of primary restless legs syndrome: Results of a general population survey in the United States. *Mov Disord* 2011; 26: 114–120.

6. Seidel S, Böck A, Schlegel W et al. Increased RLS prevalence in children and adolescents with migraine: A case–control study. *Cephalalgia* 2012; 32: 693–699.

7. Wesstrom J, Nilsson S, Sundstrom-Poromaa I, Ulfberg J. Restless legs syndrome among women: Prevalence, co-morbidity and possible relationship to menopause. *Climacteric* 2008; 11: 422–428.

8. Aigner M, Prause W, Freidl M et al. High prevalence of restless legs syndrome in somatoform pain disorder. *Eur Arch Psychiatry Clin Neurosci* 2007; 257: 54–57.

9. Gupta R, Lahan V, Goel D. A study examining depression in restless legs syndrome. *Asian J Psychiatr* 2013; 6: 308–312.

10. Gupta R, Lahan V, Goel D. Primary headaches in restless legs syndrome patients. *Ann Indian Acad Neurol* 2012; 15: S104–S108.

11. Provini F, Antelmi E, Vignatelli L et al. Increased prevalence of nocturnal smoking in restless legs syndrome (RLS). *Sleep Med* 2010; 11: 218–220.

12. Hening WA, Allen RP, Chaudhuri KR et al. Clinical significance of RLS. *Mov Disord* 2007; 22(Suppl 1): S395–S400.

13. Cuellar NG, Strumpf NE, Ratcliffe SJ. Symptoms of restless legs syndrome in older adults: Outcomes on sleep quality, sleepiness, fatigue, depression, and quality of life. *J Am Geriatr Soc* 2007; 55: 1387–1392.

14. Hornyak M, Kopasz M, Berger M, Riemann D, Voderholzer U. Impact of sleep-related complaints on depressive symptoms in patients with restless legs syndrome. *J Clin Psychiatry* 2005; 66: 1139–1145.

15. Pullen SJ, Wall CA, Angstman ER, Munitz GE, Kotagal S. Psychiatric comorbidity in children and adolescents with restless legs syndrome: A retrospective study. *J Clin Sleep Med* 2011; 7: 587–596.

16. Winkelmann J, Prager M, Lieb R et al. "Anxietas tibiarum." Depression and anxiety disorders in patients with restless legs syndrome. *J Neurol* 2005; 252: 67–71.

17. Turkel Y, Oguzturk O, Dag E, Buturak SV, Ekici MS. Minnesota Multiphasic Personality Inventory profile in patients with restless legs syndrome. *Asia Pac Psychiatry* 2015; 7: 153–156.

18. Kalaydjian A, Bienvenu OJ, Hening WA, Allen RP, Eaton WW, Lee HB. Restless legs syndrome and the five-factor model of personality: Results from a community sample. *Sleep Med* 2009; 10: 672–675.

19. Brand S, Beck J, Hatzinger M, Holsboer-Trachsler E. Patients suffering from restless legs syndrome have low internal locus of control and poor psychological functioning compared to healthy controls. *Neuropsychobiology* 2013; 68: 51–58.

20. Altunayoglu Cakmak V, Gazioglu S, Can Usta N et al. Evaluation of temperament and character features as risk factors for depressive symptoms in patients with restless legs syndrome. *J Clin Neurol* 2014; 10: 320–327.

21. Ozder A, Eker HH. Anxiety levels among Turkish public transportation drivers: A relation to restless legs syndrome? *Int J Clin Exp Med* 2014; 7: 1577–1584.

22. Wesström J, Skalkidou A, Manconi M, Fulda S, Sundström-Poromaa I. Pre-pregnancy restless legs syndrome (Willis–Ekbom disease) is associated with perinatal depression. *J Clin Sleep Med* 2014; 10: 527–533.

23. Li Y, Mirzaei F, O'Reilly EJ et al. Prospective study of restless legs syndrome and risk of depression in women. *Am J Epidemiol* 2012; 176: 279–288.

24. Froese CL, Butt A, Mulgrew A et al. Depression and sleep-related symptoms in an adult, indigenous, North American population. *J Clin Sleep Med* 2008; 4: 356–361.

25. Szentkiralyi A, Molnar MZ, Czira ME et al. Association between restless legs syndrome and depression in patients with chronic kidney disease. *J Psychosom Res* 2009; 67: 173–180.

26. Driver-Dunckley E, Connor D, Hentz J et al. No evidence for cognitive dysfunction or depression in patients with mild restless legs syndrome. *Mov Disord* 2009; 24: 1840–1842.

27. Becker PM, Novak M. Diagnosis, comorbidities, and management of restless legs syndrome. *Curr Med Res Opin* 2014; 30: 1441–1460.

28. Brown LK, Dedrick DL, Doggett JW, Guido PS. Antidepressant medication use and restless legs syndrome in patients presenting with insomnia. *Sleep Med* 2005; 6: 443–450.

29. Hornyak M. Depressive disorders in restless legs syndrome: Epidemiology, pathophysiology and management. *CNS Drugs* 2010; 24: 89–98.

30. American Academy of Medicine. *International Classification of Sleep Disorders*, 3rd edition. Darien, IL: American Academy of Sleep Medicine, 2014.

31. Kim J Bin, Koo YS, Eun M-Y, Park K-W, Jung K-Y. Psychosomatic symptom profiles in patients with restless legs syndrome. *Sleep Breath* 2013; 17: 1055–1061.

32. Crönlein T, Wagner S, Langguth B, Geisler P, Eichhammer P, Wetter TC. Are dysfunctional attitudes and beliefs about sleep unique to primary insomnia? *Sleep Med* 2014; 15: 1463–1467.

33. Anderson K, Jones DEJ, Wilton K, Newton JL. Restless leg syndrome is a treatable cause of sleep disturbance and fatigue in primary biliary cirrhosis. *Liver Int* 2013; 33: 239–243.

34. Ferri R, Cosentino FII, Manconi M, Rundo F, Bruni O, Zucconi M. Increased electroencephalographic high frequencies during the sleep onset period in patients with restless legs syndrome. *Sleep* 2014; 37: 1375–1381.

35. Allen RP, Barker PB, Horská A, Earley CJ. Thalamic glutamate/glutamine in restless legs syndrome: Increased and related to disturbed sleep. *Neurology* 2013; 80: 2028–2034.

36. Lin S-W, Chen Y-L, Kao K-C et al. Diseases in patients coming to a sleep center with symptoms related to restless legs syndrome. *PLoS One* 2013; 8: e71499.

37. Gamaldo CE, Benbrook AR, Allen RP, Oguntimein O, Earley CJ. A further evaluation of the cognitive deficits associated with restless legs syndrome (RLS). *Sleep Med* 2008; 9: 500–505.

38. Bayard S, Yu H, Langenier MC, Carlander B, Dauvilliers Y. Decision making in restless legs syndrome. *Mov Disord* 2010; 25: 2634–2640.

39. Galbiati A, Marelli S, Giora E, Zucconi M, Oldani A, Ferini-Strambi L. Neurocognitive function in patients with idiopathic restless legs syndrome before and after treatment with dopamine-agonist. *Int J Psychophysiol* 2015; 95: 304–309.

40. Chen P-K, Fuh J-L, Chen S-P, Wang S-J. Association between restless legs syndrome and migraine. *J Neurol Neurosurg Psychiatry* 2010; 81: 524–528.

41. Sabayan B, Bagheri M, Borhani Haghighi A. Possible joint origin of restless leg syndrome (RLS) and migraine. *Med Hypotheses* 2007; 69: 64–66.

42. Fernández-Matarrubia M, Cuadrado ML, Sánchez-Barros CM et al. Prevalence of migraine in patients with restless legs syndrome: A case–control study. *Headache* 2014; 54: 1337–1346.

43. d'Onofrio F, Cologno D, Petretta V et al. Restless legs syndrome is not associated with migraine with aura: A clinical study. *Neurol Sci* 2011; 32(Suppl 1): S153–S156.

44. Gozubatik-Celik G, Benbir G, Tan F, Karadeniz D, Goksan B. The prevalence of migraine in restless legs syndrome. *Headache* 2014; 54: 872–877.

45. Schürks M, Winter A, Berger K, Kurth T. Migraine and restless legs syndrome: A systematic review. *Cephalalgia* 2014; 34: 777–794.

46. d'Onofrio F, Bussone G, Cologno D et al. Restless legs syndrome and primary headaches: A clinical study. *Neurol Sci* 2008; 29(Suppl 1): S169–S172.

47. Suzuki S, Suzuki K, Miyamoto M et al. Evaluation of contributing factors to restless legs syndrome in migraine patients. *J Neurol* 2011; 258: 2026–2035.

48. d'Onofrio F, Barbanti P, Petretta V et al. Migraine and movement disorders. *Neurol Sci* 2012; 33(Suppl 1): S55–S59.

49. Clauw DJ. Fibromyalgia: A clinical review. *JAMA* 2014; 311: 1547–1555.

50. Ray Fleming, O. of C. and P. L. *Questions and Answers About Fibromyalgia: National Institute of Arthritis and Musculoskeletal and Skin and Diseases.* 2014. Retrieved September 17, 2014, from http://www.niams.nih.gov/Health_Info/Fibromyalgia/default.asp.

51. Stehlik R, Arvidsson L, Ulfberg J. Restless legs syndrome is common among female patients with fibromyalgia. *Eur Neurol* 2009; 61: 107–111.

52. Civelek GM, Ciftkaya PO, Karatas M. Evaluation of restless legs syndrome in fibromyalgia syndrome: An analysis of quality of sleep and life. *J Back Musculoskelet Rehabil* 2014; 27: 537–544.

53. Viola-Saltzman M, Watson NF, Bogart A, Goldberg J, Buchwald D. High prevalence of restless legs syndrome among patients with fibromyalgia: A controlled cross-sectional study. *J Clin Sleep Med* 2010; 6: 423–427.

54. Stehlik R, Ulfberg J, Hedner J, Grote L. High prevalence of restless legs syndrome among women with multi-site pain: A population-based study in Dalarna, Sweden. *Eur J Pain* 2014; 18: 1402–1409.

55. Goulart LI, Delgado Rodrigues RN, Prieto Peres MF. Restless legs syndrome and pain disorders: What's in common? *Curr Pain Headache Rep* 2014; 18: 461.

56. Chatterjee S, Mitra S, Guha P, Chakraborty K. Prevalence of restless legs syndrome in somatoform pain disorder and its effect on quality of life. *J Neurosci Rural Pract* 2015; 6: 160.

57. Picchietti DL, Stevens HE. Early manifestations of restless legs syndrome in childhood and adolescence. *Sleep Med* 2008; 9: 770–781.

58. Oner P, Dirik EB, Taner Y, Caykoylu A, Anlar O. Association between low serum ferritin and restless legs syndrome in patients with attention deficit hyperactivity disorder. *Tohoku J Exp Med* 2007; 213: 269–276.

59. Wagner ML, Walters AS, Fisher BC. Symptoms of attention-deficit/hyperactivity disorder in adults with restless legs syndrome. *Sleep* 2004; 27: 1499–1504.

60. Steinlechner S, Brüggemann N, Sobottka V et al. Restless legs syndrome as a possible predictor for psychiatric disorders in parents of children with ADHD. *Eur Arch Psychiatry Clin Neurosci* 2011; 261: 285–291.

61. Philipsen A, Hornyak M, Riemann D. Sleep and sleep disorders in adults with attention deficit/hyperactivity disorder. *Sleep Med Rev* 2006; 10: 399–405.

62. England SJ, Picchietti DL, Couvadelli BV et al. L-dopa improves restless legs syndrome and periodic limb movements in sleep but not attention-deficit–hyperactivity disorder in a double-blind trial in children. *Sleep Med* 2011; 12: 471–477.

63. Gagliano A, Aricò I, Calarese T et al. Restless leg syndrome in ADHD children: Levetiracetam as a reasonable therapeutic option. *Brain Dev* 2011; 33: 480–486.

64. Inoue Y. Sleep-related eating disorder and its associated conditions. *Psychiatry Clin Neurosci* 2015; 69: 309–320.

65. Antelmi E, Vinai P, Pizza F, Marcatelli M, Speciale M, Provini F. Nocturnal eating is part of the clinical spectrum of restless legs syndrome and an underestimated risk factor for increased body mass index. *Sleep Med* 2014; 15: 168–172.

66. Provini F, Antelmi E, Vignatelli L et al. Association of restless legs syndrome with nocturnal eating: A case–control study. *Mov Disord* 2009; 24: 871–877.

67. Santin J, Mery V, Elso MJ et al. Sleep-related eating disorder: A descriptive study in Chilean patients. *Sleep Med* 2014; 15: 163–167.

68. Howell MJ, Schenck CH. Restless nocturnal eating: A common feature of Willis–Ekbom syndrome (RLS). *J Clin Sleep Med* 2012; 8: 413–419.

69. Kobayashi N, Yoshimura R, Takano M. Successful treatment with clonazepam and pramipexole of a patient with sleep-related eating disorder associated with restless legs syndrome: A case report. *Case Rep Med* 2012; 2012: 893681.

70. Kang S-G, Lee H-J, Jung SW et al. Characteristics and clinical correlates of restless legs syndrome in schizophrenia. *Prog Neuropsychopharmacol Biol Psychiatry* 2007; 31: 1078–1083.

71. Kang S-G, Lee H-J, Choi J-E et al. Association study between antipsychotics-induced restless legs syndrome and polymorphisms of dopamine D1, D2, D3, and D4 receptor genes in schizophrenia. *Neuropsychobiology* 2008; 57: 49–54.

72. Cho C-H, Kang S-G, Choi J-E, Park Y-M, Lee H-J, Kim L. Association between antipsychotics-induced restless legs syndrome and tyrosine hydroxylase gene polymorphism. *Psychiatry Investig* 2009; 6: 211–215.

73. Kang S-G, Park Y-M, Choi J-E et al. Association study between antipsychotic-induced restless legs syndrome and polymorphisms of monoamine oxidase genes in schizophrenia. *Hum Psychopharmacol* 2010; 25: 397–403.

74. Kang S-G, Lee H-J, Park Y-M et al. The *BTBD9* gene may be associated with antipsychotic-induced restless legs syndrome in schizophrenia. *Hum Psychopharmacol* 2013; 28: 117–123.

75. Jung J-S, Lee H-J, Cho C-H et al. Association between restless legs syndrome and *CLOCK* and *NPAS2* gene polymorphisms in schizophrenia. *Chronobiol Int* 2014; 31: 838–844.

7

Narcolepsy and psychosomatic illnesses

MEETA GOSWAMI, KENNETH BUTTO,
AND SEITHIKURIPPU R. PANDI-PERUMAL

THE SYMPTOMS OF NARCOLEPSY

The earliest reported description of narcolepsy appears to be recorded by Oliver in 1704.[1] Following reports of cases with symptoms of narcolepsy by Graves in 1851,[2] Caffé in 1862,[3] and Westphal in 1877,[4] it was Gelineau in 1881[5] who ascribed the term *narcolepsie* to describe a condition characterized by brief episodes of irresistible sleep and by falls (*astasias*) associated with emotional stimuli.

Narcolepsy is a chronic, debilitating neurological disorder, the hallmarks of which are hypersomnia and cataplexy (a sudden and transient decrement of muscle tone and loss of deep tendon reflexes, leading to muscle weakness, paralysis, or postural collapse, usually in response to an external stimulus).[6] Persons with narcolepsy often have rapid eye movement (REM) sleep intrusion into wakefulness. Characteristically, patients report relief of discomfort from episodes of sleep attacks upon taking short naps. Cataplectic attacks—another form of REM intrusion—are usually evoked by a strong or deeply felt outburst of intense emotional expression such as laughter, joy, anger or fright, surprise, amusement or excitement.[7,8] Yoss and Daly[9] described a classic tetrad of symptoms for the diagnosis of narcolepsy: excessive daytime somnolence (EDS), which is usually the first symptom; cataplexy (70%–80%); hypnagogic or hypnopompic hallucinations (which can occur just prior to sleep onset or upon arousal) (70%); and sleep paralysis (loss of muscle tone at sleep onset or on awakening) (60%). However, each of these symptoms can appear at various stages of the disorder and in different combinations and degrees of severity.[8]

Cataplexy is characterized by a sudden bilateral loss of postural muscle tone (i.e., weakness with loss of deep tendon reflex) and often occurs in response to emotional triggers. During a cataplectic attack, a person loses muscular control but is aware of the environment. Hypnagogic hallucinations (vivid, dream-like experiences), sleep paralysis, and automatic behaviors (the performance of routine tasks without awareness; e.g., talking, driving, or writing) are auxiliary symptoms; fatigue, cognitive impairment, and disturbed nocturnal sleep are common complaints of narcolepsy patients.[10]

In most cases, sleepiness rather than cataplexy is the more perplexing and debilitating symptom; people with narcolepsy go through life feeling the way most of us would feel if we had been awake for 24 hours.[11] Nocturnal sleep is fragmented, with a reduction in slow-wave sleep (SWS); naps may offer some help, but sleepiness occurs following refreshing naps.[11] Thus, both nocturnal sleep and EDS pose a problem to these patients.[12,13]

DIAGNOSIS

If narcolepsy is suspected, a polysomnographic (PSG) and Multiple Sleep Latency Test (MSLT) are often used to establish a diagnosis of narcolepsy.[14] If sleep apnea and other excessive sleep problems can be ruled out, a daytime mean sleep onset latency (SOL) of less than 8 minutes—typically below 5 minutes—and the presence of REM sleep in two or more of five daytime nap periods (MSLT) are abnormal and confirm the clinical laboratory diagnosis of narcolepsy. According to Stores,[8] a combination of EDS and definite cataplexy can be considered pathognomonic of narcolepsy.

PATHOPHYSIOLOGY

Narcolepsy is a neurological condition, but its etiology and pathogenesis remain obscure,[15] although there have been recent advances in the understanding of the disease. A major milestone in narcolepsy research was the discovery of sleep-onset REM period sleep,[16] which led to a school of thought that argued that cataplexy, hypnagogic hallucinations, and sleep paralysis might be due to alterations in REM sleep regulatory mechanisms.[17,18] In 1975, the dysfunction of REM sleep in narcolepsy was emphasized.[19] However, others supported the idea that narcolepsy is characterized by abnormalities of REM and non-REM sleep activity.[20–22] The pathophysiology of narcolepsy appears to be related to the abnormal expression and manifestations of REM sleep that intrude upon periods of wakefulness.[16,23–25]

Earlier studies of the genetics of narcolepsy showed that 98%–100% of patients with narcolepsy are *HLA-DR2* positive.[26–30] It is important to note that 20%–25% of the normal Caucasian

population may have this gene. In Israel, where only 11% of the control subjects in one study were *DR2* positive, the prevalence rate of narcolepsy is about 1/100 of the rate of narcoleptics quoted in other studies.[31] Researchers in Japan demonstrated that *HLA-DR2* and *DQW1* were positive in most narcolepsy patients, and 30 patients were found to have the *DRW15* subspecificity.[32] However, the incidence of *HLA-DR15* varies among ethnic groups and is lower in the African-American population. Later, it was discovered that *HLA-DQ6* and *DQB1*0602* occur more frequently in persons with narcolepsy with cataplexy than those without cataplexy, and subjects were also found to be *DQA1*0102* positive. In addition, there is a positive correlation between *DQB1*0602* positivity and severity of cataplexy.[33,34] This gene appears to confer susceptibility to narcolepsy; however, several *DQB1*0602*-negative families have narcolepsy with cataplexy.[35] A multifactorial model in which exogenous factors may precipitate symptoms in predisposed individuals may be helpful in explaining the development of the symptoms. Experiments carried out on monozygotic twins point to environmental factors.[35] However, the appearance of symptoms at puberty may indicate a hormonal effect. Some authors have proposed an autoimmune mechanism of narcolepsy.[36–38] The role of the neuropeptide hypocretin/orexin in the pathophysiology of narcolepsy is a recent discovery. These neuropeptides were deficient in some cases of narcolepsy,[39] and the brains of persons with narcolepsy have shown absence of hypocretin neurons in the hypothalamus.[40,41] Hypocretin neurons project to the locus coeruleus and are excitatory. They also project to regions of the brain that are important in producing and maintaining arousal. Changes in neurotransmission in these areas due to a defect in the hypocretin system may explain the EDS of persons with narcolepsy.[42]

PREVALENCE

The prevalence of narcolepsy is estimated at 0.05% and 0.067% in the San Francisco Bay area and Los Angeles county, respectively, and about 250,000 are estimated to have narcolepsy in the United States.[43,44] Recent epidemiological studies suggest that narcolepsy occurs in approximately 1 in 2000 individuals.[8,45,46] It is not as rare as was noted in earlier literature, and may in fact affect as many people in the United States as multiple sclerosis.[47] Similar prevalence rates have been observed in Europe.[48]

Gender does not appear to be a discriminating variable. A higher rate for men noted by some researchers may reflect a different pattern of utilization of services or the inclusion of cases of sleep apnea, which is reported to be more prevalent among men than women. Socioeconomic status and race have not been reported as differentiating variables in the United States. However, a low prevalence rate (0.0002%) has been observed in Israel, probably because of a concurrent low rate of the *DR2* gene in the population.[31]

DISEASE ONSET

While frequently undiagnosed or misdiagnosed, the peak age of onset of narcolepsy is in adolescence and young adulthood, and is generally between 15 and 30 years.[49,50] However, it can occur at any age, including early childhood or late adulthood. For example, some authors have reported the onset of narcolepsy before the age of 11.[51–53]

TREATMENT (NON-PHARMACOLOGICAL AND PHARMACOLOGICAL)

Presently, there is no cure for this condition; however, the symptoms of EDS and cataplexy can be ameliorated with medications. A comprehensive treatment plan must be adopted to help allay the clinical symptoms as well as improve the health-related quality of life (HRQOL) of affected individuals. Individual needs of patients and their reactions to medications play important roles in the treatment strategy. Nonmedical management includes modifying sleep–wake schedules, napping at strategic times of the day, balancing nutritional intake, and exercising to reduce excessive weight and to increase subjective feelings of alertness. Support and counseling may be needed—as we will see in the ensuing discussion—to cope with the effects of narcolepsy on the lives of patients.

The pharmacological management of narcolepsy entails the use of stimulants for EDS, tricyclic antidepressants for secondary symptoms, and benzodiazepines to consolidate disturbed nocturnal sleep.[54,55] Other sleep disorders, if present, must also be treated. Stimulants such as amphetamines, methylphenidate, mazindol, and wakefulness-promoting medications such as modafinil are prescribed for EDS. Commonly reported side effects of these medications include tremors, irritability, sweating, gastrointestinal symptoms, headache, nausea and, in rare cases, dyskinesia, hypertension, tachycardia, and psychosis. Tricyclic antidepressants such as imipramine, clomipramine, and protriptyline are prescribed for cataplexy. Side effects include anticholinergic effects, weight gain, sexual dysfunction, orthostatic hypotension, and anti-histiminic symptoms. Selective serotonin reuptake inhibitors such as fluoxetine and venlafaxine are effective in reducing cataplexy and have fewer side effects than tricyclics.[42]

Research shows that gamma-hydroxybutyrate, an endogenous hypnotic chemical, consolidates nighttime sleep and improves sleep disruption, cataplexy, daytime sleepiness, and other secondary symptoms of narcolepsy. It is reported to promote both REM sleep and SWS.[56] A multicenter study was conducted on 136 narcolepsy patients who experienced 3–249 (median 21) cataplexy attacks weekly. Three doses of sodium oxybate (3, 6, or 9 g) or placebo were taken by the subjects in equally divided doses at bedtime. Compared to placebo, a dose-related response was observed in the occurrence of cataplexy, sleep attacks, and naps. Weekly cataplexy attacks were reduced at the 6-g dose and significantly at the 9-g dose ($p = 0.0008$). The results for the Epworth Sleepiness Scale showed a reduction at all doses, and was significant at the 9-g dose ($p = 0.0001$). The Clinical Global Impression change was significant at the 9-g dose ($p = 0.0002$). Naps and sleep attacks were also significantly reduced at the 9-g dose ($p = 0.0035$). Sodium oxybate was well tolerated at all three doses. Commonly reported side effects were nausea, headache, dizziness, and enuresis.[57] Another study of 25 patients with narcolepsy–cataplexy demonstrated dose-related decreases in nocturnal awakenings and increases in daytime SOL. There was an increase in SWS and delta power, and a reduction in REM sleep. Significant improvements in sleep architecture coincided with significant improvements in daytime functioning.[58]

An unsettling observation made by several investigators is that, in many cases, even with medications, the symptoms of EDS cannot be controlled. Consequently, patients must learn to live with it and its profound effects on their lives.[54,59,60]

IMPACT OF NARCOLEPSY ON HEALTH AND WELFARE: QUALITY OF LIFE ISSUES

It is significant that, whereas strides have been made in the study of the biomedical aspects of narcolepsy, it is only recently that scientific consideration has been given to the social and psychological aspects of narcolepsy. There is a lack of knowledge about narcolepsy among professionals and lay people. Narcolepsy may be misdiagnosed as depression or hypothyroidism, while the symptoms of hypnagogic hallucinations may be mistaken for schizophrenia. In many cases, narcolepsy may remain undiagnosed for several years, and schoolteachers or parents may mistake the symptoms of EDS for laziness, lack of interest, or drug addiction. Thus, contact with an appropriate professional may be delayed by as long as 10 years from the time of the inception of symptoms. During this time, the affected individual may drop out of school or become unemployed because of their inability to stay awake.

A strong denial mechanism is often operative, in which case the affected individual may delay an initial contact with a professional for diagnosis and treatment. Denial also poses difficulty in making a role transition from a well to an impaired state, thereby precipitating personal and interaction problems for the individual with their family and friends.

The socioeconomic effect impinging on the lives of patients has been reported.[61] International studies, which included Canada, Japan, and former Czechoslovakia, have shown that work (performance, income, and job loss), education, driving (accidents and loss of license), interpersonal relationships, social life, and personality were adversely affected.[62] Patients attributed most of these effects to EDS, although some were due to adverse reactions to medications (e.g., loss of libido and impotence). Based on the clinical presentation and the natural history of narcolepsy, it is suggested that these effects were not influenced by ethnic, cultural, or genetic factors, but were attributable to disease. EDS is characterized in most patients by its unrelenting chronicity, marked intensity, and poor response to treatment.[63] Even in treated patients, their functional status is poor compared to continuous positive airway pressure-treated patients and untreated obstructive sleep apnea/hypopnea syndrome patients.[60] The compound effects of narcolepsy were noted in one study in New York City: unemployment and divorce rates were higher in narcolepsy in comparison with the U.S. national rates.[64]

A study in Germany[65] documents the economic burden of narcolepsy on 75 patients diagnosed with the disease. Information on the symptoms of narcolepsy and their economic impacts was obtained through a standardized telephone interview. A mailed questionnaire was utilized to assess HRQOL (36-item Short Form Survey [SF-36] and European Quality of life instrument to measure five dimensions [EQ-5D]). The study measured direct and indirect costs in 2002 Euros. All values were converted from Euros to 2002 U.S. dollars ($1 U.S. = €0.96) and are presented in U.S. dollars. Direct medical costs covered medications, inpatient care (hospital), outpatient care (doctor visits), and diagnostic testing. Direct nonmedical costs included treatment such as physiotherapy and home equipment. Nursing and sick benefits were included in this category. In Germany, employees receive sick benefits if they are out of work for more than 30 days. None of the patients needed nursing care and none was out of work for more than 30 days. Indirect costs included days off from work due to narcolepsy and early retirement.

Direct costs amounted to $3310. Drug costs were $1060, more than 50% of which was attributed to new wake-promoting medications. Annual indirect costs amounted to $11,860 per patient due to early retirement because of narcolepsy. Out of 75 patients, 32 reported narcolepsy as the cause of unemployment. The high economic burden of narcolepsy is comparable to diseases such as Parkinson's disease, Alzheimer's disease, epilepsy, and stroke (annual direct costs [$3310] plus indirect costs [$11,860] equals $15,170).[65]

If we examine this issue of impact from a public health perspective, it is pertinent to note that, according to public health reports, automobile accidents are the third-leading cause of death and injury in the United States.[66] It has been reported that sleepiness is a risk factor for automobile accidents.[67-71]

A study of the hazards of sleep-related motor vehicle accidents showed that the proportion of individuals with sleep-related accidents was 1.5–4 times greater in the hypersomnolent patient group than in the control group. Patients with sleep apnea and narcolepsy accounted for 71% of all sleep-related accidents. The proportion of patients with sleep-related accidents was highest in the narcolepsy group.[72] Motor vehicle crash drivers show significantly more driver sleepiness, slower reaction times, and a trend for greater objective sleepiness compared with matched controls.[73] A recent analysis of responses from 10,870 drivers showed that 23% of all respondents experienced ≥1 accident. Among respondents who reported ≥4 accidents, a strong association existed for the most recent accident to include injury (p < 0.0001). Sleep disorders were reported by 22.5% of all respondents, with a significantly higher prevalence (35%, p = 0.002) for drivers who had been involved in ≥3 accidents. Thus, sleepiness was strongly associated with a greater risk of automobile accidents.[74] Early diagnosis and education about the impact of sleepiness could reduce automobile accidents; however, most physicians receive little education in the diagnosis and management of sleep disorders.[75]

A comparison study with epilepsy showed that narcolepsy patients were more affected at work and had poorer driving records, higher accident rates from smoking, and greater problems in planning recreation. Epilepsy patients showed greater problems in educational achievement and maintaining a driver's license.[76] The negative impact of narcolepsy due to EDS and impaired alertness was reaffirmed in 1990.[77] In 1982, McMahon et al.[78] used the Human Service Scale to study need satisfaction in 114 persons with narcolepsy compared to 2406 disabled individuals in order to gain information for rehabilitation planning. This scale is designed to measure physiological (health), emotional, family, social needs, economic security, economic self-esteem, and vocational self-actualization. A high proportion of those who had narcolepsy scored below the average of the disabled group in needs satisfaction: social needs (35%), physiological needs (25%), emotional needs (23%), and family needs (21%). On the other hand, persons with narcolepsy had higher scores than the disabled group in vocational self-actualization, economic security, and economic self-esteem. The authors concluded that the feasibility of employment of persons with narcolepsy is high; however, the psychological aspects of narcolepsy require treatment. Thus, the wide-ranging ramifications of having narcolepsy are evident and devastating.

In addition to the social- or community-level influence of narcolepsy on health, many studies have examined the role of narcolepsy on psychological and mental health. Difficulty with concentration, memory and depression, high anxiety levels, and apathetic demeanor have been described by several investigators.[62,64,79,80] Sours[49] described problems of adjustment in overall functioning in narcolepsy. Krishnan et al.[81] reviewed the charts of 24 ambulatory male veterans with narcolepsy and found that these patients showed poor adjustment to illness, high unemployment rates, and disturbed family relationships. Sixteen of the patients had adjustment disorder, depression, alcohol dependence, and personality disorder. There were no psychotic diagnoses in the group. Although the sample is highly selected, the devastating impact of narcolepsy is obvious. However, psychosis can occur during treatment with amphetamines. The risk of psychosis appears to be dose related, although it is likely that some subjects are more susceptible to this complication than others.[82] Psychosis is usually predominately paranoid. Psychosis symptoms in narcolepsy may also appear during sodium oxybate treatment. Hallucinations resemble those seen in schizophrenia. However, the insight that the symptoms are delusional is usually preserved.[83]

In 1982, Kales et al.[84] found obsessive–compulsiveness, depression, and anxiety, a tendency to internalize and control feelings, feelings of frustration in emotional fulfillment and impaired interpersonal relationships in 50 patients with narcolepsy.

In 1986, Baker et al.[85] reported high scores of hypochondriasis, depression, and hypomania. It was suggested by the authors that these high scores were probably due to difficulties in coping with the symptoms of hypersomnolence. High scores on schizophrenia may have been related to symptoms of hypnagogic hallucinations and other auxiliary symptoms of narcolepsy.[86]

Another devastating effect of narcolepsy is the complaint of decreased libido or impotence or both among male patients with narcolepsy.[62,81,87] It is reported that impotence is a well-documented side effect of tricyclic antidepressants that are prescribed for cataplexy.[87] Depression in narcolepsy may be another factor accounting for sexual dysfunction. Cataplexy or sleep attacks during sex are likely to impair the sexual relationships of partners.

The problem of untoward side effects of medications poses a dilemma for many patients. Adverse drug reactions of stimulant medications may occur, especially if prescribed in high doses for prolonged periods of time for severe cases of narcolepsy. Amphetamines may cause irritability, mood changes, headaches, palpitations, sweating and tremors,[88] or paranoia or psychosis with visual hallucinations and paranoid delusions.[89-92] Dyskinesias may also result from prolonged use of amphetamines.[93] Thus, patients may cease treatment because of the negative effects of medications on the quality of their personal and social lives.[94] Cessation of treatment may in turn render the person totally incapacitated, being unable to work or to participate in meaningful social activities.

The frequency of alcohol abuse or dependence was reported in 1989.[95] Patients with disorders of excessive somnolence (DOES) abused alcohol more frequently than did patients with other sleep disorders or controls. A Self-Administered Alcohol Screening Test showed that 26.8% in the narcolepsy group scored greater than 7%, which indicates possible or probable alcoholism, compared to 13.2% of the entire DOES group or 5.4% of the controls. The cause for this dependence on alcohol is not known and may be correlated with personality type or frustrations in coping with the negative life effects of narcolepsy. Management must include careful screening, treatment, and support from appropriate professionals.

Despite the well-known and long-recognized impact of narcolepsy on psychosocial functioning and health in general,[96,97] only limited studies address HRQOL issues.[98] The first clinical trial in narcolepsy to include HRQOL was conducted to study the effects of modafinil (Provigil).[99] Data were collected in two similar 9-week double-blind studies. A total of 558 subjects from 38 centers were randomized into one of three groups: placebo, 200 mg modafinil, or 400 mg modafinil. Several validated instruments were used to measure the extent of sleepiness and alertness. A questionnaire comprised of the SF-36 and supplemental narcolepsy-specific scales was administered in order to assess quality of life changes with treatment. These instruments were pretested to narcolepsy patients in two sleep centers.[100]

Compared to the general population, subjects with narcolepsy were more affected in terms of vitality, social functioning, and ability to perform usual activities due to physical and emotional problems. People with narcolepsy experienced HRQOL effects that were as bad as or worse than those with Parkinson's disease and epilepsy in several HRQOL areas. HRQOL effects were worse among people with narcolepsy than among those with migraine headaches with one exception: bodily pain.

The 400 mg modafinil group showed improvement over placebo in 10 of the 17 HRQOL scales. When compared to the placebo group, the 400 mg modafinil group had fewer limitations in everyday activities due to physical or emotional problems, more energy and less interference with normal social activities due to health. In addition, the 400 mg modafinil group had fewer narcolepsy symptoms and higher attention/concentration, productivity, self-esteem, and overall health perceptions compared to placebo. The 200 mg modafinil group showed better results than placebo in the following 9 of the 17 HRQOL scales: physical functioning, role limitations due to physical problems, vitality, mental health summary scale, narcolepsy symptoms, attention/concentration, productivity, self-esteem, and driving limitations.

The profound effect of narcolepsy on children warrants special attention. Alertness levels and performance in young people are significantly compromised when sleepiness results in sleep attacks during task performance. Attention, memory, motor and cognitive skills are affected by sleepiness.[101,102] Due to the inability to wake up early, school-aged youth often miss the morning hours at school or their school performance may be inconsistent. Quite often, narcolepsy or other sleep disorders are not considered, and the child is viewed by teachers and school administrators as lazy or having a behavioral problem. Many children suffering from symptoms of excessive daytime sleepiness get placed in special education classes, are seen as hyperactive, and have conflicts with teachers. Socially, these youngsters suffer because they want to hide their illness from peers and teachers. The falls that can occur in cataplexy are especially embarrassing. These children benefit by meeting others with the same sleep disorder. Family members often do not understand the condition or its symptoms, and this results in interpersonal problems with the child. Prompt diagnosis and management will reduce the frustrations, self-doubt, and lack of self-confidence that are often experienced by those who are afflicted with this impairment.

NEEDS OF PERSONS WITH NARCOLEPSY

Although an estimated 250,000 people have narcolepsy in the United States, no study has attempted to assess their total health service needs and access to relevant services. Traditionally, the needs of patients are determined by medical professionals who focus on the physical aspects of a disorder and its medical management. This highly commendable service by physicians in the face of rising pressures from patients and third-party payment systems must be supported and respected. Today, however, complex arrays of factors operate that make it necessary to reexamine the traditional methods of patient care and their beneficial effects on the patient. It must be noted that with the increasing prevalence of chronic illness and an aging population, coupled with uncertain and unanticipated consequences of long-term use of medications, alternative modes of management have gained increasing significance. The proliferation of self-help groups and support groups across the country in the management of various diseases is testimony that, in addition to the medical management of disease, the nonmedical aspects of care are a needed component of treatment. The beneficial effects of social support in stressful life events have been examined by several authors[103–105]; its buffering effect has been demonstrated in pregnancy,[106] mental health,[107,108] and unemployment.[109,110]

In view of this demand by patients for supplemental/alternative care in addition to the medications prescribed by physicians, it was felt that a survey to assess the perceived needs of patients would yield useful information for program development and patient management. The Narcolepsy Project conducted such a survey in 1986 to assess the employment status and psychosocial support needs of persons with narcolepsy in New York State who had registered with the Project.[64] Of the 120 patients who received the questionnaire (80 were returned because of changes of address), 68 completed responses were received, eliciting a response rate of 57%. When respondents and nonrespondents were compared, no statistically significant differences were observed in age, sex, or marital status (p < 0.05). Therefore, age, sex, and marital status do not appear to bias the results.[10] The following

perceived needs were reported in descending order: information and referral center for narcolepsy (84%); group counseling (65%); assistance in making career transitions (50%); transportation (34%); homemaker services (25%); and homecare services (18%).

With regards to employment status, 16% were unemployed and seeking work—a rate higher than the U.S. average of 7.5%.[111]

In personal interviews, respondents expressed fear of traveling alone or in the dark, embarrassment about symptoms of daytime sleepiness, and automatic behavior and fear of being mistaken for a drug addict or being labeled as "lazy." Untreated patients manifested a low level of motivation and an apathetic demeanor. Problems with memory, a low level of energy, and pervasive feelings of tiredness or weakness were common complaints.[10]

Thus, the major perceived needs of persons with narcolepsy were psychosocial support services, assistance in making career transitions, and transportation services, indicating the need to develop programs tailored to these patients' expressed needs.

IS NARCOLEPSY A DISABILITY OR A DEVELOPMENTAL DISABILITY?

Let us now consider the definition of narcolepsy in terms of its functional status. What is the difference between a disability and impairment? According to the American Medical Association's *Guide to the Evaluation of Permanent Impairment* (page x of the Foreword), permanent medical impairment is related to the health status of the individual, whereas disability may be determined within the context of the personal, social, or occupational demands or the statutory or regulatory requirements that the individual is unable to meet as a result of the impairment.[112] Again, on p. 215, the Guide states "Impairment is a medical determination. It involves any anatomical or functional abnormality or any clinically significant behavior changes," whereas "disability" refers to "the functional, social, or vocational level of an individual that has been altered by an impairment."

Because narcolepsy does not cause physical disfigurement or a discernable handicap, many professionals, as well as patients, do not realize that the federal and state governments classify narcolepsy legally as a "disability."

The laws and regulations that protect people with disabilities, and specifically narcolepsy, are described in detail by Sundram and Johnson.[113] Here, we briefly present these laws. The Rehabilitation Act of 1973, S 701,[114] prohibits discrimination on the basis of disability under any program or activity receiving federal financial assistance and defines an "individual with handicaps" as a person who has

a. A physical or mental impairment that substantially limits one or more of the major life activities of such individuals;
b. A record of such impairments; or
c. Been regarded as having such an impairment.[115]

Narcolepsy, being a neurological disorder, falls within this definition. The Americans with Disabilities Act (ADA) of 1990 uses the same definition of disability as individuals with handicaps under the Rehabilitation Act, which includes narcolepsy.[113] The ADA protects individuals with disabilities and provides equal opportunities in employment, public accommodations, transportation, state and local government services, and telecommunication.[111] Narcolepsy is also considered a disability under the New York State Human Rights Law that bars discrimination against the disabled. Thus, individuals with narcolepsy are eligible for services through the state developmental disabilities programs. Most providers of care as well as consumers are unaware of the rights of individuals with a disability such as narcolepsy.

What is a developmental disability? An advisory from the New York State Office of Mental Retardation and Developmental Disabilities states that under Section 1.03(22) of the New York State Mental Hygiene Law, which is the legal base for eligibility determination, a developmental disability is defined as a disability of a person that:

1. Is attributable to mental retardation, cerebral palsy, epilepsy, neurological impairment, or autism;
2. Is attributable to any other condition of a person found to be closely related to mental retardation because such condition results in similar impairment of general intellectual functioning or adaptive behavior to that of mentally retarded persons or requires

treatment and services similar to those required for such persons; or
3. Is attributable to dyslexia resulting from a disability described in (1) or (2);
   a. Originates before such person attains age twenty-two;
   b. Has continued or can be expected to continue indefinitely; and
   c. Constitutes a substantial handicap to such person's ability to function normally in society.

Functional limitations constituting a substantial handicap are defined in this document as significant limitations in adaptive functioning that are determined from the findings of assessment by using a nationally normed and validated, comprehensive, and individual measure of adaptive behavior administered by a qualified practitioner. Some patients with narcolepsy qualify for services and benefits following this definition of a developmental disability.

IMPLICATIONS FOR PATIENTS, FAMILY, AND COMMUNITY

The consequences of narcolepsy on the life of the patient are severe and often pervasive. Because people with narcolepsy look normal, their tendency to fall asleep in socially unacceptable situations is perceived by others as lack of drive or interest. This causes relationship problems with family members, teachers, and employers. The individual, often unaware of the problem, strives unusually hard to meet the expectations of a highly success-oriented society. Unable to fulfill these expectations, the person experiences a considerable amount of role strain and role conflict. Most of the person's time and energy are expended in resolving these conflicts, with little time for rewarding social activities. Repeated failures in fulfilling role obligations lead to disrupted relationships, poor self-esteem, lack of confidence, and a sense of vulnerability to all kinds of stressful situations. Alienation (a sense of isolation) is often observed in personal interviews with patients.

Individuals with narcolepsy, it was noted earlier, also tend to manifest symptoms of depression and anxiety. Cataplectic attacks are avoided by keeping away from situations that precipitate such attacks, thus depriving the individual of many enriching activities. Anticipation of such attacks is likely to lead to high levels of anxiety. Depression may result from dysfunctional relationships and frustrations in achieving life goals.

In personal interviews, patients with narcolepsy report fear of being mistaken for drug addicts by others; therefore, they are afraid of revealing the nature of their illness to anyone, especially in the work environment. In severe and prolonged uncontrolled cases, these patients may find themselves without a family or a job and quite destitute and lonely.

At a global level, an undiagnosed case of narcolepsy is a loss to the community since most individuals with narcolepsy, when proper treatment is instituted, can function normally within the framework of the family and the work environment. Lack of awareness of narcolepsy's impact and management can result in ineffective utilization of human resources with a subsequent drain on the national economy.

Reviews of studies on quality of life in narcolepsy in the United States and other countries document the extensive negative effects of narcolepsy on the physical health, mental health, work, and social health of patients. Keeping in mind that there are differences in methodology in different studies, a pattern of reduced function in narcolepsy is evident. Emerging studies indicating the presence of medical and sleep comorbidities in narcolepsy present further challenges to its management. Furthermore, the developmental aspects of narcolepsy and its comorbidities pose another set of issues in the diagnosis and management of narcolepsy. Pharmacological management of symptoms is not sufficient, although it is necessary in most cases. Successful management of objective and observable clinical features may be affected by other complaints of narcolepsy patients, such as a low level of energy and fatigue, problems with memory, depressed mood, adjustment problems, and a perception of a lack of security in the environment. Medical treatment should be supplemented with behavior modification, lifestyle changes and psychosocial support and counseling. A person-centered approach in the management strategy would consider the person's perceived needs and developing goals with a choice of medical and psychosocial interventions. The challenge is to study the variables that are manipulable so that meaningful management strategies can be instituted.

In conclusion, a person-centered and family-centered comprehensive management approach must be designed in order to address narcolepsy, its disabling effects, and the roles of social stressors and social supports on the quality of life of the individual and the family. Following sound scientific principles in conjunction with humanistic care will maintain the dignity of the individual and enhance the quality of life of our patients.

REFERENCES

1. Oliver W. A relation of an extraordinary sleepy person, at Tinsbury, near Bath. *Philosophical Transactions* 1753; 24: 2177–2182.
2. Graves RJ. Observations on the nature and treatment of various diseases. *Dublin Q J Med* 1851; 11: 1–20.
3. Caffé M. Maladie du sommeil. *J Connais Med Pharmaceut* 1862; 29: 323.
4. Westphal C. Eigenthumliche mit Einschafen verbundene Anfalle. *Arch Psychiat Nervenkr* 1877; 7: 631–635.
5. Gélineau JB. De la narcolepsie. *Tessier, imprimerie Surgères (Charente-Inferieure)* 1881:63.
6. Bassetti C. Narcolepsy. *Curr Treat Options Neurol* 1999; 1: 291–298.
7. Bassetti C, Aldrich MS. Narcolepsy. *Neurol Clin* 1996; 14(3): 545–571.
8. Stores G. The protean manifestations of childhood narcolepsy and their misinterpretation. *Dev Med Child Neurol* 2006; 48: 307–310.
9. Yoss RE, Daly D. Criteria for the diagnosis of the narcoleptic syndrome. *Mayo Clin Proc* 1957; 3: 320–328.
10. Goswami M. The influence of clinical symptoms on quality of life in patients with narcolepsy. *Neurology* 1998; 50(2 Suppl 1): S31–S36.
11. Siegel JM. Narcolepsy: A key role for hypocretins (orexins). *Cell* 1999; 98: 409–412.
12. Guilleminault C, Brooks SN. Excessive daytime sleepiness: A challenge for the practicing neurologist. *Brain* 2001; 124: 1482–1491.
13. van den Pol AN. Narcolepsy: A neurodegenerative disease of the hypocretin system? *Neuron* 2000; 27(3): 415–418.

14. Roth T, Roehrs TA. Etiologies and sequelae of excessive daytime sleepiness. *Clin Ther* 1996; 18: 562–576.

15. Dauvilliers Y, Billiard M, Montplaisir J. Clinical aspects and pathophysiology of narcolepsy. *Clin Neurophysiol* 2003; 114: 2000–2017.

16. Vogel G. Studies in psychophysiology of dreams III. The dream of narcolepsy. *Arch Gen Psychiatry* 1960; 3: 421–428.

17. Rechtschaffen A, Wolpert EA, Dement WC, Mitchell SA, Fischer C. Nocturnal sleep in narcoleptics. *Electroencephalogr Clin Neurophysiol* 1963; 15: 599–609.

18. Takahashi H, Jimbo M. Polygraphic study of narcoleptic syndrome with special reference to hypnogogic hallucinations and cataplexy. *Folia Psychiatr Neurol Japonica* 1963; 7: 343–347.

19. Guilleminault C, Passouant P, Dement WC (Eds.). *Narcolepsy*. New York, NY: Spectrum, 1976, p. 689.

20. Dement WC, Rechtschaffen A. Narcolepsy: Polygraphic aspects, experimental and theoretical considerations. In: Gastaut H, Lugaresi E, Berti-Ceroni G, Coccagna G (Eds.). *The Abnormalities of Sleep in Man*. Bologna: Aulo Gaggi, 1968, pp. 147–164.

21. Broughton RJ. Narcolepsy [letter to the editor]. *Can Med Assoc J* 1974; 110(100): 7.

22. Montplaisir J, DeChamplain J, Young S, Missala K, Sourkes T. Narcolepsy and idiopathic hypersomnia: Biogenic amines and related compounds in CSF. *Neurology* 1982; 32: 1299–1302.

23. Aldrich MS. The neurobiology of narcolepsy-cataplexy syndrome. *Int J Neurol* 1991/1992; 25/26: 29–40.

24. Krahn LE, Black JL, Silber MH. Narcolepsy: New understanding of irresistible sleep. *Mayo Clin Proc* 2001; 76: 185–194.

25. Scammell TE. The neurobiology, diagnosis, and treatment of narcolepsy. *Ann Neurol* 2003; 53: 154–166.

26. Honda Y, Juji T, Matsuki K et al. *HLA-DR2* and *DW2* in narcolepsy and in other disorders of excessive somnolence without cataplexy. *Sleep* 1986; 9(1): 133–142.

27. Langdon N, Welsh KI, Dam WV, Vaughan RW, Parkes JD. Genetic markers in narcolepsy. *Lancet* 1984; 2: 1178–1180.

28. Billiard M, Seignalet J, Besset A, Cadilhac J. *HLA-DR2* and narcolepsy. *Sleep* 1986; 9(1): 149–152.

29. Pollack S, Peled R, Gideoni O, Lavie P. *HLA-DR2* in Israeli Jews with narcolepsy–cataplexy. *Isr J Med Sci* 1988; 24: 123–125.

30. Lock CB, So AK, Welsh KI, Parkes JD, Trowsdale J. MHC class II sequences of an *HLA-DR2* narcoleptic. *Immunogenetics* 1988; 27: 449–455.

31. Peled R, Lavie P. Narcolepsy–cataplexy: An extremely rare disorder in Israel. *Sleep Res* 1987; 16: 404–406.

32. Honda Y, Matsuki K. Genetic aspects of narcolepsy. In: Thorpy MJ (Ed.). *Handbook of Sleep Disorders*. New York, NY: Marcel Dekker, Inc., 1990, pp. 217–234.

33. Mignot E; Hayduk R; Black J et al. *HLA DQB1*0602* is associated with cataplexy in 509 narcoleptic patients. *Sleep* 1997; 20: 1012–1020.

34. Mignot E, Ling L, Rogers R et al. Complex *HLA-DR* and *-DQ* interactions confer risk of narcolepsy–cataplexy in three ethnic groups. *Am J Hum Genet* 2001; 68: 686–699.

35. Mignot E. Genetic and familial aspects of narcolepsy. *Neurology* 1998; 50(2 Suppl. 1): S16–S22.

36. Lin L, Hungs M, Mignot E. Narcolepsy and the *HLA* region. *J Neuroimmunol* 2001; 117: 9–20.

37. Black JL 3rd. Narcolepsy: A review of evidence for autoimmune diathesis. *Int Rev Psychiatry* 2005; 17: 461–469.

38. Overeem S, Verschuuren JJ, Fronczek R et al. Immunohistochemical screening for autoantibodies against lateral hypothalamic neurons in human narcolepsy. *J Neuroimmunol* 2006; 6: 6.

39. Nishino S, Ripley B, Overeem S, Lammers GJ, Mignot E. Hypocretin (orexin) deficiency in human narcolepsy. *Lancet* 2000; 355(9197): 39–40.

40. Peyron C, Faraco J, Rogers W et al. A mutation in a case of early onset narcolepsy and a generalized absence of hypocretin peptides in human narcoleptic brains. *Nat Med* 2000; 6: 991–997.

41. Thannical TC, Moore RY, Nienhuis R et al. Reduced number of hypocretin neurons

in human narcolepsy. *Neuron* 2000; 27: 469–474.

42. Brooks SN, Mignot E. Narcolepsy and idiopathic hypersomnolence. In: Lee-Chiong TL Jr, Sateia MJ, Carskadon MA (Eds). *Sleep Medicine*. Philadelphia, PA: Hanley and Belfus, Inc., 2002, pp. 193–202.

43. Dement WC, Zarcone V, Varner V et al. The prevalence of narcolepsy. *Sleep Res* 1972; 1: 148.

44. Dement WC, Carskadon M, Ley R. The prevalence of narcolepsy II. *Sleep Res* 1973; 2: 147.

45. Hublin C, Kapiro J, Partinen M et al. The prevalence of narcolepsy: An epidemiological study of the Finnish twin cohort. *Ann Neurol* 1994; 35: 709–716.

46. Silber MH, Krahn LE, Olson EJ et al. The epidemiology of narcolepsy in Olmsted County, Minnesota: A population based study. *Sleep* 2002; 25: 197–202.

47. Uryu N, Maeda M, Nagata Y et al. No difference in the nucleotide sequence of the DQB B1 domain between narcoleptic and healthy individuals with *DR2 DW2*. *Hum Immunol* 1989; 24: 175–181.

48. Ohayon MM, Priest RG, Zulley J, Smirne S, Paiva T. Prevalence of narcolepsy symptomatology and diagnosis in the European general population. *Neurology* 2002; 58(12): 1826–1833.

49. Sours JA. Narcolepsy and other disturbances in the sleep–waking rhythms. A study of 115 cases with review of the literature. *J Nerv Ment Dis* 1963;137:523–542.

50. Passouant P, Billiard M. The evolution of narcolepsy with age. In: Guilliminault C, Dement WC, Passouant P (Eds). *Narcolepsy*. New York, NY: Spectrum, 1976, pp. 179–197.

51. Roth B. *Narcolepsy and Hypersomnia*. Basel: S Karger, 1980.

52. Guilleminault C, Anders TF. Sleep disorders in children. *Adv Pediat* 1976; 22: 155–174.

53. Guilleminault C. Narcolepsy syndrome. In: Kryger MH, Roth TA, Dement WC (Eds). *Principles and Practice of Sleep Medicine*. Philadelphia, PA: WB Saunders, 1989, p. 338.

54. Thorpy MJ, Goswami M. Treatment of narcolepsy. In: Thorpy MJ (Ed.). *A Handbook of Sleep Disorders*. New York, NY: Marcel Dekker, 1990, pp. 235–258.

55. Houghton WC, Scammell TE, Thorpy M, Houghton WC. Pharmacotherapy for cataplexy. *Sleep Med Rev* 2004; 8(5): 355–366.

56. Lammers GJ, Arends J, Declerk AC et al. Gamma-hydroxybuterate and narcolepsy: A double-blind placebo-controlled study. *Sleep* 1993; 16: 216–220.

57. The U.S. Xyrem Multicenter Study Group. A randomized, double blind, placebo-controlled multicenter trial comparing the effects of three doses of orally administered sodium oxybate with placebo for the treatment of narcolepsy. *Sleep* 2002; 25(1): 42–49.

58. Mamelak M, Black J, Montplaisir J, Ristanovich R. A pilot study on the effects of sodium oxybate on sleep architecture and daytime alertness in narcolepsy. *Sleep* 2004; 27(7): 1327–1334.

59. Mitler MM, Hajdukovic R, Erman M, Koziol JA. Narcolepsy. *J Clin Neurophysiol* 1990; 7(1): 93–118.

60. Teixeira VG, Faccenda JF, Douglas NJ. Functional status in patients with narcolepsy. *Sleep Med* 2004; 5(5): 477–483.

61. Broughton R, Ghanem Q. The impact of compound narcolepsy on the life of the patient. In: Guilleminault C, Dement WC, Passouant P (Eds). *Narcolepsy*. New York, NY: Spectrum, 1976, pp. 201–219.

62. Broughton R, Ghanem Q, Hishikawa Y, Sugita Y, Nevsimalova S, Roth B. Life effects of narcolepsy in 180 patients from North America, Asia and Europe compared to matched controls. *Can J Neurol Sci* 1981; 8(4): 299–304.

63. Broughton R, Valley V, Aguirre M, Roberts J, Suwalski W, Dunham W. Excessive daytime sleepiness and the pathophysiology of narcolepsy–cataplexy: A laboratory perspective. *Sleep* 1986; 9(1): 205–215.

64. Goswami M, Glovinsky P, Thorpy MJ. Needs assessment and sociodemographic characteristics in narcolepsy. 5th *International Congress of Sleep Research*, Abstract No. 805.Copenhagen, Denmark. June 1987.

65. Dodel R, Peter H, Walbert T et al. The socioeconomic impact of narcolepsy. *Sleep* 2004; 27(6): 1123–1128.

66. *US Bureau of the Census Statistical Abstract of the United States: 1990*. Washington, DC: U.S. Department of Commerce, 1990, Vol. 79, pp. 606.

67. Findley LJ, Fabrizio M, Thommi G, Surratt PM. Severity of sleep apnea and automobile crashes. *N Engl J Med* 1989; 320: 868–869.

68. Mitler M, Carskadon M, Czeisler C, Dement WC, Dinges D, Graeber R. Catastrophies, sleep and public policy: Consensus report. *Sleep* 1988; 11: 100–109.

69. *National Commission on Sleep Disorders Research report. Vol 1. Executive summary and executive report*. Bethesda, MD: National Institutes of Health, 1993.

70. Horne J, Reyner L. Vehicle accidents related to sleep: A review. *Occup Environ Med* 1999; 56(5): 289–294.

71. Maclean AW, Davies DR, Thiele K. The hazards and prevention of driving while sleepy. *Sleep Med Rev* 2003; 7(6): 507–521.

72. Aldrich MS. Automobile accidents in patients with sleep disorders. *Sleep* 1989; 12(6): 487–494.

73. Kingshott RN, Jones DR, Smith AD, Taylor DR. The role of sleep-disordered breathing, daytime sleepiness, and impaired performance in motor vehicle crashes—A case control study. *Sleep Breath* 2004; 8: 61–72.

74. Powell NB, Schechtman KB, Riley TW, Li K, Guilleminault C. Sleepy driving: Accidents and injury. *Otolaryngol Head Neck Surg* 2002; 126(3): 217–227.

75. Rosen RC, Rosekind M, Rosevear C, Cole WE, Dement WC. Physician education in sleep and sleep disorders. *Sleep* 1993; 16: 249–254.

76. Broughton R, Guberman A, Roberts J. Comparison of the psychosocial effects of epilepsy and of narcolepsy/cataplexy: A controlled study. *Epilepsia* 1984; 25: 423–433.

77. Rosenthal L, Merlotti L, Young DK et al. Subjective and polysomnographic characteristics of patients diagnosed with narcolepsy. *Gen Hosp Psychiatry* 1990; 12: 191–197.

78. McMahon ST, Walsh JK. Sexton K, Smitson SA. Need satisfaction in narcolepsy. *Rehabil Lit* 1982; 43(3–4): 82–85.

79. Ganado W. The narcolepsy syndrome. *Neurology* 1958; 8(6): 487–496.

80. Roth TA, Merlotti L. Advances in the diagnosis of narcolepsy. *3rd International Symposium on Narcolepsy*. San Diego, CA. June, 1988.

81. Krishnan RR, Volow MR, Miller PP, Carwile ST. Narcolepsy: Preliminary retrospective study of psychiatric and psychosocial aspects. *Am J Psychiatry* 1984; 141: 428–431.

82. Vourdas A, Shneerson JM, Gregory CA et al. Narcolepsy and psychopathology: Is there an association? *Sleep Med* 2002; 3: 353–360.

83. Sarkanen, T., Nirmela V., Landtblom A-M, Partinen M. Psychosis in patients with narcolepsy as an adverse effect of sodium oxybate. *Front Neurol* 2014; 5; 126.

84. Kales A, Soldatos CR, Bixler EO et al. Narcolepsy–cataplexy II. Psychosocial consequences and associated psychopathology. *Arch Neurol* 1982; 139: 169–171.

85. Baker TL, Guilleminault C, Nino-Murcia G, Dement WC. Comparative polysomnographic study of narcolepsy and idiopathic central nervous system hypersomnia. *Sleep* 1986; 9(1): 232–242.

86. Kishi Y, Konishi S, Koizumi S, Kudo Y, Kurosawa H, Kathol RG. Schizophrenia and narcolepsy: A review with a case report. *Psychiatry Clin Neurosci* 2004; 58: 117–124.

87. Karacan I. Erectile dysfunction in narcoleptic patients. *Sleep* 1986; 9: 227–231.

88. Parkes JD. Amphetamines and other drugs in the treatment of daytime drowsiness and cataplexy. In: Parkes JD (Ed.). *Sleep and its Disorders*. London: WB Saunders, 1985, pp. 459–482.

89. Zarcone V. Narcolepsy. *N Engl J Med* 1973; 288, 1156–1166.

90. Pfefferbaum A, Berger P. Narcolepsy, paranoid psychosis, and tardive dyskinesia: A pharmacological dilemma. *J Nerv Ment Dis* 1977; 164: 293–297.

91. Schrader G, Hicks EP. Narcolepsy, paranoid psychosis, major depression, and tardive dyskinesia. *J Nerv Ment Dis* 1984; 172: 439–441.

92. Young D, Scoville WB. Paranoid psychosis in narcolepsy and possible danger of benzedrine treatment. *Med Clin North Am* 1938; 22: 637–646.

93. Mattson RH, Calverley JR. Dextroampheta-mine-sulfate-induced dyskinesias. *JAMA* 1968; 204(5): 400–402.

94. Gillin JC, Horowitz D, Wyatt RJ. Pharmacologic studies of narcolepsy involving serotonin, acetylcholine and monoamine oxidase. In: Guilleminault C, Dement WC, Passouant P (Eds). *Narcolepsy*. New York, NY: Spectrum, 1976, pp. 585–604.

95. Fredrickson PA, Kaplan J, Richardson J, Esther M. Prevalence of alcohol abuse in narcolepsy and other disorders of excessive somnolence. *Arch Found Thanatol* 1989; 16(1): 12.

96. Broughton WA, Broughton RJ. Psychosocial impact of narcolepsy. *Sleep* 1994; 17: S45–S49.

97. Bruck D. The impact of narcolepsy on psychosocial health and role behaviours: Negative effects and comparisons with other illness groups. *Sleep Med* 2001; 2: 437–446.

98. Reimer MA, Flemons WW. Quality of life in sleep disorders. *Sleep Med Rev* 2003; 7: 335–349.

99. Beusterien KM, Rogers AE, Walsleben JA et al. Health-related quality of life effects of modafinil for treatment of narcolepsy. *Sleep* 1999; 22(6): 757–765.

100. Stoddard RB, Goswami M, Ingalls KK. The development and validation of an instrument to evaluate quality of life in narcolepsy patients. *Drug Information Journal* 1996; 30(6): 850.

101. Kashden J, Wise M, Avarado I et al. Neurocognitive functioning in children with narcolepsy. *Sleep Res* 1996; 25: 262.

102. Nevšímalová S. Narcolepsy in children. *Sleep Res Online* 1999; 2(Suppl 1): 410.

103. Cohen S, Wills TA. Stress, social support and the buffering hypothesis. *Psychol Bull* 1985; 98: 310–357.

104. Leavy RL. Social support and psychological disorder: A review. *J Community Psychol* 1983; 11: 3–21.

105. Gottlieb BH. *Social Networks and Social Support*. Beverly Hills, CA: Sage Publications, 1981.

106. Nuckolls KG, Gassell J, Kaplan BH. Psychosocial assets, life crises and the prognosis of pregnancy. *Am J Epidemiol* 1972; 95: 431–441.

107. Wilcox BL. Social support, life stress and psychological adjustment: A testing of the buffering hypothesis. *Am J Community Psychol* 1981; 9: 371–387.

108. Lin N, Woefel MW, Light SC. The buffering effect of social support subsequent to an important life event. *J Health Soc Behav* 1985; 26: 247–263.

109. Gore SL. The effects of social support in moderating the health consequences of unemployment. *J Health Soc Behav* 1978; 19: 157–165.

110. Pearlin LI, Lieberman M, Menaghan E, Mullen J. The stress process. *J Health Soc Behav* 1981; 22: 337–357.

111. U.S. Bureau of the Census. Statistical abstract of the US 1986. In: *Characteristics of the Civilian Labor Force, by State: 1984* (106th ed.). Washington, DC, 1985, Vol. 662, p. 393.

112. American Medical Association. *Guide to the Evaluation of Permanent Impairment* (2nd ed.). Chicago, IL: American Medical Association, 1984.

113. Sundram CJ, Johnson PW. The legal aspects of narcolepsy. In: Goswami M, Pollak PC, Cohen FL, Thorpy MJ, Kavey NB, Kutscher AH (Eds). *Psychosocial Aspects of Narcolepsy*. New York, London: Haworth Press, 1992, pp. 175–192.

114. The Rehabilitation Act of 1973, 29, U.S.C. § 701 et seq.

115. The Rehabilitation Act of 1973, 29, U.S.C. § 706 (7) (b).

8

Sleep and headache disorders

JEANETTA C. RAINS, J. STEVEN POCETA, AND DONALD B. PENZIEN

INTRODUCTION

The relationship between sleep and headache was recognized well over a century ago in early medical texts and journals.[1,2] Historically, evidence of a sleep–headache relationship comprised anecdotal observations and small clinical studies, which led to inconsistencies in outcome and interpretation. The last quarter century has yielded growth and maturation of the scientific literature, with an increased number of controlled studies. Although reviewers agree that the specific nature, magnitude, and physiological mechanisms of the relationship are not well defined, hypothetical associations have been proposed to account for the relationships between sleep and headache. Knowledge of current research and theory may facilitate management of primary headache, especially migraine, cluster, and tension-type headache (TTH), as well as the identification of secondary headache (i.e., sleep apnea headache [SAH]). This chapter describes the classification of sleep-related headache and the models and evidence for the reciprocal association of headache and sleep processes, and presents a clinical algorithm to guide sleep evaluation and management, a key component of head pain management for a substantial portion of headache sufferers.

DIAGNOSIS AND CLASSIFICATION

Classification of headaches follows the *International Classification of Headache Disorders—3rd Edition, Beta Version* (ICHD-3).[3] The subset of headaches comorbid with sleep disorders have also been addressed in a general manner in the most recent revision of the *International Classification of Sleep Disorders—3rd Edition* (ICSD-3).[4] The ICSD-3 recently added a five-page appendix that describes the diagnosis and clinical course under a section entitled "Sleep-related medical and neurological disorders." Finally, sleep-related headaches may be diagnosed from the *Diagnostic and Statistical Manual of Mental Disorders—5th Edition* (DSM-5).[5] The DSM-5 includes diagnoses for 10 major sleep–wake disorders consistent with ICSD-3. Psychiatric disorders are comorbid with both sleep disorders and headache and the symptom constellation of sleep–headache–mood disorder is a common and clinically challenging case. Thus, familiarity with the DSM-5 may facilitate a differential diagnosis for cases in which psychological symptoms, distress, and impairment are prominent.

Primary headaches linked to sleep

HYPNIC HEADACHE

Hypnic headache is the sole example of a pure sleep-related headache (Table 8.1). Headaches tend to occur in the mid to latter portion of the sleep period and patients are abruptly awakened with pain (a.k.a. "alarm clock" headache due to the tendency to occur at approximately the same time each night). A review pooled data from all published reports of hypnic headache between 1988 and 2014 (250 cases).[6] Hypnic headache is relatively rare and accounts for less than 0.1% of all headaches and only 1.4% of geriatric headaches. The majority of cases were female (65%). The mean age of headache onset was 61 years, although younger patients were also found. Five cases of hypnic headache have been reported in children.

Table 8.1 *International Classification of Headache Disorders—3rd Edition, Beta Version*[3] diagnostic criteria for hypnic and sleep apnea headache

Hypnic headache
  A. Recurrent headache attacks fulfilling criteria B–E
  B. Developing only during sleep and causing wakening
  C. Occurring on ≥10 days per month for >3 months
  D. Lasting ≥15 minutes and for up to 4 hours after waking
  E. No cranial autonomic symptoms or restlessness
  F. Not better accounted for by another ICHD-3 diagnosis

Sleep apnea headache
  A. Headache present on awakening after sleep and fulfilling criterion C
  B. Sleep apnea (Apnea–Hypopnea Index ≥5) has been diagnosed
  C. Evidence of causation demonstrated by at least two of the following:
    1. Headache has developed in temporal relation to the onset of sleep apnea
    2. Either or both of the following:
      a. Headache has worsened in parallel with worsening of sleep apnea
      b. Headache has significantly improved or remitted in parallel with improvement in or resolution of sleep apnea
    3. Headache has at least one of the following three characteristics:
      a. Recurs on >15 days per month
      b. All of the following:
        i. Bilateral location
        ii. Pressing quality
        iii. Not accompanied by nausea, photophobia, or phonophobia
      c. Resolves within 4 hours
  D. Not better accounted for by another ICHD-3 diagnosis

The average attack duration was 162 ± 74 minutes, with a frequency of 21 ± 10 days per month, 68% were bilateral, 94% were moderate or severe, and a third of cases also had migraine. Two of the studies had also assessed sleep apnea and identified a high incidence of obstructive sleep apnea (73% and 83%). A few studies have reported an association between hypnic headaches and sleep disorders such as decreased sleep efficiency, restless legs, snoring, and sleep apnea.[7-10]

CLUSTER HEADACHE

Cluster headache is characterized by a circadian pattern, and sufferers have been found to be at substantial risk for sleep apnea. Circadian patterns were reported, with 75% of attacks presenting from 9:00 p.m. to 10:00 a.m. Headaches last for weeks or months, followed by remissions that could last for several months to years.[3,11,12] Prevalence is less than 1% of the population, and the majority of sufferers are male.[3] Attacks are accompanied by cranial autonomic features (e.g., ipsilateral conjunctival injection, lacrimation, nasal congestion, rhinorrhea, forehead and facial sweating, miosis, ptosis, or eyelid edema) (Table 8.1).

Cluster headache patients are at risk for sleep apnea. A study of 37 cluster headache patients who underwent polysomnography identified a fourfold increase in the incidence of sleep apnea relative to age- and gender-matched controls (58% versus 14%, respectively), and this risk increased to 24-fold in overweight and obese patients (body mass index [BMI] ≥ 25 kg/m^2).[11] Another report found sleep apnea in 80% of cluster patients (n = 25/31) who underwent polysomnography.[12] Management is described below, but the diagnosis of cluster headache should raise suspicion for sleep apnea.

CHRONIC PAROXYSMAL HEMICRANIA

Chronic paroxysmal hemicrania (CPH) is a variant of cluster headache, except with female rather than male preponderance and higher-frequency and shorter-duration attacks.[3] Headache symptoms closely resemble cluster, with severe and unilateral orbital, supraorbital, or temporal pain associated with one or more cranial autonomic features. Conversely, attacks usually last only 2–30 minutes, but may recur often, with at least five attacks per day. Attacks are more likely to occur at night, similarly to cluster.[13] In the 1980s, polysomnographic studies found that attacks often occurred in rapid eye movement (REM) sleep, and thus the term "REM-locked" headache disorder became synonymous with CPH.[14,15]

MIGRAINE HEADACHE

Migraine is the most common neurological disorder, occurring in 12% of the population worldwide with a 3:1 female:male preponderance.[3] Headaches are often unilateral, pulsatile, and associated with nausea and/or vomiting. One in five migraineurs experiences visual or other sensory auras prior to the onset of headache. Chronobiological patterns and sleep-related triggers (presented below) are well evidenced in migraine. Prospective longitudinal studies showed migraine attacks occur following a monophasic sinusoidal 24-hour cycle, peaking during morning[16] or midday[17] (Figure 8.1).

Chronic migraine

Sleep has been implicated in the progression of episodic to chronic migraine. The diagnosis of chronic migraine is employed when headache frequency equals or exceeds 15 days each month, and is a common diagnosis presenting to neurology and multidisciplinary headache clinics and a leading cause of disability.[18-20] Often, chronic migraine presents as daily (often unremitting) or near-daily headache. A history of transformation from episodic to chronic migraine over time is a frequent clinical scenario (a.k.a. transformed migraine or chronic daily headache). Progression or *chronification* of migraine has been well evidenced in the literature, and snoring and sleep disturbance have been identified as being among the risk factors for progression (as well as medication overuse, psychiatric comorbidity, stress, obesity, and others).[21,22]

TENSION-TYPE HEADACHE

TTH is by far the most common headache, distinguished from migraine by its non-pulsatile, pressing, often bilateral pain, which is mild to moderate and less often severe.[3] Unlike migraine, tension is not usually exacerbated by physical activity and vomiting, although mild nausea, photophobia, or phonophobia may occur. Headache is subclassified by the presence or absence of pericranial muscle tenderness. Headaches are subclassified by frequency as episodic or chronic (15 or more headache days per month or >180 days per year) and may also be unremitting. The 1-year prevalence of episodic forms is 42% of adults, and the prevalence

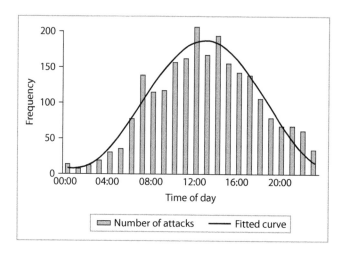

Figure 8.1 Chronobiological pattern in migraine: 24-hour temporal distribution of headache attacks. A 24-hour temporal distribution of migraine, plotting the time of headache onset from 2302/2314 headache attacks (12 cases missing data for time of onset) and showing peak onset after noon. (Reproduced with permission from Alstadhaug K, Salvesen R. *Headache* 2007; 48: 95–100.)

of chronic forms is 2%–3% of adults.[20] Chronic TTH is the most common headache symptom pattern to present secondary to sleep apnea and other sleep-related breathing disorders,[23] and as described below, sleep is a recognized headache trigger.

Secondary headache

SLEEP APNEA HEADACHE

SAH is the only formal diagnosis of headache *secondary* to a sleep disorder, coded within "Headache attributed to hypoxia or hypercapnia," although its mechanisms and criteria have not been validated (see Table 8.1).[3] The headache is defined by proximity to sleep in the context of a confirmed sleep apnea diagnosis and is often called "awakening headache." Headache is present in the morning upon awakening from sleep. Pain location, quality and associated features often resemble tension headache, although headache is often distinct from tension and migraine in its high-frequency, short-duration attacks. Alberti et al.[24] demonstrated that SAH presents commonly as bilateral (53%) versus unilateral (47%); frontal (33%), frontotemporal, (28%) or temporal in location (16%); usually with pressing/tightening pain (79%); and of mild (47%) or moderate intensity (37%), but can be severe (16%). SAH may present as a new-onset headache or otherwise as an exacerbation of primary headache,

especially those described above. Effective treatment of apnea may resolve headache or revert it to an earlier form of primary headache.

Mechanisms of SAH

Mechanisms of SAH remain a subject of debate. SAH is classified as a secondary condition attributable to a disorder of homeostasis.[3] However, the authors of the diagnostic criteria note that it is unknown whether headache is a function of "hypoxia, hypercapnia, or disturbance in sleep." A comprehensive literature review concluded that it was unclear whether the mechanisms underlying SAH were hypoxemia or hypercapnia, or instead another non-respiratory consequence of sleep apnea (i.e., sleep disturbance, autonomic arousal, cervical/cranial muscle tension, or intracranial cerebrospinal fluid pressure changes).[25] Objective polysomnographic evidence supporting metabolic mechanisms of SAH was published by Loh et al.,[26] who observed a dose–response relationship between apnea severity (e.g., apneic events and oxygen desaturation) and severity of morning headache.

Persuasive recently published evidence found nocturnal intermittent hypoxemia to be an independent risk factor for headache, as well as other pain complaints.[27] Data from the largest longitudinal cohort study to date examining sleep—the Cleveland Family Study—compared spontaneous pain reports with recurrent nocturnal hypoxemia versus sleep. The authors concluded that nocturnal hypoxemia increased the risk for both headache and chest pain, independent of sleep disturbance; the association remained significant after controlling for systemic inflammation. When comparing the lowest quartile of oxygen saturation (saturation less than 75%) with the highest quartile (saturation greater than 92%), odds ratios (ORs) doubled for self-reported morning headache (OR = 2.1 [CI: 1.2–3.7]), headache disrupting sleep (OR = 1.8 [CI: 1.3–2.7]), and chest pain in bed (OR = 2.2 [CI: 1.3–3.6]).

Although the evidence regarding hypoxemia is compelling, significant experimental evidence has been published that shows that hypoxemia and respiratory measures did not account for SAH.[28-31] For example, Chen et al.[32] found headache to be more common in apneics (27%) compared to primary snorers (16%) among 268 consecutive snorers undergoing sleep studies, but headache was not accounted for by oxygen desaturation or apnea severity. Furthermore, the OR (adjusted for gender, age, BMI, and smoking) for sleep apnea (adjusted OR = 2.6) was actually lower than for other predictors, including migraine diagnosis (6.5), insomnia (4.2), and psychological distress (3.9). Finally, evidence supporting multiple mechanisms was published by Alberti et al.,[24] who compared patients with sleep apnea versus insomnia and found that approximately half of each group exhibited awakening headache, although the likelihood was greater with apnea, especially for more severe sleep apnea.

Primary sleep disorders

EXPLODING HEAD SYNDROME

The parasomnia exploding head syndrome (a.k.a. sensory sleep starts or sensory sleep shocks) can present as sleep-related headache. Although not associated with pain, exploding head syndrome may present to either the headache or sleep specialist. Exploding head syndrome is a parasomnia (Table 8.2) rather than a headache disorder, and is classified by the ICSD-3.[4] The condition is not included in the ICHD-3 because of the absence of pain, although many papers and reviews include exploding head syndrome among papers on unusual and short-lived headaches.

The syndrome is considered to be a rare and benign disorder of the wake-to-sleep or

Table 8.2 *International Classification of Sleep Disorders—3rd Edition*[4] diagnostic criteria for exploding head syndrome

Diagnostic criteria (A–C must be met)

A. There is a complaint of a sudden loud noise or sense of explosion in the head either at the wake–sleep transition or upon waking during the night.
B. The individual experiences abrupt arousal following the event, often with a sense of fright.
C. The experience is not associated with significant complaints of pain.

sleep-to-wake transition. The condition appears to be a sensory variant of the better-known transient motor phenomenon of sleep starts or hypnic jerks occurring at wake–sleep transition. The neurophysiologic mechanisms underlying these hypnagogic phenomena are unknown. There are few studies on exploding head syndrome.[33] In one study of nine patients with a history of the syndrome using polysomnographic recordings, five patients reported the sensation of "explosions" during the recording period, and in each case, the electroencephalogram (EEG) demonstrated that the patients were awake and relaxed. Two attacks were characterized by EEG arousals, while no EEG changes were observed in the remaining three. No epileptiform activity was recorded in any case.[34]

COMORBIDITY OF HEADACHE AND SLEEP DISORDERS

Population studies

Observational studies document relationships between chronic headache and sleep disorders, particularly insomnia. Using data collected by the Nord-Trondelag Health Study (HUNT-3), a cross-sectional analysis of 297 individuals found severe sleep disturbance was threefold more likely among those with TTH (n = 135) and fivefold more likely among those with migraine (n = 51) than those without headache.[35] Those with chronic headache were 17-fold more likely to have severe sleep disturbance.[36]

Boardman et al.[37] conducted a cross-sectional study within the United Kingdom and identified a relationship between headache severity and sleep (i.e., trouble falling asleep, waking up several times, trouble staying asleep, or waking after usual amount of sleep feeling tired or worn out). Among 2662 respondents, headache frequency was associated with slight (age and gender adjusted OR = 2.4 [1.7–3.2]), moderate (OR = 3.6 [2.6–5.0]) and severe (OR = 7.5 [4.2–13.4]) sleep complaints. Likewise, a European study using 18,980 telephone interviews found that "chronic morning headache" was associated with increased rates of insomnia (OR = 2.14 [1.79–2.57]).[38]

In a cross-sectional study of 1000 Copenhagen residents, Rasmussen[39] found that "sleep problems" were more common in TTH than migraine or the general population; waking non-refreshed was more common among women with TTH and with migraineurs than in the general population. The 12-year follow-up survey of the Lyngberg et al.[40] cohort determined sleep problems to be associated with a poorer headache outcome (i.e., at least 180 headache days/year at follow-up due to increased frequency from episodic to chronic or unremitting) in those individuals with sleep complaints (OR = 2.7 [1.1–6.3]) for TTH but not migraine. Poorer prognosis for TTH was predicted by fewer hours of sleep (OR = 1.4 [1.1–2.0]), waking unrefreshed (OR = 2.0 [1.1–3.7]), and fatigue (OR = 2.5 [1.3–4.6]).[41]

Clinical studies

Sleep disorders are increased in migraine and TTH. Sancisi et al.[42] matched chronic headache patients from a neurology service with episodic headache controls. The sample of 105 chronic headache sufferers included 50 probable migraine with medication overuse, 30 probable tension headache with medication overuse, 4 chronic migraine, and 21 chronic tension headache; patients with chronic headache were at substantially greater risk of insomnia every day than episodic headache controls (54% vs. 24%; OR = 2.71 [1.2–6.4]), sleep apnea and/or snoring (49% vs. 37%), daytime sleepiness (49% vs. 24%), and anxiety and/or depression (43% vs. 27%; OR = 2.11 [1.2–3.8]). Risk of insomnia was especially high; 68% of the chronic headache sample reported insomnia, including 23% reporting daily use of hypnotics (vs. 10% of the episodic headache sample). Further multivariate analysis showed that the presence of insomnia was independently associated with chronic headache (OR = 5.01 [2.3–10.9]).

Other variables independently associated with chronic headache were lower education, lower age of headache onset, and antidepressant therapy. While antidepressant use is probably a marker for more severe psychiatric comorbidity, they are also commonly employed in headache prophylaxis.

Prospective population studies

Insomnia has been shown to be a risk factor for new-onset headache. Longitudinal research has identified insomnia as a risk factor for TTH. The HUNT was one of the largest longitudinal population health studies in history, and headache was included in the collection of the health histories of 75,000 individuals in Norway, beginning in 1984–1986 (HUNT-1) and assessed again at regular intervals. Insomnia (HUNT-2, 1995–1997) was found to predict new-onset headache 11 years later (HUNT-3, 2006–2008).[36] Of 15,268 headache-free individuals in HUNT-2, a total of 2323 individuals developed new-onset headache in HUNT-3, most of which was TTH (n = 1299 vs. n = 388 for migraine, n = 251 for probable migraine and n = 370 for unclassifiable headache). After controlling for gender, age, and sleep medication, insomnia was associated with increased risk during the 11-year interval for TTH (Relative Risk [RR] = 1.4 [1.2–1.8]) and both migraine (overall RR = 1.4 [1.0–1.9]) and unclassifiable headache (RR = 1.4 [1.0–2.0]). For those with insomnia who reported impairment in ability to work, there was a 60% increased risk (RR = 1.6 [1.3–2.1]) for developing headache compared to those with insomnia but without work impairment.

The pattern of progression or transformation from episodic migraine and TTH into chronic headache subforms (i.e., "chronification") has been documented.[43] Risk factors for this phenomenon include medication overuse, psychiatric comorbidity, stress, obesity, other pain and sleep variables (habitual snoring and sleep disturbance).[44] Regulation of sleep and early detection of sleep disorders could eventually be used as tools of primary or secondary prevention in order to halt progression, although research is needed to explore this possibility. At this time, however, there are few drawbacks in headache patients pursuing efforts to regulate sleep, as these strategies are highly compatible with and easily integrated into usual headache care.[45]

Sleep dysregulation triggers headache

Sleep dysregulation (e.g., sleep disturbance, sleep loss, or oversleeping) is one of the most frequent acute "headache triggers" for migraine and TTH. Stress, menstruation, and fasting are other commonly cited triggers.[46,47] Both short or long sleep patterns have also been associated with chronic migraine. The Kelman and Rains study[48] described the association of sleep duration and migraine severity in 1283 patients; 398 short sleepers (sleeping less than 6 hours per night) exhibited greater headache frequency than 573 normal sleepers, who slept 6–8 hours per night (Table 8.3). Headache frequency for a small group of 73 long sleepers (sleeping greater than 8 hours per night) was 18 headache days per month, which was similar to short sleepers, although the trend was not statistically significant.

PROSPECTIVE TIME-SERIES ANALYSIS

Sleep triggers for migraine and TTH were confirmed prospectively using time-series analysis. Houle et al.[49] examined relationships between daily stress, sleep duration, and headache severity (recorded four times daily) over 28 days of self-monitoring. The relationship between sleep and headache intensity was nonlinear, with both extremes of the sleep period distribution (short = <6 hours; long = >8.5 hours) associated with increased headache. The lowest headache severity was associated with traditional sleep periods lasting 7–8 hours. Interestingly, there was an interaction effect with stress that suggested that sleep might be a moderator in the stress/headache relationship; headache was most severe when high stress occurred in conjunction with low sleep duration over a 2-day period. Data suggest consistent sleep durations of 7–8 hours may have a stress-buffering effect, supporting the regulation of sleep in headache management, especially during periods of stress.

Kikuchi et al.[50] compared sleep over a 7-day interval using autography, and headache intensity was recorded in electronic diaries at 6-hour intervals (four times daily) in 27 tension headache patients (74% chronic), the majority of whom were female. Oversleeping was associated with headache, although insufficient sleep, which earlier studies identified as a frequent headache trigger, was not associated with subsequent headache. Potentially, a longer assessment period, such as the 28-day interval used by Houle et al.,[49] would have increased the power to identify an effect of sleep loss.

MODELS OF POTENTIAL MECHANISMS FOR SLEEP-RELATED HEADACHE

Shared neuroanatomy

Several recent papers have discussed the potential underlying mechanisms of sleep-related headache.[51-54] The convergence of sleep and headache disorders is generally believed to have its basis in neuroanatomical connections and neurophysiological mechanisms, particularly involving the hypothalamus, serotonin, and melatonin. Wakefulness depends principally on the functioning of the reticular activating system in the brainstem, which is maintained by influences of cortical neurotransmitters such as norepinephrine, dopamine, and acetylcholine. Non-REM sleep is primarily controlled by influences from the basal forebrain, with the functions of non-REM sleep

Table 8.3 Increased headache frequency with short and long sleep duration

Headache days per month	Average sleep duration		
	(n = 398) Short sleep (<6 hours/night)	(n = 573) Normal sleep (6–8 hours/night)	(n = 73) Long sleep (>8 hours/night)
All headache days	17.6[a]	15.1[a]	17.5
Severe headache days	7.3[a]	5.9[a]	6.6
Headache-free days	8.7[a]	11.0[a]	9.3

Source: Data from Kelman L, Rains JC. *Headache* 2005; 45(7): 904–910.
[a] p < 0.001; n = 1044.

maintained by gamma-amino-butyric acid from basal forebrain neurons. REM sleep-generating processes have been localized within the dorsolateral pontine tegmentum. REM sleep is initiated by the release of acetylcholine, which activates pontine neurons. Serotonin is abundant in the dorsal raphe nuclei and has a well-established but incompletely delineated role in acute migraine. The trigeminal nucleus caudalis in the pons and midbrain has been considered to be a potential "migraine generator" by some researchers, since there appears to be activation of the vascular structures that are supplied by this nucleus during migraine attacks. However, many of the migraine symptoms, especially those associated with prodrome and aura, are more likely to be the result of hypothalamic or cerebral cortical activity and include clinical features such as yawning, hunger, cravings, fatigue, mood changes, and sensory and visual distortions. The hypothalamus, which is the location of the suprachiasmatic nuclei, has extensive connections, some of which include connections to the limbic system, pineal gland (a source of neuronal melatonin), and brainstem nuclei involved in autonomic efferent control (nucleus tractus solitarius), sleep stage and motor control (locus ceruleus), and pain modulation (periaqueductal gray matter). Melatonin is well established as a factor in circadian rhythmicity and might have therapeutic efficacy in cluster headache. Further study of headache syndromes that exhibit chronobiological patterns, such as cluster headache, have the most potential to provide a clearer understanding of the anatomical and physiological links between headache and sleep.

Functional bio-behavioral model

Sleep disturbance is a trigger for migraine and TTH; this is well evidenced by prospective studies, including the time-series studies described above. Interestingly, there is evidence that the reverse is also true—headache triggers sleep disturbance. Headache sufferers tend to rest, nap, and sleep as a palliative response to pain. Haque et al.[55] examined headache triggers and relieving factors in 500 patients presenting for treatment to a large neurological practice who were diagnosed with migraine and TTH. Sleeping (52% TTH and 58% migraine) and medication taking (50% TTH and 61% migraine) were the most commonly endorsed

means of headache relief. Similarly, in a young non-clinical sample of individuals with TTH, Ong and colleagues[56] identified sleeping (81%) and medication taking (75%) as the most common responses to pain.

The bidirectional nature of sleep and headache is the basis of the biobehavioral model that has been proposed to explain the vicious cycle of sleep dysregulation and chronic headache.[57] Using a prevailing theory of basic sleep processes (two-process model of sleep/wake regulation, depicting the interplay of homeostatic sleep drive and circadian processes) and conceptual understanding of the development of chronic insomnia ("3-P model" for insomnia including predisposing, precipitating, and perpetuating factors), Ong and Park[57] proposed three tenets: (1) the palliative behavioral response to headache of sleeping/resting can precipitate and perpetuate sleep disturbance over time; (2) the subsequent disruption in sleep physiology increases the propensity for headache; and (3) over time, this vicious cycle may serve to transform episodic headache into chronic headache. These tenets parallel the well-known phenomenon in which daytime napping undermines the nocturnal sleep drive and causes increased night awakenings, which in turn leads to increases in daytime tiredness and more napping, which further disintegrates nocturnal sleep. This cycle undermines the basic physiological sleep/wake processes, which also happens to be critical in the regulation of pain and headache.

MEASUREMENT OF SLEEP

Polysomnography

The gold standard of objective sleep measurement remains polysomnography, or technologist-attended overnight assessment of multiple physiologic parameters including EEG, respiration, cardiac activity, and movement.[58] Testing may be laboratory based, but may be unattended using ambulatory equipment. Polysomnography is necessary in order to document complex sleep-disordered breathing, narcolepsy, and parasomnias. When headache is evaluated in relation to sleep, polysomnography can identify specific sleep-stage and headache antecedents. Limited-channel EEG cassette recordings may be used to assess sleep parameters only.[59] Unattended cardiorespiratory monitoring is increasingly utilized in order to assess obstructive sleep apnea.[60]

Actigraphy

This activity monitor or motion detector has been used to infer sleep and waking states based on activity level.[61] Industry is increasingly promoting the use of smartphones and other devices that contain accelerometers and could assess correlates of movement that suggest sleep. Actigraphy requires patients to wear a small wristband recording device that records and stores activity data, which are later downloaded for evaluation. Sleep is inferred by extended periods of inactivity so that approximate sleep time can be determined for patients who are relatively inactive during sleep and maintain a normal activity level while awake. Actigraphy may be a cost-effective tool for validating self-report diary data and measuring the timing and duration of sleep in circadian rhythm disorders and insomnia over days or weeks. Statistical correlations for actigraphy with one night of polysomnography varied by the sleep variable of interest, such as time in bed ($r = 0.99$), total sleep time (0.68), sleep onset (0.87), wake after sleep onset (0.69), total wake time (0.74), and sleep efficiency (0.67).[62] Actigraphy has been employed in migraine research,[63,64] although rarely. To our knowledge, commercially available accelerometers have not been validated against polysomnography.

Questionnaires

A wide variety of questionnaires are available to assess sleep disorders, sleep quality, daytime sleepiness, sleep-related psychosocial functioning, impairment, and quality of life. These questionnaires vary in level of psychometric development, and are reviewed elsewhere.[65-68] The Berlin Sleep Questionnaire[68] identified high- versus low-risk patients for obstructive sleep apnea (OSA) in the primary care environment (sensitivity 86%, specificity 77%, positive predictive value 89%, likelihood ratio 3.79) based on patient's neck circumference, habitual snoring, or witnessed apnea and hypertension. Questionnaires are available for a wide range of conditions (e.g., insomnia, restless legs, daytime sleepiness, headache, and sleep quality of life). Interestingly, validated questionnaires have been rarely utilized in studies of sleep-related headaches.

Mnemonics

In adults, simple screening questions can direct the sleep history to identify patients who are "at risk" of sleep disorders. Inquiring about the *Restorative* nature of the patient's sleep, *Excessive* daytime sleepiness, tiredness or fatigue, the presence of habitual *Snoring,* and whether the *Total* sleep time is sufficient can be revealing. The mnemonic *REST* can help the clinician to remember these four key questions in the screening history. Likewise, the STOP-BANG questionnaire has been validated in the general medical population for quickly screening for sleep apnea with the variables of *Snoring, Tiredness/ sleepiness, Observed apnea, blood Pressure, Body mass index, Age, Neck circumference,* and *Gender.*[69]

Sleep history

The basic elements of a 24-hour sleep history include: presleep routine, sleep period (sleep latency, duration of sleep relative to time in bed, and mid-cycle and early morning awakenings), nocturnal perturbations (e.g., respiratory, movement, and waking), awakening headache features if reported (e.g., location, ancillary symptoms, severity, and duration untreated), daytime function (e.g., napping, alertness vs. sleepiness, and fatigue) and substances or behavioral measures to promote sleep (e.g., hypnotics, alcohol, bedtime rituals, or white noise) or wake (e.g., caffeine, nicotine, snacks, boost activity, or other stimulation). The history may be supplemented with bed partner report and questionnaires in order to assess sleep apnea, insomnia, restless legs, etc.

Sleep diary

Paper-and-pencil and electronic sleep diaries are probably the most commonly used systematic self-report tools for sleep assessment. With once-a-day monitoring, subjective estimates can be obtained over time in terms of the regularity, duration, and quality of sleep. Typical findings of general interest are latency to sleep onset, number and duration of nocturnal awakenings, total sleep time relative to time in bed (i.e., sleep efficiency), sleep quality ratings, daytime sleepiness and alertness, and napping. Monitoring can also include other specific variables that are potentially related to sleep, such as headache (combined headache/sleep diary) (Figure 8.2).[70] Interestingly, sleep diaries are seldom employed in headache research, although

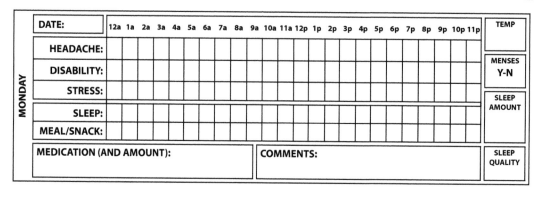

Figure 8.2 Headache diary. Daily headache monitoring in relation to precipitants: sleep, stress, meals, and menstruation. (Reproduced with permission from Rhudy JL, Penzien DB, Rains JC. *Self-Management Training Program for Chronic Headache: Therapist Manual.* Available at: https://www.apa.org/pubs/videos/4310731-diary.pdf)

headache diaries are commonly used and are familiar tools in headache research.

CLINICAL IMPLICATIONS

While there are no empirically established algorithms to guide clinical practice, there are at least a few empirically supported tenets. The review provided in this chapter supports the following recommendations for identifying potentially important sleep disorders in common headache disorders.

The headache practitioner would be encouraged to approach sleep in a systematic manner as presented below (Figure 8.3).[71]

Diagnose headache using ICHD-3 criteria

Recommendations are provided according to ICHD-3 primary and secondary headache diagnoses and are based upon diagnosis-specific risk for sleep and psychiatric disorders.

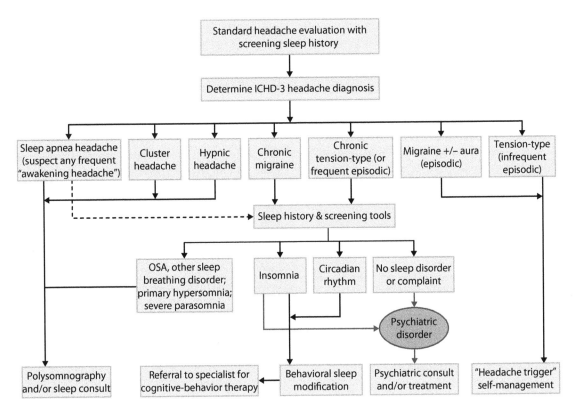

Figure 8.3 Algorithm for sleep-related headache. (Reproduced with permission from Rains JC, Poceta JS. *Curr Treat Options Neurol* 2010; 12: 1–15.)

Collect sleep history

The basic sleep history as described above may be supplemented by standardized questionnaires to assess "awakening headache" or other headache proximally related to sleep. The sleep history includes circadian rhythms, sleep habits, daytime sleepiness or fatigue, and abnormal behaviors of sleep. Responses to behavioral and pharmacological treatments are monitored in the headache diary over time. Sleep patterns may also help to determine the choice of prophylactic headache treatments, with more-sedating agents preferred in cases of insomnia, while more-alerting or neutral agents are preferred for hypersomnolent headache patients.[71]

Rule out sleep apnea headache in high-risk diagnoses (suspect any awakening headache)

The diagnosis of cluster, hypnic, chronic migraine or TTH or any headache that usually emerges during

sleep or at the termination of sleep—also known as awakening headache warrants screening for sleep apnea and other major sleep disorders. As noted above, SAH is suspected in cases of new-onset headache or exacerbation of a pre-existing primary headache. A diagnosis of cluster headache carries an 8.4-fold increased risk of sleep apnea and 24-fold increased risk in overweight patients[11]; thus, while a review of symptoms and risk factors is encouraged, the threshold for polysomnography is probably met based on diagnosis alone and the potential to improve cluster headache control with treatment for apnea. The available evidence indicates a marked increased incidence of sleep-disordered breathing among cluster headache patients, and that treatment of the apnea can improve this form of headache.[72,73] Likewise, hypnic headache probably warrants polysomnography based on anecdotal evidence implicating sleep-disordered breathing.

When headache frequently occurs during or after sleep onset or upon awakening, it is prudent to screen for the presence of significant sleep disorder or disturbance. The identification of obstructive sleep apnea or other sleep-related breathing abnormalities is particularly important because of the potential for headache to improve with treatment of the apnea,[74] as well as to avert the significant morbidity and mortality associated with sleep apnea.

Patients with suspected sleep apnea warrant diagnostic testing and treatment according to sleep medicine evidence-based guidelines.[58] The clinical symptoms, risk factors, and common treatments for obstructive sleep apnea are presented (Table 8.4).[71]

Table 8.4 Identification and treatment of obstructive sleep apnea

Clinical symptoms
- Habitual snoring
- Wake gasping
- Witnessed apnea
- Morning headache
- Hypersomnia or insomnia
- Night sweats
- Nocturia

Risk factors
- Obesity (↑ body mass index, neck, chest, waist, and hips)
- Male gender (male preponderance reduced in elderly)
- Age (positive correlation)
- Family history
- Craniofacial morphology and oral anatomy
- Neuromuscular disorders
- Substances (e.g., tobacco, alcohol, and sedatives)

Treatment options
- Positive airway pressure (continuous, bi-level, or adaptive servo-ventilation for central or complex apnea)
- Oral appliances (mandibular advancement devices and tongue retaining devices)
- Surgery (e.g., uvulopalatopharyngoplasty, tonsillectomy, and tracheostomy)
- Hypoglossal nerve stimulation
- Oxygen case by case
- Conservative measures
- Positional therapy to avoid supine sleep
- Weight loss
- Avoid alcohol, muscle relaxants, anxiolytics, sedatives, and hypnotics

The probable presence of obstructive sleep apnea as well as hypersomnia and other major sleep disorders warrants referrals for diagnostic testing and treatment. The identification and management of primary sleep disorders can be crucial not only for optimal head pain management, but also for managing the often substantial medical consequences of the sleep disorders themselves. Reevaluation of the headache at 1 month following the initiation of treatment for the sleep disorder is recommended.

Behavioral sleep reregulation for chronic headache

Among patients with chronic migraine and TTH, insomnia is the most common sleep complaint, being reported by half to two-thirds of clinic patients. Chronic headache sufferers should be screened for sleep disorders, particularly in cases that are refractory to standard treatments (e.g., prophylaxis or withdrawal from medication overuse). Episodic migraine and TTH patients may benefit from screening for sleep disorders when the history suggests a significant sleep complaint or daytime sleepiness.

Chronic migraine and TTH sufferers should be encouraged to employ behavioral sleep regulation, which has been shown to significantly improve chronic migraine (Table 8.5).[75] More extensive behavioral insomnia treatments are available by referral. Although there are no studies demonstrating that headache improves solely with pharmacologic treatment for insomnia, pharmacologic treatment may be appropriate on a case-by-case

Table 8.5 The five-component behavioral sleep intervention of Calhoun and Ford

1. Schedule consistent bedtime that allows 8 hours in bed.
2. Eliminate watching television, reading, or listening to music in bed.
3. Use visualization techniques to shorten time to sleep onset.
4. Move supper to at least 4 hours before bedtime and limit fluids within 2 hours of bedtime.
5. Discontinue naps.

Source: Adapted from Calhoun AH, Ford S. *Headache* 2007; 47: 1178–1183.

basis. There are no data to recommend a particular sleep aid for headache patients from the available hypnotic, anxiolytic, or sedating antidepressants, but tailoring treatment to symptom patterns would be in order.

Screen for psychiatric comorbidity in chronic headache

Individuals with chronic migraine and TTH are at increased risk for psychiatric disorders.[76] Sleep disturbance and daytime sleepiness are symptoms of a number of psychiatric disorders, and they occur in the majority of patients with depression, anxiety, and chemical dependencies.[19] Thus, the threshold for screening is met when either insomnia or hypersomnia is present in the headache sufferer. Psychiatric referral or treatment may be indicated. While there is little research to direct treatment, recognition of insomnia with depression or anxiety may guide headache prophylaxis towards sedating antidepressants or anticonvulsants, while hypersomnia would call for neutral or more-alerting medications.[71]

Manage sleep-related triggers of episodic headache

All headache sufferers, and particularly patients with episodic headache, may benefit from the inclusion of sleep variables in trigger management. Triggers are probably most accurately identified by prospective self-monitoring because patients are often unaware of the behavioral and psychosocial precipitants. Headache diaries can be highly informative for patients as well as clinicians for identifying previously unrecognized headache patterns. Diaries track the regularity, duration, and quality of sleep as headache triggers. Headache may be linked to triggers such as sleep, stress, mood, and diet when comprehensive diaries are used (see Figure 8.1 above).[70]

RESEARCH CRITIQUE

There are methodological limitations in the sleep/headache literature, particularly concerning headache classification, research design and methods, and reporting. Many published studies, including nearly all studies on SAH, have reported no formal headache diagnosis for validating existing criteria or enabling comparison of the phenomenon across studies. The popular terminology of chronic, morning, and awakening headache likely encompasses different forms of headache with varying pathophysiologies and introduces substantial variance into the research equation. Consistent terminology such as that from the ICHD-3 and ICSD-3, coupled with additional validation studies, would foster meta-analyses and generalization across studies.

Research methods have varied widely across studies. Many studies employed small and selected diagnostic groups rather than larger, unselected samples of headache patients or the general population. A small number of subjects might be unavoidable in certain circumstances because of the rarity of certain disorders (e.g., hypnic headache). However, clinical studies most commonly employ samples of convenience in which recruitment methods and sample characteristics are not reported; thus, the stability of outcomes and generalizability are unknown. Likewise, the significance of uncontrolled group outcome studies cannot be determined.

Many studies have utilized non-standardized and subjective outcome measures, although validated questionnaires are available for a wide range of sleep and headache variables.[1] Studies that have reported objective polysomnographic data to quantify measures of sleep have tended to include small numbers of subjects. No doubt this limitation is an unfortunate consequence of the substantial costs and labor involved in conducting polysomnography. However, even low-cost tools, such as standardized questionnaires and sleep diaries, have rarely been utilized. Likewise, many published polysomnographic studies date back several years and would not meet the current standards for assessing, scoring, and reporting respiratory events and the microstructure of sleep. Methodological improvements that have been recommended for future research include the use of controlled research designs, random sampling methods, ICHD-3 classification of headache, objective or validated outcome measures, and improved reporting. Intervention studies to assess effects on headache that might manipulate sleep duration or utilize hypnotic sedatives and cognitive–behavioral therapy to improve sleep quality or otherwise treat specific sleep disorders would be of value.

CONCLUSION

The association between headache and sleep has been long recognized in the medical literature, but is still not fully understood. Although there is good evidence supporting the comorbidity of headache and sleep disorders, the nature, extent, and causes of this comorbidity are debated. The empirical literature has yielded inconsistencies and, in some cases, highly divergent outcomes, no doubt due in part to the methodological limitations of the studies comprising the evidence base (e.g., imprecise diagnoses, small numbers of subjects, varied sampling methods, lack of objective sleep measures, other non-standardized measures, and inconsistent reporting). It is highly probable that the pathogenic sources of sleep-related headache are multifactorial and stem from the involvement of common neuroanatomical structures and neurophysiological processes. With the rapid acquisition of scientific advancements and knowledge regarding the complex interrelationships of functional anatomical, neurological, and molecular processes occurring over the last decade, there is great promise that future research that is specifically designed to examine and better understand the associations between headache and sleep disorders is close at hand. Regardless, there are now at least a few empirically supported tenets, and this chapter forwards a preliminary algorithm to guide clinicians in identifying headache that is a consequence of primary sleep disorders and identifying sleep variables that may impact the headache threshold, and thereby providing a means by which to improve headache management.

REFERENCES

1. Wright H. *Headaches: Their Causes and Their Cures*. Philadelphia, PA: Lindsay & Blakiston, 1871.
2. Liveing E. *On Megrim, Sick-Headache, and Some Allied Disorders: A Contribution to the Pathology of Nervous Storms*. London: Churchill, 1873.
3. Headache Classification Committee of the International Headache Society (IHS). The international classification of headache disorders, 3rd edition (beta version). *Cephalalgia* 2013; 33(9): 629–808.
4. American Academy of Sleep Medicine. *International Classification of Sleep Disorders* (3rd ed.). Darien, IL: American Academy of Sleep Medicine, 2014.
5. American Psychiatric Association. *Diagnostic and Statistical Manual of Mental Disorders* (5th ed.). Arlington, VA: American Psychiatric Association, 2013.
6. Liang JF, Wang SJ. Hypnic headache: A review of clinical features, therapeutic options and outcomes. *Cephalalgia* 2014; 34(10): 795–805.
7. Evers S, Goadsby PJ. Hypnic headache: Clinical features, pathophysiology, and treatment. *Neurology* 2003; 60(6): 905–909.
8. Lanteri-Minet M. Hypnic headache. *Headache* 2014; 54(9): 1556–1559.
9. Silva-Néto RP, Bernardino SN. Ambulatory blood pressure monitoring in patient with hypnic headache: A case study. *Headache* 2013; 53(7): 1157–1158.
10. Ruiz M, Mulero P, Pedraza MI et al. From wakefulness to sleep: Migraine and hypnic headache association in a series of 23 patients. *Headache* 2015; 55(1): 167–173.
11. Nobre ME, Leal AJ, Filho PM. Investigation into sleep disturbance of patients suffering from cluster headache. *Cephalalgia* 2005; 25(7):488–492.
12. Graff-Radford SB, Newman A. Obstructive sleep apnea and cluster headache. *Headache* 2004; 44(6): 607–610.
13. Barloese M, Lund N, Jensen R. Sleep in trigeminal autonomic cephalalgias: A review. *Cephalalgia* 2014; 34(10): 813–822.
14. Kayed K, Sjaastad O. Nocturnal and early morning headaches. *Ann Clin Res* 1985; 17(5): 243–246.
15. Newman LC, Goadsby PJ. Unusual primary headache disorders. In: Silberstein SD, Dalessio, DJ (Eds). *Wolff's Headache and other Head Pain*. New York, NY: Oxford University Press, 2001, pp. 310–324.
16. Fox AW, Davis RL. Migraine chronobiology. *Headache* 1998; 38: 436–441.
17. Alstadhaug K, Salvesen R, Bekkelund S. 24-hour distribution of migraine attacks. *Headache* 2007; 48: 95–100.
18. Smitherman TA, Burch R, Sheikh H, Loder E. The prevalence, impact, and treatment of migraine and severe headaches in the

United States: A review of statistics from national surveillance studies. *Headache* 2013; 53(3): 427–436.

19. Vos T, Flaxman AD, Naghavi M et al. Years lived with disability (YLDs) for 1160 sequelae of 289 diseases and injuries 1990–2010: A systematic analysis for the Global Burden of Disease Study 2010. *Lancet* 2012; 280: 2163–2196.

20. Stovner L, Hagen K, Jensen R et al. The global burden of headache: A documentation of headache prevalence and disability worldwide. *Cephalagia* 2007; 27: 193–210.

21. Bigal ME, Lipton RB. Migraine chronification. *Curr Neurol Neurosci Rep* 2011; 11(2): 139–148.

22. Scher AI, Lipton RB, Stewart WF. Habitual snoring as a risk factor for chronic daily headache. *Neurol* 2003; 60(8): 1366–1368.

23. Rains JC, Davis RE, Smitherman TA. Tension-type headache and sleep. *Curr Neurol Neurosci Rep* 2015; 15(2): 520.

24. Alberti A, Mazzotta G, Gallinella E, Sarchielli P. Headache characteristics in obstructive sleep apnea and insomnia. *Acta Neurol Scand* 2005; 111: 309–316.

25. Provini F, Vetrugno R, Lugaresi E et al. Sleep-related breathing disorders and headache. *Neurol Sci* 2006; 27(Suppl 2): S149–S152.

26. Loh NK, Dinner DS, Foldvary N et al. Do patients with obstructive sleep apnea wake up with headaches? *Arch Intern Med* 1999; 159(15): 1765–1768.

27. Doufas AG, Tian L, Davies MF, Warby SC. Nocturnal intermittent hypoxia is independently associated with pain in subjects suffering from sleep-disordered breathing. *Anesthesiology* 2013; 119(5): 1149–1162.

28. Russell MB, Kristiansen HA, Kværner KJ. Headache in sleep apnea syndrome: Epidemiology and pathophysiology. *Cephalagia* 2014; 34: 752–755.

29. Göder R, Friege L, Fritzer G et al. Morning headaches in patients with sleep disorders: A systematic polysomnographic study. *Sleep Med* 2003; 4: 385–391.

30. Greenough GP, Nowell PD, Sateia MJ. Headache complaints in relation to nocturnal oxygen saturation among patients with sleep apnea syndrome. *Sleep Med* 2002; 3: 361–364.

31. Sand T, Hagen K, Schrader H. Sleep apnoea and chronic headache. *Cephalagia* 2003; 23:90–95.

32. Chen PK, Fuh JL, Lane HY et al. Morning headache in habitual snorers: Frequency, characteristics, predictors and impacts. *Cephalagia* 2011; 31(7): 829–836.

33. Frese A, Summ O, Evers S. Exploding head syndrome: Six new cases and review of the literature. *Cephalagia* 2014; 34(10): 823–827.

34. Sachs C, Svanborg E. The exploding head syndrome: Polysomnographic recordings and therapeutic suggestions. *Sleep* 1991; 14(3): 263–266.

35. Odegard SS, Engstrom M, Sand T et al. Associations between sleep disturbance and primary headaches: The third Nord-Trøndelag Health Study. *J Headache Pain* 2010; 11: 197–206.

36. Odegard SS, Sand T, Engstrom M et al. The long-term effect of insomnia on primary headaches: A prospective population-based cohort study (HUNT-2 and HUNT-3). *Headache* 2011; 51: 570–580.

37. Boardman HF, Thomas E, Millson DS, Croft PR. Psychological, sleep, lifestyle, and comorbid associations with headache. *Headache* 2005; 45: 657–669.

38. Ohayon MM. Prevalence and risk factors of morning headaches in the general population. *Arch Intern Med* 2004; 164: 97–102.

39. Rasmussen BK. Migraine and tension-type headache in the general population: Precipitating factors, female hormones, sleep pattern and relation to lifestyle. *Pain* 1993; 53: 65–72.

40. Lyngberg AC, Rasmussen BK, Jorgensen T, Jensen R. Incidence of primary headache: A Danish epidemiologic follow up study. *Am J Epidemiol* 2005; 161: 1066–1073.

41. Lyngberg AC, Rasmussen BK, Jorgensen T, Jensen R. Prognosis of migraine and tension-type headache: A population-based follow-up study. *Neurol* 2005; 65(4): 580–585.

42. Sancisi E, Cevoli S, Vignatelli L et al. Increased prevalence of sleep disorders in chronic headache: A case-control study. *Headache* 2010; 50: 1464–1472.

43. Saper JR. Chronic daily headache: Transformational migraine, chronic migraine, and related disorders. *Curr Neurol Neurosci Rep* 2008; 8: 100–107.

44. Scher AI, Midgette LA, Lipton RB. Risk factors for headache chronification. *Headache* 2008; 48: 16–25.

45. Rains JC. Chronic headache and potentially modifiable risk factors: Screening and behavioral management of sleep disorders. *Headache* 2008; 48: 32–39.

46. Wöber C, Wöber-Bingöl C. Triggers of migraine and tension-type headache. *Handb Clin Neurol* 2010; 97: 161–172.

47. Wang J, Huang Q, Li N et al. Triggers of migraine and tension-type headache in China: A clinic-based survey. *Eur J Neurol* 2013; 20: 689–696.

48. Kelman L, Rains JC. Headache and sleep: Examination of sleep patterns and complaints in a large clinical sample of migraineurs. *Headache* 2005; 45(7): 904–910.

49. Houle TT, Butschek RA, Turner DP et al. Stress and sleep predict headache severity in chronic headache sufferers. *Pain* 2012; 153: 2432–2440.

50. Kikuchi H, Yoshiuchi K, Yamamoto Y et al. Does sleep aggravate tension-type headache?: An investigation using computerized ecological momentary assessment and actigraphy. *Biopsychosoc Med* 2011; 5: 10.

51. Holland PR. Headache and sleep: Shared pathophysiological mechanisms. *Cephalalgia* 2014; 34(10): 725–744.

52. Brennan KC, Charles A. Sleep and headache. *Semin Neurol* 2009; 29(4): 406–418.

53. Evers S. Sleep and headache: The biological basis. *Headache* 2010; 50(7): 1246–1251.

54. Rains JC, Poceta JS. Sleep-related headache. In: Vaughn B (Ed.). *Neurologic Clinics*. Philadelphia, PA: Elsevier, 2012, Vol. 30, pp. 1285–1298.

55. Haque B, Rahman KM, Hoque A et al. Precipitating and relieving factors of migraine vs tension type headache. *BMC Neurol* 2012; 12: 82.

56. Ong JC, Stepanski EJ, Gramling SE. Pain coping strategies for tension-type headache: Possible implications for insomnia? *J Clin Sleep Med* 2009; 5(1): 52–56.

57. Ong JC, Park M. Chronic headaches and insomnia: Working toward a biobehavioral model. *Cephalalgia* 2012; 32: 1059–1070.

58. Kushida CA, Littner MR, Morgenthaler T et al. Practice parameters for the indications for polysomnography and related procedures: An update for 2005. *Sleep* 2005; 28(4): 499–521.

59. Drake ME Jr, Pakalnis A, Andrews JM, Bogner JE. Nocturnal sleep recording with cassette EEG in chronic headaches. *Headache* 1990; 30(9): 600–603.

60. Collop NA, Anderson WM, Boehlecke B et al. Clinical guidelines for the use of unattended portable monitors in the diagnosis of obstructive sleep apnea in adult patients. *J Clin Sleep Med* 2007; 3(7): 737–747.

61. Saadeh A, Hauri PJ, Kripke DF, Lavie P. The role of actigraphy in the evaluation of sleep disorders. *Sleep* 1995; 18: 288–302.

62. Edinger JD, Means MK, Stechuchak KM, Olsen MK. A pilot study of inexpensive sleep-assessment devices. *Behav Sleep Med* 2004; 2(1): 41–49.

63. Bruni O, Russo PM, Violani C, Guidetti V. Sleep and migraine: An actigraphic study. *Cephalalgia* 2004; 24: 134–139.

64. Capuano A, Vollono C, Rubino M et al. Hypnic headache: Actigraphic and polysomnographic study of a case. *Cephalalgia* 2005; 25(6): 466–469.

65. Devine EB, Hakim Z, Green J. A systematic review of patient-reported outcome instruments measuring sleep dysfunction in adults. *Pharmacoeconomics* 2005; 23 (9): 889–912.

66. Moule DE, Hall M, Pilkonis PA, Buysse DJ. Self-report measures of insomnia in adults: Rationales, choices, and needs. *Sleep Med Rev* 2004; 8: 177–198.

67. Harding S. Prediction formulae for sleep-disordered breathing. *Curr Opin Pulm Med* 2001; 7: 381–385.

68. Flemons WW, Whitelaw WA, Brant R, Remmers JE. Likelihood ratios for a sleep apnea clinical prediction rule. *Am J Respir Crit Care Med* 1994; 150(5 Pt 1): 1279–1285.

69. Ong TH, Raudha S, Fook-Chong S et al. Simplifying STOP-Bang: Use of a simple questionnaire to screen for OSA in an Asian population. *Sleep Breath* 2010; 14: 371–376.

70. Rhudy J, Penzien DB, Rains JC. Daily headache self-monitoring form. Available at: https://www.apa.org/pubs/videos/4310731-diary.pdf

71. Rains JC, Poceta JS. Sleep and headache. *Curr Treat Option Neurol* 2010; 12: 1–15.

72. Chervin RD, Zallek SN, Lin X et al. Sleep disordered breathing in patients with cluster headache. *Neurology* 2000; 54(12): 2302–2306.

73. Buckle P, Kerr P, Kryger M. Nocturnal cluster headache associated with sleep apnea. A case report. *Sleep* 1993; 16(5): 487–489.

74. Johnson KG, Ziemba AM, Garb JL. Improvement in headaches with continuous positive airway pressure for obstructive sleep apnea: A retrospective analysis. *Headache* 2013; 53(2): 333–343.

75. Calhoun AH, Ford S. Behavioral sleep modification may revert transformed migraine to episodic migraine. *Headache* 2007; 47: 1178–1183.

76. Smitherman TA, Walters AB, Ambrose CE et al. Randomized controlled trial of behavioral insomnia treatment for chronic migraine with comorbid insomnia: Preliminary results of a sham-controlled pilot study. *Headache* 2014; 54: 33–34.

9

Fibromyalgia and the neurobiology of sleep

DANIEL J. WALLACE

Fibromyalgia (FM) is not a disease, but a syndrome that is characterized by centralized sensitization of afferent inputs into the spinal cord from tactile, chemical, thermal, and nociceptive stimuli, which leads to amplified pain. Sleep pathology is a unifying feature of the syndrome, and improvement in sleep architecture is the principal treatment goal.

CLINICAL ASPECTS OF FM

Evolution of the concept

References to the symptoms of FM date back to biblical times and can be found in the books of Job and Jeremiah among individuals whose insomnia was associated with musculoskeletal discomfort.[1] Modern understanding stems from observations by Sir William Gowers in 1904 who coined the term "fibrositis" to depict tender points in patients with lumbago (back pain). Serious investigation into tender points first took place in the early 1800s, but its connection with fatigue and systemic symptoms is credited to Hugh Smythe and Harvey Moldofsky at the University of Toronto in the 1970s, who described alpha-wave intrusion into delta-wave sleep on polysomnograms and correlated it with the presence of tender points.[2] Work by Muhammad Yunus in 1981 statistically correlated tender points with fatigue, functional bowel disease, sleep pathology, tension headache, and other systemic symptoms.[3] The American College of Rheumatology published statistically validated working criteria for the syndrome in 1990, which were followed by epidemiologic surveys to define what constituted FM (Figure 9.1).[4]

Epidemiology

Population-based studies using the 1990 American College of Rheumatology criteria by Wolfe et al. and White et al. showed that FM had a prevalence of 2%–4% and was 8–9 times more prevalent in women.[5-7] Onset is primarily

The 1990 ACR Criteria for Fibromyalgia

1. *History of widespread pain.*

   Definition: Pain is considered widespread when all of the following are present: pain in the left side of the body, pain in the right side of the body, pain above the waist and pain below the waist. In addition, axial skeletal pain (cervical spine or anterior chest or thoracic spine or low back) must be present. In this definition shoulder and buttock pain is considered as pain for each involved side. "Low back" pain is considered lower segment pain.

2. *Pain in 11 of 18 tender point sites on digital palpation.*

   Definition: Pain, on digital palpation, must be present in at least 11 of the following 18 tender point sites:

   Occiput: bilateral, at the suboccipital muscle insertions.

   Low cervical: bilateral, at the anterior aspects of the inter-transverse spaces at C5–C7.

   Trapezius: bilateral, at the midpoint of the upper border.

   Supraspinatus: bilateral, at origins, above the scapula spine near the medial border.

   2nd rib: bilateral, at the second costochondral junctions, just lateral to the junctions on upper surfaces.

   Lateral epicondyle: bilateral, 2 cm distal to the epicondyles.

   Gluteal: bilateral, in upper outer quadrants of buttocks in anterior fold of muscle.

   Greater trochanter: bilateral, posterior to the trochanteric prominence.

   Knees: bilateral, at the medial fat pad proximal to the joint line.

   For a tender point to be considered "positive" the subject must state that the palpation was painful. "Tender" is not to be considered painful.

Note: For classification purposes patients will be said to have fibromyalgia if both criteria are satisfied. Wide-spread pain must have been present for at least 3 months. The presence of a second clinical disorder does not exclude the diagnosis of fibromyalgia.

Fibromyalgia tender points. (Adapted from "The Three Graces," Louvre Museum, Paris.)

Figure 9.1 1990 American College of Rheumatology criteria for fibromyalgia. (Updated and revised with permission from Wallace DJ, Wallace JB. *All about Fibromyalgia*. New York/London: Oxford University Press, 2003.)

between the ages of 20 and 50. Approximately 6 million people in the United States have FM, and another 6 million have FM-related complaints but never seek medical attention for them (termed "community FM"). It is the third most common reason for referral to a rheumatologist. In 2010, a committee expanded the diagnostic variables to include a widespread pain index and categorical scales for cognitive symptoms, unrefreshed sleep, fatigue, and a number of body symptoms.[8] A severity index was proposed that allows for longitudinal evaluation of patients. It is demonstrated in Table 9.1. This new metric can be used along with or instead of the 1990 criteria and is over 90% sensitive and specific.

Clinical overview

The key feature of FM is pain due to central sensitization, which is an exaggeration of sensory stimuli and the perception of touch and pressure as pain.[9] From an operational standpoint, FM requires 3 months of discomfort in individuals, with 11 of 18 specified tender points in all four quadrants of the body. The presence of fewer FM-like tender points in one to three quadrants is termed "myofascial pain syndrome" or "regional myofascial pain." "Primary FM" is of unknown cause, but patients with "secondary FM" can trace the onset of their symptoms to an inciting event or cumulative emotional or physical trauma. The most common include inception after a trauma, a viral process, untreated inflammatory arthritis, or lifting heavy loads with poor body mechanics. Most of these individuals have preexisting risk factors such as psychosocial stressors, poor sleep habits, or myofascial pain syndrome. FM is a subset of chronic widespread pain.[10] This differs from the 1990 FM criteria in that it requires more diffuse limb pain that is present in two or more sections

Table 9.1 2010 preliminary American College of Rheumatology diagnostic criteria for fibromyalgia

Fibromyalgia can be diagnosed based on a healthcare professional-administered questionnaire if all three of the following conditions are met:

1. Widespread pain index: the number of painful body regions (scored 0–19, where the patient has had pain over the previous week with a score of at least 5).
2. Symptom severity scale that assesses fatigue, waking unrefreshed and cognitive symptoms, and quantifies the occurrence of other somatic symptoms. Each category is rated on a 0–3 scale with a final score of 0–12. The somatic symptom category includes over 20 complaints, including muscle pain, numbness, tingling, depression, dry mouth, bladder spasm, etc.
3. Pain and symptoms present for 3 months or longer with no other disorder to explain it.

Table 9.2 Prevalence of frequently observed symptoms and signs in fibromyalgia patients

Symptom	Prevalence (%)
Widespread pain with tender points	100
Generalized weakness, muscle, and joint aches	80
Unrefreshing sleep	75
Fatigue	75
Stiffness	70
Anxiety	60
Psychological stress	60
Dizziness/vertigo	55
Tension headache	55
Cognitive impairment or "fibro fog"	50
Painful periods	40
Irritable colon	40
Subjective numbness, burning, or tingling	35
Subjective complaints of swelling or edema	35
Depression	35
Skin redness or lace-like skin mottling	30
Complaints of fever	20
Complaints of swollen glands	20
Complaints of non-medication dryness or dry eyes	18
Posttraumatic stress disorder	18
Nocturnal myoclonus or restless legs syndrome	15
Raynaud's disease	15
Irritable bladder or interstitial cystitis	12
Chronic pelvic pain	12
Reflex sympathetic dystrophy	5
Bipolar illness	5

of contralateral limbs, as well as axial pain present for 3 months. These individuals have more severe disability and higher levels of associated symptoms. Chronic widespread pain can also often be present without tender points.

The principal symptoms and signs of FM are listed in order of prevalence in Table 9.2.[11,12] Fatigue is a near-universal complaint and results in decreased mental and physical endurance. Cardiopulmonary disease, hypothyroidism, and inflammation must be excluded as potential causes of fatigue. Although some patients complain of fevers, the "feverish" sensation reflects autonomic or hormonal imbalances, and true FM patients are always afebrile unless infected. Tender lymph glands without lymphadenopathy are a frequent complaint. FM patients have tenderness in their muscle regions without inflammation, weakness, or myositis. The most tender areas are in the neck, upper back, anserine bursa, proximal hip, and shoulder girdle region. Similarly, joint discomfort or stiffness is common, always without synovitis. Stiffness or aching is worse in the late afternoons and tends to spare the hands or feet, the opposite of what is reported in rheumatoid arthritis or systemic lupus.

Cognitive, memory, and psychological disturbances include short-term memory impairment and inability to concentrate or multitask; sensory and cognitive overload are common as side effects of medication or due to an evolving dementia or cerebrovascular disease. Tension headaches are common and cervical osteoarthritis, sinusitis, or migraine need to be ruled out. Migraine is more frequent in FM patients due to vasomotor insta-bility. Alterations in blood flow patterns to the brain as documented by neuroimaging account for some of the headache-related complaints, as well as intermittent symptoms of cognitive impairment (also termed "fibro fog"). The sympathetic nervous system is dysfunctional in FM, which leads to vaso-motor hyper-reactivity and is manifested by a rest-ing decrease in oncotic pressure, higher prevalence of Raynaud's disease, mitral valve prolapse, livedo reticularis, self-reported edema, and complaints of numbness, burning, and tingling. Dysautonomia can also be associated with "hypervigilance syn-dromes" where sensitivity to loud noises, bright lights, and the sensation of dizziness are common, along with anxiety disorders. Numbness, tingling, and swelling is part of a small-fiber neuropathy (and electromyograms are normal).[13,14] A total of 1% of people with FM evolve dramatic sympathetic pathology associated with frank swelling, burning, and tenderness, which is termed "reflex sympa-thetic dystrophy" or "regional complex pain syn-drome, type 1."

Bladder and bowel disturbances include pain, which is present during urination and when the bladder is full, urge and stress incontinence ensue. Bowel symptoms include pain, which may be con-stant or colicky and may be present in any quad-rant. Hypersensitivity of the bowel due to irritable bowel syndrome is thought to be responsible for the pain and may coexist with constipation and diarrhea. Abdominal bloating after eating can be observed and may be due to the presence of bacte-rial overgrowth in the small intestine. Changes in barometric pressure, poor sleep, and anxiety can exacerbate symptoms.

Association with other central sensitization syndromes

Patients with FM-like complaints often have symp-toms that are more bothersome than musculoskel-etal ones, and so consult physicians who diagnose them with conditions that are also associated with central sensitization.[9] A listing of these conditions and their prevalence is found in Table 9.3. Chronic fatigue syndrome has its own statistically vali-dated criteria, but differs from FM in that fatigue is more prominent than myofascial discomfort, and many more chronic fatigue patients had doc-umented evidence that an infectious process had

Table 9.3 Examples of some central sensitization syndromes

Fibromyalgia

Myofascial pain syndrome
  Temporomandibular disorder
  Whiplash
  Repetitive strain disorder
  Chronic idiopathic low-back pain

Chronic fatigue syndrome; post-infectious fatigue syndromes

Gastrointestinal syndromes
  Non-ulcer dyspepsia
  Esophageal dysmotility
  Irritable bowel syndrome
  Biliary dyskinesia; post-cholecystectomy syndrome

Cardiac region syndromes
  Syndrome X
  Noncardiac chest pain
  Costochondritis
  Mitral valve prolapse

Headache: tension-type or migraine

Gynecologic syndromes
  Primary dysmenorrhea
  Chronic pelvic pain
  Dyspareunia, vulvodynia, vulvar vestibulitis
  Endometriosis

Urologic syndromes
  Irritable bladder/painful bladder
  Interstitial cystitis
  Chronic prostatitis

Psychiatric conditions
  Depression/anxiety
  Posttraumatic stress disorder
  Bipolar illness
  Obsessive compulsive disorder

Multiple chemical sensitivities (a form of anxiety)

Periodic limb movement disorder

Dysautonomias

induced their symptoms. Post-infectious fatigue syndrome patients (e.g., Epstein–Barr and Lyme diseases) often manifest myofascial symptomatol-ogy. There is considerable overlap between FM and conditions associated with visceral hyperalgesia, such as irritable bowel syndrome, noncardiac chest pain, non-ulcer dyspepsia, and esophageal spasm. Pathophysiologically, there is greater emphasis on parasympathetic dysfunction and smooth muscles as opposed to sympathetic dysfunction and striated muscle involvement in FM. Dysmenorrhea, chronic pelvic pain, vulvodynia, vaginismus, interstitial cystitis, and irritable bladder are found in a minority of patients with FM, but once mechanical problems or hormonal imbalances are taken into account, there is an increased prevalence of sexual abuse, rape experiences, pelvic trauma, or guilt surrounding sexual feelings. Other regional (not necessarily four-quadrant)-associated syndromes include repetitive strain, temporomandibular joint dysfunction, and scoliosis.

Clinical evaluation and differential diagnosis

FM patients have normal blood chemistry panels, complete blood counts, immune profiles, imaging studies, and electrodiagnostic testing. It is often a diagnosis of exclusion. Many individuals diagnosed with FM turn out not to have the syndrome. Other disorders and conditions are associated with myofascial symptoms and need to be differentiated from FM. These include multiple sclerosis, hypothyroidism, rheumatoid arthritis, bipolar illness, early pregnancy, allergies, nutritional deficiencies, anorexia, cancer, substance (e.g., steroid, alcohol, heroin, or cocaine) withdrawal, and opportunistic infections.

Psychological profiles

The majority of FM patients have a history of depression, but only 18% are depressed at any given visit. Individuals with FM tend to have more anxiety, poorer coping skills and psychosocial stressors than control populations. A total of 20%–30% of people with FM have no psychologic problems (e.g., those with scoliosis) influencing the syndrome, but two personality profiles predominate. Posttraumatic stress disorder is found in 20% of people with FM, and another 20% are above-average intelligence females with perfectionistic tendencies, chronic anxiety, and hypervigilance symptoms who find it difficult to relax and often have no hobbies.[10,11]

THE ETIOPATHOGENESIS OF FM

Pain pathways in healthy and FM patients

What causes the "pain without purpose" of FM? FM affects chronic but not acute pain. Chronic pain states consist of psychogenic or organic pain. Within the realm of organic pain, the discomfort can be localized (which plays a minor role in FM) or central. The sources of central pain are either neuropathic (not part of FM), nociceptive, or non-nociceptive. In FM, nociceptive pain—or pain that is associated with discomfort when amplified by repetitive inputs—leads to hyperalgesia, neuroplasticity, hyperpathia, and/or rarely causalgia. Non-nociceptive inputs (e.g., gentle stroking) that should not be uncomfortable become painful, a phenomenon known as allodynia.

In FM and healthy individuals, thin non-myelinated C-fibers in the skin are easily activated by chemical, mechanical, or thermal stimuli. Even without noxious stimuli, signals can arise spontaneously that are converted into neural impulses. Once sensitized by this stimulus, the C-fiber nerves convey this afferently to the dorsal root ganglion of the spinal cord. The constant bombardment of noxious inputs by C-fibers produces a "wind-up" phenomenon in the substantia gelatinosa, which leads to central sensitization and, ultimately, FM. Large, myelinated A-delta fibers, which normally transmit very noxious signals, start carrying some of the signals that are usually carried by the C-fibers. Even autonomic B fibers start carrying nociceptive stimuli in order to handle the overload. Non-nociceptive fibers begin to carry nociceptive signals. In the dorsal root ganglion, increased discharges of second- and third-rung neurons take place via the secretion of nerve growth factor and substance P. Numerous studies have documented that cerebrospinal levels of substance P are increased in FM. Excitatory amino acids (e.g., glutamate) result in N-methyl-D-aspartate receptors in the spinal column—which are normally dormant—enhancing electrical depolarization and thus calcium influx into nerve cells, which makes them more excitable. These impulses ascend via the spinothalamic tract to the thalamus (and autonomic fibers via the spinoreticular tract to the limbic system). The brain now responds with inhibitory actions via neurotransmitters (e.g., dopamine, norepinephrine, epinephrine, serotonin, and opioids) in the descending system. However, in FM, the responses are diminished (e.g., serotonin products are decreased in cerebrospinal fluid). Recently, it has been shown that glial cells make cytokines, substance P, and other chemicals that may perpetuate this process. These factors are influenced by hormones, emotional stress, cytokines, and sleep disorders. The end result is amplified pain.[12–15] Figure 9.1 summarizes these interactions.

THE ROLE OF SLEEP IN THE ETIOPATHOGENESIS OF FM

Between 2% and 15% of any given population and 60% and 90% of FM patients have non-restorative sleep.[16] First reported with FM in the 1970s by Moldofsky's group in Toronto, alpha-wave intrusion into delta-wave sleep is noted on polysomnograms during stages 2–4 of non-rapid eye movement (non-REM) sleep.[2] Other findings noted include increases in stage 1 sleep, reductions in delta sleep, and increases in the number of arousals.[17,18] This leads to one being in bed for 8 hours or so, but waking up not feeling rested. Moldofsky's group has identified three distinct patterns of alpha sleep activity in FM: phasic alpha–delta activity (50%); tonic alpha continuous throughout non-REM sleep (20%); and low alpha in the remaining 30%.[19] The phasic group had the greatest number of symptoms and lowest sleep time. Increased fragmented sleep has also been documented by greater numbers of arousals and alpha–K complexes (which promote arousal, fatigue, and muscular symptoms) in the syndrome. Electrocardiograms demonstrate increased sympathetic nervous system activity overnight, while healthy individuals report a decline with sleep. FM and pain have been associated with a higher proportion of stage 1 non-REM sleep, fewer sleep spindles, and less sleep spindle frequency activity (usually seen in phase 2 sleep), suggesting that the mechanism relates to thalamocortical mechanisms of spindle, generation. Cyclic altering patterns during sleep, which express the instability of the level of vigilance that manifests as the brain's fatigue in preserving and regulating the macrostructure of sleep, is more often present in FM.[20]

Non-restorative sleep is felt to derive from decreases in growth hormone secretion as measured by insulin-like growth hormone.[21] We have 640 muscles in our body that undergo micro-trauma during our daytime activities. Once asleep, growth hormone and melatonin secretion is increased, which heals the micro-trauma experienced by our muscles.[22] In other words, abnormal electrical activity interferes with a sound sleep. These changes are more pronounced with menstruation, stress, pain, trauma, infection, nocturia, and barometric changes. Forced awakenings (fragmented sleep) lead to the loss of diffuse noxious inhibitory controls (DNIC), causing its associated increased achiness.[23]

A total of 10%–30% of FM patients exhibit another pattern of sleep pathology that can be documented via a sleep study: restless legs syndrome, also known as sleep myoclonus or periodic limb movement syndrome.[24,25] These patients experience an alpha-wave burst followed by limb movement, and may have excess sympathetic tone, more movement arousals, and less stage 3 and 4 sleep. They do not respond to usual sleep aids and report that their legs shoot out, lift, jerk, or go into spasm. Bed partners are often the first to alert one that this is present. Respiratory flow dynamics during sleep in FM are just beginning to be surveyed. While initial reports suggested that these may correlate with sleep apnea (especially in men), its true prevalence is only 5%. One group has associated upper-airway resistance syndrome, rather than sleep apnea or hypopnea, in the overwhelming majority of FM patients.[26,27]

Hormones and cytokines play important roles in the disturbances reported in FM.[28–30] Behavioral makeup, stress, estrogen release, sympathetic nerve activity, and interleukin-1β (IL-1β) all lead to the release of corticotropin-releasing hormone, which indirectly blocks growth hormone secretion. IL-1 independently promotes fatigue, sleep, and muscle aches and blocks the release of substance P. Chronic insomnia is associated with a shift of IL-6 and tumor necrosis factor secretion from the nighttime to daytime, as well as hypersecretion of cortisol. This leads to daytime fatigue and difficulty sleeping (see Figure 9.2).

THE IMPACT OF DISORDERED SLEEP IN FM

Altered or poor sleep in FM is associated with increased somatic symptoms, pain, anxiety, hyperalgesia, altered quality of life, fatigue, depression, mood changes, and higher medical costs.[31–37]

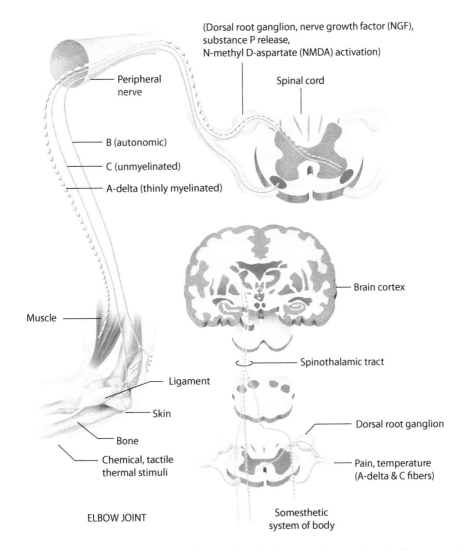

(Dorsal root ganglion, nerve growth factor (NGF), substance P release, N-methyl D-aspartate (NMDA) activation)

Peripheral nerve

Spinal cord

B (autonomic)

C (unmyelinated)

A-delta (thinly myelinated)

Brain cortex

Muscle

Ligament

Spinothalamic tract

Skin

Bone

Dorsal root ganglion

Chemical, tactile thermal stimuli

Pain, temperature (A-delta & C fibers)

ELBOW JOINT

Somesthetic system of body

Figure 9.2 Ascending pain pathways in healthy individuals. (Updated and revised with permission from Wallace DJ, Wallace JB. *All about Fibromyalgia*. New York/London: Oxford University Press, 2003.)

THE MANAGEMENT OF SLEEP DISORDERS IN FM

Before treating sleep problems in a FM patient, a medical work-up should classify the nature of the problem. Does the patient have classic FM? Is periodic limb movement syndrome part of the picture? Is sleep apnea, hypopnea, upper-airway resistance syndrome, or bruxism present? Are there psychiatric considerations? Are medical comorbidities such as hypothyroidism or inflammatory arthritis present? Is the patient taking medications (especially for fatigue) that keep them up during the day or make it more difficult to sleep? Is the patient taking analgesic agents that interfere with sleep? These aspects may influence the medications or treatment regimens advocated.

Ascertainment methodologies

Improved quality of sleep in FM has been correlated with decreased musculoskeletal pain, better quality of life, and less fatigue. A review of studies examining sleep in FM found that multiple methodologies, most of which were being validated for non-FM conditions, were used as methods of ascertainment.[38]

These included interviewing patients, keeping a sleep–wake diary, self-rating scales (Beck Depression Inventory, Multidimensional Fatigue Inventory, and Fatigue Severity Scale), sleepiness scales (e.g., Stanford Sleepiness Scale and Epworth Sleepiness Scale), qualitative and quantitative self-rating scales of sleep (e.g., Pittsburgh Sleep Quality Index, Sleep Assessment Questionnaire, and Karolinska Sleep Diary), and performance tasks. The Jenkins Sleep Scale is a four-item questionnaire that evaluates the frequency and intensity of certain sleep difficulties in respondents. The Medical Outcomes Sleep Scale measures six dimensions of sleep, including initiation, maintenance (e.g., staying asleep), quantity, adequacy, somnolence (e.g., drowsiness), and respiratory impairments (e.g., shortness of breath and snoring). These metrics and inventories are complemented by polysomnography.[39,40]

Sleep hygiene

The rules of sleep hygiene reviewed in Chapters 16 and 19 are applicable here. Briefly, they include making sure the room is dark, the mattress firm, that bed partners do not snore, that children and/or pets are not in the bedroom, taking a hot shower before sleeping, not napping during the day, creating a restful environment during the hour before going to sleep, eliminating alcohol or caffeine after 6 p.m., not exercising in the evenings, and going to sleep and waking up at the same times, among other actions. If one is up in the middle of the night and unable to sleep, it is desirable to do an activity or read for less than 1 hour and then return to bed and wake up at the same time as usual. In FM, any actions that reduce anxiety or improve body mechanics are appropriate additional considerations.

Medication

An Internet survey of 2596 patients with self-reported FM revealed that 30%–40% reported ongoing use of either cyclobenzaprine, amitriptyline, zolpidem, or a benzodiazepine for insomnia.[41] Since few controlled studies have evaluated sleep therapies for FM, textbooks and experience from rheumatic disease practices have generated a set of general concepts, which are listed below[42–63]:

1. Medication should only be used in patients who have failed implementation of sleep hygiene regimens. Elderly patients are at an increased risk of falls.
2. Is the diagnosis correct, and are there other medical or psychiatric considerations that apply?
3. If medication is to be used, a tricyclic antidepressant that promotes sleep with or without muscle relaxation can be prescribed. These include cyclobenzaprine, trazodone, amitriptyline, or doxepin in low doses given 1–2 hours before going to bed. Tizanidine also has a place in the management of FM-disordered sleep.
4. Selective serotonin reuptake inhibitors may make sleep more problematic, but can be used with tricyclics. Nonsteroidal anti-inflammatory agents have no effect on sleep. Diphenhydramine and other sedating antihistamines do not adequately address the problem.
5. Failure to respond or a partial response to the above regimens is followed by the introduction or addition of a benzodiazepine. These agents are effective and may ameliorate anxiety, but can tolerize (e.g., diazepam) and sometimes lead to depression (e.g., clonazepam). The most commonly prescribed drugs used in FM patients include temazepam or zolpidem. Rarely, sodium oxybate may be indicated.
6. Severe anxiety may warrant innovative interventions such as atypical antipsychotics (e.g., quetiapine), concomitant burning, and tingling agents such as gabapentin or pregabalin.
7. Selected patients may benefit from over-the-counter supplements such as melatonin, valerian root, or St John's wort.
8. Non-medicinal methods of promoting stress reduction and encouraging restful sleep include hypnotherapy, acupuncture, music, manual therapy, electro-acupuncture, meditation, behavioral therapy, and supervised exercise (38–42).
9. Failure to respond to usual measures warrants a psychiatric evaluation and polysomnogram with a sleep center consultation.

SUMMARY

FM is a pain amplification syndrome produced by persistent afferent sensory stimulation and

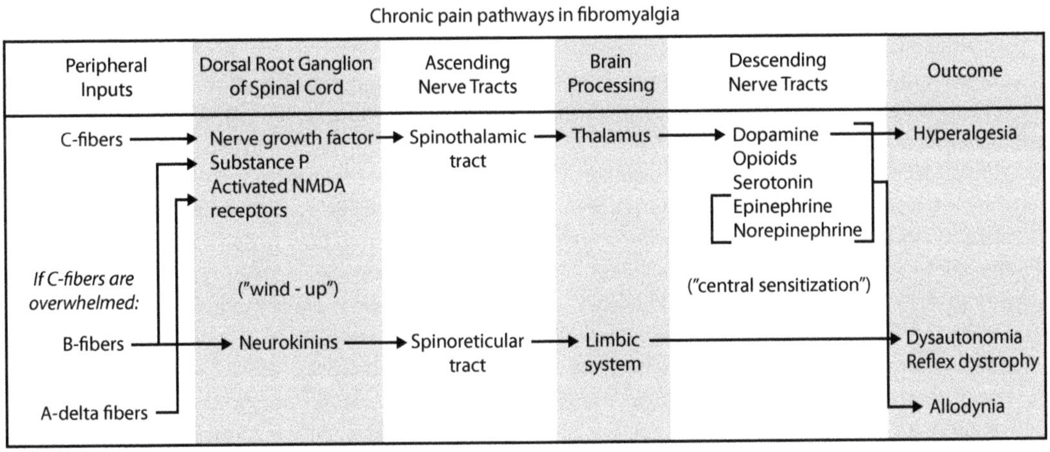

Chronic pain pathways in fibromyalgia
Peripheral Inputs
Dorsal Root Ganglion of Spinal Cord
Ascending Nerve Tracts
Brain Processing
Descending Nerve Tracts
Outcome
C-fibers
Nerve growth factor
Substance P
Activated NMDA receptors
Spinothalamic tract
Thalamus
Dopamine
Opioids
Serotonin
Epinephrine
Norepinephrine
Hyperalgesia
If C-fibers are overwhelmed:
("wind - up")
("central sensitization")
B-fibers
Neurokinins
Spinoreticular tract
Limbic system
Dysautonomia
Reflex dystrophy
A-delta fibers
Allodynia

Figure 9.3 Ascending and descending pathways in fibromyalgia. (Updated and revised with permission from Wallace DJ, Wallace JB. *All about Fibromyalgia*. New York/London: Oxford University Press, 2003.)

manifested as a central sensitization syndrome. It is not a disease, but is present in a variety of medical and behavioral conditions. FM is modified by hormonal, cytokine, neurotransmitter, and autonomic influences. The overwhelming majority of people with FM have sleep disorders, with the alpha–delta abnormality being the principal pathology. Managing sleep pathology in FM appropriately ameliorates the symptoms and signs of the syndrome more than almost any other intervention (Figures 9.3 and 9.4).

Sympathetic nerve activity
Substance P
Nerve growth factor
Estrogen
Prolactin
CRH
Interleukin - 1 beta
Interleukin - 6
Behavioral makeup stress
Serotonin
ACTH
Endorphins
Cortisol
Pain stimulus
Melatonin
Sleep
↑Growth hormone
Cell and muscle growth and repair

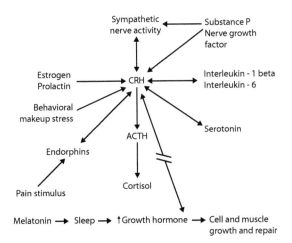

Figure 9.4 How sleep is influenced by hormones, neurotransmitters, and cytokines. (Updated and revised with permission from Wallace DJ, Wallace JB. *All about Fibromyalgia*. New York/London: Oxford University Press, 2003.)

REFERENCES

1. Wallace DJ. The history of fibromyalgia. In: Wallace DJ, Clauw DJ (Eds). *Fibromyalgia and Other Central Pain Syndromes*. Philadelphia, PA: Lippincott, Williams & Wilkins, 2005, pp. 1–8.
2. Smythe HA, Moldofsky H. Two contributions to understanding of the fibrositis syndrome. *Bull Rheum Dis* 1977–1978; 28: 928–931.
3. Yunus M, Masi AT, Calabro JJ et al. Primary fibromyalgia (fibrositis): Clinical study of 50 patients with matched normal controls. *Semin Arthritis Rheum* 1981; 11: 151–170.
4. Wolfe F, Smythe HA, Yunus MB et al. The American College of Rheumatology 1990 criteria for the classification of fibromyalgia. Report of the Multicenter Criteria Committee. *Arthritis Rheum* 1990; 33: 160–172.
5. Wallace DJ, Wallace JB. *Making Sense of Fibromyalgia: An Essential Guide for Patients and Their Families*. New York/London: Oxford University Press, 2014.
6. Wolfe F, Ross K, Anderson J et al. The prevalence and characteristics of fibromyalgia in the general population. *Arthritis Rheum* 1995; 38: 19–28.
7. White KP, Speechley M, Harth M et al. The London Fibromyalgia Epidemiology Study: Comparing the demographic and clinical characteristics in 100 random community cases versus controls. *J Rheumatol* 1999; 26: 885–889.

8. Wolfe F, Clauw F, Fitzcharles M-A et al. The American College of Rheumatology preliminary diagnostic criteria for fibromyalgia and measurement of symptom severity. *Arthritis Care Res* 2010; 62: 600–610.

9. Yunus MB. Central sensitivity syndromes: A new paradigm and group nosology for fibromyalgia and overlapping conditions, and the related issue of disease versus illness. *Semin Arthritis Rheum* 2008; 37: 339–352.

10. Clauw DJ, Crofford LJ. Chronic widespread pain: What we know and what we need to know. *Best Pract Res Clin Rheumatol* 2003; 17: 685–701.

11. Hallegua DS, Wallace DJ. Managing fibromyalgia: A comprehensive approach. *J Musculoskelet Med* 2005; 22: 382–390.

12. Yunus M, Masi AT, Calabro JJ et al. Primary fibromyalgia (fibrositis): Clinical study of 50 patients with matched normal controls. *Seminars Arthritis Rheum* 1981; 11: 151–171.

13. Abeles AM, Pillinger MH, Soltiar BM, Abeles M. Narrative review: The pathophysiology of fibromyalgia. *Ann Intern Med* 2007; 146: 726–734.

14. Caro XJ, Winter EF. Evidence of abnormal epidermal nerve fiber density in fibromyalgia: Clinical and immunologic implications. *Arthritis Rheum* 2014; 66: 1945–1954.

15. Staud R, Vierck CJ, Cannon RL et al. Abnormal sensitization and temporal summation of second pain (wind up) in patients with fibromyalgia syndrome. *Pain* 2001; 91: 165–175.

16. Ohayon MM. Prevalence and correlates of nonrestorative sleep complaints. *Arch Intern Med* 2005; 165: 35–41.

17. Landis CA, Lentz MJ, Rothermel J, Buchwald D, Shaver JL. Decreased sleep spindles and spindle activity in midlife women with fibromyalgia and pain. *Sleep* 2004; 27: 741–750.

18. Moldofsky H. The significance of the sleeping–waking brain for the understanding of widespread musculoskeletal pain and fatigue in fibromyalgia syndrome and allied conditions. *Joint Bone Spine* 2008; 75: 397–402.

19. Roizenblatt S, Moldofsky H, Benedito-Silva AA, Tufik S. Alpha sleep characteristics in fibromyalgia. *Arthritis Rheum* 2001; 44: 222–230.

20. Rizzi M, Sarzi-Puttini P, Atenzi F et al. Cyclic alternating pattern: A new marker of sleep alteration in patients with fibromyalgia. *J Rheumatol* 2004; 31: 1193–1199.

21. Ruhr UD, Herold J. Melotonin deficiencies in women. *Mauritas* 2002; 41(Suppl 1): S85–S104.

22. Bagge E, Bengstsson BA, Carlsson L, Carlsson J. Low growth hormone secretion in patients with fibromyalgia—A preliminary report on 10 patients and 10 controls. *J Rheumatol* 1998; 25: 145–148.

23. Besteiro Gonzalez JL, Suarez-Fernandez TV, Arboleva Rodrigues L, Muniz J, Lemos Giraldez S, Alverez Fernandez A. Sleep architecture in patients with fibromyalgia. *Psiotherma* 2011; 23: 368–373.

24. Bara-Jimeniz W, Aksu M, Graham B et al. Periodic limb movements in sleep: State-dependent excitability of the spinal flexor reflex. *Neurology* 2000; 54: 1609–1616.

25. Civelek GM, Cifkaya PO, Karatas M. Evaluation of restless legs syndrome in fibromyalgia syndrome: An analysis of quality of sleep and life. *J Back Musculoskelet Rehab* 2014; 27:537–544.

26. Gold A, Dipalo F, Gold MS, Broderick J. Respiratory airflow dynamics during sleep in women with fibromyalgia. *Sleep* 2004; 27: 459–466.

27. Chen LX, Baqir M, Schumacher HR et al. Increased incidence of sleep apnea in fibromyalgia patients. *Arthritis Rheum* 2004; 50: S494.

28. Bennett RM, Clark SR, Campbell SM, Burkhardt CS. Low levels of somatomedin C in patients with the fibromyalgia syndrome. A possible link between sleep and muscle pain. *Arthritis Rheum* 1992; 35: 1113–1116.

29. Landis CA, Lentz MJ, Rothermel J et al. Decreased nocturnal levels of prolactin and growth hormone in women with fibromyalgia. *J Clin Endocrinol Metab* 2001; 86: 1672–1678.

30. Wikner J, Hirsch Y, Wetterberg L, Rojdmark S. Fibromyalgia—A syndrome associated with decreased nocturnal melatonin secretion. *Clin Endcrinol* 1998; 49: 179–183.

31. Diaz-Piedra C, Catena A, Miro E, Pilar Martinez M, Sanchez AI, Buela-Casal G. The impact of pain on anxiety and depression is mediated by objective and subjective sleep characteristics in fibromyalgia patients. *Clin J Pain* 2014; 30: 852–859.

32. Ablin JN, Clauw DJ, Lyden AK et al. Effects of sleep restriction and exercise deprivation on somatic symptoms and mood in healthy adults. *Clin Exp Rheumatol* 2013; 6(Suppl 79): S53–S59.

33. Schuh-Hofer S, Wodarski R, Pfau DB et al. One night of total sleep deprivation promotes a state of generalized hyperalgesia: A surrogate pain model to study the relationship of insomnia and pain. *Pain* 2013; 154: 1613–1621.

34. Wagner JS, DiBonaventura MD, Chandran AB, Cappilleri JC. The association of sleep difficulties with health-related quality of life among patients with fibromyalgia. *BMC Musculoskeletal Disord* 2012; 13: 199.

35. Ulus Y, Akyol Y, Tander B, Durmus D, Bilgici A, Kuru O. Sleep quality in fibromyalgia and rheumatoid arthritis: Associations with pain, fatigue, depression and disease activity. *Clin Exp Rheumatol* 2011; 29(Suppl 69): S92–S96.

36. Okifuji A, Donaldwon GW, Barck L, Fine PG. Relationship between fibromyalgia and obesity in pain, function, mood and sleep. *J Pain* 2010; 11: 1329–1337.

37. Wagner JS, Chandran A, DiBonaventura M, Cappellieri JC. The costs associated with sleep symptoms among patients with fibromyalgia. *Expert Rev Pharmacoecon Outcomes Res* 2013; 13: 131–139.

38. Moldofsky H, MacFarlane JG. Sleep and its potential role in chronic pain and fatigue. In: Wallace DJ, Clauw DJ (Eds). *Fibromyalgia and Other Central Pain Syndromes*. Philadelphia, PA: Lippincott, Williams & Wilkins, 2005, pp. 115–124.

39. Williams DA, Arnold LM. Measures of fibromyalgia: Fibromyalgia Impact Questionnaire (FIQ), Brief Pain Inventory (BPI), Multidimensional Fatigue Inventory (MFI-20), Medical Outcomes Study (MOS) Sleep Scale, and Multiple Ability Self-Report Questionnaire (MASQ). *Arthritis Care Res (Hoboken)* 2011; 63: S86–S97.

40. Crawford BK, Piault EC, Lai C, Sarzi-Puttini P. Assessing sleep in fibromyalgia: Investigation of an alternative scoring method in the Jenkins Sleep Scale based on data from randomized controlled studies. *Clin Exp Rheumatol* 2010; 6(Suppl 63): S100–S109.

41. Bennett RM, Jones J, Turk DC, Russell JI, Matallana L. An internet survey of 2596 people with fibromyalgia. *BMC Musculoskelet Disord* 2007; 8: 27.

42. Citera G, Arias MA, Maldonado-Cocco JA et al. The effect of melatonin in patients with fibromyalgia: A pilot study. *Clin Rheumatol* 2000; 19: 9–13.

43. Clark S, Tindall E, Bennett RM. A double blind crossover trial of prednisone versus placebo in the treatment of fibrositis. *J Rheumatol* 1985; 12: 908–983.

44. Goldenberg DL, Felson DT, Dinerman H. A randomized controlled trial of amitryptiline and naproxen in the treatment of patients with fibromyalgia. *Arthritis Rheum* 1986; 29: 1371–1377.

45. McCain GA, Bell DA, Mai FM, Halliday PD. A controlled study of the effects of a supervised cardiovascular fitness training program on the manifestations of primary fibromyalgia. *Arthritis Rheum* 1988; 31: 1135–1141.

46. Caruso I, Sarzi Puttini P, Cazzola M et al. Double blind study of 5-hydroxytryptophan versus placebo in the treatment of primary fibromyalgia syndrome. *J Int Med Res* 1990; 18: 201–209.

47. Haanen HC, Hoenderdos HT, van Rommunde RK et al. Controlled trial of hypnotherapy in the treatment of refractory fibromyalgia. *J Rheumatol* 1991; 18: 72–75.

48. Reynolds WJ, Moldofsky H, Saskin P, Lue FA. The effects of cyclobenzaprine on sleep physiology and symptoms in patients with fibromyalgia. *J Rheumatol* 1991; 18: 452–454.

49. Fossaluzza V, de Vita S. Combined therapy with cyclobenzaprine and ibuprofen in primary fibromyalgia syndrome. *Int J Clin Pharmacol Res* 1992; 12: 99–102.

50. Deluze C, Bosia L, Zirbs A et al. Electroacupuncture in fibromyalgia: Results of a controlled trial. *Br Med J* 1992; 305: 1249–1252.

51. Santandrea S, Montrone F, Sarzi-Puttini P et al. A double-blind crossover study of two cyclobenzaprine regimens in primary fibromyalgia syndrome. *J Int Med Res* 1993; 21: 74–80.

52. Carette S, Oakson G, Guimont C, Steriade M. Sleep electroencephalography and the clinical response to amitriptyline in patients with fibromyalgia. *Arthritis Rheum* 1995; 38: 1211–1217.

53. Kaplan KH, Goldenberg DL, Galvin-Nadeau M. The impact of a mediation-based stress reduction program on fibromyalgia. *Gen Hosp Psychiatry* 1993; 15: 284–289.

54. Carette S, Oakson G, Guimont C, Steriade M. Sleep electroencephalography and the clinical response to amitriptyline in patients with fibromyalgia. *Arthritis Rheum* 1995; 38: 1211–1217.

55. Moldofsky H, Lue FA, Mously C et al. The effect of zolpidem in patients with fibromyalgia: A dose ranging, double blind, placebo controlled, modified crossover study. *J Rheumatol* 1996; 23: 529–533.

56. Goldenberg D, Mayiskiy M, Mossey C et al. A randomized, double-blind crossover trial of fluoxetine and amitriptyline in the treatment of fibromyalgia. *Arthritis Rheum* 1996; 39: 1852–1859.

57. Citera G, Arias MA, Maldonado-Cocco JA et al. The effect of melatonin in patients with fibromyalgia: A pilot study. *Clin Rheumatol* 2000; 19: 9–13.

58. Scharf MB, Baumann M, Berkowitz DV. The effects of sodium oxybate on clinical symptoms and sleep patterns in patients with fibromyalgia. *J Rheumatol* 2003; 30: 1070–1074.

59. Crofford LJ, Rowbotham MC, Mease PJ et al. Pregabalin for the treatment of fibromyalgia syndrome: Results of a randomized, double-blind, placebo-controlled trial. *Arthritis Rheum* 2005; 52: 1264–1273.

60. Edinger JD, Wohlgemuth WK, Krystal AD et al. Behavioral insomnia therapy for fibromyalgia patients: A randomized clinical trial. *Arch Intern Med* 2005; 165: 2527–2535.

61. Martinez MP, Miro E, Sanchez AI. Cognitive-behavioral therapy for insomnia and sleep hygiene in fibromyalgia: A randomized controlled trial. *J Behav Med* 2014; 37: 683–697.

62. Castro-Sanchez AM, Aguillar-Ferrandiz ME, Mataran-Penarocha GA, Sanchez-Joya Mdel M, Arroyo-Morales M, Fernandez-de-Las-Penas C. Short term effects of a manual therapy protocol on pain, physical function, quality of sleep, depressive symptoms and pressure sensitivity in women and men with fibromyalgia syndrome: A randomized controlled trial. *Clin J Pain* 2014; 30: 589–597.

63. Castro-Sanchez AM, Mataran-Penaroccha GA, Granero-Molina J, Aquilera-Manrique G, Quesada-Rubio JM, Moreno-Lorenzo C. Benefits of massage–myofascial release therapy on pain, anxiety, and quality of life in patients with fibromyalgia. *Evid Based Complement Alternat Med* 2011; 2011: 561753.

10

Sleep in chronic fatigue syndrome

ZOE MARIE GOTTS, JASON ELLIS, AND JULIA NEWTON

INTRODUCTION

Most people feel overly tired at some time or another, and in 10% of people who see their general practitioner, fatigue presents as the principle symptom. However, 1%–2% of the population experience severely disabling and ongoing fatigue. It is this significant minority of patients enduring profound and unexplained fatigue that has been the topic of debate for the past 30 years. Naming the illness has been a topic of controversy dating back to the early 1980s, when patients were considered to be having a reaction to stressors from modern society, otherwise labelled as "yuppie flu."[1] Conflicting arguments advocating for an organic cause of the illness advanced a series of names to reflect this, such as post-viral fatigue syndrome and myalgic encephalomyelitis (ME).

In response to the debates surrounding the use of terminology with the condition, and in an attempt to define a homogenous patient group for the purpose of research, a renaming of the condition to chronic fatigue syndrome (CFS) was coined by the Centers for Disease Control (CDC) in Atlanta, who also published the first standardized diagnostic criteria for the condition.[2] Following this, a number of related definitions were published from the United Kingdom, Australia, and Canada, and these case definitions are discussed further later in the chapter.

Fukuda and colleagues endeavored to standardize the diagnosis of CFS across countries in their publication of a consensus definition for the illness.[4] The definition specified that fatigue is the primary symptom, which should be of definite onset and cause significant disruption to the person's life. In addition to fatigue, at least four other key symptoms are required to fulfill these diagnostic criteria, including muscle and joint pain, headache, cognitive dysfunction, and unrefreshing sleep.

The National Health Service (NHS) estimated that 250,000 people in the United Kingdom have CFS. Sleep disturbances are frequently reported in CFS, and these complaints have been shown to persist throughout the course of the illness.[8] The symptom presentation and issues with the diagnostic criteria are discussed further in the following sections.

CFS AND ITS CLINICAL PRESENTATION

CFS is a chronic, complex, systemic disease that often can profoundly affect the lives of patients. With no consistently identifiable biomarkers, diagnosis relies upon symptom report criteria. Patients with CFS experience a multitude of symptoms. These range from those of a physical nature (i.e., severe malaise and fatigue following physical activity and muscle and joint pain/myalgia), to those that suggest ongoing abnormalities in immune system function (i.e., sore throat, swollen/painful glands, headaches, temperature control, and intermittent flu-like feelings), to brain and central nervous system symptoms (i.e., dizziness, mental fatigue, cognitive dysfunction, palpitations, and symptoms associated with low blood pressure/postural hypotension and fainting). Other symptoms include sleep disturbances (often increased requirements for sleep at illness onset followed by problems with sleep maintenance or onset and waking unrested) and irritable bowel symptomology. Over time, patients can also develop emotional lability and mood disturbances. Not all symptoms are experienced by all patients, and besides the widespread symptoms described, there are also a myriad of "minor" ones. Fluctuation in symptom severity is common and patients often report as having "good" and "bad" days. Nonetheless, individuals experience marked functional disability[9] and face significant reductions in their quality of life.[10]

DIAGNOSTIC CHALLENGES

In order to accurately diagnose an illness or disease, it is important to have a reliable set of criteria for researchers and clinicians. That said, diagnosing CFS can be complicated by a number of factors. There is no generally accepted diagnostic test to reliably diagnose or exclude CFS (i.e., no laboratory test or biomarker for CFS), and fatigue and other symptoms of CFS are common to many illnesses.

Further, the illness has a pattern of remission and relapse, and symptoms vary from person to person in type, number, and severity.

Whilst CFS affects at least 250,000 people in the United Kingdom, the variance in the rates reported from epidemiological studies (0.23%–2.6%[11,12]) are a likely result of the differing published criteria and the guidance that they set out. A meta-analysis to examine variability among prevalence estimates for CFS suggested that the observed heterogeneity in CFS prevalence may also be due to differences in the method of assessment used.[13]

Case definitions

There are many aspects to CFS that are controversial, from its etiology through to its pathophysiology, treatments, and even to the naming of the condition. Classification, including the case definitions and the problems surrounding the varied clinical descriptions of CFS, is the key issue when it comes to agreement on naming the condition, and this disagreement occurs among researchers, medical practitioners, and patients. This is problematic for a field of research that is attempting to redress uncertainty. Despite there being some overlap of symptoms between the clinical descriptions, the definitions differ.

Sleep-related symptoms in case definitions and diagnostic criteria for CFS/ME

Issues with classification add to the contentious nature of the illness. With disagreements already existing between patients, clinicians, and researchers over its etiology, treatment, and name, classification of this condition is an area that warrants further work. That said, the 1994 CDC case definition (Fukuda et al.[4]) appears to be the most reliable clinical assessment tool available at the current time,[13] and has been recommended for use in the UK clinical services.[7] It specifies that in addition to being present for at least 6 months, fatigue must have a definite onset, cause substantial disruption to the individual's day-to-day activities, and should not be caused by continual exertion. At least four additional key symptoms, such as muscle and joint pain, headaches, unrefreshing sleep, and cognitive dysfunction, need to be reported. There is also a final requirement that other known causes of chronic fatigue must have been ruled out, specifically clinical depression, side effects of medication, eating disorders, and substance abuse. However, improving clinical case definitions and their adoption internationally will enable better comparisons of findings and inform healthcare systems about the true burden of CFS.

Minor case definition criteria: The pitfalls

The lack of regularity in working case definitions and guidelines for CFS creates uncertainty regarding the role of sleep in CFS. Specifically, complication exists as to where sleep complaints fit within the available case definitions of CFS. For example, the Holmes et al. (CDC) definition includes sleep disturbances such as hypersomnia or insomnia in its minor criteria,[2] whereas the Fukuda et al. (CDC) definition considers sleep disorders such as sleep apnea and narcolepsy as exclusionary and regards unrefreshing sleep as a minor criterion (Table 10.1).[4] Given that some available criteria regard sleep problems as minor—whilst sleep complaints are consistently reported in CFS—together with the considerable variation in the use of terminology, there is scope to investigate sleep disturbances in CFS in much more detail and, certainly, there is a need to establish more definitive criteria with regards to sleep. This raises the issue of the overlap/confusion between sleep disorders and CFS.

NEW UPDATES TO THE CRITERIA FOR CFS

In 2015, the Institute of Medicine (IOM) proposed new diagnostic criteria for CFS in order to help facilitate timely diagnosis and care and enhance understanding among healthcare providers and the public. The IOM also proposed a new name for CFS,[14] recommending replacing the term CFS/ME with "systemic exertion intolerance disease" (SEID). In the new definition, the IOM outlines three core symptoms, of which sleep disturbance features as one of these requirements ("unrefreshing sleep") (Box 10.1).

PATIENT DESCRIPTIONS OF SLEEP-RELATED SYMPTOMS IN CFS

Patients with CFS describe sleep to be a vital process for health and well-being, which has a direct bearing on the course and progression of their illness.[15] They often experience sleep-related problems such as insomnia, unrefreshing sleep, non-restorative sleep and sleep disturbances, and these sleep-related symptoms can be helpful in attempting to operationalize sleep in the CFS/ME diagnosis.[14] However, the nature (and severity) of sleep problems changes over the illness course,[15] but typically, the complaint of "unrefreshing sleep" remains universal among patients with CFS/ME. Some terms commonly used by patients to describe their "unrefreshing sleep" symptoms (and which have the potential to alert clinicians to the diagnosis) are: "Feeling like I never slept," "Cannot fall asleep or stay asleep," and "After long/normal hours' sleep, I still don't feel good in the morning" (Table 10.2).[14]

DOES SLEEP PLAY A ROLE IN THE ETIOPATHOGENESIS OF CFS?

The etiology and pathophysiology of CFS remain as disputed as the nosology, with several theories proposed ranging from viral infections[16] and immunological and neurobiological factors[17,18] to psychological stress.[19] Given that CFS is not likely to be explained by one single etiological mechanism, it has been proposed that it is the interaction of multiple factors that serves to precipitate and/or maintain CFS. This more generic bio-psychosocial model has been proposed by various authors.[20,21] This "3-P" model incorporates predisposing, precipitating, and perpetuating factors, and ultimately seeks to explain the phenomenology of the condition as arising through the interaction of biological, affective, behavioral, and cognitive factors.[20,22] In a recent review of explanatory models of functional somatic symptoms,[23] this multifactorial model was distinguished from other single-modality models (which propose that symptoms are the result of one pathogenic mechanism) as being a meta-model that provides a coherent theoretical framework for describing how the interaction of such physiological, behavioral, cognitive, and affective factors can cause and/or exacerbate physical symptoms. The 3-P model hypothesizes that the combination of predisposing, precipitating, and perpetuating factors serves to keep the condition going. This model is particularly important when considering factors that might come into play in terms of mediating the exacerbation of sleep-related complaints in CFS. The perpetuating factors range from the physical through to behavioral, physiological, and cognitive contributors, which may all serve to maintain symptoms, particularly disturbed sleep.

Table 10.1 Overview of the different case definitions for CFS

	Holmes	Fukuda	Reeves	Oxford	ICC
Minimum duration of illness	6 months	6 months	NA	6 months	NA
Onset type	Distinct	New or definite	NA	Distinct	Infectious or gradual
Laboratory tests used	Minimum battery of standard laboratory screening tests looking for known cause of fatigue	Minimum battery of standard laboratory screening tests looking for known cause of fatigue	Routine analysis of blood and urine	None	None listed
Exclusions	Clinical conditions that would produce similar symptoms	Unless clinically indicated, no additional tests are required to exclude other diagnosis Findings, laboratory, or imaging test suggesting the presence of a condition that may explain chronic fatigue must be resolved (meaning is not clear) before further classification	A list of permanent medical and psychiatric exclusions is given, as well as possible exclusions	Medical conditions that cause chronic fatigue Range of mental health disorders Organic brain disease	Unless clinically indicated, no additional tests are required to exclude other diagnoses Primary psychiatric disorders, somatoform disorder, and substance use are excluded
Depression and anxiety	Not excluded	Not excluded, only major depressive disorder with psychotic or melancholic feature is excluded	Not excluded, only major depressive disorder with psychotic or melancholic feature is excluded for 5 years before onset of illness	Not excluded	Not excluded; reactive depression is

(Continued)

Table 10.1 (Continued) Overview of the different case definitions for CFS

	Holmes	Fukuda	Reeves (2005)	Oxford	ICC
PEM	Increased symptoms of fatigue as a result of exercise (previously tolerated) Recovery 24 hours or longer	PEM not required for this diagnosis, but can be included as a minor symptom Increased symptom of malaise after exertion Recovery 24 hours or longer	Increased symptoms of fatigue as a result of any activity that is not a demanding schedule	No version of PEM included in the criteria	Increased symptoms of fatigue as a result of any level of activity No duration of recovery required
Fatigue	Debilitating fatigue or fatigability	Persistent or relapsing chronic fatigue that is not the result of ongoing exertion and not substantially alleviated by rest that substantially reduces activity level	Fatigue is incorporated into the three self-report scales	Fatigue of psychiatric or idiopathic origin	Fatigue is included under the term PENE: a pathological inability to produce sufficient energy on demand
Minor symptoms	6 or more of the 11 symptom criteria and 2 or more of the 3 physical criteria; or 8 or more of the 11 symptoms listed (minimum of 6–8 symptoms)	4 or more of the 8 symptoms listed	NA	May be present	1 symptom from each of the 3 symptom categories of pain, sleep disturbance, and cognitive symptoms 3 symptoms from a mix of immune and neuroendocrine/autonomic symptoms 1 symptom from autonomic symptoms (minimum of 7 symptoms)

(Continued)

Table 10.1 (*Continued*) Overview of the different case definitions for CFS

	Holmes	Fukuda	Reeves (2005)	Oxford	ICC
Pain	New headaches, muscle discomfort, or myalgia Migratory arthralgia without joint swelling or redness	New headaches Muscle pain Multi-joint pain without swelling or redness	NA	NA	Headaches Noninflammatory muscle pain or joint pain Abdomen or chest pain
Sleep disturbance symptoms	Sleep disturbance	Unrefreshing sleep	NA	NA	Sleep disturbance Unrefreshing sleep
Cognitive/neurological symptoms	Neuropsychological complaints Muscle weakness	Symptoms related to cognitive impairment	NA	NA	Symptoms related to cognitive impairment Perceptual and sensory disturbances Ataxia Muscle weakness Fasciculations Sensory overload
Autonomic symptoms	Fever (temperature 37.5–38.6°C) or chills	NA	NA	NA	Symptoms related to blood pressure, gastric and urinary systems, cardiac involvement
Neuroendocrine symptoms	NA	NA	NA	NA	Symptoms related to temperature Genitourinary symptoms
Immune symptoms	Painful lymph nodes, sore throat	Painful lymph nodes, sore throat	NA	NA	Symptoms such as painful lymph nodes, sore throat, influenza-like symptoms, sensitivities to food, medicine and/or chemicals

Source: Adapted from Morris G and Maes M. *BMC Medicine* 2013; 11(1): 205.
NA, not applicable; ICC, international consensus criteria for myalgic encephalomyelitis; PEM, post-exertional malaise; PENE, post-exertional neuroimmune exhaustion.

BOX 10.1: Diagnostic criteria for CFS/ME (SEID)

PROPOSED DIAGNOSTIC CRITERIA FOR CFS/ME

Diagnosis requires that the patient have the following three symptoms:

1. A substantial reduction or impairment in the ability to engage in pre-illness levels of occupational, educational, social, or personal activities, that persists for more than 6 months and is accompanied by fatigue, which is often profound, is of new or definite onset (not lifelong), is not the result of ongoing excessive exertion, and is not substantially alleviated by rest
2. Post-exertional malaise*
3. Unrefreshing sleep*

At least one of the two following manifestations is also required:

1. Cognitive impairment*
2. Orthostatic intolerance

---

* Frequency and severity of symptoms should be assessed.

The diagnosis of ME/CFS (SEID) should be questioned if patients do not have these symptoms at least half of the time with moderate, substantial, or severe intensity.

Adapted from IOM (Institute of Medicine). Beyond Myalgic Encephalomyelitis/Chronic Fatigue Syndrome: Redefining an Illness. *Washington, DC: The National Academies Press; 2015. http://www.iom.edu/mecfs.*

The role of sleep in existing CFS models

We should be mindful of a number of factors that interact throughout the course of illness and ultimately keep disturbed sleep as an ongoing problem in patients. In individuals who are more vulnerable to developing CFS (i.e., increased arousal), sleep may act as a mediating factor between stress and disability, and thus precipitate illness onset. A combination of socio-emotional factors (i.e., lack of support and understanding of the condition) in the early stages of the illness, and socio-economic factors (i.e., not being able to work and adapting to living on disability) that occur during the illness course may contribute to CFS and feed into symptoms such as fatigue. A consequence of these factors is an irregular sleeping pattern, which contributes to reductions

Table 10.2 Operationalizing sleep in the chronic fatigue syndrome/myalgic encephalomyelitis diagnosis

Symptom	Patient descriptions	Questions to ask (explore frequency and severity)	Observations to make; tests to conduct
Unrefreshing sleep	• "Feeling like I never slept" • "Cannot fall asleep or stay asleep" • "After long or normal hours of sleep, I still don't feel good in the morning"	• Do you have any problems getting to sleep or staying asleep? • Do you feel rested in the morning or after you have slept? • Tell me about the quality of your sleep • Do you need too much sleep? • Do you need to take more naps than other people? (There may be other disruptors as well)	There is no evidence that currently available sleep studies contribute to the diagnosis of chronic fatigue syndrome/myalgic encephalomyelitis

Source: Adapted from IOM (Institute of Medicine). *Beyond Myalgic Encephalomyelitis/Chronic Fatigue Syndrome: Redefining an Illness.* Washington, DC: The National Academies Press; 2015. http://www.iom.edu/mecfs.

in sleep efficiency (SE), less deep (slow-wave) sleep, circadian dysregulation, and homeostatic dysregulation. Disruptions in these four features of sleep may lead to a sleep disorder that, in turn, reinforces the existing fatigue. The disordered sleep may also feed into hypothalamic-pituitary-adrenal axis (HPA) irregularities and autonomic dysregulation, and also have an impact on the cognitive and behavioral components of sleep with regards to a patient's beliefs about their required amounts of sleep and their resulting sleep behaviors (i.e., sleep state misperception and napping). This reinforces the symptom of fatigue and thus the continuation of an irregular sleeping pattern (Figure 10.1).

Where other (bio)psychosocial models have been proposed,[21,24] these emphasize the role of precipitants and the eventual perpetuation of symptoms in CFS. They focus on explaining how precipitating and perpetuating factors (i.e., prior stress and personality traits) induce the biological pathophysiology that accounts for specific symptoms. The proposed model (Figure 10.1) therefore acknowledges such vulnerability factors that may predispose individuals to CFS. Prospective studies may wish to explore these further using birth cohorts, given that these may infer arousability as a generic risk factor for ill health.

Considered together, the theoretical model of the role of sleep in CFS may explain the process by which fatigue develops, is maintained and continues via a cycle of irregular sleeping patterns, biological dysregulation, and cognitive and behavioral responses. This preliminary model integrates sleep into existing models of CFS. The model might be considered as a useful basis for CFS research and clinical work, helping to generate more multidisciplinary, mixed-methodology work in order to further understand this complex illness.

THE PRESENCE OF SLEEP DISORDERS IN CFS

CFS/ME has the largest prevalence of a diagnosable sleep disorder in any single illness population, with

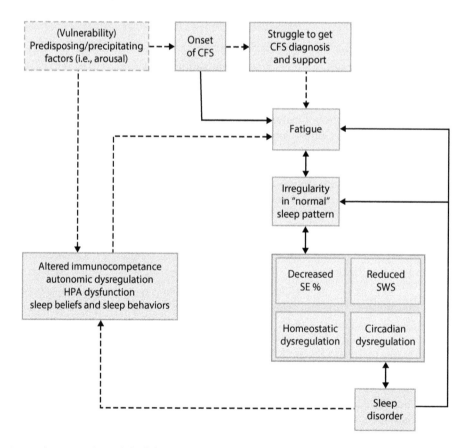

Figure 10.1 A theoretical model of sleep in chronic fatigue syndrome.

rates as high as 69%.[25–28] Furthermore, studies have shown the presence of sleep apnea in up to 65% of patients with CFS. Sleep-disordered breathing can in itself result in arousals during sleep, which can create daytime symptoms of sleepiness and fatigue.[29] There are effective interventions available for sleep apnea (e.g., continuous positive airway pressure); however, there is little evidence for the effectiveness of such treatments in reducing CFS symptoms.[30] Even after excluding individuals with diagnosable sleep disorders, population-based studies report that between 87% and 95% of patients continue to report unrefreshing sleep as a principle complaint.[12,31]

The presence of sleep disorders has been highlighted in sleep studies with CFS patients. In a single-night polysomnography (PSG) and self-report observational study,[32] it was found that, of the 37 CFS patients, 58% fulfilled the criteria for a diagnosable sleep disorder, 11 (42%) had apnea, and 4 (16%) had restless legs syndrome/periodic limb movement disorder (RLS/PLMD). High rates of self-reported insomnia (86%) were also evident in the CFS group. In another study that combined PSG, actigraphy, and self-reports, 42 (86%) of the 49 CFS patients had a diagnosable sleep disorder, with 32 (65%) meeting *Diagnostic and Statistical Manual of Mental Disorders—4th Edition* (DSM-IV) criteria for chronic insomnia and 10 (20%) with apnea.[25] In a recent large-scale study of 205 patients (n = 410, combined data over two recorded nights), a third of the sample had suspected apnea, based on their respiratory disturbance index (average number of episodes of apnea, hypopnea, and respiratory event-related arousal per hour of sleep).[33] In a recent survey of a specialist CFS service in London,[34] it was reported that of the 377 patients referred to the CFS service, almost half (49%) had alternative diagnoses made, based on an assessment that included a detailed history and physical and mental state examination. Of those assessed (n = 250), a sleep disorder was the most common diagnosis (28%), which suggests an alternative diagnosis may be warranted.

These findings highlight the fact that sleep disorders (i.e., insomnia, apnea, RLS, and PLMD) may well be comorbid, overlooked, or misdiagnosed in CFS. Furthermore, some authors have suggested that CFS is a primary sleep disorder in itself. However, this questions the nature of causal relationships between physiology, symptoms, and behavior in this population (indeed, in any chronic

condition) and how the appellation "primary" must be used with caution. Hypersomnia and insomnia are common features of CFS and are as likely to be effects as they are causes of fatigue. To date, no treatment studies have sought to exclusively target sleep in order to ascertain how this impacts on other symptoms. Only experimental designs would afford causality. Given the lack of consistency in findings from studies of sleep architecture and multiple sleep latency times, it could also be considered that symptoms such as unrefreshing sleep and unremitting fatigue may not reflect a sleep disorder per se, but rather impaired sleep homeostasis,[35,36] the body's natural ability to regulate the sleep/wake cycle,[37] but again this would have to be experimentally tested. Practically, what this work highlights is the necessity of performing thorough sleep assessments in this population.

SLEEP SUBTYPES IN CFS

Sleep problems are not homogenous in CFS/ME, and four sleep-specific phenotypes have been described in a Dutch CFS/ME patient cohort (Figure 10.2).[26] After excluding a third of individuals for an objectively verifiable sleep disorder (obstructive sleep apnea or

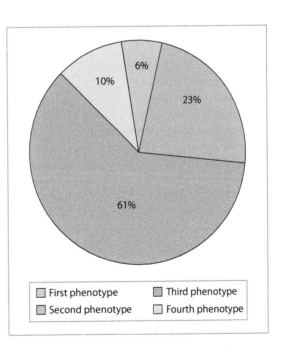

Figure 10.2 Proportion of Dutch chronic fatigue syndrome cohort in each of the four sleep phenotypes.

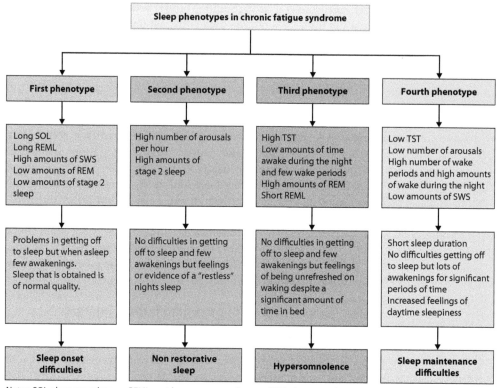

Figure 10.3 Key features and subjective presentation of the four chronic fatigue syndrome sleep phenotypes.

PLMD) that could explain the CFS/ME diagnosis, 89.1% of the CFS/ME sample (n = 239) met quantitative criteria for insomnia or hypersomnolence disorders. In this single-night, case-controlled observational PSG study, groups 1 and 4 are characterized by insomnia symptoms (difficulties initiating or maintaining sleep), groups 2 and 3 share overlapping characteristics of disorders characterized by poor sleep quality, and group 3 in particular shares characteristics of hypersomnolence (longer total sleep duration or shorter sleep-onset latencies). Importantly, these different sleep profiles can be defined in clinical practice based on subjective reports of sleep from patients after ruling out objective sleep disorders, as information from a detailed sleep interview and sleep diary provides a subjective account that can be matched to the phenotypes (Figure 10.3).[26]

First phenotype

The first phenotype is characterized by long sleep-onset and rapid eye movement (REM) latencies,

and a high percentage of slow-wave sleep (SWS). Moreover, this group has low percentages of both stage 2 sleep and REM sleep. Statistically, this phenotype differs from the other three groups in terms of longer sleep-onset and REM latencies and a lower percentage of REM.

Second phenotype

The second phenotype has the highest percentage of stage 2 sleep and the highest number of arousals per hour, although neither of these variables statistically separated them from the other three phenotypes.

Third phenotype

The third phenotype comprised the largest proportion (61%) of the patient group, with high amounts of total sleep time (TST) and a high percentage of REM. Additionally, this group demonstrated short sleep-onset and REM latencies, low amounts of wake after sleep onset (WASO), percentages of wake time and stage 1 sleep, and a low number of awakenings. Statistically, TST, percentage wake, and WASO differentiated this phenotype from each of the others.

Fourth phenotype

The fourth phenotype demonstrates the highest amount of WASO, percentages of wake time and stage 1 sleep, and the highest number of awakenings. This group has low amounts of TST, numbers of arousals per hour and percentages of SWS. Statistically, only WASO and percentages of wake time differentiated this group from each of the other groups.

EVIDENCE-BASED REVIEW OF SLEEP STUDIES IN CFS

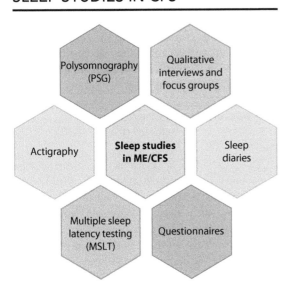

Subjective sleep assessment

QUALITATIVE FINDINGS

The qualitative literature in CFS typically describes themes from both patients and physicians, outlining their perspectives of the illness and identifying the struggle to understand and manage CFS.[10,15,38,39] These studies utilize interview-based techniques with relatively small samples of patients, and differ in their use of case definitions of CFS (i.e., Fukuda, Oxford, or Holmes). Although they discuss quality of life and social factors in CFS—with extreme fatigue, pain, cognitive dysfunction,

and unrefreshing sleep being described as patients' key illness experiences—the studies rarely explore below the surface of these symptoms.[40,41] The main focus of these studies is rather on themes of gender differences in the illness experience, illness beliefs regarding development and potential causes. A recent comprehensive review of the qualitative literature on CFS highlighted that a large proportion of the qualitative studies in CFS and fibromyalgia describe perspectives from healthcare professionals regarding medical practice,[42] rather than focusing on patients' experience. As such, there are no real in-depth data in these qualitative accounts on the role of sleep in CFS. This is an area of research that needs to be addressed, as qualitative work can, amongst other things, indicate the direction needed in order to move forward with hypothesis-driven studies, larger cohort studies, objective measures, and intervention development. In sum, a potentially rich source of data on sleep and how sleep-related disturbances may play a role in maintaining the other symptoms experienced in CFS is being overlooked.

SELF-REPORT DIARIES AND QUESTIONNAIRES

Self-report techniques have been employed to determine the perceptual role of sleep in CFS and include the use of sleep diaries and sleep questionnaires, such as the Pittsburgh Sleep Quality Index (PSQI)[43] and the Epworth Sleepiness Scale (ESS).[44] Studies assessing sleep through self-report methods show higher than normal ESS scores[33,45] and PSQI scores.[45] Mariman and colleagues reported ESS and PSQI scores in 415 Fukuda-defined CFS patients. Excessive sleepiness was observed in 53% (ESS scores greater than 10) and poor sleep quality (global PSQI scores above 5) was observed in 86% of CFS patients. Further, when patients were divided into groups based on ESS results, these scores corresponded to a clinical profile of insomnia (complaints of sleep disturbance associated with increased alertness) or hypersomnia (excessive daytime sleepiness persisting despite normal nocturnal sleep).[45] This provides an indication that sleep problems may be heterogeneous in this population. Conversely, in a study of 339 CFS patients from the Wichita population,[46] ESS scores and factors from the Sleep Assessment Questionnaire (SAQ) showed that, while fatigued, CFS subjects tended not to report excessive sleepiness. This study also showed that most (81.4%) CFS patients had an abnormal score in at least one of the five possible SAQ sleep factors (sleep apnea, restlessness, non-restorative sleep, insomnia, and excessive daytime somnolence). Interestingly, those with sleep abnormalities also had significantly lower wellness scores but unchanged fatigue severity scores compared to those with no abnormalities.[46]

Morriss et al.[47] included 69 Oxford-diagnosed CFS patients without a psychiatric disorder, 58 CFS patients with a psychiatric disorder, 45 controls, and 38 psychiatric outpatients with chronic depressive disorders in their study. A specially designed sleep questionnaire was constructed in order to measure self-rated sleep complaints according to the International Classification of Sleep Disorders[48] over 4 weeks.[47] The study found a higher prevalence of sleepiness and daytime naps in CFS patients compared to healthy controls and depressed subjects, and CFS patients were also significantly more likely to wake up because of temperature problems and pain than healthy controls or depressed patients. Further, restless legs were more frequently reported by CFS (41%) and depressed patients (40%) than by controls (4%). This study suggests that difficulty in maintaining sleep is the principle sleep complaint in CFS patients with or without a psychiatric disorder, and showed nocturnal waking and restless legs to be significantly associated with global disability in CFS patients, as assessed by the Medical Outcomes Survey.[49]

Krupp et al.[50] assessed 68 CFS patients diagnosed according to Holmes criteria and 20 non-illness controls using a modified version of the St. Mary's Sleep Questionnaire.[51] They showed patients' sleep to be more disrupted than that of healthy controls; 37% of CFS patients reported sleeping lightly compared to 20% of controls, and upon awakening, 76% of CFS patients reported feeling drowsy compared to 48% of controls.[50] Additionally, patient sleep diaries have shown that, compared to controls, those with CFS report a significantly longer time in bed at night, take longer to fall asleep, and wake more frequently during the night, whilst also feeling less refreshed on waking.[52,53]

There are methodological issues with these studies. There is a lack of standardized assessment for sleep disorders, and subjective self-report is not corroborated by any objective indicators such as actigraphy or PSG assessment. In addition, a range of self-report scales is utilized to assess the degree to which sleep disturbances occur in this patient group. This lack of consistency is further compounded by inconsistent criteria for patient selection and atypical CFS patient groups,[50,54] small patient samples,[50,53,55] and the examination of only a limited number of sleep complaints.[53,54] Given these caveats, probably the most robust conclusion that can be drawn from the self-report data is that sleep disturbances are commonly reported and they appear analogous to a range of sleep disorders, including hypersomnia and both sleep onset and sleep maintenance insomnia; in short, we can say that despite methodological issues, sleep problems are variable but significant in this population. There are also indications of how this might fit into the multifactorial model outlined above. Duncan[56] highlighted the possible interaction of daytime behavioral and lifestyle factors driving sleep disturbance in CFS. He has suggested that a sedentary lifestyle and daytime sleep (napping) may serve to maintain disturbed nighttime sleep, thus establishing a vicious circle of poor nighttime sleep and compensatory daytime sleep.

Objective assessment

ACTIGRAPHY

Actigraphy offers a method of characterizing gross objective measures of sleep continuity (the quantity and timing of sleep episodes) without interfering with sleep or daytime functioning. Actigraphy is usually measured by an unobtrusive device on the subject's body (often placed like a wristwatch) and allows for 24-hour recording of wake and sleep activity in a person's natural environment. A combination of actigraphy and symptom measurement offers real-time prospective activity–symptom relationships to be examined.[57]

Actigraphy studies in CFS differ in sample selection and their reporting practices. One assessment was carried out with children,[58] another with a small group of CFS patients,[59] and a third study was conducted with no comparative control group.[25] With regards to the findings, an assessment of 12 children with CFS showed that they had longer sleep durations and lower levels of daytime physical activity compared to age-matched controls. The actigraph also identified an interaction between disrupted sleep–wake and daytime napping.[58] Conversely, a more recent sleep study of 15 CFS patients and 15 controls found no significant actigraphy differences between CFS patients and healthy controls in daytime activity levels, fragmentation during sleep, SE, duration of sleep, or duration of napping. Interestingly, in this study, CFS patients still reported poorer sleep quality than healthy control subjects.[59] This sample, however, is particularly small and perhaps unrepresentative of the general CFS population, given that these were patients undergoing an intervention program of graded activity-oriented cognitive–behavioral therapy at a tertiary referral clinic. As such, they were treatment-motivated patients who were also likely to be well informed in terms of behavioral strategies for sleep and symptom management. Notably, Creti et al.[25] identified difficulties with actigraphy as an objective measurement modality in this population, showing that it underestimated sleep-onset latency (SOL; how long it took to get to sleep following retiring to bed) in patients, and was also not able to accurately or consistently identify nocturnal wakefulness.

Further, it is difficult to elucidate actual sleep versus inactivity (lying still in bed, but being awake) in actigraphy, and the actigraph alone cannot provide specific information on sleep architecture (i.e., the progression and timing of sleep–wake stage transitions). As a sole method for examining sleep, actigraphy is considered problematic.[60] However, when combined with other objective measurement modalities, actigraphy can help differentiate CFS individuals who have chronic insomnia from those without insomnia.[25]

MULTIPLE SLEEP LATENCY TESTS

CFS patients often use the terms tired, sleepy, and fatigued interchangeably, which is why it is important—albeit difficult—to separate the symptoms of fatigue and daytime sleepiness.[61] This is also important as the two complaints have different implications for diagnosis and treatment. Whereas fatigue relates to a lack of available energy and a loss of ability to exert mental and physical effort, sleepiness is a tendency to fall asleep, and only becomes problematic if it occurs at an inappropriate time or situation.[62] The Multiple Sleep Latency Test (MSLT) objectively assesses daytime sleepiness by measuring the amount of time it takes people to fall asleep during the day, given the opportunity. In the case of an MSLT, four daytime sleep opportunities

are offered, separated by 2 hours each time. The results from MSLTs carried out in CFS have generally been inconsistent. Several studies show that CFS patients do not differ from healthy controls on MSLT values.[63–65] A twin study has shown that CFS twins, despite reporting significantly more subjective sleepiness than their healthy co-twins, did not differ in their mean sleep latencies.[66] This indicates that CFS and biological sleepiness are not associated, and also points towards a heightened sense of sleepiness among CFS twins.

Where clinically significant MSLT latencies (<10 minutes) have been identified in CFS patients, these again are not consistent; one study revealed 41% of CFS patients had an abnormal MSLT, yet their scores did not differ significantly from those of non-fatigued controls.[64] Another study showed that a quarter of a CFS sample had an abnormal MSLT.[50] In a recent study of CFS patients from the Wichita population (n = 225), 59.5% of patients had an abnormal MSLT. Interestingly, in the same sample, 61.7% had an ESS within the normal range.[33] Moreover, in a study assessing CFS patients, apnea patients and healthy controls, the CFS group showed significantly smaller MSLT scores than controls during two of the four sleep opportunities provided; however, these still fell within a normal range. Overall, comparisons of scores on the MSLT showed that CFS patients presented with intermediate values between apnea patients (demonstrating the most excessive sleepiness) and healthy controls.[61]

The key shortcoming of the reported results of MSLTs carried out with CFS patients is that they are based on tests usually following a single night of PSG. As such, MSLTs after a single night of PSG may not be accurate indications of a patient's daytime somnolence, given that the first night of PSG (usually conducted in an unfamiliar environment) is commonly associated with a phenomenon known as the "first-night effect." The first-night effect is the set of differences in sleep parameters observed on the first night of recording in comparison to consecutive ones. The main characteristics of the first-night effect include a reduction in sleep time, more time awake during the night, and reduced SE parameters.[67,68] Moreover, PSG studies are often criticized for their artificial sleep laboratory conditions that do not reflect normal sleep at home and do not allow for habituation to cumbersome equipment.[69] As such, patients are likely to be sleepy during the next day. The functional impact of the PSG study night may also explain the lack of an observed relationship between MSLT scores, sleepiness scores, and fatigue scales.

Examinations of individuals who have been deprived of or restricted from sleep consistently demonstrate deteriorations in mood, cognition, and performance.[70] The purpose of each different sleep stage is also unclear, although it is generally agreed that the lighter stages of sleep (stage 1 and stage 2 sleep) afford transitions between wakefulness and sleep, and then between SWS and REM sleep. SWS and REM sleep are believed to confer recuperative, restorative, and learning properties on the individual (e.g., the secretion of growth hormone and consolidation of memory) (Table 10.3).[71] Therefore, the proportion of each sleep stage and timing of entry into each sleep stage—SWS and REM sleep in particular—are important for the long-term maintenance of human physical and mental health. PSG studies allow us to measure these sleep parameters, and the next section will review the PSG studies that have been carried out in CFS.

PSG STUDIES IN CFS

Overnight PSG is an all-night recording of sleep physiology, including both sleep continuity and sleep architecture (the progression and timing of sleep–wake stage transitions) and involves attaching electrodes to the scalp, forehead, and chin in order to record electroencephalogram (EEG) eye movements or electro-oculogram and muscle activity from the submental muscle (electromyogram [EMG]). In addition, variables such as electrocardiogram, EMG from leg muscles, and a range of respiratory variables and body movements may be measured at the same time. Overall, PSG sleep findings have not shown any clear pattern of sleep abnormality in this population. One characteristic of the PSG research in CFS may account for this: there is a high degree of variability in terms of the sleep continuity and sleep architecture features that are reported in research studies (Table 10.4). This makes comparisons between PSG studies of sleep in CFS difficult, and meta-analyses virtually impossible. There is also a fair degree of variability in the protocols, adding

Table 10.3 Summary of sleep stages and corresponding normal architectural parameters

Sleep stage		Activity	Normal architectural parameters
Wake		Eyes open, responsive to external stimuli, can hold intelligible conversation	<5%
NREM sleep	N1	Transition between wakefulness and sleep. Eyes closed, breathing slows, muscles relax, and brain starts to produce alpha waves. Many people notice the falling sensation during this stage of sleep, which may cause a sudden muscle contraction (called hypnagogic jerk)	2%–5%
	N2	Onset of sleep, heart rate slows, and the body temperature drops. The brain produces bursts of rapid, rhythmic brain wave activity known as sleep spindles	45%–55%
	N3 (SWS)	Deep sleep or SWS, delta brain waves occur. Deepest, most restorative sleep, muscles relaxed, blood pressure drops and breathing slows. Blood supply to muscles increases, tissue growth and repair occurs	13%–23%
REM sleep	REM	Body becomes immobile and muscles relax. Energy provided to brain and body. Heart rate and breathing become more variable. REM and dreaming (dreaming can also occur in other stages of sleep)	20%–25%

NREM, non-rapid eye movement; REM, rapid eye movement; N1, stage 1 sleep; N2, stage 2 sleep; N3, stage 3 sleep; SWS, slow-wave sleep.

Table 10.4 Previous polysomnography in chronic fatigue syndrome/myalgic encephalomyelitis

Sleep variable	Normal sleep parameters	Range from CFS studies
TST (minutes)	>360	304–495
SE (%)	>85	68–90
SOL (minutes)	<30	6–69
WASO (minutes)	<30	43–75
N1 (%)	2–5	4–36
N2 (%)	45–55	21–58
N3 (SWS) (%)	13–23	13–42
REM (%)	20–25	7–27
Wake (%)	<5	11–46
REM latency (Minutes)	90	63–149

CFS, chronic fatigue syndrome; TST, total sleep time; SE, sleep efficiency; SOL, sleep-onset latency; WASO, wake after sleep onset; N1, stage 1 sleep; N2, stage 2 sleep; SWS, slow-wave sleep (stages 3 and 4); REM, rapid eye movement.

Note: Significant ranges reported across all sleep variables and normal continuity and architectural parameters.

to the complexity. PSG studies have mostly been carried out over one night[25,32,64,72–75] or two,[27,66,76–81] which may present potential "first-night effects."

Does the "first-night effect" exist in CFS?

A great deal of the assumptions regarding the variability and lack of consistency in the objective sleep patterns of patients with CFS rests on the idea that people with CFS experience a first-night effect. In a seminal study,[79] this phenomenon was studied in 83 CFS patients without an objectively verifiable sleep disorder. Le Bon and colleagues observed clear differences between the first- and second-night sleep parameters. On night 1, there was less TST and REM sleep, a longer REM latency, more intermittent wake time, and a reduced number of sleep cycles,[79] all indicating poorer first-night sleep, in comparison to night 2. That said, and rather complicating the issue, a quarter of Le Bon et al.'s[79] sample demonstrated an "inverse first-night effect," with patients experiencing better sleep on night 1 than night 2. These issues may highlight the need for at least a three-night assessment for PSG research.

Continuity

There is significant variation between PSG studies on reported sleep continuity variables; for example, in one single-night PSG assessment study, Sharpley and colleagues showed a mean SOL of 69 minutes,[82] whereas Togo and colleagues showed a SOL of 31 minutes.[74] Further, Morriss and colleagues reported a mean SOL of 12.2 minutes in their CFS patient cohort.[53] Given that a SOL of longer than 30 minutes is considered to indicate the presence of potential "sleep initiation difficulty,"[83] these findings are highly variable and relate to both problematic and non-problematic sleep values. A two-night PSG assessment—arguably a more representative indication of "typical" sleep—identified a significantly longer SOL (39.9 minutes) on night 2 in CFS patients compared to healthy controls (21.5 minutes).[78]

Further single-night PSG assessments have shown other types of sleep disruption in CFS. After exclusion for medical illness, psychiatric disorders, apnea, hypersomnia, and PLMD, 26 CFS patients with coexisting fibromyalgia had significantly reduced TST and reduced SE than controls.[74] Home-based PSG studies have also shown CFS patients to sleep less efficiently than controls, with patients spending more time in bed and significantly more time awake during the night.[53,82] However, results in terms of percentage time awake during the night are also highly variable between studies, with ranges from 11.7%[27,65] to 31.9%,[53] and even as much as 46.28%.[81]

Architecture

Architectural findings of PSG studies in CFS are equally equivocal. Discrepancies between studies in their reporting of sleep stages and abbreviations used may account for this, based on the advancements that have been made in the visual scoring of sleep stages. Since 2007, the American Academy of Sleep Medicine (AASM) manual replaced Rechtschaffen and Kales's (R&K) rules that had originally divided the sleep stages into wakefulness, stage 1–4 (non-REM [NREM]), or REM. Sleep stages were now defined as N1–N3 (NREM) and stage R (REM), the key difference being stage 3 and stage 4 in the old R&K rules being abbreviated to stage N3 (SWS) in the new rules, as it is considered that no physiological basis exists for a difference between stages 3 and 4. This creates discrepancies with regards to some studies reporting on stages 3 and 4, with others referring to SWS. In addition to this, few studies report a full characterization of sleep architectural variables (amount of each sleep or wake stage and the timing of transitions to each sleep stage), making any conclusive statements about sleep abnormalities in this patient population difficult. The first report, from a group of 49 CFS patients compared to 20 matched healthy controls, found a significantly lower percentage of NREM N3 (formerly considered stage 3 and stage 4 sleep) in CFS.[78] Conversely, Le Bon et al.[84] showed significantly increased NREM N3 in the sleep of CFS patients (free of medical illness and psychiatric or primary sleep disorders), compared to apnea patients and healthy controls, thus drawing the conclusion that there is an increase in the amount of deep sleep over light sleep in CFS. This latter finding was corroborated in 2009 by Neu and colleagues, who reported architectural differences from the second night of PSG assessment, with CFS patients exhibiting less light (N1 and N2) sleep and more deep (slow-wave) sleep than healthy controls or apnea patients.[80]

Large variations also exist between studies in terms of reported percentages of REM sleep in CFS cohorts. Normal adults spend approximately 20%–25% of their TST in REM.[85] Some CFS studies have shown reduced REM (7.6%[73]; normal range: 22.3%[27]), and others have observed increased REM (27.7%).[76] Moreover, reported REM latencies (length of time to the first REM cycle), which in normal adult sleep is around 90 minutes,[85] are equally varied in individuals with CFS. Some report latencies that are as short as 63.5 minutes,[76] and others report much longer latencies (e.g., 149 minutes).[74]

These results relating to the proportion of REM are based on either a single night[73,74] or a second night[27,76] of recording, so perhaps, again, the first-night effect explains this variance. Indeed, much of the variance in the reported structural and architectural sleep variables in these studies of CFS patients' sleep may well be explained by the number of study assessment nights. Very few studies have conducted a three-night protocol of PSG assessment.[35,78,86] This protocol inconsistency is further complicated by the fact that reporting practices differ, making interpretation and comparisons difficult. Some studies report the percentage of each sleep and wake stage as an index of sleep period time (amount of the whole sleep period), TST (amount of time once sleep has been initiated), or even amount of time in bed,[35,65,75,76,78,81,86–88] whilst others report the minutes of each stage.[53,73,79,82,89] However, collectively, these studies do not demonstrate any consistent architectural differences in the sleep of patients and healthy controls/co-twins. Consequently, the results should be treated carefully. The variability in measures of sleep parameters could be the result of either the heterogeneity of sleep problems within the CFS populations, a variability in measurement protocols, and/or the result of still often-neglected primary sleep disorders in these studies (i.e., mild obstructive sleep apnea syndrome or periodic limb movement disorder).

There have also been indications of alterations in the transition patterns of sleep stages in CFS; in particular, significantly fewer transitions from REM to NREM sleep over the night in CFS patients than in controls.[73,90] This suggests a potential disruption in the normal circadian regulation of sleep–wake patterns, abnormalities that could lead to unrefreshing sleep.

Power spectral analysis

EEG brain wave activities can be categorized into frequency bands, and they can to some extent provide a gross indication of a particular sleep–wake state. For example, alpha activity (8–12 Hz) is the dominant rhythm in a relaxed wake state (eyes closed) in posterior regions of the scalp. That said, alpha waves can also be observed (although they do not predominate) during the various stages of sleep. When these alpha waves intrude into deep sleep, it is suggested that the brain is not resting like it should (also known as alpha–delta intrusion), indicating that a wakeful period during sleep and high levels of alpha intrusion into sleep are often associated with complaints of non-refreshing sleep. Power spectral analysis (PSA) deconstructs the amount and density of each frequency band over each phase of wake and sleep. PSA integrates the amount of energy (power) and respective density in each frequency band and their potential overlap or "intrusion." In some cases, these intrusions can result in a change of stage shift (e.g., going from one sleep stage to another, including wake). In terms of the PSA evidence, Armitage et al.[35] assessed 13 twin pairs (from the University of Washington CFS Twin Registry) of the 22 who originally underwent PSG assessments. Having applied PSA, using fast Fourier transformation (FFT; a widely applied linear modeling method to obtain EEG power spectra), delta power was observed to be slightly elevated overall in the CFS twins, although this was not enough to significantly differentiate them from their healthy co-twins.[91] Moreover, there have been new perspectives in what the homeostatic impairment in SWS in CFS might be. Le Bon and colleagues looked at delta activity in the very slow end of the frequency band, which had mainly been overlooked in studies. They showed lower ultra-slow delta power in the sleep of 10 young females with CFS in comparison with healthy controls.[36] This underlines the importance of looking beyond the conventional gross EEG.

Decker et al.[77] also used FFT, and analysed the PSG recordings of 35 CFS patients from the Wichita population study, comparing these to 40 non-fatigued controls. Overall, there was significantly reduced spectral power of alpha activity in CFS subjects during stage 2 sleep and SWS, along with the greatest reduction observed during REM sleep. CFS patients also showed significantly reduced delta power activity in SWS, which would concur with the common symptoms found in CFS, as reduced delta power is associated with reported fatigue and the perception of pain.[92] However, this delta power was increased during stage 1 and REM sleep. This latter finding has been corroborated elsewhere.[72,80] This rebound might reflect an impairment of sleep-related homeostatic functions in CFS.

However, alpha–delta sleep or alpha intrusions are not entirely specific to CFS. Other disorders, such as rheumatoid arthritis, fibromyalgia, and lupus erythematosus, also feature differences in alpha-range EEG frequencies compared to control subjects.[93] Additionally, there are reports of this alpha–delta sleep in stages 2 and 3 and NREM sleep in fibromyalgia patients who present with excessive daytime somnolence and chronic fatigue.[94–97] Furthermore, changes in alpha can also be found in several primary sleep disorders (PLMD, sleep apnea, and narcolepsy), and occasionally in patients with no complaints of fatigue.[98] The role of alpha–delta sleep in the development of CFS remains questionable, with some studies of CFS patients observing alpha intrusions in SWS,[86,87] and others failing to support this.[45,91,99]

Co-twin control methodology

Despite offering a powerful method by which to control for genetic factors, PSG twin studies have not provided strong evidence for sleep abnormalities in CFS. In one two-night PSG assessment study, and following the exclusion of psychiatric and medical disorders, CFS twins did not appear to differ from their healthy co-twins on any sleep parameter.[66,76,91] However, delaying sleep onset by 4 hours resulted in CFS twins showing less slow-wave activity than their healthy co-twins.[35] This finding may be indicative of a potential impairment in homeostatic sleep pressure in CFS, and supports the notion that the daytime complaints observed in people with CFS may be associated with potential issues in sleep regulation.

Summary of PSG

Overall, the PSG studies report that CFS patients have abnormal sleep; however, such disturbances are variable and not found across all patients. There is also no standardization in protocol, selection criteria, or reporting practices, making interpretation and comparisons between studies difficult. Moreover, different studies exclude different groups and, whilst most exclude medical illness, psychiatric disorders, and some sleep disorders (primary hypersomnias, sleep apnea, and PLMD), they tend not to exclude insomnia.[25,82] This is highly problematic in terms of reporting sleep findings in CFS, where patients are likely to encounter symptoms that are similar to those experienced in insomnia, such as problems getting off to sleep or staying asleep. Further, given the considerable overlap in the existing diagnostic criteria (used in the reported studies) between CFS and insomnia (namely non-restorative sleep), it is highly important to tease these conditions apart in order to be able to understand CFS exclusively.

A COMBINATION OF METHODS

Several of the studies mentioned earlier have used a combination of objective sleep assessment with subjective measures of patient's sleep.[25,65,66] These triangulation studies demonstrate interesting discrepancies between what emerges in subjective and objective measures. Overall, CFS patients report poorer sleep quality and more non-restorative sleep than healthy and non-fatigued controls, but objectively they appear to have close to normal sleep architecture (structure and pattern of sleep) or macrostructure (temporal organization of sleep). Similarly, CFS patients report more subjective sleepiness, yet objective measures (MSLTs) of sleepiness do not tend to differ between CFS twins and their healthy co-twins.[66] These discrepancies between subjective daytime complaints and objectively measured sleep are also common in individuals with insomnia, which is often described as sleep-state misperception (SSM; i.e., perceiving sleep as wakefulness/overestimating sleep). Such sleep misperception has been explained by the neurocognitive model of insomnia, emphasizing that brain cortical arousal is a central component whereby both physiological and cognitive arousal arises from increased cortical arousal around the sleep-onset period.[100]

Neu et al.[89] specifically demonstrated this difference between subjective and objective sleep. After exclusion of psychiatric disorders and certain sleep disorders (apnea, PLMD, and hypersomnia), their 28 "pure" CFS patients reported significantly poorer subjective sleep quality—as demonstrated by PSQI scores—compared to healthy age- and gender-matched controls, but there was no evidence of structural PSG abnormalities. This might suggest that CFS patients negatively perceive their sleep quality, even though they may sleep well. One suggestion is that they may over-monitor their sleep, and this perhaps contributes to perceived sleep problems, a phenomenon that has also been observed in insomnia.[101] Additionally, in CFS, the lack of explanation and guidance surrounding the condition may increase patient anxiety, symptom experience and, consequently, symptom focus.[20] This in turn could cause an increased monitoring of sleep duration and quality. Again, this emphasizes the importance of considering the interaction between cognitions, behaviors, physiology, and symptoms in this condition, and also the importance of combining assessment techniques.

SLEEP MANAGEMENT IN CFS

It may be possible to identify specific contributors to sleep problems in people with CFS (i.e., medications and activities). Likewise, it is also possible to establish certain sleeping behaviors that may contribute to patients' experiences of daytime symptoms. Irregularity in sleeping patterns can have a significant impact on daytime functioning. For example, it has been shown that in some CFS patients, daytime sleep can lead to cognitive impairments and increased levels of daytime sleepiness.[102] It may be beneficial to monitor sleep

patterns in CFS patients using a sleep diary in order to record nighttime (and daytime) sleep and wake, with adjunct actigraphy in order to identify any irregularities in sleeping patterns. It may also be beneficial to avoid napping for extended periods (i.e., >30 minutes) and later in the day in order to prevent entering into sleep cycles and weakening the sleep drive.

CFS patients who report symptoms of a primary sleep disorder should have a detailed evaluation carried out in order to identify and/or rule out a primary sleep disorder, as identifying these in patients may mean that they can be treated therapeutically. Sleep problems may be appropriately identified via information obtained from a detailed sleep interview and sleep diary, which provide subjective accounts that can be matched to different sleep profiles.[26] There are evidence-based treatment approaches; for example, light therapy is effective in reducing sleep and circadian rhythm problems; scheduled napping and modafinil are recommended for hypersomnolence disorder; and cognitive–behavioral therapy for insomnia (CBT-I) is recommended. Efficacy has also been shown to alleviate sleep problems in related conditions that share similar features with CFS/ME; for example, CBT-I is effective for people with fibromyalgia,[103,104] and bright light therapy has been shown to improve cancer-related fatigue.[105,106] However, there are no reports of whether these strategies are beneficial for people with CFS/ME. As such, there is a need to explore different sleep treatments in this patient group.

Why is identification of sleep problems relevant in CFS patients?

It is important to establish a relationship between objective sleep and sleep complaints in this patient population, because we do not know the extent to which disordered sleep may be influencing levels of disability in CFS. When considering (a) the level of disability associated with CFS, (b) our negligible understanding of its pathophysiology or associated biomarkers, (c) the lack of effective and efficacious therapies, and (d) its poor recognition and management in primary care, this "significant" complaint actually represents a considerable burden in terms of present and future healthcare utilization, quality of life, and employability. Treating the most disruptive symptoms first is the priority that is facing clinicians in primary care. It is therefore important that disturbed sleep is identified and treated early on in this illness in order to prevent the maintenance or worsening of other symptoms associated with CFS (fatigue, poor short-term memory, and concentration difficulties).

The ability to quickly identify an objective sleep problem in patients with CFS is clinically relevant in a population where symptoms can be maintained and/or worsened by poor sleep. Taking a complete and detailed assessment of sleep complaints is important (i.e., clinical interview with a complete sleep history, sleep screening questionnaires, and sleep diaries) will ensure that the clinician can determine the nature of the sleep problem and offer appropriate sleep-based interventions to help patients with their sleep-related symptoms. Specifically, appropriate treatment should be tailored for the individual patient, given that sleep problems are not homogenous in CFS. Whilst PSG is not required to diagnose CFS, it may be appropriate to use it when screening for treatable sleep disorders.

The importance of such assessments is based on the premise that there is potential for sleep disorders to develop over the illness course,[27] that sleep-related symptoms may change over time in patients with CFS,[15,47] and that they differ in frequency and/or severity.[8] Importantly, however, sleep-related changes may be a result of changes in medication use or environmental/behavioral adaptation.

CFS is more complex than just one factor (i.e., sleep), as otherwise it would have been explained by now, and we must consider that this condition has a multifactorial biopsychosocial etiology, and within that, one may be able to identify factors that we can modify in order to help decrease the impact of CFS (e.g., educating about daytime management, cognitive–behavioral therapy for maladaptive coping strategies, etc.). The biopsychosocial etiology of CFS is borne out in the quantitative and qualitative literature on the biopsychosocial impacts of CFS. Management should therefore be tailored to the individual.

Based on findings from studies that have assessed the impact of poor sleep on daytime functioning in people with CFS, it is important that clinicians are aware of the factors that may serve to maintain the condition and/or prevent people from being able to effectively manage their CFS:

- Irregular sleeping patterns (i.e., delayed sleep phase/advanced sleep phase)
- Daytime napping that occurs in the afternoon/evening (closer to sleep period)
- Daytime napping that exceeds 30 minutes (resulting in entering sleep cycles)
- Medications that cause nocturnal arousal/interference (i.e., selective serotonin reuptake inhibitor (SSRI)/serotonin-norepinephrine reuptake inhibitor (SNRI) antidepressants, analgesics, antianxiety medications, antihistamines, and blood pressure-controlling medications)
- Symptom "flare-up" (i.e., pain, virus, and "crash" from overexertion)

SUMMARY

Despite the absence of objective architectural differences in people with CFS and healthy controls for determining CFS-specific sleep pathologies, the complaint of unrefreshing sleep remains universal among CFS patients. Sleep disturbance closely follows the principal complaint of fatigue; to recap, 87%–95% of CFS patients report unrefreshing sleep as a key symptom complaint.[8] More specifically, patients tend to report fragmented sleep and sleep-onset difficulties, despite feeling tired. Continuing sleep disturbance can cause fatigue, myalgia, and poor concentration in healthy individuals,[71,107] and therefore sleep disruption may not be just a consequence, but also a cause of symptoms in CFS. Despite its frequent presentation in primary care, clinicians face difficulty in objectively verifying sleep complaints due to a lack of standardized assessments. Unidentified sleep difficulties early on may delay appropriate therapy; likewise, early identification will ensure timely and appropriate treatment that may help to alleviate other symptoms. Improving symptoms in this population—namely fatigue and unrefreshing sleep—has the potential to improve the health and quality of life of individuals with CFS and significantly reduce the costs (both at individual and societal levels) associated with this disabling condition.

THE FUTURE DIRECTION OF RESEARCH INTO SLEEP IN CFS

Sleep studies in CFS have shown very mixed results, particularly with regards to PSG. Not only do these studies significantly differ in their results, but their protocol and exclusion criteria are also equally inconsistent. As a result, no consistent picture of sleep disturbance emerges from the data. The most consistent findings are in the subjective reports of sleep quantity and quality. However, rather than dismissing this inconsistency as an artefact of inconsistent methodologies, we would suggest that a more plausible conclusion is that there is significant heterogeneity of sleep phenotypes in the CFS population. This needs to be further investigated.

There is a need for more in-depth qualitative accounts from patients about their experiences of sleep. This could highlight the proposed heterogeneity and also guide or complement subsequent objective studies to further investigate which specific components of sleep are disturbed and how they might play a role in maintaining symptoms such as fatigue. There is a need for more mixed-method studies, combining and comparing objective and subjective data. This would afford an examination of SSM as a principle problem in this population. Moreover, it is important that research moves forward in measuring sleep according to more stringent protocols, avoiding single-night recordings, selecting well-matched control subjects, and accounting for sleep disorders such as apnea.

More research is also needed to tease apart the causative and consequential role of sleep in CFS and to explore whether the daytime fatigue that is attributed to CFS is related to a sleep disturbance or something else (i.e., autonomic dysregulation, activity patterns, or homeostatic dysregulation). Changes have now been made in the DSM-5 and *International Classification of Sleep Disorders— 3rd Edition*, where non-restorative sleep has been removed from the criteria for insomnia disorder, which may afford a greater differentiation between CFS and insomnia. In clinical practice, it is important that CFS patients are screened for the presence of a sleep disorder in order to identify any incorrect or comorbid diagnoses, as currently, complete sleep testing is not part of routine CFS evaluation.[2,4,108]

Ultimately, sleep comprises specific brain activities and physiological systemic adaptations, which are implicated in various functions of brain and body restoration, learning processes, memory consolidation, and mood regulation.[85] Investigating sleep is thus highly relevant to this population, given the overlap of these features with symptom presentation in CFS patients (loss of physical functioning and impairments in memory, attention, and concentration).[109] Currently, however, there is no standardized method to measure sleep in this patient cohort. Practically, there is a need to address and standardize the technical aspects of assessing sleep efficiently by employing a standardized three-night protocol in order to observe sleep continuity, architecture, and microstructure in CFS patients. This would help to determine: (1) sleep disorders in this patient group; (2) the overlap between sleep disorders and CFS; and (3) the distinct sleep characteristics of CFS patients. It may then be possible to clarify the precise relationship between sleep, behavior, cognition, physiology, and the physical symptoms of CFS.

REFERENCES

1. Wessely S. Chronic fatigue syndrome: A 20th century illness? *Scand J Work Environ Health* 1997; 23: 17–34.
2. Holmes GP et al. Chronic fatigue syndrome: A working case definition. *Ann Intern Med* 1988; 108(3): 387–389.
3. Sharpe MC. A report—Chronic fatigue syndrome: Guidelines for research. *J R Soc Med* 1991; 84(2): 118–121.
4. Fukuda K et al. The chronic fatigue syndrome: A comprehensive approach to its definition and study. *Ann Intern Med* 1994; 121(12): 953–959.
5. Jain AK et al. Fibromyalgia syndrome: Canadian clinical working case definition, diagnostic and treatment protocols—A consensus document. *J Musculoskel Pain* 2003; 11(4): 3–107.
6. Reeves WC et al. Chronic fatigue syndrome—A clinically empirical approach to its definition and study. *BMC Med* 2005; 3(1): 19.
7. NICE clinical guideline 53. Chronic fatigue syndrome/myalgic encephalomyelitis (or encephalopathy): Diagnosis and management of CFS/ME in adults and children. London: National Institute for Health and Clinical Excellence, 2007. Available from: http://www.nice.org.uk/nicemedia/pdf/ CG53QuickRefGuide.pdf.
8. Nisenbaum R et al. A population-based study of the clinical course of chronic fatigue syndrome. *Health Qual Life Outcomes* 2003; 1: 49.

9. Tiersky LA et al. Longitudinal assessment of neuropsychological functioning, psychiatric status, functional disability and employment status in chronic fatigue syndrome. *Appl Neuropsychol* 2001; 8(1): 41–50.

10. Anderson JS, Ferrans CE. The quality of life of persons with chronic fatigue syndrome. *J Nerv Ment Dis* 1997; 185(6): 359–367.

11. Jason LA et al. CFS: A review of epidemiology and natural history studies. *Bull IACFS ME* 2009; 17(3): 88–106.

12. Reyes M et al. Prevalence and incidence of chronic fatigue syndrome in Wichita, Kansas. *Arch Intern Med* 2003; 163(13): 1530–1536.

13. Johnston S et al. The prevalence of chronic fatigue syndrome/myalgic encephalomyelitis: A meta-analysis. *Clin Epidemiol* 2013; 5: 105.

14. Clayton EW. Beyond myalgic encephalomyelitis/chronic fatigue syndrome: An IOM report on redefining an illness. *JAMA* 2015; 313(11): 1101–1102.

15. Gotts ZM et al. The experience of sleep in chronic fatigue syndrome: A qualitative interview study with patients. *Br J Health Psychol* 2015 [Epub ahead of print] doi: 10.1111/bjhp.12136.

16. Wessely S, Powell R. Fatigue syndromes— A comparison of chronic postviral fatigue with neuromuscular and affective-disorders. *J Neurol Neurosurg Psychiatry* 1989; 52(8): 940–948.

17. Cho HJ et al. Chronic fatigue syndrome: An update focusing on phenomenology and pathophysiology. *Curr Opin Psychiatry* 2006; 19(1): 67–73.

18. Cleare AJ. The HPA axis and the genesis of chronic fatigue syndrome. *Trends Endocrinol Metab* 2004; 15(2): 55–59.

19. Van Houdenhove B et al. Daily hassles reported by chronic fatigue syndrome and fibromyalgia patients in tertiary care: A controlled quantitative and qualitative study. *Psychother Psychosom* 2002; 71(4): 207–213.

20. Deary V, Chalder T, Sharpe M. The cognitive behavioural model of medically unexplained symptoms: A theoretical and empirical review. *Clin Psychol Rev* 2007; 27(7): 781–797.

21. Harvey SB, Wessely S. Chronic fatigue syndrome: Identifying zebras amongst the horses. *BMC Med* 2009; 7: 58.

22. Moss-Morris R, Deary V, Castell B. Chronic fatigue syndrome. In: Michael PB, David CG (Eds). *Handbook of Clinical Neurology*. Edinburgh: Elsevier, 2013, pp. 303–314.

23. Van Ravenzwaaij J et al. Explanatory models of medically unexplained symptoms: A qualitative analysis of the literature. *Ment Health Fam Med* 2010; 7(4): 223–231.

24. Maes M, Twisk FN. Why myalgic encephalomyelitis/chronic fatigue syndrome (ME/CFS) may kill you: Disorders in the inflammatory and oxidative and nitrosative stress (IO&NS) pathways may explain cardiovascular disorders in ME/CFS. *Neuro Endocrinol Lett* 2009; 30(6): 677–693.

25. Creti L et al. Impaired sleep in chronic fatigue syndrome: How is it best measured? *J Health Psychol* 2010; 15(4): 596–607.

26. Gotts ZM et al. Are there sleep-specific phenotypes in patients with chronic fatigue syndrome? A cross-sectional polysomnography analysis. *BMJ Open* 2013; 3(6): e002999.

27. Reeves WC et al. Sleep characteristics of persons with chronic fatigue syndrome and non-fatigued controls: Results from a population-based study. *BMC Neurol* 2006; 6: 41.

28. Le Bon O et al. How significant are primary sleep disorders and sleepiness in the chronic fatigue syndrome? *Sleep Res Online* 2000; 3(2): 43–48.

29. Cowan DC et al. Predicting sleep disordered breathing in outpatients with suspected OSA. *BMJ Open* 2014; 4(4): e004519.

30. Libman E et al. Sleep apnea and psychological functioning in chronic fatigue syndrome. *J Health Psychol* 2009; 14(8): 1251–1267.

31. Jason LA et al. A community-based study of chronic fatigue syndrome. *Arch Intern Med* 1999; 159(18): 2129–2137.

32. Fossey M et al. Sleep quality and psychological adjustment in chronic fatigue syndrome. *J Behav Med* 2004; 27(6): 581–605.

33. Decker MJ et al. Validation of ECG-derived sleep architecture and ventilation in sleep apnea and chronic fatigue syndrome. *Sleep Breath* 2010; 14(3): 233–239.

34. Devasahayam A et al. Alternative diagnoses to chronic fatigue syndrome in referrals to a specialist service: Service evaluation survey. *JRSM Short Rep* 2012; 3(1): 4.

35. Armitage R et al. The impact of a 4-hour sleep delay on slow wave activity in twins discordant for chronic fatigue syndrome. *Sleep* 2007; 30(5): 657–662.
36. Le Bon O et al. Ultra-slow delta power in chronic fatigue syndrome. *Psychiatry Res* 2012; 200(2–3): 742–747.
37. Borbely AA, Achermann P. Sleep homeostasis and models of sleep regulation. *J Biol Rhythms* 1999; 14(6): 557–568.
38. Lovell DM. Chronic fatigue syndrome among overseas development workers: A qualitative study. *J Travel Med* 1999; 6(1): 16–23.
39. Soderlund A, Skoge AM, Malterud K. "I could not lift my arm holding the fork…"—Living with chronic fatigue syndrome. *Scand J Prim Health Care* 2000; 18(3): 165–169.
40. Clarke JN. Chronic fatigue syndrome: Gender differences in the search for legitimacy. *Aust N Z J Ment Health Nurs* 1999; 8(4): 123–133.
41. Schoofs N et al. Death of a lifestyle: The effects of social support and healthcare support on the quality of life of persons with fibromyalgia and/or chronic fatigue syndrome. *Orthop Nurs* 2004; 23(6): 364–374.
42. Anderson VR et al. A review and meta-synthesis of qualitative studies on myalgic encephalomyelitis/chronic fatigue syndrome. *Patient Educ Couns* 2012; 86(2): 147–155.
43. Buysse DJ et al. The Pittsburgh Sleep Quality Index: A new instrument for psychiatric practice and research. *Psychiatry Res* 1989; 28(2): 193–213.
44. Johns MW. A new method for measuring daytime sleepiness: The Epworth Sleepiness Scale. *Sleep* 1991; 14(6): 540–545.
45. Mariman A et al. Subjective sleep quality and daytime sleepiness in a large sample of patients with chronic fatigue syndrome (CFS). *Acta Clin Belg* 2012; 67(1): 19–24.
46. Unger ER et al. Sleep assessment in a population-based study of chronic fatigue syndrome. *BMC Neurol* 2004; 4: 6.
47. Morriss RK, Wearden AJF, Battersby L. The relation of sleep difficulties to fatigue, mood and disability in chronic fatigue syndrome. *J Psychosom Res* 1997; 42(6): 597–605.
48. Thorpy MJ. Classification of sleep disorders. *J Clin Neurophysiol* 1990; 7(1): 67–81.
49. Stewart AL, Hays RD, Ware JE Jr. The MOS short-form general health survey. Reliability and validity in a patient population. *Med Care* 1988; 26(7): 724–735.
50. Krupp LB et al. Sleep disturbance in chronic fatigue syndrome. *J Psychosom Res* 1993; 37(4): 325–331.
51. Ellis BW et al. The St. Mary's Hospital Sleep Questionnaire: A study of reliability. *Sleep* 1981; 4(1): 93–97.
52. Fossey M et al. Sleep quality and psychological adjustment in chronic fatigue syndrome. *J Behav Med* 2004; 27(6): 581–605.
53. Morriss R et al. Abnormalities of sleep in patients with the chronic fatigue syndrome. *Br Med J* 1993; 306(6886): 1161–1164.
54. Vercoulen JH et al. Dimensional assessment of chronic fatigue syndrome. *J Psychosom Res* 1994; 38(5): 383–392.
55. Moldofsky H. Nonrestorative sleep and symptoms after a febrile illness in patients with fibrositis and chronic fatigue syndromes. *J Rheumatol Suppl* 1989; 19: 150–153.
56. Duncan I. Insomnia in chronic fatigue syndrome: Due to daytime dozing. *Br Med J* 1993; 306(6890): 1480.
57. Yoshiuchi K et al. A real-time assessment of the effect of exercise in chronic fatigue syndrome. *Physiol Behav* 2007; 92(5): 963–968.
58. Ohinata J et al. Actigraphic assessment of sleep disorders in children with chronic fatigue syndrome. *Brain Dev* 2008; 30(5): 329–333.
59. Rahman K, Burton A, Galbraith S, Lloyd A, Vollmer-Conna U. Sleep–wake behavior in chronic fatigue syndrome. *Sleep* 2011; 34(5): 671–678.
60. Sadeh A. The role and validity of actigraphy in sleep medicine: An update. *Sleep Med Rev* 2011; 15(4): 259–267.
61. Neu D et al. Are patients with chronic fatigue syndrome just 'tired' or also 'sleepy'? *J Sleep Res* 2008; 17(4): 427–431.
62. Shen JH, Barbera J, Shapiro CM. Distinguishing sleepiness and fatigue: Focus on definition and measurement. *Sleep Med Rev* 2006; 10(1): 63–76.

63. Bailes S et al. Brief and distinct empirical sleepiness and fatigue scales. *J Psychosom Res* 2006; 60(6): 605–613.

64. Buchwald D et al. Sleep disorders in patients with chronic fatigue. *Clin Infect Dis* 1994; 18(S1): S68–S72.

65. Majer M et al. Perception versus polysomnographic assessment of sleep in CFS and non-fatigued control subjects: Results from a population-based study. *BMC Neurol* 2007; 7: 40.

66. Watson NF et al. Subjective and objective sleepiness in monozygotic twins discordant for chronic fatigue syndrome. *Sleep* 2004; 27(5): 973–977.

67. Agnew HW Jr, Webb WB, Williams RL. The first night effect: An EEG study of sleep. *Psychophysiology* 1966; 2(3): 263–266.

68. Rechtschaffen A, Verdone P. Amount of dreaming: Effect of incentive, adaptation to laboratory, and individual differences. *Percept Mot Skills* 1964; 19(3): 947–958.

69. Newell J et al. Assessment of polysomnography: First-night effect and night-to-night variability: Relationships to diagnosis and intra-individual differences. *J Sleep Res* 2012; 21: 216.

70. Turner TH et al. Effects of 42 Hr of total sleep deprivation on component processes of verbal working memory. *Neuropsychology* 2007; 21(6): 787–795.

71. Van Cauter E, Leproult R, Plat L. Age-related changes in slow wave sleep and REM sleep and relationship with growth hormone and cortisol levels in healthy men. *JAMA* 2000; 284(7): 861–868.

72. Guilleminault C et al. Chronic fatigue, unrefreshing sleep and nocturnal polysomnography. *Sleep Med* 2006; 7(6): 513–520.

73. Kishi A et al. Dynamics of sleep stage transitions in healthy humans and patients with chronic fatigue syndrome. *Am J Physiol Regul Integr Comp Physiol* 2008; 294(6): R1980–R1987.

74. Togo F et al. Sleep structure and sleepiness in chronic fatigue syndrome with or without coexisting fibromyalgia. *Arthritis Res Ther* 2008; 10(3): R56.

75. Stores G, Fry A, Crawford C. Sleep abnormalities demonstrated by home polysomnography in teenagers with chronic fatigue syndrome. *J Psychosom Res* 1998; 45(1): 85–91.

76. Ball N et al. Monozygotic twins discordant for chronic fatigue syndrome—Objective measures of sleep. *J Psychosom Res* 2004; 56(2): 207–212.

77. Decker MJ et al. Electroencephalographic correlates of chronic fatigue syndrome. *Behav Brain Funct* 2009; 5: 43.

78. Fischler B et al. Sleep anomalies in the chronic fatigue syndrome—A comorbidity study. *Neuropsychobiology* 1997; 35(3): 115–122.

79. Le Bon O et al. First-night effect in the chronic fatigue syndrome. *Psychiatry Res* 2003; 120(2): 191–199.

80. Neu D et al. High slow-wave sleep and low-light sleep: Chronic fatigue syndrome is not likely to be a primary sleep disorder. *J Clin Neurophysiol* 2009; 26(3): 207–212.

81. Van Hoof E et al. Defining the occurrence and influence of alpha–delta sleep in chronic fatigue syndrome. *Am J Med Sci* 2007; 333(2): 78–84.

82. Sharpley A et al. Do patients with "pure" chronic fatigue syndrome (neurasthenia) have abnormal sleep? *Psychosom Med* 1997; 59(6): 592–596.

83. Edinger JD et al. Derivation of research diagnostic criteria for insomnia: Report of an American Academy of Sleep Medicine Work Group. *Sleep* 2004; 27(8): 1567–1596.

84. Le Bon O, Neu D, Valente F, Linkowski P. Paradoxical NREMS distribution in "pure" chronic fatigue patients: A comparison with sleep apnea–hypopnea patients and healthy control subjects. *J Chron Fatigue Syndr* 2007; 14(2): 45–60.

85. Kryger MH, Roth T, Dement WC. *Principles and Practice of Sleep Medicine* (5th ed.). Philadelphia, PA: Saunders/Elsevier, 2011.

86. Whelton CL, Salit I, Moldofsky H. Sleep, Epstein–Barr virus infection, musculoskeletal pain, and depressive symptoms in chronic fatigue syndrome. *J Rheumatol* 1992; 19(6): 939–943.

87. Manu P et al. Alpha–delta sleep in patients with a chief complaint of chronic fatigue. *South Med J* 1994; 87(4): 465–470.

88. Watson NF et al. Comparison of subjective and objective measures of insomnia in monozygotic twins discordant for chronic fatigue syndrome. *Sleep* 2003; 26(3): 324–328.

89. Neu D et al. Sleep quality perception in the chronic fatigue syndrome: Correlations with sleep efficiency, affective symptoms and intensity of fatigue. *Neuropsychobiology* 2007; 56(1): 40–46.

90. Kishi A et al. Sleep-stage dynamics in patients with chronic fatigue syndrome with or without fibromyalgia. *Sleep* 2011; 34(11): 1551–1560.

91. Armitage R et al. Power spectral analysis of sleep EEG in twins discordant for chronic fatigue syndrome. *J Psychosom Res* 2009; 66(1): 51–57.

92. Lentz MJ et al. Effects of selective slow wave sleep disruption on musculoskeletal pain and fatigue in middle aged women. *J Rheumatol* 1999; 26(7): 1586–1592.

93. Macfarlane JG, Moldofsky H. Fibromyalgia and chronic fatigue syndromes. In: Kryger MH, Roth T, Dement CW (Eds). *Principles and Practice of Sleep Medicine* (5th ed.). St. Louis, MO: Elsevier/Saunders, 2011, pp. 1422–1434.

94. Branco J, Atalaia A, Paiva T. Sleep cycles and alpha–delta sleep in fibromyalgia syndrome. *J Rheumatol* 1994; 21(6): 1113–1117.

95. Drewes AM et al. Sleep intensity in fibromyalgia: Focus on the microstructure of the sleep process. *Br J Rheumatol* 1995; 34(7): 629–635.

96. Moldofsky H et al. Musculosketal symptoms and non-REM sleep disturbance in patients with "fibrositis syndrome" and healthy subjects. *Psychosom Med* 1975; 37(4): 341–351.

97. Roizenblatt S et al. Alpha sleep characteristics in fibromyalgia. *Arthritis Rheum* 2001; 44(1): 222–230.

98. MacFarlane JG et al. Periodic K-alpha sleep EEG activity and periodic limb movements during sleep: Comparisons of clinical features and sleep parameters. *Sleep* 1996; 19(3): 200–204.

99. Flanigan MJ, Morehouse RL, Shapiro CM. Determination of observer-rated alpha activity during sleep. *Sleep* 1995; 18(8): 702–706.

100. Perlis ML et al. Which depressive symptoms are related to which sleep electroencephalographic variables? *Biol Psychiatry* 1997; 42(10): 904–913.

101. Harvey AG, Payne S. The management of unwanted pre-sleep thoughts in insomnia: Distraction with imagery versus general distraction. *Behav Res Ther* 2002; 40(3): 267–277.

102. Gotts ZM et al. The association between daytime napping and cognitive functioning in chronic fatigue syndrome. *PLoS One* 2015; 10(1): e0117136.

103. Edinger JD et al. Behavioral insomnia therapy for fibromyalgia patients: A randomized clinical trial. *Arch Intern Med* 2005; 165(21): 2527–2535.

104. Martínez MP et al. Cognitive–behavioral therapy for insomnia and sleep hygiene in fibromyalgia: A randomized controlled trial. *J Behav Med* 2013: 1–15.

105. Ancoli-Israel S et al. Light treatment prevents fatigue in women undergoing chemotherapy for breast cancer. *Support Care Cancer* 2012; 20(6): 1211–1219.

106. Redd WH et al. Systematic light exposure in the treatment of cancer-related fatigue: A preliminary study. *Psychooncology* 2014; 23(12): 1431–1434.

107. Stickgold R. Sleep-dependent memory consolidation. *Nature* 2005; 437(7063): 1272–1278.

108. Carruthers BM et al. Myalgic encephalomyelitis: International consensus criteria. *J Intern Med* 2011; 270(4): 327–338.

109. Afari N, Buchwald D. Chronic fatigue syndrome: A review. *Am J Psychiatry* 2003; 160(2): 221–236.

110. Morris G, Maes M. Myalgic encephalomyelitis/chronic fatigue syndrome and encephalomyelitis disseminata/multiple sclerosis show remarkable levels of similarity in phenomenology and neuroimmune characteristics. *BMC Medicine* 2013; 11(1): 205.

111. IOM (Institute of Medicine). *Beyond Myalgic Encephalomyelitis/Chronic Fatigue Syndrome: Redefining an Illness*. Washington, DC: The National Academies Press; 2015. http://www.iom.edu/mecfs.

11

Sleep and coronary heart disease

ELIZABETH BRUSH, RIDHWAN Y. BABA, AND NEOMI SHAH

Coronary heart disease (CHD) is the leading cause of death in the United States. According to the American Heart Association (AHA), one out of every six deaths in the United States is caused by CHD.[1] The incidence of CHD in our aging population is on the rise due to improved healthcare and subsequent prolonged life expectancy. As a result, the AHA projects that the direct and indirect costs of CHD will nearly double from $110 billion in 2010 to approximately $220 billion in 2030.[1] It is therefore clear that effective new strategies to prevent and manage CHD need to be devised and implemented in our healthcare system.

Like CHD, sleep-related breathing disorders such as obstructive sleep apnea (OSA) are exceedingly prevalent in the United States. It is estimated that amongst the western population, 24% of men and 9% of women have OSA, diagnosed as an Apnea–Hypopnea Index (AHI) $\geq$ 5.[2] OSA occurs when the upper airway collapses during sleep, resulting in a cycle of hypoxemia, increased respiratory effort, frequent arousals, and increased sympathetic activity. It has been associated with numerous cardiovascular conditions including hypertension,[3] CHD,[4,5] cardiac arrhythmias,[6] heart failure,[7,8] stroke,[9] and sudden death.[10] Further, OSA has been linked with increased cardiovascular mortality.

In this chapter, we focus on the relationship between sleep and CHD. Specifically, we will focus on (1) normal sleep, (2) extremes of sleep duration, and (3) OSA on CHD events. We describe several mechanisms through which sleep duration and OSA contribute to CHD. Lastly, we discuss the effect of continuous positive airway pressure (CPAP) treatment of OSA on CHD and CHD-related outcomes.

NORMAL SLEEP AND THE CARDIOVASCULAR SYSTEM

Normal sleep provides a period of physiologically reduced workload for the cardiovascular system for nearly a third of the average human lifespan. There are two markedly different stages of sleep; non-rapid eye movement (NREM) and rapid eye movement (REM) sleep. Normal sleep architecture consists of NREM/REM cycles, which repeat on average four to five times a night. Overall, NREM and REM stages occupy 70%–80% and 20%–25%

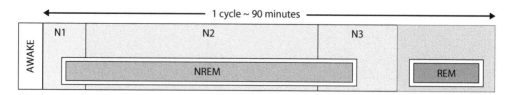

Figure 11.1 Normal sleep architecture. NREM: non-rapid eye movement; REM: rapid eye movement.

of total sleep time, respectively (Figure 11.1). These two sleep phases are associated with dramatically different cardiovascular responses. Generally, NREM sleep is associated with a reduction in blood pressure, heart rate, systemic vascular resistance, and cardiac output.[11-13] Alternatively, during REM sleep, the blood pressure undergoes frequent oscillations, at times reaching the lowest levels of the circadian cycle and then rising above waking values.[12] In addition to blood pressure, the heart rate also experiences greater variability during REM sleep. In an elegant study by Somers et al., sympathetic nerve traffic was recorded in eight patients during wakefulness and sleep. The resulting data provided clear, metric evidence that sympathetic activity is reduced by more than half from wakefulness to stage 4 NREM sleep, but increases to levels above waking values during REM sleep.[14]

SLEEP DURATION AND CHD

A paradigm shift in sleep duration has been observed over the past 50 years. This change coincides with the progression of the westernized lifestyle, which includes distractors such as TV and mobile devices along with longer working hours. The amount of sleep needed varies from person to person; however, a recent National Sleep Foundation survey found that, on average, most adults need about 6.5–7 hours to feel alert and well rested.[15] The National Sleep Foundation also reported that 16% of Americans were sleeping 6 hours or less and 24% were sleeping 6–6.9 hours a night. Further, 38% of those respondents who sleep less than 6 hours per night reported getting less sleep than they need for optimal daytime functioning.[15] This is a cause for concern as multiple studies have examined the role of sleep duration and quality as a risk factor for comorbid conditions and premature mortality.[16,17]

Evidence has emerged that the extremes of sleep duration are associated with adverse health outcomes, including increased incidence of obesity,[18] metabolic dysregulation including diabetes mellitus,[18,19] hypertension,[20,21] atherogenic lipid profile,[22] respiratory disorders,[23] poor self-rated health,[24] and all-cause mortality in both adults and children.[16,17] It has been suggested that a U-shaped association exists between sleep duration and cardiovascular events and all-cause mortality.[16,17,25] Numerous cross-sectional observational studies have reported a temporal association between sleep duration and cardiovascular disease.[26-32] Further, a meta-analysis of prospective cohort studies described a higher incidence of CHD, CHD-related mortality, and stroke in subjects who reported sleeping for 5–6 hours or less per night at a follow-up of at least 3 years.[16] A similar increase was reported in the long sleep group who had sleep durations of greater than 8–9 hours per night.[16]

Various theories have been investigated to understand the mechanisms underlying the correlation between sleep duration and CHD. An increase in the circulating levels of the anorexigenic hormone leptin and a reduction in orexigenic factor ghrelin have been observed in patients after sleep restriction.[33,34] This in turn causes an increase in appetite,[33,34] calorie-dense food intake,[33] obesity,[34] and impaired glycemic control.[33] Leptin and ghrelin levels are typically regulated by sympatheticovagal balance, and a reciprocal change in sympathetic nervous system and vagal activity due to sleep loss may explain the potential underlying mechanism.[35-39] An increase in cortisol secretion after sleep deprivation and subsequent glucocorticoid excess has also been suggested.[34,39,40] Additionally, low-grade inflammation activated during short sleep with possible implications for CHD has been reported.[41]

To date, limited data exist on the potential mechanisms for long sleep causing cardiovascular disease or death. The Nurses' Health Study II attempted to identify potential correlates to explain increased mortality in women with prolonged sleep. They reported a significant impact of psychiatric, lifestyle, socioeconomic, medical, gynecologic, and sleep factors on the association between long sleep and mortality.[42] As such, the presence of long sleep duration may be considered an indication to investigate the underlying mental or physiological disorder, instead of a cause of these chronic disorders.

SLEEP FRAGMENTATION AND CHD

Whether sleep fragmentation in addition to sleep duration affects cardiovascular morbidity and mortality is unclear. Multiple studies have provided evidence to support the association of disturbed sleep with increased CHD risk in both adults and children.[32,43,45–47] The concomitant effect of sleep disturbance and short sleep duration on the incidence of CHD was examined in a prospective study using the Whitehall II Study cohort data. Both sleep disturbance and sleep duration were measured using standardized questionnaires, and after adjusting for confounding risk factors, the authors concluded that the risk of CHD and CHD-related mortality was highest among those who report sleep disturbance in addition to sleep duration of less than 6 hours per night.[45] It is plausible that sleep disturbance and extremes of sleep duration act synergistically with regards to their impact on cardiovascular disease. This remains to be investigated.

OSA: PREVALENCE AND ASSOCIATION WITH CHD

Stable CHD

The prevalence of OSA , as defined by the American Academy of Sleep Medicine,[44] is high in patients with stable CHD. The estimates vary depending on the study design, yet the most commonly reported estimate is from Mooe et al., in which the authors examined 142 men with CHD and found a 37% prevalence of OSA, defined as an AHI ≥ 10.[48] Similarly, in a case–control study of women with CHD, it was found that 54% had an AHI ≥ 5 and 30% had an AHI ≥ 10.[49] Both studies also reported that OSA was a significant predictor of CHD after adjustment for confounding factors like age, body mass index, hypertension, smoking habits, and diabetes.[48,49] In a similar study, Schafer and colleagues report a high prevalence (31%) of OSA in patients with angiographically proven CHD.[50] Maekawa et al. also investigated the prevalence of CHD (documented by cardiac stress test, radionuclide myocardial scintigraphy, and/or coronary angiography) in 386 subjects suspected of OSA with heavy snoring. They found the prevalence of CHD among these patients with untreated OSA (defined as an AHI ≥ 10) to be 23.8%.[51] These results are noteworthy, as evidence suggests that OSA (AHI ≥ 10) is associated with a worse long-term prognosis.[52]

Acute CHD

The prevalence of OSA is even higher in patients with acute CHD events. In a study by Konecny et al., OSA was present in 69% of patients hospitalized with acute myocardial infarction.[53] Interestingly, only 12% of patients had documentation of diagnosed or suspected OSA. The authors suggest a lack of awareness and recognition of OSA generally, and specifically in the acute phase of a myocardial infarction, in order to explain this discrepancy.

Circadian effect on CHD mortality

There appears to be a circadian effect on CHD mortality. Reports in the world literature going back as far as 1960 have described a circadian periodicity in the onset of myocardial infarction, and have documented a peak incidence between the hours of 6 a.m. and noon, with a nadir between midnight and 6 a.m.[54] As such, the risk of sudden death from cardiac causes in the general population is significantly greater during the morning hours after waking (i.e., from 6 a.m. to noon) than during any other 6-hour interval of the day.[55] The presumed mechanism underlying this trend involves the disturbance of vulnerable atherosclerotic plaque. It is believed that the plaque within arteries is disrupted due to alterations in blood pressure, thereby creating shearing forces within the coronary blood vessels and promoting a thrombogenic event.

Patients with OSA, however, show a significant deviation from this well-established day–night pattern. A study by Gami et al. demonstrated a marked nocturnal peak (between 10 p.m. and 6 a.m.) in sudden death from cardiac causes in patients with OSA.[10] This is likely due to the cyclical mini-arousals (increases in blood pressure, heart rate, etc.) that are characteristic of OSA, which occur repeatedly throughout the night. However, beyond simply shifting time of onset, OSA has been shown to be an independent risk factor for CHD-related events as well.[45,56] In the subsequent section, we will discuss potential mechanisms for this association.

OSA AND CHD: POTENTIAL MECHANISMS

Various theories on the physiological cascades linking OSA and CHD have been examined. These mechanisms include, but are not limited to: (1) endothelial dysfunction/systemic inflammation, (2) sympathetic activation, (3) metabolic dysregulation, and (4) mechanical load effects.

Endothelial dysfunction and systemic inflammation

Systemic inflammation and oxidative stress resulting from intermittent hypoxia observed in OSA are gaining attention as fundamental mechanisms leading to endothelial dysfunction.[57–60] Endothelial dysfunction in turn is a dynamic and progressive disease that can lead to atherosclerosis—the underlying pathogenetic mechanism of cardiovascular and cerebrovascular events (Figure 11.2).[60]

Examples of systemic inflammation and oxidative stress resulting from OSA-induced intermittent hypoxia are numerous. For instance, it has been shown that the oxygenation–reoxygenation process, which produces reactive oxygen species due to

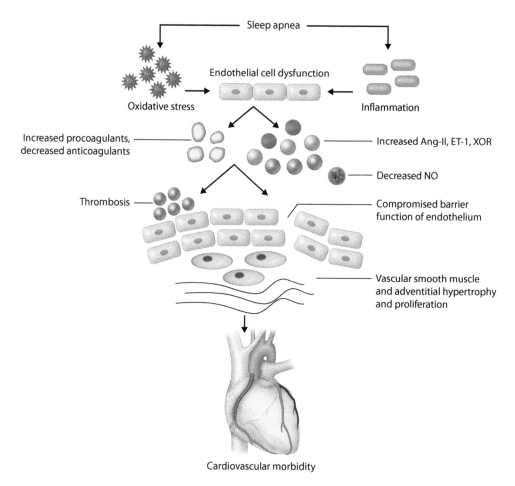

Figure 11.2 Schematic showing the role of endothelial dysfunction in sleep apnea as a plausible mechanism for cardiovascular morbidity. Ang-II: angiotension II; ET-1: endothelin-1; XOR: xanthine oxidoreductase; NO: nitric oxide. (Reprinted with permission from Quan S. *J Clin Sleep Med* 2007; 3: 409–415.)

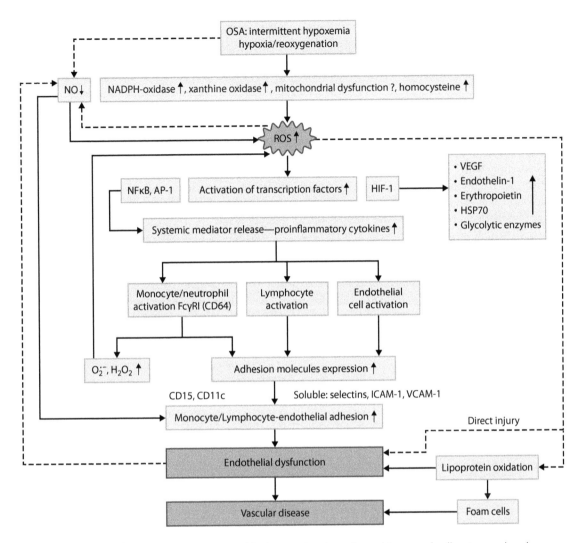

Figure 11.3 The inflammatory response with the production of cytokines and adhesion molecules secondary to OSA-induced intermittent hypoxemia resulting in endothelial dysfunction and vascular disease. NADPH: reduced form of nicotinamide adenine dinucleotide phosphate; NFκB: nuclear factor κB; AP-1: activator protein-1; HIF-1: hypoxia inducible factor-1; VEGF: vascular endothelial growth factor; HSP70: heat-shock protein 70; O_2^-: superoxide radical anion; H_2O_2: hydrogen peroxide; ICAM-1: intercellular adhesion molecule-1; VCAM-1: vascular cell adhesion molecule-1. (Reprinted from *Sleep Med Rev*, 7, Lavie L., 35–51, Copyright 2003, with permission from Elsevier.)

recurrent obstructive apneas, initiates an inflammatory response. The resulting cascade produces cytokines and adhesion molecules, which have been shown to promote endothelial dysfunction (Figure 11.3).[61] Further, it is known that biomarkers of inflammation, such as C-reactive protein and interleukin-6, are important risk factors for endothelial dysfunction and atherosclerosis. Studies have shown that these indicators are also elevated in patients with OSA.[61] It is therefore evident that OSA is associated with the mechanistic progression

of endothelial dysfunction and atherosclerosis via oxidative stress and inflammation, which in turn correlate to a higher incidence of CHD. Another example of impaired endothelial function resulting from OSA is seen in the nitric oxide (NO) pathway. Patients with OSA have enhanced endothelial dysfunction via a lack of NO-mediated vascular homeostasis.[62] Endothelial NO is a key regulator of vascular homeostasis. Impaired NO release by dysfunctional endothelial cells is regarded as an initiator and promoter of atherosclerosis.

On an observable level, noninvasive scanning techniques like B-mode imaging can be used to directly identify and monitor endothelial dysfunction and preclinical atherosclerosis in arteries. Carotid intima–media thickness (IMT) has been recognized as a surrogate measure of atherosclerosis and a useful index of subclinical cardiovascular disease. Epidemiological studies have shown that carotid IMT is a strong independent risk factor for predicting future CHD and stroke, even after adjusting for traditional risk factors.[63–66] Silvestrini et al. found that the IMT of the common carotid arteries of patients with OSA was significantly greater than that of control subjects.[67] Further, a study by Friedlander et al.[68] indicated that individuals with OSA (AHI ≥ 15) have a greater prevalence of calcified carotid artery atheromas, as measured by radiographs, than healthy individuals. A beneficial impact of both short-term and long-term CPAP therapy on subclinical atherosclerosis using carotid IMT and arterial pulse wave velocity has been demonstrated.[69–72]

Sympathetic activation

While the pathophysiology of OSA is not fully delineated, the downstream consequences of OSA—such as sympathetic activation—have been studied extensively. As a result of obstructed breathing, patients with OSA experience repeated and prolonged episodes of arterial oxygen desaturation and carbon dioxide retention. During recurrent apneas, hypoxemia and retained carbon dioxide can stimulate chemoreceptors, leading to an increase in sympathetic tone. The duration of apnea and the level of oxygen desaturation are key determinants in the degree of sympathetic activation. The sympathetic tone increases progressively during the obstructive event, reaching its peak after arousal.

Narkiewicz and Somers found that the blood pressure surge at the end of the apneic event can reach levels as high as 250/110 mmHg in a subject who is normotensive during wakefulness.[73] Other factors, such as increased muscle tone and arousal, may also contribute to the increased blood pressure at the end of an apnea. Oxyhemoglobin saturation decreases during the apnea and recovers only slowly after termination. Therefore, the sudden increase in cardiac output at arousal with increased myocardial oxygen demand is met with low oxygen delivery to the myocardium and therefore increased risk of cardiac ischemia. Studies have also shown that in addition to the heightened

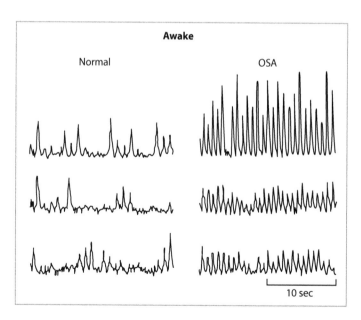

Figure 11.4 Recordings of sympathetic nerve activity during wakefulness in patients with obstructive sleep apnea and matched controls showing high levels of sympathetic nerve activity in patients with sleep apnea. (Reprinted from Somers V et al., *J Clin Invest* 1995; 96(4): 1897–1904, with permission from RightsLink 2014.)

nocturnal levels, sympathetic drive remains elevated in OSA patients during awake, normoxic conditions (Figure 11.4).[74,75] Further, with CPAP therapy, a significant reduction in sympathetic activity to levels below wakefulness has been observed (Figure 11.5).[75]

Norepinephrine is the primary neurotransmitter for the post-ganglionic sympathetic nervous system. Baylor et al. analyzed plasma samples from patients who were highly suspected of OSA by history over a 5.5–hour sleeping period.[76] The team found that as oxygen saturation decreases, the variability in plasma norepinephrine increases. They also reported that the patients with the most severe degree of cyclic desaturation had the greatest variability in plasma norepinephrine levels; specifically, the surges in norepinephrine concentrations were greatest in those with the greatest desaturations. It should be noted that changes in norepinephrine levels across the night are related to the degree of cyclic oxygen desaturation and not to AHI. Regardless, there is strong evidence supporting the link between OSA-associated sympathetic activation (via norepinephrine).

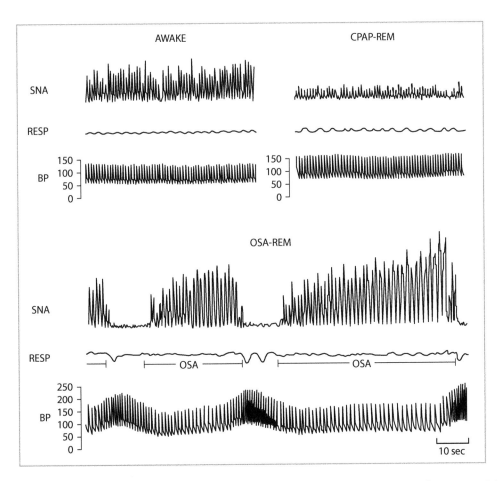

Figure 11.5 Recordings of sympathetic nerve activity (SNA), respirations (RESP), and intra-arterial blood pressure (BP) in the same subject when awake with obstructive sleep apnea during rapid eye movement (REM) sleep and with elimination of obstructive apnea by continuous positive airway pressure (CPAP) therapy during REM sleep. SNA is very high during wakefulness, but increases even further secondary to obstructive apnea during REM sleep. BP increases from 130/65 mmHg when awake to 256/110 mmHg at the end of apnea. Elimination of apneas by CPAP results in decreased nerve activity and prevents BP surges during REM sleep. (Reprinted from Somers V et al., *J Clin Invest* 1995; 96(4): 1897–1904, with permission from RightsLink 2014.)

Metabolic dysregulation

Patients with OSA have increased prevalence of dyslipidemia, insulin resistance, and glucose intolerance, which are associated with increased CHD morbidity and mortality. It is important to recognize that OSA and obesity frequently coexist, and therefore the independent role of each of these conditions in metabolic dysregulation is challenging to delineate.

DYSLIPIDEMIA

Dyslipidemia—an abnormal amount or subtype of lipids in the systemic circulation—is a known risk factor for the development and progression of atherosclerosis and subsequent cardiovascular events.[77] The prevalence of dyslipidemia among those with OSA has been reported to be as high as 55% (hypertriglyceridemia) and 61% (hypercholesterolemia).[78] Greater OSA severity, particularly degree of nocturnal intermittent hypoxemia, has been associated with higher fasting triglyceride levels and lower high-density lipoprotein levels.[79] Further, short-term treatment studies have shown that OSA treatment with CPAP may reduce total cholesterol, low-density lipoprotein, and triglyceride levels.[80] Interestingly, studies of the impact of CPAP on high-density lipoprotein levels appear to be inconsistent, with some showing an increase and others reporting no effect.[80,81] Thus, conclusive evidence regarding the long-term effect of OSA treatment on lipids is lacking, and so this requires further investigation.

INSULIN RESISTANCE AND GLUCOSE INTOLERANCE

Insulin resistance, glucose intolerance, and diabetes are well-established risk factors for CHD. OSA has been associated with insulin resistance in animal and human investigations spanning cross-sectional and prospective studies, as well as clinical trials. It appears that OSA is associated with insulin resistance independent of obesity, which is an important confounder on this relationship.[82] Furthermore, even in otherwise healthy individuals, OSA has been associated with a twofold increase in the risk of impaired or diabetic glucose tolerance.[83] Similar to dyslipidemia, hypoxic stress appears to be the driving factor for impaired glucose tolerance in individuals with OSA.[84] Results from the Sleep Heart Health Study demonstrate that sleep-related hypoxemia is associated with glucose intolerance independent of age, gender, and obesity.[85] In this study, increasing levels of nocturnal hypoxemia were associated in a "dose-dependent" manner with increasing glucose levels. This finding was observed even within nonobese individuals. Further, several investigations have assessed the impact of CPAP therapy on glucose control. It has been suggested that among patients with severe OSA, improvement in glucose metabolism may occur after 2 months of CPAP.[86]

In summary, numerous studies have revealed an association between OSA, dyslipidemia, insulin resistance, and glucose intolerance. In fact, OSA has been identified as an independent risk factor for type II diabetes.[87] It also appears that CPAP therapy improves insulin resistance and glucose intolerance. Nevertheless, larger studies with long-term follow-up are necessary in order to better assess the independent impact of CPAP therapy on glucose intolerance and dyslipidemia across patients with a range of OSA severity. These studies should address obesity, especially visceral obesity, as an important confounder. Further, they should also work to identify optimal levels of CPAP therapy, which will result in significant improvement in metabolic dysregulation.

Mechanical load

Swings in mechanical load via intra-thoracic pressure can impose significant strain and stress on the cardiovascular system. Intrathoracic pressure as low as -80 cm H_2O can result from inspiratory efforts during apneic events, similarly to the Mueller maneuver.[88,89] Studies by Virolainen et al. and Buda et al. showed that during a sustained Mueller maneuver, most of the intrathoracic pressure drop is transmitted into the pericardial cavity, increasing the left ventricular (LV) transmural pressure gradient and LV afterload.[90,91] Virolainen et al. also demonstrated that the LV relaxation rate becomes impaired during this maneuver, which further impedes LV filling and preload.[90] Increased afterload and reduced preload together lead to a reduction in stroke volume and cardiac output during the apnea. With arousal and resumption of breathing, venous return increases, potentially distending the right ventricle and causing the interventricular septum to shift to the left. This results in impaired LV compliance and LV diastolic filling.

Additionally, myocardial hypertrophy from OSA may be associated with compression of endocardial capillaries, impaired myocardial perfusion, and ischemia. In a study by Cloward et al., 83% of the patients who were normotensive and had underlying OSA were found to have LV hypertrophy.[92] Similarly, Hedner et al. compared 61 men with OSA to 61 control subjects, and reported that LV mass and LV mass index (obtained by dividing the LV mass by the body surface area) were significantly higher among the patients with OSA.[93] Specifically, they found the LV mass index to be approximately 15% higher in normotensive OSA patients compared to normotensive control subjects. Therefore, it is apparent that OSA-induced mechanical load effects induce myocardial hypertrophy and consequently increase the risk for ischemia and therefore CHD.

EFFECT OF TREATMENT OF OSA ON CHD

The impact of the treatment of OSA with CPAP on long-term cardiovascular outcomes has been investigated in numerous studies. These studies have been largely observational in nature. A landmark study pertinent to this area was from Marin et al. (Figure 11.6).[94] The investigators recruited 1387 men from a sleep clinic-based sample and an additional 264 healthy men (age- and body mass index-matched with untreated severe OSA subgroup) from a population-based sample. All patients underwent in-laboratory attended polysomnogram studies. Patients with an AHI > 30 per hour and those with an AHI between 5 and 30 in addition to daytime sleepiness or cardiac failure were offered CPAP therapy. The rest of the patients were offered conservative advice including weight loss, avoidance of alcohol, smoking, sedatives, and appropriate sleep hygiene. All patients were followed forward in time for 10 years with yearly clinic appointments. The major endpoints for the study were fatal or nonfatal cardiovascular events including myocardial infarction, stroke, and acute coronary syndromes.

Objective CPAP compliance was ascertained at each visit. Of the 1347 patients, 377 had simple snoring, 403 had mild to moderate OSA (untreated), 235 had severe OSA (untreated), and 372 had treated OSA with CPAP. The results from this study are noted in Table 11.1. Patients with untreated severe OSA were found to have an increased risk of both fatal and nonfatal cardiovascular events (after adjusting for the confounding variables listed in Table 11.1). Patients with no sleep apnea (snorers), untreated mild to moderate OSA, and treated OSA with CPAP had no significant increase in the risk of either fatal or nonfatal cardiac events. The authors of this study do not provide outcome-specific odds ratios (i.e., coronary or cerebrovascular event-specific outcomes). Therefore, the ability to infer the impact of CPAP therapy on CHD-specific events (fatal or nonfatal) is limited. Further, it is important to note that the observational nature of this study makes it at risk for residual confounding. Despite these limitations, this study was the first of its kind to demonstrate a significant impact of CPAP therapy on cardiovascular outcomes over a long-term period.

Another study by Milleron et al. investigated the long-term effect of treating OSA on cardiovascular events in patients with underlying CHD.[95] The authors found that OSA treatment significantly reduced the number of cardiovascular events (cardiovascular death, acute coronary syndrome, hospitalization for heart failure, or need for coronary revascularization) to 24% in treated group versus 58% in untreated group (p < 0.01). This was demonstrated over a period of 86.5 ± 39 months. The two groups were similar in risk factors for cardiovascular diseases at baseline, and treatment of risk factors other than OSA was similar in the two groups. A majority of the patients in the treatment group (n = 25) was treated with CPAP (n = 21), with a small fraction treated with upper-airway surgery (n = 4).

Similarly, others have shown that treatment of OSA improves cardiovascular outcomes in the setting of acute or chronic CHD. A recent case–control study of 192 acute myocardial infarction patients and 96 matched control subjects without CHD (ratio 2:1) revealed that treated OSA patients (AHI ≥ 5 per hour) had a lower risk of recurrent myocardial infarction (adjusted hazard ratio: 0.16 [95% CI 0.03–0.76, p = 0.021]) and revascularization (adjusted hazard ratio: 0.15 [95% CI 0.03–0.79, p = 0.025]) compared to untreated OSA patients.[96]

Despite the evidence suggesting a beneficial role of the treatment of OSA on cardiovascular outcomes—either in the setting of CHD[95] or outside of the setting of acute coronary syndromes[39,41,94]—there has been a lack of large-scale randomized

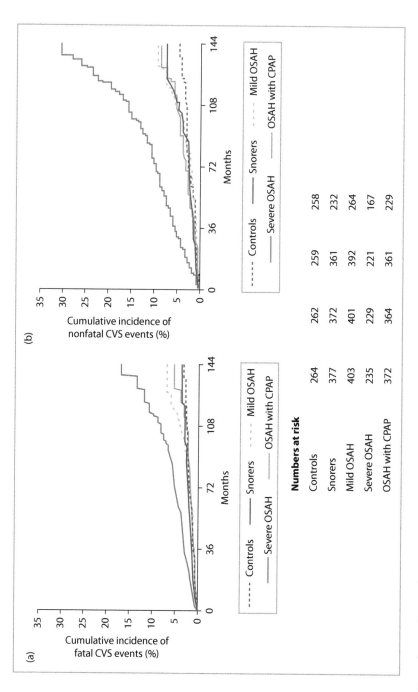

Figure 11.6 Data from a large observational cohort study that compared rates of fatal and nonfatal cardiovascular events (including myocardial infarction, stroke, coronary artery bypass surgery, and percutaneous transluminal coronary angiography) in patients with obstructive sleep apnea (OSA) to simple snorers and healthy participants. The presence of severe OSA (Apnea–Hypopnea Index > 30) was significantly associated with a significant risk of both fatal (a) and nonfatal (b) cardiovascular events. (Reprinted from *Lancet*, 365, Marin JM et al., 1046–1053, Copyright 2005, with permission from Elsevier.)

Table 11.1 Adjusted odds ratio for nonfatal and fatal cardiovascular events

Diagnostic group	Adjusted OR* (CI, p-value) nonfatal cardiovascular events	Adjusted OR* (CI, p-value) fatal cardiovascular events
Snoring	1.32 (0.64–3.01, 0.38)	1.03 (0.31–1.84, 0.88)
Mild to moderate OSA	1.57 (0.62–3.16, 0.22)	1.15 (0.34–2.69, 0.71)
Severe OSA	3.17 (1.12–7.52, 0.001)	2.87 (1.17–7.51, 0.025)
CPAP	1.42 (0.52–3.40, 0.29)	1.05 (0.39–2.21, 0.74)

OR, odds ratio; CI, confidence interval; OSA, obstructive sleep apnea; CPAP, continuous positive airway pressure.

* Variables in the fully adjusted mode include: age, diagnostic group, presence of cardiovascular disease, hypertension, diabetes, lipid disorders, smoking status, alcohol use, systolic and diastolic blood pressure, blood glucose, total cholesterol, triglycerides and current use of antihypertensive, lipid-lowering and anti-diabetic drugs.

studies indicating that treatment of OSA reduces cardiovascular disease-related outcomes. Notably, in the last few years, there has been an emergence of randomized clinical trials that are investigating the impact of OSA treatment on the risk of cardiovascular disease. One such trial is currently underway in Europe. It is a prospective randomized intervention study (n = 400 patients) called "RICCADSA" (Randomized Intervention with CPAP in CAD and OSA).[97] It will assess the impact of CPAP on a composite endpoint of myocardial infarction, new revascularization, stroke, and cardiovascular mortality among those with both CHD and OSA. Similarly, in the United States, a multicenter study called the Heart Biomarker Evaluation in Apnea Treatment (HeartBEAT) was recently completed (August 2012). This study randomized patients with OSA and CHD or CHD risk factors to CPAP, nocturnal oxygen, and health lifestyle instruction. The major goal of this trial was to determine whether CPAP or oxygen altered cardiac biomarkers including (but not limited to) markers of systemic inflammation and oxidative stress, cardiac rhythm, impulse generation, ischemia and myocardial stress. Finally, another clinical trial—Continuous Positive Airway Pressure Treatment of Obstructive Sleep Apnea to Prevent Cardiovascular Disease (SAVE)—is underway and will be the largest clinical trial to date in the sleep apnea field (http://www.savetrial.org/results).[98] The goal of SAVE is to determine whether CPAP treatment reduces the risk of heart attack, stroke, or heart failure for people with OSA. This trial plans to recruit 5000 participants and is a multicenter study involving various countries, including China, India, Australia, and New Zealand. Similar to the RICCADSA trial, it will assess the impact of OSA treated with CPAP on incident serious cardiovascular events among those with established cardiovascular disease.

Clinical trials such as the ones noted above are required in order to assess the impact of CPAP therapy on important clinical outcomes, specifically those pertaining to the cardiovascular system. However, ethical and regulatory topics continue to be a major focus of concern for investigators in the field of sleep medicine. Although the ethics of clinical trials are outside of the scope of this chapter, we encourage interested readers to follow up with an article by Brown et al., which reviews important information on ethical apprehensions pertaining to the design and conduct of clinical trials in OSA.[99]

SUMMARY

It is clear that OSA, via endothelial dysfunction, sympathetic activation, metabolic dysregulation, and mechanical load effects, initiates a complex cascade of cardiovascular-deleterious responses. Epidemiological evidence from numerous cross-sectional observational studies suggests an independent association between OSA and CHD and CHD-related mortality. Further research from longitudinal and interventional studies is necessary in order to more accurately define the link between sleep and the sleep-related breathing disorder OSA and CHD. Given the prevalence of OSA in the general population and the public health burden of CHD, this is an area of great importance for continued investigation.

REFERENCES

1. Go AS, Mozaffarian D, Roger VL et al.; American Heart Association Statistics Committee and Stroke Statistics Subcommittee. Heart disease and stroke statistics—2014 update: A report from the American Heart Association. *Circulation* 2014; 129: e28–e292.

2. Young T, Palta M, Dempsey J, Skatrud J, Weber S, Badr S. The occurrence of sleep-disordered breathing among middle-aged adults. *N Engl J Med* 1993; 328: 1230–1235.

3. Peppard P, Young T, Palta M, Skatrud J. Prospective study of the association between sleep-disordered breathing and hypertension. *N Engl J Med* 2000; 342: 1378–1384.

4. Peker Y, Kraiczi H, Hedner J, Loth S, Johansson A, Bende M. An independent association between obstructive sleep apnoea and coronary artery disease. *Eur Respir J* 1999; 14: 179–184.

5. Shah NA, Yaggi HK, Concato J, Mohsenin V. Obstructive sleep apnea as a risk factor for coronary events or cardiovascular death. *Sleep Breath* 2010; 14: 131–136.

6. Mehra R, Benjamin EJ, Shahar E et al. Sleep Heart Health Study. Association of nocturnal arrhythmias with sleep-disordered breathing: The Sleep Heart Health Study. *Am J Respir Crit Care Med* 2006; 173: 910–916.

7. Shahar E, Whitney C, Redline S et al. Sleep-disordered breathing and cardiovascular disease: Cross-sectional results of the Sleep Heart Health Study. *Am J Respir Crit Care Med* 2001; 163: 19–25.

8. Gottlieb DJ, Yenokyan G, Newman AB et al. Prospective study of obstructive sleep apnea and incident coronary heart disease and heart failure: The Sleep Heart Health Study. *Circulation* 2010; 122: 352–360.

9. Yaggi H, Concato J, Kernan W, Lichtman J, Brass L, Mohsenin V. Obstructive sleep apnea as a risk factor for stroke and death. *N Engl J Med* 2005; 353: 2034–2041.

10. Gami AS, Howard DE, Olson EJ, Somers VK. Day–night pattern of sudden death in obstructive sleep apnea. *N Engl J Med* 2005; 352: 1206–1214.

11. Parish JM. Cardiovascular effects of sleep disorders. *Chest* 1990; 97: 1220–1226.

12. Khatri IM, Freis ED. Hemodynamic changes during sleep. *J Appl Physiol* 1967; 22: 867–873.

13. Baccelli G, Guazzi M, Mancia G, Zanchetti A. Neural and non-neural mechanisms influencing circulation during sleep. *Nature* 1969; 223: 184–185.

14. Somers V, Dyken M, Mark A, Abboud F. Sympathetic nerve activity during sleep in normal adults. *N Engl J Med* 1993; 328: 303–307.

15. National Sleep Foundation. *Adult Sleep Habits and Styles, [Sleep in America poll]*. Arlington, VA: National Sleep Foundation; 2005. http://www.sleepfoundation.org/article/sleep-america-polls/2005-adult-sleep-habits-and-styles/.

16. Cappuccio FP, D'Elia L, Strazzullo P, Miller MA. Sleep duration and all-cause mortality: A systematic review and meta-analysis of prospective studies. *Sleep* 2010; 33: 585–592.

17. Gallicchio L, Kalesan B. Sleep duration and mortality: A systematic review and meta-analysis. *J Sleep Res* 2009; 18: 148–158.

18. Spiegel K, Tasali E, Leproult R, Van Cauter E. Effects of poor and short sleep on glucose metabolism and obesity risk. *Nat Rev Endocrinol* 2009; 5: 253–261.

19. Cappuccio FP, D'Elia L, Strazzullo P, Miller MA. Quantity and quality of sleep and incidence of type 2 diabetes: A systematic review and meta-analysis. *Diabetes Care* 2010; 33: 414–420.

20. Gangwisch JE, Heymsfield SB, Boden-Albala B et al. Short sleep duration as a risk factor for hypertension: Analyses of the first National Health and Nutrition Examination Survey. *Hypertension* 2006; 47: 833–839.

21. Gottlieb DJ, Redline S, Nieto FJ et al. Association of usual sleep duration with hypertension: The Sleep Heart Health Study. *Sleep* 2006; 29: 1009–1014.

22. Kaneita Y, Uchiyama M, Yoshiike N, Ohida T. Associations of usual sleep duration with serum lipid and lipoprotein levels. *Sleep* 2008; 31: 645–652.

23. Bliwise DL, Pascualy RA. Sleep-related respiratory disturbance in elderly persons. *Compr Ther* 1984; 10: 8–14.

24. Steptoe A, Peacey V, Wardle J. Sleep duration and health in young adults. *Arch Intern Med* 2006; 166: 1689–1692.

25. Knutson KL, Turek FW. The U-shaped association between sleep and health: The 2 peaks do not mean the same thing. *Sleep* 2006; 29: 878–879.

26. King CR, Knutson KL, Rathouz PJ, Sidney S, Liu K, Lauderdale DS. Short sleep duration and incident coronary artery calcification. *JAMA* 2008; 300: 2859–2866.

27. Wolff B, Volzke H, Schwahn C, Robinson D, Kessler C, John U. Relation of self-reported sleep duration with carotid intima–media thickness in a general population sample. *Atherosclerosis* 2008; 196: 727–732.

28. Mallon L, Broman JE, Hetta J. Sleep complaints predict coronary artery disease mortality in males: A 12-year follow-up study of a middle-aged Swedish population. *J Intern Med* 2002; 251: 207–216.

29. Amagai Y, Ishikawa S, Gotoh T et al. Sleep duration and mortality in Japan: The Jichi Medical School Cohort Study. *J Epidemiol* 2004; 14: 124–128.

30. Ayas NT, White DP, Manson JE et al. A prospective study of sleep duration and coronary heart disease in women. *Arch Intern Med* 2003; 163: 205–209.

31. Meisinger C, Heier M, Lowel H, Schneider A, Doring A. Sleep duration and sleep complaints and risk of myocardial infarction in middle-aged men and women from the general population: The MONICA/KORA Augsburg cohort study. *Sleep* 2007; 30: 1121–1127.

32. Ikehara S, Iso H, Date C et al.; JACC Study Group. Association of sleep duration with mortality from cardiovascular disease and other causes for Japanese men and women: The JACC study. *Sleep* 2009; 32: 295–301.

33. Spiegel K, Tasali E, Penev P, Van Cauter E. Brief communication: Sleep curtailment in healthy young men is associated with decreased leptin levels, elevated ghrelin levels, and increased hunger and appetite. *Ann Intern Med* 2004; 141: 846–850.

34. Taheri S, Lin L, Austin D, Young T, Mignot E. Short sleep duration is associated with reduced leptin, elevated ghrelin, and increased body mass index. *PLoS Med* 2004; 1: e62.

35. Rayner DV, Trayhurn P. Regulation of leptin production: Sympathetic nervous system interactions. *J Mol Med (Berl)* 2001; 79: 8–20.

36. van der Lely AJ, Tschop M, Heiman ML, Ghigo E. Biological, physiological, pathophysiological, and pharmacological aspects of ghrelin. *Endocr Rev* 2004; 25: 426–457.

37. Sugino T, Yamaura J, Yamagishi M et al. Involvement of cholinergic neurons in the regulation of the ghrelin secretory response to feeding in sheep. *Biochem Biophys Res Commun* 2003; 304: 308–312.

38. Heath RB, Jones R, Frayn KN, Robertson MD. Vagal stimulation exaggerates the inhibitory ghrelin response to oral fat in humans. *J Endocrinol* 2004; 180: 273–281.

39. Spiegel K, Leproult R, L'hermite-Baleriaux M, Copinschi G, Penev PD, Van Cauter E. Leptin levels are dependent on sleep duration: Relationships with sympathovagal balance, carbohydrate regulation, cortisol, and thyrotropin. *J Clin Endocrinol Metab* 2004; 89: 5762–5771.

40. Spiegel K, Leproult R, Van Cauter E. Impact of sleep debt on metabolic and endocrine function. *Lancet* 1999; 354: 1435–1439.

41. Miller MA, Cappuccio FP. Inflammation, sleep, obesity and cardiovascular disease. *Curr Vasc Pharmacol* 2007; 5: 93–102.

42. Patel SR, Malhotra A, Gottlieb DJ, White DP, Hu FB. Correlates of long sleep duration. *Sleep* 2006; 29: 881–889.

43. Ekstedt M, Akerstedt T, Soderstrom M. Microarousals during sleep are associated with increased levels of lipids, cortisol, and blood pressure. *Psychosom Med* 2004; 66: 925–931.

44. Berry RB, Budhiraja R, Gottlieb DJ et al.; American Academy of Sleep Medicine. Rules for scoring respiratory events in sleep: Update of the 2007 AASM Manual for the Scoring of Sleep and Associated Events. Deliberations of the Sleep Apnea Definitions Task Force of the American Academy of Sleep Medicine. *J Clin Sleep Med* 2012; 8: 597–619.

45. Chandola T, Ferrie JE, Perski A, Akbaraly T, Marmot MG. The effect of short sleep duration on coronary heart disease risk is greatest among those with sleep disturbance: A prospective study from the Whitehall II cohort. *Sleep* 2010; 33: 739–744.

46. Narang I, Manlhiot C, Davies-Shaw J et al. Sleep disturbance and cardiovascular risk in adolescents. *CMAJ* 2012; 184: E913–E920.

47. Hoevenaar-Blom MP, Spijkerman AM, Kromhout D, van den Berg JF, Verschuren WM. Sleep duration and sleep quality in relation to 12-year cardiovascular disease incidence: The MORGEN study. *Sleep* 2011; 34: 1487–1492.

48. Mooe T, Rabben T, Wiklund U, Franklin KA, Eriksson P. Sleep-disordered breathing in men with coronary artery disease. *Chest* 1996; 109: 659–663.

49. Mooe T, Rabben T, Wiklund U, Franklin KA, Eriksson P. Sleep-disordered breathing in women: Occurrence and association with coronary artery disease. *Am J Med* 1996; 101: 251–256.

50. Schafer H, Koehler U, Ewig S, Hasper E, Tasci S, Luderitz B. Obstructive sleep apnea as a risk marker in coronary artery disease. *Cardiology* 1999; 92: 79–84.

51. Maekawa M, Shiomi T, Usui K, Sasanabe R, Kobayashi T. Prevalence of ischemic heart disease among patients with sleep apnea syndrome. *Psychiatry Clin Neurosci* 1998; 52: 219–220.

52. Mooe T, Franklin KA, Holmstrom K, Rabben T, Wiklund U. Sleep-disordered breathing and coronary artery disease: Long-term prognosis. *Am J Respir Crit Care Med* 2001; 164: 1910–1913.

53. Konecny T, Kuniyoshi FH, Orban M et al. Under-diagnosis of sleep apnea in patients after acute myocardial infarction. *J Am Coll Cardiol* 2010; 56: 742–743.

54. Muller JE, Stone PH, Turi ZG et al. Circadian variation in the frequency of onset of acute myocardial infarction. *N Engl J Med* 1985; 313: 1315–1322.

55. Cohen MC, Rohtla KM, Lavery CE, Muller JE, Mittleman MA. Meta-analysis of the morning excess of acute myocardial infarction and sudden cardiac death. *Am J Cardiol* 1997; 79: 1512–1516.

56. Shahar E, Whitney C, Redline S et al.; Sleep Heart Health Study Research Group. Sleep-disordered breathing and cardiovascular disease: Cross-sectional results of the Sleep Heart Health Study. *Am J Respir Crit Care Med* 2001; 163: 19–25.

57. Libby P. Inflammation in atherosclerosis. *Nature* 2002; 420: 868–874.

58. Lefer DJ, Granger DN. Oxidative stress and cardiac disease. *Am J Med* 2000; 109: 315–323.

59. Babior BM. Phagocytes and oxidative stress. *Am J Med* 2000; 109: 33–44.

60. Davignon J, Ganz P. Role of endothelial dysfunction in atherosclerosis. *Circulation* 2004; 109: III27–III32.

61. Lavie L. Obstructive sleep apnoea syndrome—An oxidative stress disorder. *Sleep Med Rev* 2003; 7: 35–51.

62. Ohike Y, Kozaki K, Iijima K et al. Amelioration of vascular endothelial dysfunction in obstructive sleep apnea syndrome by nasal continuous positive airway pressure—Possible involvement of nitric oxide and asymmetric NG, NG-dimethylarginine. *Circ J* 2005; 69: 221–226.

63. O'Leary DH, Polak JF. Intima–media thickness: A tool for atherosclerosis imaging and event prediction. *Am J Cardiol* 2002; 90: 18L–21L.

64. O'Leary DH, Polak JF, Kronmal RA, Manolio TA, Burke GL, Wolfson SK Jr. Carotid-artery intima and media thickness as a risk factor for myocardial infarction and stroke in older adults. Cardiovascular Health Study Collaborative Research Group. *N Engl J Med* 1999; 340: 14–22.

65. Kuller LH, Shemanski L, Psaty BM et al. Subclinical disease as an independent risk factor for cardiovascular disease. *Circulation* 1995; 92: 720–726.

66. Bots ML, Hoes AW, Koudstaal PJ, Hofman A, Grobbee DE. Common carotid intima–media thickness and risk of stroke and myocardial infarction: The Rotterdam Study. *Circulation* 1997; 96: 1432–1437.

67. Silvestrini M, Rizzato B, Placidi F, Baruffaldi R, Bianconi A, Diomedi M. Carotid artery

wall thickness in patients with obstructive sleep apnea syndrome. *Stroke* 2002; 33: 1782–1785.

68. Friedlander AH, Yueh R, Littner MR. The prevalence of calcified carotid artery atheromas in patients with obstructive sleep apnea syndrome. *J Oral Maxillofac Surg* 1998; 56: 950–954.

69. Drager LF, Bortolotto LA, Figueiredo AC, Krieger EM, Lorenzi GF. Effects of continuous positive airway pressure on early signs of atherosclerosis in obstructive sleep apnea. *Am J Respir Crit Care Med* 2007; 176: 706–712.

70. Sharma SK, Agrawal S, Damodaran D et al. CPAP for the metabolic syndrome in patients with obstructive sleep apnea. *N Engl J Med* 2011; 365: 2277–2286.

71. Hui DS, Shang Q, Ko FW et al. A prospective cohort study of the long-term effects of CPAP on carotid artery intima–media thickness in obstructive sleep apnea syndrome. *Respir Res* 2012; 13: 22.

72. Buchner NJ, Quack I, Stegbauer J, Woznowski M, Kaufmann A, Rump LC. Treatment of obstructive sleep apnea reduces arterial stiffness. *Sleep Breath* 2012; 16: 123–133.

73. Narkiewicz K, Somers VK. Cardiovascular variability characteristics in obstructive sleep apnea. *Auton Neurosci* 2001; 90: 89–94.

74. Narkiewicz K, van de Borne PJ, Montano N, Dyken ME, Phillips BG, Somers VK. Contribution of tonic chemoreflex activation to sympathetic activity and blood pressure in patients with obstructive sleep apnea. *Circulation* 1998; 97: 943–945.

75. Somers V, Dyken M, Clary M, Abboud F. Sympathetic neural mechanisms in obstructive sleep apnea. *J Clin Invest* 1995; 96: 1897–1904.

76. Baylor P, Mouton A, Shamoon HH, Goebel P. Increased norepinephrine variability in patients with sleep apnea syndrome. *Am J Med* 1995; 99: 611–615.

77. Arsenault BJ, Boekholdt SM, Kastelein JJ. Lipid parameters for measuring risk of cardiovascular disease. *Nat Rev Cardiol* 2011; 8: 197–206.

78. Chou YT, Chuang LP, Li HY et al. Hyperlipidaemia in patients with sleep-related breathing disorders: Prevalence & risk factors. *Indian J Med Res* 2010; 131: 121–125.

79. Trzepizur W, Le Vaillant M, Meslier N et al. Independent association between nocturnal intermittent hypoxemia and metabolic dyslipidemia. *Chest* 2013; 143: 1584–1589.

80. Dorkova Z, Petrasova D, Molcanyiova A, Popovnakova M, Tkacova R. Effects of continuous positive airway pressure on cardiovascular risk profile in patients with severe obstructive sleep apnea and metabolic syndrome. *Chest* 2008; 134: 686–692.

81. Steiropoulos P, Tsara V, Nena E et al. Effect of continuous positive airway pressure treatment on serum cardiovascular risk factors in patients with obstructive sleep apnea–hypopnea syndrome. *Chest* 2007; 132: 843–851.

82. Ip MS, Lam B, Ng MM, Lam WK, Tsang KW, Lam KS. Obstructive sleep apnea is independently associated with insulin resistance. *Am J Respir Crit Care Med* 2002; 165: 670–676.

83. Punjabi NM, Bandeen-Roche K, Marx JJ, Neubauer DN, Smith PL, Schwartz AR. The association between daytime sleepiness and sleep-disordered breathing in NREM and REM sleep. *Sleep* 2002; 25: 307–314.

84. Sulit L, Storfer-Isser A, Kirchner HL, Redline S. Differences in polysomnography predictors for hypertension and impaired glucose tolerance. *Sleep* 2006; 29: 777–783.

85. Punjabi N, Shahar E, Redline S, Gottlieb D, Givelber R, Resnick H. Sleep-disordered breathing, glucose intolerance, and insulin resistance: The Sleep Heart Health Study. *Am J Epidemiol* 2004; 160: 521–530.

86. Weinstock TG, Wang X, Rueschman M et al. A controlled trial of CPAP therapy on metabolic control in individuals with impaired glucose tolerance and sleep apnea. *Sleep* 2012; 35: 617–625B.

87. Botros N, Concato J, Mohsenin V, Selim B, Doctor K, Yaggi HK. Obstructive sleep apnea as a risk factor for type 2 diabetes. *Am J Med* 2009; 122: 1122–1127.

88. Shamsuzzaman AS, Gersh BJ, Somers VK. Obstructive sleep apnea: Implications for cardiac and vascular disease. *JAMA* 2003; 290: 1906–1914.

89. Shiomi T, Guilleminault C, Stoohs R, Schnittger I. Leftward shift of the intraventricular septum and pulsus paradoxus in obstructive sleep apnea syndrome. *Chest* 1991; 100: 894–902.

90. Virolainen J, Ventila M, Turto H, Kupari M. Effect of negative intrathoracic pressure on left ventricular pressure dynamics and relaxation. *J Appl Physiol* 1995; 79: 455–460.

91. Buda AJ, Pinsky MR, Ingels NB Jr, Daughters GT 2nd, Stinson EB, Alderman EL. Effect of intrathoracic pressure on left ventricular performance. *N Engl J Med* 1979; 301: 453–459.

92. Cloward TV, Walker JM, Farney RJ, Anderson JL. Left ventricular hypertrophy is a common echocardiographic abnormality in severe obstructive sleep apnea and reverses with nasal continuous positive airway pressure. *Chest* 2003; 124: 594–601.

93. Hedner J, Ejnell H, Caidahl K. Left ventricular hypertrophy independent of hypertension in patients with obstructive sleep apnoea. *J Hypertens* 1990; 8: 941–946.

94. Marin JM, Carrizo SJ, Vicente E, Agusti AG. Long-term cardiovascular outcomes in men with obstructive sleep apnoea–hypopnoea with or without treatment with continuous positive airway pressure: An observational study. *Lancet* 2005; 365: 1046–1053.

95. Milleron O, Pilliere R, Foucher A et al. Benefits of obstructive sleep apnoea treatment in coronary artery disease: A long-term follow-up study. *Eur Heart J* 2004; 25: 728–734.

96. Garcia-Rio F, Alonso-Fernandez A, Armada E et al. CPAP effect on recurrent episodes in patients with sleep apnea and myocardial infarction. *Int J Cardiol* 2013; 168: 1328–1335.

97. Peker Y, Glantz H, Thunstrom E, Kallryd A, Herlitz J, Ejdeback J. Rationale and design of the Randomized Intervention with CPAP in Coronary Artery Disease and Sleep Apnoea—RICCADSA trial. *Scand Cardiovasc J* 2009; 43: 24–31.

98. McEvoy RD, Anderson CS, Antic NA et al. The Sleep Apnea Cardiovascular Endpoints (SAVE) trial: Rationale and start-up phase. *J Thorac Dis* 2010; 2: 138–143.

99. Brown DL, Anderson CS, Chervin RD et al. Ethical issues in the conduct of clinical trials in obstructive sleep apnea. *J Clin Sleep Med* 2011; 7: 103–108.

12

Sleep and personality disorders

LAMPROS PEROGAMVROS AND C. ROBERT CLONINGER

The sleep–personality relationship has been studied in several ways. The personality traits of patients with sleep disorders have been reported in one body of literature. A separate literature has examined the sleep features of patients with personality disorders. The role of sleep in the development of personality in childhood is not reviewed here, nor is the role of sleep in the exacerbation of personality diatheses.

PERSONALITY TRAITS IN PATIENTS WITH SLEEP DISORDERS

Only a few studies of personality assessment in patients with sleep disorders have been conducted. One main reason motivating these studies is that the pathophysiology of these disorders has remained largely unknown until now. Defining the personality profile of these patients would thus at least partly elucidate the role of psychological factors in the etiology, perpetuation, and treatment of sleep disorders. The main hypothesis is that there exists a vulnerability temperament predisposing these patients to developing their sleep disorder. Longitudinal studies before and after the appearance of the sleep symptoms are certainly needed in order to confirm the causality (and not simply an association or risk factor) of personality traits in sleep disorders.

Personality traits of patients with insomnia

Early studies assessing the personalities of insomnia patients have mainly used the Minnesota Multiphasic Personality Inventory and reported that these patients are more self-preoccupied, defensive, and sensation avoiding than controls.[1,2] In accordance with these results, some authors proposed that insomnia is related to an internalization of psychological distress.[3,4] More recent studies have used the Temperament and Character Inventory (TCI), a well-validated instrument based on the unified biosocial theory of personality developed by Cloninger and colleagues.[5,6] The TCI is a true–false questionnaire that measures four dimensions of temperament—novelty seeking (NS), harm avoidance (HA), reward dependence (RD), and persistence (P)—and three dimensions of character—self-directedness (SD), self-transcendence (ST), and cooperativeness (C).[6] There is evidence to suggest that these features have

Table 12.1 Temperament and character contributions to sleep-related disorders

	Insomnia	Obstructive sleep apneas	Parasomnias
Novelty seeking	–	↑	↑
Harm avoidance	↑	–	↑*
Reward dependence	–	–	↑*
Persistence	–	–	–
Self-directedness	↓	–	↓

* High anticipatory worry (HA1) and dependence on social attachment (RD3) together indicate sensitivity to social separation or rejection in patients with sleepwalking or nightmares

neurochemical correlates in the brain and clinical correlates with regards to personality disorders, as well as a differential response to the treatment of depression and anxiety.[7,8] de Saint Hilaire and colleagues[9] have used the TCI to investigate personality components in primary insomnia. This study of 32 nondepressed patients with primary insomnia revealed significantly higher HA scores and lower SD scores in insomniacs versus normal controls (Table 12.1). These were correlated with polysomnographic sleep variables related to insomnia. Sleep latency was positively correlated with HA, while anticipatory worry (a subset of the HA scale) was negatively correlated with rapid eye movement (REM) latency. Insomniacs were higher in all subscales of HA, including anticipatory worry, fear of uncertainty, shyness, and fatigability. Patients and controls were not significantly different in the other aspects of temperament: RD, NS, or P (Table 12.1). There was also a positive correlation between HA and Hospital Anxiety and Depression Scale anxiety scores. A total of 43% of the patients with high HA scores demonstrated a passive–aggressive profile (i.e., high on HA, NS, and RD), while 17% tended toward a passive–dependent (cautious/avoidant) personality type (i.e., high on HA and RD and low on NS). These findings are consistent with the hypothesis that insomnia may be related to sensation-avoiding behavior,[1] which may be similar to TCI harm avoidance. In addition, they suggest that HA may be a reliable psychobiological vulnerability marker for primary insomnia and that psychotherapeutic treatment aiming at this temperament dimension could be used for the management of this sleep disorder.

Personality traits of patients with sleep apneas

Initial investigations suggested higher rates of personality and affective disorders in patients with sleep apneas.[10,11] Associations have been suggested between sleep apneas and hypochondriasis, conversion–hysteria, and possibly psychosis; however, other studies have not identified apnea density as a risk factor for personality disorders or affective illness.[12,13] Pillar and Lavie[14] studied psychological sequelae in over 2000 patients with sleep apneas and failed to show a connection between the severity or existence of sleep apneas and the development of depression or anxiety symptoms. A recent study, however, showed that snorting/stopping breathing ≥5 nights/week compared to never was strongly associated with major depression in men and women.[15]

To further elucidate a possible connection between sleep apneas and personality changes, Sforza and colleagues[11] administered the TCI to 60 patients with reduced daytime alertness and snoring who presented to clinic for evaluation of possible obstructive sleep apnea (OSA). Apart from an increase in NS in patients with OSA compared to controls (Table 12.1), no other relationships between apnea density and other TCI variables were found. Is there direct causality between NS and apneas? This seems unlikely, as there was no significant correlation between NS and the Apnea-Hypopnea Index, nocturnal hypoxemia, or diurnal sleepiness. The authors hypothesized that as NS is higher in persons with dependence on nicotine and alcohol and in obesity, this finding reflects a genetic personality trait favoring risk factors that increase comorbidity for apneas. It is also possible that symptoms associated with sleep apneas, including daytime sleepiness, reduced vigor and energy, or reduced lust for life, may be misinterpreted as symptoms of affective illness or personality disorder rather than primary sequelae of sleep apneas.[11]

Personality traits of narcoleptic patients

Early studies demonstrated that narcoleptic patients are characterized by anxiety and social introversion,[12,16] although it has been noted that there may also be a coupling between the genetics of narcolepsy and biologically based personality traits. Narcoleptic persons often present a unique personality change named "narcoleptoid personality" characterized by denial of sleepiness, decreased psychic tension, low self-esteem, and passivity.[17] These characteristics do not seem to change with treatment.

Personality traits of parasomnia patients

A recent study (Perogamvros et al.[18]) demonstrated that, compared to controls, parasomnia patients (12 sleepwalkers and 12 patients with idiopathic nightmares) scored higher on NS and in particular on the exploratory excitability/curiosity (NS1) subscale, and lower on SD, suggesting a general increase in reward sensitivity and impulsivity. Furthermore, parasomnia patients tended to worry about social separation persistently, as indicated by greater anticipatory worry (HA1) and dependence on social attachment (RD3). Moreover, exploratory excitability correlated positively with the severity of parasomnia (i.e., the frequency of self-reported occurrences of nightmares and sleepwalking), and with time spent in REM sleep in patients with nightmares. These results suggest that patients with parasomnia might share common waking personality traits associated to reward-related brain functions.

SLEEP FEATURES IN PATIENTS WITH PERSONALITY DISORDERS

There is limited formal investigation of the sleep characteristics of patients with personality disorders. The most investigated personality disorders to date are borderline personality disorder (BPD) and antisocial personality disorder (ASP).

Sleep in BPD

Sleep complaints are more frequent in BPD than in healthy controls.[19] However, as rates of substance abuse and depression are increased in patients with BPD, the data from sleep studies can be confounded.[20] Several polysomnographic studies have been conducted in patients with BPD.[21,22] The majority of the earlier studies were not controlled for comorbid depressive illness, and those that acknowledged depression in their research subjects often did not do so with standardized instruments.[23] The most common finding among these studies was shortened REM latency, increased REM density, and overall disturbance of sleep continuity.[24-26] Similar findings have been reported in non-European and non-American populations, with similar methodological shortcomings.[19] Several studies have controlled for comorbid depressive illness with mixed results. Akiskal et al.[27] examined 24 non-depressed patients with BPD and found shortened REM latency in these patients versus non-BPD personality-disordered controls and healthy controls. This finding contributed to the hypothesis that affective disorders and BPD may have a common biological origin or similar biological pathways, or that at least a subset of BPD patients may be at elevated risk for affective illness.[27] Battaglia et al.[28] examined ten never-depressed patients with BPD and correlated shortened REM latency with familial risk for depression in these patients. In a similar follow-up study, the same group found that increased REM density in the first period of REM sleep was related to risk for depressive illness in patients with BPD.[29] These authors have advanced these data in order to hypothesize a link between BPD and vulnerability for affective illness.

Current research has more effectively controlled for depressive illness and, to some degree, for alcohol usage. De la Fuente et al.[30,31] were unable to differentiate between BPD, major depressive disorder, and normal controls on the basis of REM latency. Philipsen et al.[22] studied conventional polysomnographic parameters as well as electroencephalographic (EEG) spectral power analysis in 20 nondepressed, unmedicated female patients with

BPD. No severe sleep problems were noted in the subjects versus healthy controls, but a significant discrepancy was noted between objective sleep measurements and subjective sleep quality. This raises the possibility that sleep disturbance in BPD may be more closely related to the psychological correlates and interpretive behaviors that are characteristic of BPD than to a primary neurobiological mechanism. Indeed, complaints of subjectively poor sleep quality may be more consistent with the constellation of somatoform and psychoform symptoms seen in comorbid BPD and somatization disorder, with sleep-state misperception being part of a broader disturbance in self-awareness. A second finding of the study by Philipsen et al.[22] was that delta power in non-REM (NREM) sleep appears to be elevated in BPD. This is also consistent with findings in patients with ASP.[31,32] As will be discussed below, this similarity in NREM sleep between patients with BPD and ASP may be indicative of dysregulated impulse control and aggression, and lends itself to a serotonergic and dopaminergic hypothesis of impulse control, NREM sleep disturbance, and personality disorder. It should be remembered that BPD and ASP are clinically similar; patients with BPD are higher in HA than those with ASP, but are otherwise the same with regards to other aspects of personality.[34]

Two more recent studies, that excluded a psychotic or mood disorder during the past year, showed that BPD patients experience a higher rate of nightmares, dream anxiety, and disturbed sleep than controls.[35,36] Dream anxiety was positively correlated with early traumatic experiences and impaired subjective sleep quality. Patients with nightmares had a more severe clinical profile than those without nightmares. In addition, BPD and primary insomnia patients seem to share common objective sleep modifications (longer sleep onset, shorter sleep time, and lower sleep efficiency than controls), but different subjective complaints (subjectivity of sleep quality and quantity in BPD more strongly resembles that of controls than of insomniacs). The authors conclude that BPD patients suffer from insomnia symptoms.[37] Other studies have shown that treatment for BPD improves sleep quality/quantity in these patients.[38]

Overall, the most common sleep architecture change reported in early polysomnographic studies of patients with BPD is shortened REM latency. This is an abnormality that is commonly seen in mood disorders.[39] As later studies excluded depressive disorder and found normal REM latencies, it seems that these findings are confounded by comorbid depression. Further studies in homogeneous samples of BPD patients and without other psychiatric comorbidities are certainly needed in order to clarify the sleep abnormalities in BPD.

Sleep in ASP and aggression

There is a limited yet significant body of research regarding sleep architecture and ASP. In a sleep EEG study of 19 habitually violent and drug-free subjects with ASP, Lindberg et al.[33] found an increase in slow-wave sleep, with decreased sleep efficiency in the form of more frequent awakenings than controls. In contrast to other studies of psychiatric disorders, this appears to be the only instance of increased slow-wave sleep among patients with psychopathology, although this finding may be consistent with observational findings in preadolescent boys with conduct disorder.[31] Lindberg et al.[33] also found that subjects with ASP (and, by definition, a history of conduct disorder) had a variable degree of impulsiveness that was correlated with variations in sleep architecture. More recently, this group found that in 14 habitually violent men with ASP, increases in the amount and delta power of slow-wave sleep correlated with a retrospective measure of childhood attention deficit–hyperactivity disorder (ADHD).[40] From these findings, Lindberg et al.[40] hypothesized a shared deficit related to sleep architecture among children with ADHD and adults who develop ASP. In addition, Philipsen et al.[22] hypothesized a common pathway between ASP and BPD, wherein impulse control and aggression are related to sleep architecture, perhaps through a serotonergic pathway with or without a dopaminergic component. While the $5-HT_{2C}$ receptor may play a role in slow-wave sleep regulation, and treatment with serotonergic medications may have some benefit in sleep regulation, it seems premature to declare a causal relationship.[41]

Complicating factors

Patients with personality disorders or pathologic disruptions of temperament and character have more somatic complaints overall, tend to utilize additional medical resources, and are prone to making poor health-related lifestyle choices.[42] In addition to their primary illness, these patients have elevated incidences of substance abuse and may be at elevated risk for major depressive disorder or other mood disorders. The importance of acknowledging or controlling for depression in study designs has been discussed above. To our knowledge, limited efforts have been made at incorporating comorbid substance abuse, particularly alcoholism, into study designs. Substance abuse may be a significant confounding factor in studies of personality disorders and sleep. On the other hand, "self-medication" with alcohol or other drugs of abuse may initially develop in some patients as a response to a sleep disturbance. Furthermore, substance abuse may also impact the subjective experience or reporting of sleep quality, particularly in outpatient or ambulatory survey instruments.

SUMMARY

Progress, challenges, and limitations in the investigation of a relationship between sleep and personality factors have largely followed those of their two component fields. Sleep research, in particular with regards to insomnia, is continually faced with the challenges of definition, validation, and confounding variables. Personality research is often limited by its measurement instruments and the heterogeneity of diagnostic schemes. Nevertheless, investigations from both sides of a sleep–personality research perspective have yielded interesting results, which have generated additional research questions and may also have clinical utility.

Personality assessment has revealed that HA may play a role in primary insomnia, while NS seems to increase comorbidity for sleep apneas by increasing the probability of substance abuse and obesity. Low SD may influence not only the development of insomnia, but also its perpetuation. Increased NS and reduced SD also seem to be related to parasomnias, such as sleepwalking and idiopathic nightmares. These findings have therapeutic implications, as specific personality characteristics may be targeted by psychotherapy.

While research regarding the sleep components of personality disorders is complex and the recent iterations of more carefully controlled sleep EEG studies have been largely equivocal with regards to REM latency, current evidence suggests a NREM sleep component to impulse control-related personality disorders. These findings hold potential for therapeutic modalities involving modulation of sleep, impulsivity, and character components through pharmacologic intervention, self-awareness training, or alterations of sleep patterns.

REFERENCES

1. Hauri P, Fisher J. Persistent psychophysiologic (learned) insomnia. *Sleep* 1986; 9; 38–53.
2. Marchini EJ, Coates TJ, Magistad JG, Waldum SJ. What do insomniacs do, think, and feel during the day? A preliminary study. *Sleep* 1983; 6: 147–155.
3. Kales A, Caldwell AB, Preston TA, Healey S, Kales JD. Personality patterns in insomnia. Theoretical implications. *Arch Gen Psychiatry* 1976; 33: 1128–1134.
4. Kales A, Vgontzas AN. Predisposition to and development and persistence of chronic insomnia: Importance of psychobehavioral factors. *Arch Intern Med* 1992; 152: 1570–1572.
5. Cloninger CR. A systematic method for clinical description and classification of personality variants. A proposal. *Arch Gen Psychiatry* 1987; 44: 573–588.
6. Cloninger CR, Svrakic DM, Przybeck TR. A psychobiological model of temperament and character. *Arch Gen Psychiatry* 1993; 50: 975–990.
7. Battaglia M, Przybeck TR, Bellodi L, Cloninger CR. Temperament dimensions explain the comorbidity of psychiatric disorders. *Compr Psychiatry* 1996; 37: 292–298.
8. Cloninger CR, Bayon C, Svrakic DM. Measurement of temperament and character in mood disorders: A model of fundamental states as personality types. *J Affect Disord* 1998; 51: 21–32.
9. de Saint Hilaire Z, Straub J, Pelissolo A. Temperament and character in primary insomnia. *Eur Psychiatry* 2005; 20: 188–192.
10. Kales A, Caldwell AB, Cadieux RJ, Vela-Bueno A, Ruch LG, Mayes SD. Severe obstructive sleep apnea—II: Associated psychopathology and psychosocial consequences. *J Chronic Dis* 1985; 38: 427–434.

11. Sforza E, de Saint Hilaire Z, Pelissolo A, Rochat T, Ibanez V. Personality, anxiety and mood traits in patients with sleep-related breathing disorders: Effect of reduced daytime alertness. *Sleep Med* 2002; 3: 139–145.

12. Beutler LE, Ware JC, Karacan I, Thornby JI. Differentiating psychological characteristics of patients with sleep apnea and narcolepsy. *Sleep* 1981; 4: 39–47.

13. Hudgel DW. Neuropsychiatric manifestations of obstructive sleep apnea: A review. *Int J Psychiatry Med* 1989; 19: 11–22.

14. Pillar G, Lavie P. Psychiatric symptoms in sleep apnea syndrome: Effects of gender and respiratory disturbance index. *Chest* 1998; 114: 697–703.

15. Wheaton AG, Perry GS, Chapman DP, Croft JB. Sleep disordered breathing and depression among U.S. adults: National Health and Nutrition Examination Survey, 2005–2008. *Sleep* 2012; 35: 461–467.

16. Sachs C, Levander S. Personality traits in patients with narcolepsy. *Pers Individ Dif* 1981; 2: 319–324.

17. Honda Y. *Clinical Features of Narcolepsy: Japanese Experiences*. Heidelberg: Springer, 1988.

18. Perogamvros L, Aberg C, Gex-Fabry M, Perrig S, Cloninger CR, Schwartz S. Increased reward-related behaviors during sleep and wakefulness in sleepwalking and idiopathic nightmares. *Plos One* 2015; 10(8): e0134504.

19. Asaad T, Okasha T, Okasha A. Sleep EEG findings in ICD-10 borderline personality disorder in Egypt. *J Affect Disord* 2002; 71: 11–18.

20. Joyce PR, Mulder RT, Luty SE, McKenzie JM, Sullivan PF, Cloninger RC. Borderline personality disorder in major depression: Symptomatology, temperament, character, differential drug response, and 6-month outcome. *Compr Psychiatry* 2003; 44: 35–43.

21. De la Fuente JM, Bobes J, Vizuete C, Mendlewicz J. Effects of carbamazepine on dexamethasone suppression and sleep electroencephalography in borderline personality disorder. *Neuropsychobiology* 2002; 45: 113–119.

22. Philipsen A, Feige B, Al-Shajlawi A et al. Increased delta power and discrepancies in objective and subjective sleep measurements in borderline personality disorder. *J Psychiatr Res* 2005; 39: 489–498.

23. Boutros NN, Torello M, McGlashan TH. Electrophysiological aberrations in borderline personality disorder: State of the evidence. *J Neuropsychiatry Clin Neurosci* 2003; 15: 145–154.

24. Bell J, Lycaki H, Jones D, Kelwala S, Sitaram N. Effect of preexisting borderline personality disorder on clinical and EEG sleep correlates of depression. *Psychiatry Res* 1983; 9: 115–123.

25. Lahmeyer HW, Val E, Gaviria FM et al. EEG sleep, lithium transport, dexamethasone suppression, and monoamine oxidase activity in borderline personality disorder. *Psychiatry Res* 1988; 25: 19–30.

26. Reynolds CF 3rd, Soloff PH, Kupfer DJ et al. Depression in borderline patients: A prospective EEG sleep study. *Psychiatry Res* 1985; 14: 1–15.

27. Akiskal HS, Yerevanian BI, Davis GC, King D, Lemmi H. The nosologic status of borderline personality: Clinical and polysomnographic study. *Am J Psychiatry* 1985; 142: 192–198.

28. Battaglia M, Ferini-Strambi L, Smirne S, Bernardeschi L, Bellodi L. Ambulatory polysomnography of never-depressed borderline subjects: A high-risk approach to rapid eye movement latency. *Biol Psychiatry* 1993; 33: 326–334.

29. Battaglia M, Ferini Strambi L, Bertella S, Bajo S, Bellodi L. First-cycle REM density in never-depressed subjects with borderline personality disorder. *Biol Psychiatry* 1999; 45: 1056–1058.

30. De la Fuente JM, Bobes J, Morlan I et al. Is the biological nature of depressive symptoms in borderline patients without concomitant Axis I pathology idiosyncratic? Sleep EEG comparison with recurrent brief, major depression and control subjects. *Psychiatry Res* 2004; 129: 65–73.

31. De la Fuente JM, Bobes J, Vizuete C, Mendlewicz J. Sleep-EEG in borderline patients without concomitant major depression: A comparison with major depressives and normal control subjects. *Psychiatry Res* 2001; 105: 87–95.

32. Lindberg N, Tani P, Appelberg B et al. Human impulsive aggression: A sleep research perspective. *J Psychiatr Res* 2003; 37: 313–324.

33. Lindberg N, Tani P, Appelberg B et al. Sleep among habitually violent offenders with antisocial personality disorder. *Neuropsychobiology* 2003; 47: 198–205.

34. Svrakic DM, Whitehead C, Przybeck TR, Cloninger CR. Differential diagnosis of personality disorders by the seven-factor model of temperament and character. *Arch Gen Psychiatry* 1993; 50: 991–999.

35. Schredl M, Paul F, Reinhard I, Ebner-Priemer UW, Schmahl C, Bohus M. Sleep and dreaming in patients with borderline personality disorder: A polysomnographic study. *Psychiatry Res* 2012; 200: 430–436.

36. Semiz UB, Basoglu C, Ebrinc S, Cetin M. Nightmare disorder, dream anxiety, and subjective sleep quality in patients with borderline personality disorder. *Psychiatry Clin Neurosci* 2008; 62: 48–55.

37. Bastien CH, Guimond S, St-Jean G, Lemelin S. Signs of insomnia in borderline personality disorder individuals. *J Clin Sleep Med* 2008; 4: 462–470.

38. Plante DT, Frankenburg FR, Fitzmaurice GM, Zanarini MC. Relationship between sleep disturbance and recovery in patients with borderline personality disorder. *J Psychosom Res* 2013; 74: 278–282.

39. Kupfer DJ, Ehlers CL. Two roads to rapid eye movement latency. *Arch Gen Psychiatry* 1989; 46: 945–948.

40. Lindberg N, Tani P, Porkka-Heiskanen T, Appelberg B, Rimon R, Virkkunen M. ADHD and sleep in homicidal men with antisocial personality disorder. *Neuropsychobiology* 2004; 50: 41–47.

41. Sharpley AL, Elliott JM, Attenburrow MJ, Cowen PJ. Slow wave sleep in humans: Role of 5-HT_{2A} and 5-HT_{2C} receptors. *Neuropharmacology* 1994; 33: 467–471.

42. Frankenburg FR, Zanarini MC. The association between borderline personality disorder and chronic medical illnesses, poor health-related lifestyle choices, and costly forms of health care utilization. *J Clin Psychiatry* 2004; 65: 1660–1665.

13

Sleep and dermatology

MADHULIKA A. GUPTA, ADITYA K. GUPTA, AND KATIE KNAPP

INTRODUCTION

Sleep symptoms in dermatology patients[1-4] can be conceptualized as the interplay of one or more factors that are not necessarily mutually exclusive (Figure 13.1). Dermatologic disorders may contribute to sleep–wake symptoms such as sleep disruption, insomnia, and daytime fatigue as a result of nocturnal pruritus and cutaneous pain, and possibly obstructive sleep apnea (OSA) in association with a heightened pro-inflammatory state (e.g., as a result of psoriasis). Secondly, sleep disorders and certain sleep–wake factors may contribute to the exacerbation of dermatologic diseases (Table 13.1)[5-37]; for example, sleep deprivation can contribute to a heightened pro-inflammatory state and inhibition of skin barrier function recovery, and circadian rhythm sleep disorders may aggravate pruritus by disruption of the normal diurnal pattern of cortisol secretion and transepidermal water loss (TEWL). Studies from the Danish National Patient Registry[38,39] of 19,438 patients with OSA and 755 patients with obesity hypoventilation syndrome (OHS) have reported increased comorbidity with diseases of the skin and subcutaneous tissue and other medical disorders at least 3 years prior to the OSA/OHS diagnosis (odds ratio [OR]: 1.18,

95% CI 1.07–1.30 for OSA and OR: 2.12, 95% CI 1.33–3.38 for OHS).[39] Data from 2998 children (aged 0–19 years) with OSA out of 11,974 children in the registry showed that children with OSA also have increased medical comorbidities, including skin conditions (OR: 1.32, 95% CI 1.02–1.71) 3 years before the OSA diagnosis, and a higher incidence of skin conditions 3 years after the diagnosis of OSA (OR: 1.42, 95% CI 1.06–1.89) in contrast to children without OSA.[38] OSA was diagnosed with polysomnography (PSG), polygraphy or oximetry and the specific skin disorders were not identified.[38,39] The results from these epidemiologic studies[38,39] suggest that OSA may predispose the patient to developing certain dermatologic disorders or OSA may be precipitated by certain dermatologic disorders, or both. Finally, the treatment-related side effects of the drug therapies used in dermatology can be associated with sleep-related symptoms. The "biologics" appear to be especially effective in the treatment of sleep-related complaints (Table 13.2).[40-50] Continuous positive airway pressure (CPAP) masks,[51,52] especially when they are poorly fitting, can be associated with contact dermatitis and skin abrasion.

In this paper, we have reviewed some of the neurobiological factors that play important roles in the

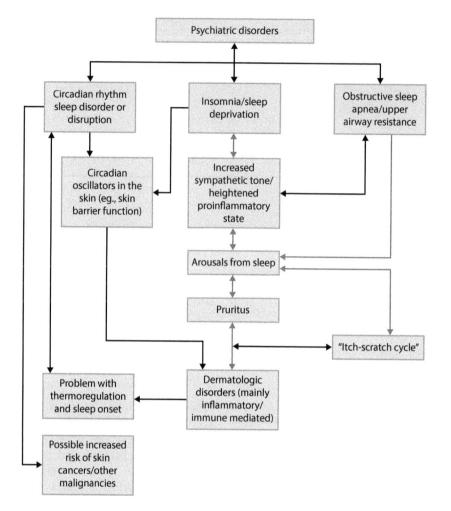

Figure 13.1 Some of the neurobiological factors involved in the interface between sleep disorders and dermatologic disease. The gray arrows indicate the most salient interactions.

interface between sleep and dermatology. We have further reviewed sleep–wake factors and sleep disorders in primary dermatologic disease. There are a relatively large number of recent epidemiologic and clinical studies that have examined the association of sleep and dermatology—whenever possible, we have tried to focus upon both epidemiologic and clinical studies that have used objective measures of sleep, such as PSG and actigraphy versus patient-rated questionnaires.

GENERAL FACTORS

Sleep deprivation/restriction

The skin generates the cutaneous permeability barrier, localized in the external stratum corneum,

which prevents entry of foreign substances and prevents excessive water loss.[53] One night of sleep deprivation can inhibit recovery (e.g., after tape stripping) of skin barrier function in humans.[53] Skin barrier function impairment is present in chronic skin disorders such as atopic dermatitis (AD), and can lead to an exacerbation of allergic and irritant contact dermatitis.[53]

Sleep restriction of 25%–50% of the normal 8-hour sleep period has been associated with an elevation of the mediators of inflammation. Various pro-inflammatory cytokines have been reported to be elevated in response to sleep deprivation, with the most commonly reported cytokines being interleukin-1β (IL-1β) and tumor necrosis factor (TNF)-α.[54] The extensive literature on sleep and immunity is outside the scope of this chapter.

Table 13.1 Overview of sleep disorders and dermatologic disorders

Insomnia

Atopic dermatitis—Polysomnographic (PSG) studies of scratching during sleep show decreased sleep efficiency, normal sleep architecture, and overall increased arousal even when disease is not active; atopic dermatitis (AD) can be a factor[5] in patients with unexplained arousals on PSG

Psoriasis—Sleep disruption from pruritus, obstructive sleep apnea (OSA) and restless legs syndrome[13]

Urticaria—Insomnia is a predisposing factor[6] that is possibly indicative of underlying increased sympathetic tone; sleep disruption from pruritus

Scabies—Skin infestation is associated with significant morbidity in economically deprived settings, intense or severe itch-related sleep disturbance in 37.5% of patients and itch intensity correlates significantly ($p < 0.001$) with sleep disturbance; with effective treatment, both itching and sleep disturbance decrease significantly ($p < 0.001$)[7]

Acne—Has been reported to be triggered by a lack of sleep in up to 75.2%[8] (n = 1236) of Korean adults; over 50% (n = 597) of Thai adolescents[9] endorsed inadequate sleep as an aggravating factor for their acne; acne has been associated with several comorbidities including insomnia in a large (n = 9417) U.S. epidemiologic study[10]

Obstructive sleep apnea

Atopic dermatitis—Association with OSA in children[11] and adults[12]

Psoriasis—Increased risk following OSA[13]; metabolic syndrome and increased levels of pro-inflammatory cytokines in both disorders

Nocturnal sweating—Results from Icelandic Sleep Apnea Cohort[14]

Acanthosis nigricans—Case study[15] of association with obesity and insulin resistance

Yellow nail syndrome—Case[16] of PSG-diagnosed OSA, yellowing of nails related to lymphedema and pleuro-pulmonary inflammatory disorders

Angioedema/urticaria—Case[17] of OSA secondary to edema of soft palate in hereditary angioedema and edema of uvula[18]

Hypertrophic burn scars—Two cases[19] of severe OSA secondary to pressure garments used to treat hypertrophic burn scars

Lipodystrophy—Case reports[20] of familial partial lipodystrophy type 2, a rare genetic disorder affecting the total body fat distribution associated with increased fat tissue around the neck; HIV-associated partial lipodystrophy resulting from treatments[1]

Cicatricial pemphigoid—Series[21] of 142 patients showing mucosal lesions affecting the oral cavity, pharynx, larynx, and stenosis of nasopharynx or larynx

Malignant melanoma—Case[22] of secondary OSA caused by malignant melanoma in the nasal cavity and paranasal sinus

Hyperpigmentation and lichenification of forehead skin—Case[23] of localized post-inflammatory skin changes secondary to resting against a wall when attempting to sleep in severe OSA

Local skin reactions to nasal continuous positive airway pressure mask[24]—E.g., abrasion of the ridge of the nose

Onychophagia—Case study[25] suggesting impulse control disorders may be unmasked by untreated OSA

Central sleep apnea

Xeroderma pigmentosum (XP)[26] group A, a variant of XP that is associated with neurological abnormalities

(Continued)

Table 13.1 (*Continued*) Overview of sleep disorders and dermatologic disorders

Parasomnias

Atopic dermatitis—Questionnaire-based controlled study[27] of 57 AD children showed increased frequency of bedwetting, sleep talking, restlessness and movement and sleep walking to be directly related to AD disease severity, attributed to "sensory hypersensitivity" and "hyperarousability" in AD[27]

Vitiligo—Study of 116 adults[28] versus non-dermatologic controls showed higher frequency of parasomnias (nocturnal enuresis, sleepwalking, sleep terrors, nightmares, and "night illusions") in childhood and adolescence before onset of vitiligo, possible common neuropathological factors involving catecholamines/serotonin implicated

Trichotillomania (TM)—TM exclusively during sleep or "sleep-isolated" TM[29]; case study[30] of PSG with videography indicated non-rapid eye movement parasomnia; cases of scratching during sleep[31] have been associated with factitious skin ulcers[32]

Restless legs syndrome

Atopic dermatitis—Study[33] of 120 AD patients, restless legs syndrome (RLS) significantly higher in active AD (40.8%) versus inactive AD (23.6%), psoriasis (18.0%) and controls (10.8%); stronger family history of RLS in AD. The paresthesias in RLS may be mistakenly attributed to a primary dermatologic disorder or delusions of parasitosis[34]

Psoriasis—Increased prevalence of 15.1%–18% in psoriasis versus 5%–10% prevalence of RLS in European and North American samples[13]

Narcolepsy

Alopecia areata (AA)—Five cases[35] of comorbid narcolepsy, with possible common autoimmune factors in both conditions. Study[36] of 105 AA patients revealed no significant difference in mean Epworth Sleepiness Scale (ESS) scores between AA and general Japanese population; 12 (11.4%) patients had ESS scores >10

Psoriasis and atopic dermatitis—Cases of immune-mediated comorbid conditions,[37] including dermatologic disorders, in 156 patients with narcolepsy with cataplexy

Autonomic nervous system activation secondary to sleep deprivation and small, subclinical shifts in basal inflammatory cytokine levels can be associated with the future development of metabolic syndrome in some dermatologic disorders that interfere with sleep, such as psoriasis.[55] In a mouse model[56] of psoriasis, 48 hours of selective paradoxical sleep deprivation was associated with significant increases in pro-inflammatory cytokines (IL-1β, IL-6, and IL-12) and decreases in the anti-inflammatory cytokine IL-10. The cytokine levels normalized after 48 hours of sleep rebound, and the authors suggest that sleep loss should be considered to be a risk factor for the development of psoriasis.[56] Resolution of dyshidrotic eczema of the palms in a patient after treatment of severe OSA with CPAP was attributed to a reduced pro-inflammatory state as a result of a reduced sympathetic tone with CPAP treatment.[57]

The pro-inflammatory cytokines tend to be somnogenic[58] and have been implicated in the pathogenesis of OSA-associated sleepiness.[58] A double-blind, placebo-controlled pilot study of the TNF-α antagonist etanercept in eight obese male OSA patients[45] observed a decrease in sleepiness. This decrease in sleepiness was about threefold higher than the reported effect of CPAP on objective sleepiness in OSA.[45] Etanercept is also used in the treatment of psoriasis; however, these preliminary findings do not necessarily suggest an underlying link between psoriasis and OSA. For example, Maari et al.[59] found no difference in improvement of OSA in psoriasis patients treated with the TNF-α antagonist adalimumab versus placebo.

Circadian rhythms, cortisol, and melatonin

The skin has a complex circadian organization, with the presence of circadian clock genes and functional oscillators. These functional oscillators function autonomously for driving rhythmic

Table 13.2 Sleep effects associated with some medications used in dermatology

Isotretinoin—used in the treatment of acne, binds to retinoic acid receptors (RARs) and, in the mouse model, the gene encoding RARs determines the contribution of the delta oscillations in the sleep electroencephalogram.[40] Polysomnographic study of 12 acne patients before and after 1 month of isotretinoin treatment demonstrated an increase in sleep efficiency from 83.5% to 89.5% (p = 0.036) and a decrease in sleep latency from 18.0 to 15.5 minutes (p = 0.023), with no significant change in other sleep parameters, including slow-wave sleep. Case studies[40] of hypersomnia, onset of Kleine–Levin syndrome and changes in dream patterns including sustained dreaming[41] have been reported

Corticosteroids—associated with a >30% incidence of sleep disturbance, especially insomnia[42]

Interferon—used in several dermatologic disorders, including malignant melanoma; can modulate the master clock[43] at the genetic level of the *Per1* gene and lead to insomnia and sleep disturbances, including decreased sleep efficiency, slow-wave sleep suppression and decreased time in rapid eye movement (REM) sleep

Efavirenz—one of the most commonly used antiretroviral medications used in HIV-infected patients, efavirenz has been associated with severe, usually transient neuropsychiatric effects,[44] including insomnia, vivid dreams and mood changes in about 50% of patients who initiate efavirenz; it is suggested that efavirenz not be used in patients with severe sleep disorders

"Biologics"—TNF-α antagonist, etanercept[45] and infliximab,[46] used in the treatment of psoriasis and psoriatic arthritis, are associated with a marked decrease in sleepiness, believed to be mediated through inhibition of circulating TNF-α levels. In a study[47] of moderate to severe psoriasis, etanercept was associated with significant improvements in sleep (measured using the Medical Outcomes Study [MOS] sleep scale) and quality of life; adalimumab[48] significantly improved sleep quality (as measured by the MOS sleep scale) in chronic plaque psoriasis patients with suboptimal response to prior therapy including etanercept; omalizumab[49] significantly improved urticaria activity and interference of urticaria with sleep in treatment-refractory chronic urticaria patients

Minocycline—decreases in short-wave sleep and no changes in REM sleep after a single dose of minocycline 200 mg, with no decreases in short-wave sleep or REM sleep after a single dose of ampicillin 500 mg; these finding were attributed to the inhibition of protein synthesis by minocycline[50]

---

skin functions and are present in the keratinocytes and melanocytes at the level of the epidermis, and present in fibroblasts at the level of the dermis.[60] There is an emerging literature on the role of the circadian clock in metabolism, immunity, and inflammation. Various skin-related factors show circadian rhythmicity, with one of the most important ones being the stratum corneum barrier of the skin.[61] There is circadian rhythmicity in TEWL, skin surface pH and skin temperature at most anatomic sites, with skin permeability being higher in the evening and night than in the morning.[62] TEWL is also associated with increased itch intensity in AD.[63] The secretion of cortisol is primarily circadian driven with a natural trough in circulating corticosteroids in the evening, which results in its anti-inflammatory effect being at a minimum level during this time. This may be the basis for the finding that up to 65% of patients with

inflammatory dermatoses, including AD, psoriasis, and chronic idiopathic urticaria, report increased pruritus during the night,[2] as well as the basis for the circadian variation of symptoms in some cutaneous sensory disorders such as burning mouth syndrome (BMS).[64] Yosipovitch et al.[61,62] discuss that skin barrier functions are predictably time dependent, and that the circadian rhythms are maintained during treatment with high-potency and medium-potency corticosteroids in healthy skin. Clinically, this could be an important consideration in the use of moisturizers and emollients during the night and use of topical corticosteroids during the late evening hours when the epidermal barrier function is decreased and there also tends to be a rise in inflammatory activity.[61,62]

Night-shift work has been associated with psoriasis comorbidities such as type 2 diabetes and coronary heart disease, as well as a significantly increased risk of psoriasis (adjusted hazard ratio: 1.19, 95% CI 1.07–1.32) after adjusting for age, body mass index (BMI), smoking and alcohol use, and physical activity.[65] Psoriasis[66] and severe AD[67] have been associated with diminished nocturnal melatonin increase, which has been possibly attributed to a reduced sensitivity of β-receptors in the pineal gland, which parallels a reduced β-receptor sensitivity in the skin.[67] In a study[68] of 42 patients with advanced melanoma who received orally administered melatonin ranging in dosage from 5 mg to 700 mg/meter[2]/day, after a median follow-up of 5 weeks, six patients had partial responses and six additional patients had stable disease; the sites of response included the central nervous system (CNS), subcutaneous tissue, and lung.

Various epidemiologic studies[69–72] have examined the possible role of night-shift work on cancer, including melanoma and other skin cancers, and come up with apparently conflicting results. It has generally been hypothesized that night-shift work may be associated with an increase in cancer risk, as melatonin has anti-mutagenic and oncostatic effects, and peak production of melatonin is typically suppressed by nighttime light exposure. Earlier studies of circadian rhythm disruption and cancer that examined flight attendants[69] and airline pilots[70] report an over twofold increase in the risk for melanoma and no increase in the risk for other skin cancers among flight attendants,[69] as well as an over twofold increased incidence of all skin cancers among pilots, including melanoma and basal and squamous cell carcinoma.[70] These increases have been subsequently largely attributed to increased exposure to cosmic radiation in these occupations. A Swedish register-based cohort study of over 3 million workers did not find an association between night-shift work and any type of cancer, including skin cancer.[71] However, results from the Nurses' Health Study database that examined 68,336 non-Hispanic white women from the United States[72] revealed that working ≥10 years on rotating night shifts was associated with a 14% decreased risk of skin cancer compared with never working night shifts. This association was strongest for cutaneous melanoma, where working for ≥10 years on rotating night shifts was associated with a 44% decreased risk for melanoma. The inverse association between ≥10 years on rotating night shifts was the strongest among women who had black or dark-brown hair color at 20 years of age.[72] The authors consider the natural hair color as a phenotypic proxy for a woman's predisposition to skin cancer, and discuss the possible role of both genetic and environmental factors in their finding.[72] A case–control study[73] of 70 patients with pathologist-confirmed skin cancer (55 basal cell carcinoma and 15 squamous cell carcinoma) and 70 healthy controls reported that the 24-hour urinary melatonin metabolite 6-sulfatoxymelatonin was significantly higher in the control group; however, the results appear to be confounded by the fact that the control group also slept for a longer duration on average. Melatonin has also been implicated in hair growth, melanoma control, and wound healing, in addition to the suppression of ultraviolet light-induced damage[74] in human skin. In a double-blind, randomized, placebo-controlled study of 40 women with diffuse alopecia or androgenetic alopecia, 0.1% melatonin solution significantly influenced hair growth by increasing anagen hair rates and decreasing telogen hair rates.[75] The protective effects of melatonin against ultraviolet radiation via its antioxidant properties and free radical-scavenging capacity is a topic with potentially important clinical implications.[76]

THERMOREGULATION AND SWEATING DURING SLEEP

The skin plays a central role in thermoregulation, which is defined as the process involved in maintaining the core body temperature (CBT).[1,2,77] The timing of the circadian component of the CBT, which plays a central role in sleep onset, is determined by the suprachiasmatic nucleus.[1,2,77] For heat loss to occur, blood that carries heat from the core is shunted—primarily as a result of a decreased sympathetic tone—from the muscles to the cutaneous vascular beds, which are rich in arteriovenous anastomoses, and heat loss occurs through vasodilation while sweating cools down the core. The frequency of sweating is highest during non-rapid eye movement (NREM) sleep, especially slow-wave sleep, and lowest during rapid eye movement (REM) periods.[78] In healthy individuals, the propensity for sleep increases when the distal skin temperature increases relative to the proximal skin temperature (i.e., when the distal to proximal gradient [DPG] is greater). The decrease in the CBT before the onset of sleep and during sleep is associated with a dilatation of peripheral blood vessels, which facilitates the dissipation of heat from the core to the periphery.[79] The propensity for sleep can be enhanced and sleep-onset latency decreased by warming the skin to the temperature that normally occurs prior to and during sleep; PSG studies have shown that the acceleration of sleep onset is attenuated in older subjects.[79] The importance of the DPG in sleep onset and sleep disorders is further illustrated by a study of 15 patients with narcolepsy with cataplexy who underwent skin temperature measurements.[80] The DPG was higher in narcoleptics than in controls through the day, and was related to a shorter sleep-onset latency during a multiple sleep latency test. The increase in DPG in narcolepsy patients was also higher than that which healthy controls achieve when asleep.

Thermoregulation and the autonomic nervous system are closely linked,[78] and sweating during sleep can be an important clinical sign of potentially serious sleep disorders. Kahn et al.[81] have reported that 25.5% out of 98 infants dying from sudden infant death syndrome (SIDS; versus 3.1% of control infants and 4.5% of siblings of SIDS infants) had been found to be sweating excessively during sleep. In a study of 258 infants using PSG and a measure of TEWL, Kahn et al.[81] observed that infants with an apparently life-threatening event (ALTE) had significantly higher TEWL evaporation rates during NREM sleep. The authors discuss that these findings may reflect differences in autonomic nervous system activity,[81] which could have potentially important clinical implications for identifying children who are at a greater risk for developing an ALTE or even SIDS.

PRURITUS

Pruritus or itch is the most common symptom of dermatologic disease and a primary symptom in the interface between sleep and dermatology (Figure 13.1). Many dermatologic disorders[4] including AD, psoriasis, urticaria, prurigo nodularis, lichen planus, dermatologic infestations such as scabies, a range of systemic disorders such as renal and hepatic failure, and malignancies such as lymphomas have been associated with significant nocturnal pruritus. The studies on pruritus and sleep have measured scratching behavior during sleep and considered it to be a proxy for itch. It is noteworthy that scratching and resultant lichenification, which is encountered in AD and considered

to be the physical indicator of itch, is typically not encountered in other pruritic dermatoses such as psoriasis and urticaria.

Over four decades ago, Savin et al.[82,83] studied sleep in pruritic skin disorders such as severe atopic eczema,[82] psoriasis, dermatitis herpetiformis, lichen planus, and urticaria[83] using PSG and observed that the distribution of scratching (a proxy for the subjective sensation of itch) during the different stages of sleep was similar in all disorders studied: in NREM sleep, scratching occurred more frequently in stage 1 than stage 2, and more frequently in stage 2 than stages 3 and 4; in REM sleep, scratching occurred with approximately the same frequency as stage 2 sleep. This finding has been essentially replicated by most subsequent studies[84–89] of scratching during sleep.

The frequency of scratching during the different sleep stages[83] appears to be proportional to the relative level of sympathetic nervous activity normally observed during the sleep stage.[83] In NREM sleep, there is a progressive decline in sympathetic nervous activity from stages 1 and 2 to slow-wave sleep stages 3 and 4, with increased and more variable sympathetic tone during REM sleep.

Scratching is typically associated with arousals and sleep stage shifts rather than awakenings from sleep, and arousals are least likely[4] if the scratching begins during slow-wave sleep. Sack and Hanifin[4] observe that scratching during sleep is more common in children and young adults who have much deeper sleep and therefore are less likely to be awakened from a bout of scratching during sleep; hence, self-reports of nocturnal itch and scratch do not correlate well with objective measures of scratching.[85,90] Most of the studies of scratching during sleep use self-reports and actigraphy. In addition to actigraphy, other objective measures of scratching during sleep include infrared videography alone[91] or videography in conjunction with PSG. Several studies of adults[85] and children[91,92] with AD have also shown that subjective measures of itch do not correlate with objective ratings of itch using PSG,[85] where scratching bouts were also confirmed with increased electromyographic activity lasting at least 3 seconds and video monitoring of scratching, infrared video monitoring alone,[91] or actigraphy.[92] Comparison of PSG and actigraphic sleep measures[85] indicated that while PSG can provide detail not measured by actigraphy, primary sleep outcomes that were measured by both actigraphy and PSG, such as sleep-onset latency ($r = 0.61$, $p < 0.005$) and sleep efficiency ($r = 0.44$, $p < 0.050$), were significantly correlated. A study[93] of 14 adult AD patients and 14 controls revealed that actigraphy provided more discriminating information about sleep fragmentation and itch in AD patients versus controls than patient self-reporting using the Pittsburgh Sleep Quality Index (PSQI).

CUTANEOUS SENSORY DISORDER

The skin is a large sensory organ[94,95] with afferent sensory nerves conveying sensations of itch, touch, pain, temperature, and other physical stimuli to the CNS and the efferent autonomic, mainly sympathetic nerves playing a role in the maintenance of cutaneous homeostasis by regulating vasomotor and pilomotor functions and the activity of the apocrine and eccrine sweat glands. The cutaneous sensory neurons transmit modality-specific information to the CNS and are associated with specialized receptors (mechanoreceptors, thermoreceptors, and chemoreceptors) and transducers for highly specific sensory functions.

Several dermatologic disorders classified under the new "Obsessive–Compulsive and Related Disorders" section of the *Diagnostic and Statistical Manual of Mental Disorders, 5th Edition* (DSM-5) such as excoriation (skin-picking) disorder and trichotillomania (hair pulling disorder) involve excessive/compulsive manipulation of the integument by the patient that helps to regulate emotions. The compulsive manipulation of the skin is often triggered by a disagreeable skin sensation (e.g., itching, burning, or stinging). Dermatologic patients with skin picking or "pathological excoriation" were noted to report more sleep complaints on the PSQI than dermatologic patients without skin picking.[96]

Skin regions that normally have a greater density of epidermal innervation (e.g., face, scalp, and perineum) tend to be more susceptible to the development of disagreeable cutaneous sensations and are most susceptible to the development of cutaneous sensory syndromes, which are often referred to in region-specific terms such as glossodynia, BMS, and vulvodynia.[94] These cutaneous sensory disorders tend to be complex and multifactorial, where local dermatologic, hormonal, autonomic nervous system, neuropsychiatric and other CNS factors, including sleep-related factors such as sleep deprivation, can all contribute to

the pathogenesis of symptoms.[94] A study[97] of 70 patients with BMS and 70 controls reported poor sleep quality (measured by the PSQI) in 67.1% of BMS patients versus 17.1% of controls. Various earlier studies[98,99] have reported similar findings. An epidemiologic study[100] from a Taiwanese health insurance database observed that patients with both sleep apnea (adjusted hazard ratio [HR]: 2.56, 95% CI 1.30–5.05) and non-apnea sleep disorders (adjusted HR: 2.89, 95% CI 2.51–3.34) had a higher risk of developing BMS in comparison to the non-sleep-disordered cohort; the HR increased with age and female gender.

DERMATOLOGIC DISORDERS

Table 13.1 provides an overview of some dermatologic disorders that have been associated with various sleep disorders. There are several large-scale questionnaire-based epidemiologic studies on sleep and dermatologic disorders, with only a few that have reported using objective measures such as PSG to diagnose OSA in patients with unspecified dermatologic disorders,[38,39] AD,[11,12] and psoriasis.[13]

AD (Atopic eczema)

DEFINITION

AD[101] or atopic eczema is a chronic, relapsing inflammatory skin disorder that is associated with persistent and severe pruritus, which is a hallmark of AD. AD is typically characterized by dry, scaly skin, which is the consequence of epidermal barrier dysfunction and an altered stratum corneum leading to increased TEWL.[101] AD can occur at any age; however, AD develops during the first 6 months of life in 45% of affected individuals and before 5 years of age in 85% of cases.[101] About 60% go into remission by 12 years of age, and in the remainder, AD persists into adolescence and adulthood. In industrialized nations, the prevalence of AD is 10%–30% among school-aged children and 2%–10% among adults. AD is a complex genetic disorder that is accompanied by other atopic disorders such as asthma and allergic rhinoconjunctivitis.[101]

STUDIES OF SLEEP AND AD (ECZEMA)

Epidemiologic studies of habitual snoring in children where the severity of sleep-disordered breathing was not evaluated by PSG have reported that atopy (asthma, allergic rhinitis, or AD) was the strongest risk factor for habitual snoring, and the effect was cumulative (i.e., the OR for habitual snoring with all three atopic diseases was 9.45 [95% CI 3.48–1.97] and the OR for AD alone was 1.80 [95% CI 1.28–2.54]).[102] However, a recent study[11] of 855 children with physician-diagnosed eczema, a mean age of 6.3 years and median obstructive Apnea–Hypopnea Index (AHI) of 2.1 episodes per hour diagnosed by PSG revealed that eczema (an index of atopy) was not related to adenoidal or tonsillar hypertrophy, and eczema did not affect OSA frequency after adjustment for adenoidal and tonsillar hypertrophy, obesity, gender, and age, with an adjusted OR of 0.82 (95% CI 0.56–1.21). Alternately, in an epidemiologic study[12] of 1222 Taiwanese individuals with newly diagnosed OSA (cases selected based upon one inpatient admission for OSA or two outpatient service claims with PSG) who were followed up over a period of 5.5 years, the adjusted OR for developing AD after controlling for confounders such as age, gender, obesity, diabetes, hypertension, allergy, allergic rhinitis, etc., was 1.5 (95% CI 1.14–2.06, p = 0.005). The hazard risk for AD was greater in male AD patients under 35 years of age.[12]

There is a large body of literature[1,3,4] on nocturnal scratching in AD, with the most consistent finding from studies using PSG being decreased sleep efficiency[103–106] due to nocturnal arousals, with no significant change in overall sleep architecture in AD. Up to 60% of children with AD have disturbed sleep, with an increase in frequency to 83% during exacerbations of AD.[3] Increased AD severity has been reported to adversely affect the sleep of the parents in over 60% of cases.[3] Objective indices of scratching correlate with objective measures of AD severity. The subjective reporting of nocturnal pruritus and sleep loss typically do not correlate with objective measures of disease severity,[107] which highlights the need for objective ratings of sleep in AD. Chang et al.[106] studied 72 pediatric AD patients (mean ± standard deviation [SD] of age: 7.5 ± 3.9 years) with physician-diagnosed AD affecting at least 5% of the total body surface area (TBSA) and 32 controls (mean ± SD of age: 8.6 ± 3.9 years) using actigraphy and PSG. Participants and their caregivers completed subjective ratings of sleep disturbance, and AD disease severity was assessed using the Scoring Atopic Dermatitis (SCORAD) index. Subjective sleep quality was poor in 54.2%

of AD patients versus 6.2% of controls (p < 0.001); difficulty falling asleep, difficulty awakening in the morning and daytime sleepiness were all more common in AD. Results of actigraphy and PSG were significantly correlated. AD patients had significantly reduced sleep efficiency (84.5% ± 9.3% versus 94.1% ± 7.5% in controls, p < 0.001) and higher limb movement indices (13.0 ± 13.4 versus 5.6 ± 2.9, p < 0.001), with no significant difference (p > 0.05) in the frequencies of stage N1 (4.8% ± 3.8% versus 4.4% ± 3.0% in controls), N2 (34.4% ± 14.3% versus 33.8% ± 16.2% in controls) and N3 (32.8% ± 14.8% versus 38.1% ± 17.2% in controls) sleep and a higher sleep fragmentation index as assessed by actigraphy (22.1 ± .8 versus 17.0 ± 6.8, p = 0.004). A SCORAD index of ≥48.7 predicted poor sleep efficiency. There was significantly more movement during sleep in the patients with AD, and the movement was highly correlated (r = 0.70, p < 0.001) with the SCORAD index and poor sleep efficiency (r = −0.73, p < 0.001). The pruritus score was correlated with lower sleep efficiency. In AD patients, lower nocturnal melatonin secretion was significantly associated with indices of disturbed sleep.[106] AD severity has been shown to be positively correlated with sleep disturbances as measured by actigraphy in some studies,[108] but not others.[109] In a study[110] of 117 children and adult eczema patients with itchy skin, there was no significant correlation between subjective ratings of itch severity on a visual analog scale and objective measures of scratching using actigraphy, reflecting a dissociation between scratch and perceived or recalled itch.

Pruritus in AD is typically worse in the evening and can interfere with nighttime sleep. The scratching in AD may become habitual and automatic[1,27] and become a conditioned response to being in bed at night, even in the absence of significant pruritus. These factors can exacerbate the AD as a result of the "itch–scratch cycle."[1] AD has been associated with a higher frequency of restless legs syndrome (RLS[33]; 40.8% in AD versus 10.8% in controls, p < 0.005), and a positive family history of RLS (24.2% of AD versus 7.2% of controls).

Bender et al.[85] studied 20 adult AD patients using two nights of PSG with videography in order to monitor scratching, actigraphy, subjective patient ratings of itch severity, ratings of the Dermatology Life Quality Index (DLQI) and PSQI, objective dermatologic ratings of AD severity and blood cytokine assays before and after PSG on the second night. All scratching events occurred only during sustained wakefulness or in association with arousal or awakening from sleep; on average, a greater proportion of time was spent scratching in stage 1 (p < 0.007) and 2 (p < 0.0001) sleep than stage 3, stage 4 or REM sleep; scratching in REM occurred only with arousals from sleep. The total scratching index correlated (Spearman correlations) directly with indices of AD severity such as TBSA affected by AD (0.33, p = 0.008) and the Rajka and Langeland skin score (0.42, p = 0.064), and inversely correlated with both PSG (−0.56, p = 0.01) and actigraphic (−0.52, p = 0.019) measures of sleep efficiency. There was an inverse correlation between sleep efficiency and inflammatory biomarkers (e.g., evening and morning IL-6 levels).[85] There was a significant association between actigraphic and PSG ratings of sleep-onset latency and sleep efficiency.[85] DLQI and PSQI scores, however, did not correlate significantly with the PSG or actigraphy scores of sleep efficiency. Self-assessment of itch did not correlate with objectively measured scratching.[85]

Using PSG, video monitoring and scratch electrodes, Reuveni et al.[104] observed that, even in remission, AD children had more arousals and awakenings per hour than controls (24.1 ± 8.1 versus 15.4 ± 6.2, p = 0.001). Scratching accounted for only 15% of arousals, while the remainder of arousals were not associated with any identifiable cause.[104] This is consistent with the underlying hyperarousability[27] and overactive sympathetic response to itch and scratching observed in AD. In a case–control PSG study of 15 patients with lichen simplex chronicus (LSC), a common pruritic disorder resulting from repeated rubbing and scratching, Koca et al.[88] observed significantly (p < 0.05) higher arousal and awakenings, higher percentages of stage 2 sleep, and lower percentages of slow-wave sleep in LSC versus controls, with no difference in sleep efficiency. Scratching episodes were also observed in stage 2 NREM sleep.[88] These findings are in contrast with AD, which is also a pruritic disorder, in which sleep architecture is typically not disturbed, but sleep efficiency is lower.

Psoriasis

DEFINITION

Psoriasis[111] is a chronic and recurrent inflammatory skin disorder that is most commonly characterized by circumscribed erythematous, dry, scaling plaques, which has been associated with an adverse effect on sleep quality[112] due to factors such as pruritus, pain, depression, and OSA. Psoriatic arthritis occurs in 5%–30% of psoriasis patients and tends to be more prevalent among patients with more severe psoriasis.[111] The worldwide prevalence of psoriasis is 2%. Psoriasis can occur at any age; in approximately 75% of patients, the onset of psoriasis is before 40 years of age, and in 35%–50%, onset is before 20 years of age.[111] Pruritus, one of the most distressing features of psoriasis, is present in over 80% of patients.[113]

STUDIES OF SLEEP AND PSORIASIS

Epidemiologic studies[13] have shown a significantly higher frequency of a sleep disorder[114] (OR: 3.89, 95% CI 2.26–6.71; sleep disorder was coded using the International Classification of Diseases, Ninth Revision, Clinical Modification code 780.57, which denotes "unspecified sleep apnea" with no mention of how the diagnosis was made) in the psoriasis patients (n = 51,800) from a national database in Taiwan, and a higher frequency of psoriasis (hazard of psoriasis: 2.30, 95% CI 1.13–4.69) in PSG-diagnosed OSA patients (n = 2258) also from a Taiwanese health insurance database.

In a study of 101 patients with severe psoriasis, sleep was reported to relieve psoriasis in 57% and aggravate pruritus in 8% of patients.[115] Pruritus[115] was associated with difficulty in falling asleep in 69% of patients; 66% reported being awakened by pruritus. The Psoriasis Area Severity Index (PASI) did not correlate with itch intensity.[115] In a survey of 420 respondents from the U.S. National Psoriasis Foundation,[55] some degree of sleep disruption was reported by 49.5% of the sample; sleep disturbance[55] was most frequently predicted by the presence of psoriatic arthritis (OR: 3.26), followed by pruritus (OR: 1.26), pain from psoriatic lesions (OR: 1.22) and decreased emotional well-being (OR: 1.18). Body surface area affected by psoriasis, BMI, and therapy were not significant predictors of sleep disturbance in psoriasis.[55]

PSG studies of psoriasis[13] have evaluated sleep-related breathing disorders in psoriasis and not itching and scratching, even though pruritus commonly interferes with sleep in psoriasis. Buslau and Benotmane[116] performed PSG on two consecutive nights in 25 psoriasis patients and 19 matched patients with chronic bronchitis, a condition that is known to be associated with OSA. The apnea index in psoriasis (14.4 ± 13.5) was greater ($p = 0.028$) than in chronic bronchitis (8.8 ± 13.4); neither group differed in other OSA risk factors (e.g., BMI or hypertension). Three patients with recalcitrant stable psoriasis and OSA that had previously not responded to standard dermatologic treatments were treated with nasal CPAP and followed prospectively for 1 year; 1 year later, all three patients were still on CPAP and had only mild psoriasis, and two patients reported that they could not remember their psoriasis being so good for years.[116] Karaca et al.[117] studied 33 psoriasis patients with PSG; 18/33 (54.5%) had OSA (11 mild with AHI scores between 5 and 15, 2 moderate with AHI between 16 and 30 and 5 severe with AHI >30). In contrast to psoriasis patients without OSA, the psoriasis patients with OSA were older ($p < 0.05$) and had a greater neck circumference ($p < 0.05$); however, they did not differ significantly with respect to BMI, duration of psoriasis, PASI scores, or Epworth Sleepiness Scale scores. The authors[116,117] discuss that OSA may act as a triggering factor for psoriasis for genetically predisposed individuals because of autonomic nervous system dysregulation and alteration of the homeostasis of the immune neuroendocrine network in the skin. In contrast to these results,[116,117] in a PSG study of 35 chronic psoriasis patients, Papadavid et al.[118] observed that only BMI and hypertension, and not psoriasis characteristics, were associated with an increased risk for OSA. Among women, OSA severity correlated ($p = 0.021$) with psoriasis duration (psoriasis duration ≥ 8 years more likely to be associated with AHI >15), and men with moderate to severe psoriasis presented with greater ($p = 0.035$) cardiovascular risk (greater BMI, hypertension, and Framingham score ≥ 10) in comparison to patients with mild psoriasis.

Maari et al.[59] studied the effect of the TNF-α antagonist adalimumab on PSG sleep parameters in 20 psoriasis patients with $\geq 5\%$ TBSA affected by psoriasis and an AHI ≥ 15 using an 8-week, randomized, double-blind, placebo-controlled study. Patients were randomized to adalimumab with a loading dose of 80 mg followed by 40 mg every other week or placebo for 8 weeks. There was no significant change in the primary outcome measure of a change in AHI between baseline and day 56.[59] The authors[59] discuss that their negative findings may in part be related to an insufficient dose of adalimumab because of the overall higher BMI of their psoriasis patients. The authors further observe that 70% of OSA patients had no previous diagnosis of OSA despite being symptomatic, suggesting that OSA may be underdiagnosed in psoriasis patients.[59]

Urticaria

DEFINITION

Urticaria or hives are characterized by transient skin or mucosal swelling, resulting in wheals due to plasma leakage.[119] Urticaria is defined as chronic after a period of time of usually ≥ 6 weeks. The lifetime occurrence of urticaria ranges from <1% to 30%.[119] In over 50% of cases, the basis for the urticaria is not found. The mast cell is the primary effector cell of urticaria. Mast cell granules contain preformed mediators of inflammation, the most important of which is histamine.[119] Histamine is a major wake-promoting neurotransmitter in the CNS, and histaminergic neurons display elevated discharge activity during increased states of vigilance.[120]

STUDIES OF SLEEP AND URTICARIA

In a questionnaire study[121] of 100 chronic idiopathic urticaria (CIU) patients, 83% experienced pruritus at night and in the evening, and 62% reported difficulty falling asleep. In a study[6] of 75 CIU patients, insomnia in the 6 months preceding CIU onset was the most important psychosomatic predisposing factor for CIU (OR: 4.9, 95% CI 1.7–14.5); insomnia may further disturb the circadian rhythm of cortisol and contribute to the vicious cycle of CIU. The authors note that treatment of insomnia[6] could aid in the treatment of urticaria, and therefore sedating antihistamines[6,121] may be more effective than nonsedating antihistamines (which are usually considered as a first-line treatment for most urticarias) in the treatment of urticaria. A recent study[122] of 24 patients with chronic spontaneous urticaria using a double-blind, crossover design examined the effectiveness and prevalence of unwanted side effects when treating with the non-sedating, second-generation H1-antihistamine levocetirizine 15 mg daily plus the sedating first-generation H1-antihistamine hydroxyzine 50 mg at night versus levocetirizine 20 mg daily, each for 5-day periods. Both regimens significantly improved urticaria-related quality of life and nighttime sleep disturbance, without significant difference between the two groups. Compared to baseline, daytime somnolence was significantly reduced with levocetirizine monotherapy (p = 0.006), but not levocetirizine plus hydroxyzine (p = 0.218). The authors conclude that the belief that nighttime sleep was aided by the addition of a sedating first-generation H1-antihistamine was not supported by the results.[122] We identified no PSG studies of urticaria. The majority of the randomized, double-blind, controlled trials for the treatment of urticaria use nonsedating histamine H1-receptor antagonists, such as desloratadine and fexofenadine, or the leukotriene receptor antagonist montelukast.[1] In all studies, sleep was not the primary outcome measure, and in most studies, it was unclear whether the improvement in sleep was sustained.[1]

REFERENCES

1. Thorburn PT, Riha RL. Skin disorders and sleep in adults: Where is the evidence? *Sleep Med Rev* 2010; 14: 351–358.
2. Gupta MA, Gupta AK. Sleep–wake disorders and dermatology. *Clin Dermatol* 2013; 31: 118–126.
3. Camfferman D, Kennedy JD, Gold M, Martin AJ, Lushington K. Eczema and sleep and its relationship to daytime functioning in children. *Sleep Med Rev* 2010; 14: 359–369.
4. Sack R, Hanifin J. Scratching below the surface of sleep and itch. *Sleep Med Rev* 2010; 14: 349–350.
5. DelRosso L, Hoque R. Eczema: A diagnostic consideration for persistent nocturnal arousals. *J Clin Sleep Med* 2012; 8: 459–460.
6. Yang HY, Sun CC, Wu YC, Wang JD. Stress, insomnia, and chronic idiopathic urticaria— A case–control study. *J Formos Med Assoc* 2005; 104: 254–263.
7. Worth C, Heukelbach J, Fengler G et al. Acute morbidity associated with scabies and other ectoparasitoses rapidly improves after treatment with ivermectin. *Pediatr Dermatol* 2012; 29: 430–436.
8. Suh DH, Kim BY, Min SU et al. A multicenter epidemiological study of acne vulgaris in Korea. *Int J Dermatol* 2011; 50: 673–681.
9. Suthipinittharm P, Noppakun N, Kulthanan K et al. Opinions and perceptions on acne: A community-based questionnaire study in Thai students. *J Med Assoc Thai* 2013; 96: 952–959.
10. Silverberg JI, Silverberg NB. Epidemiology and extracutaneous comorbidities of severe acne in adolescence: A U.S. population-based study. *Br J Dermatol* 2014; 170: 1136–1142.
11. Alexopoulos EI, Bizakis J, Gourgoulianis K, Kaditis AG. Atopy does not affect the frequency of adenotonsillar hypertrophy and sleep apnoea in children who snore. *Acta Paediatr* 2014; 103: 1239–1243.
12. Tien KJ, Chou CW, Lee SY et al. Obstructive sleep apnea and the risk of atopic dermatitis: A population-based case control study. *PLoS One* 2014; 9: e89656.
13. Gupta MA, Simpson FC, Gupta AK. Psoriasis and sleep: A systematic review. *Sleep Med Rev* 2016; 29: 63–75.
14. Arnardottir ES, Janson C, Bjornsdottir E et al. Nocturnal sweating—A common symptom of obstructive sleep apnoea: The Icelandic sleep apnoea cohort. *BMJ Open* 2013; 3: e002795.
15. Hoppin AG, Katz ES, Kaplan LM, Lauwers GY. Case records of the Massachusetts General Hospital. Case 31-2006. A 15-year-old girl with severe obesity. *N Engl J Med* 2006; 355: 1593–1602.
16. Gubinelli E, Fiorentini S, Cocuroccia B, Girolomoni G. Yellow nail syndrome associated with sleep apnoea. *J Eur Acad Dermatol Venereol* 2005; 19: 650–651.
17. Bork K, Koch P. Episodes of severe dyspnea caused by snoring-induced recurrent edema of the soft palate in hereditary angioedema. *J Am Acad Dermatol* 2001; 45: 968–969.
18. Alcoceba E, Gonzalez M, Gaig P, Figuerola E, Auguet T, Olona M. Edema of the uvula: Etiology, risk factors, diagnosis, and treatment. *J Investig Allergol Clin Immunol* 2010; 20: 80–83.
19. Hubbard M, Masters IB, Williams GR, Chang AB. Severe obstructive sleep apnoea secondary to pressure garments used in the treatment of hypertrophic burn scars. *Eur Respir J* 2000; 16: 1205–1207.

20. Patel K, Roseman D, Burbank H, Attarian H. Obstructive sleep apnea in familial partial lipodystrophy type 2 with atypical skin findings and vascular disease. *Sleep Breath* 2009; 13: 425–427.

21. Hanson RD, Olsen KD, Rogers RS 3rd. Upper aerodigestive tract manifestations of cicatricial pemphigoid. *Ann Otol Rhinol Laryngol* 1988; 97: 493–499.

22. Asai N, Ohkuni Y, Kawamura Y, Kaneko N. A case of obstructive sleep apnea syndrome caused by malignant melanoma in the nasal cavity and paranasal sinus. *J Cancer Res Ther* 2013; 9: 276–277.

23. Vorona RD. Skin pigmentation changes in a patient with a sleep disorder. *J Clin Sleep Med* 2007; 3: 535–536.

24. Pepin JL, Leger P, Veale D, Langevin B, Robert D, Levy P. Side effects of nasal continuous positive airway pressure in sleep apnea syndrome. Study of 193 patients in two French sleep centers. *Chest* 1995; 107: 375–381.

25. Nino G, Singareddy R. Severe onychophagia and finger mutilation associated with obstructive sleep apnea. *J Clin Sleep Med* 2013; 9: 379–381.

26. Kohyama J, Shimohira M, Kondo S et al. Motor disturbance during REM sleep in group A xeroderma pigmentosum. *Acta Neurol Scand* 1995; 92: 91–95.

27. Shani-Adir A, Rozenman D, Kessel A, Engel-Yeger B. The relationship between sensory hypersensitivity and sleep quality of children with atopic dermatitis. *Pediatr Dermatol* 2009; 26: 143–149.

28. Mouzas O, Angelopoulos N, Papaliagka M, Tsogas P. Increased frequency of self-reported parasomnias in patients suffering from vitiligo. *Eur J Dermatol* 2008; 18: 165–168.

29. Murphy C, Redenius R, O'Neill E, Zallek S. Sleep-isolated trichotillomania: A survey of dermatologists. *J Clin Sleep Med* 2007; 3: 719–721.

30. Murphy C, Valerio T, Zallek SN. Trichotillomania: An NREM sleep parasomnia? *Neurology* 2006; 66: 1276.

31. Schenck C, Mahowald M. Nocturnal scratching as a chronic, injurious parasomnia in patients without primary dermatologic disorders. *Sleep* 2007; 30: A277–A278.

32. Brodland DG, Staats BA, Peters MS. Factitial leg ulcers associated with an unusual sleep disorder. *Arch Dermatol* 1989; 125: 1115–1118.

33. Cicek D, Halisdemir N, Dertioglu SB, Berilgen MS, Ozel S, Colak C. Increased frequency of restless legs syndrome in atopic dermatitis. *Clin Exp Dermatol* 2012; 37: 469–476.

34. Simonetti V, Strippoli D, Pinciara B, Spreafico A, Motolese A. Ekbom syndrome: A disease between dermatology and psychiatry. *G Ital Dermatol Venereol* 2008; 143: 415–419.

35. King LE Jr, Eastham AW, Curcio NM, Schmidt AN. A potential association between alopecia areata and narcolepsy. *Arch Dermatol* 2010; 146: 677–679.

36. Inui S, Hamasaki T, Itami S. Sleep quality in patients with alopecia areata: Questionnaire-based study. *Int J Dermatol* 2014; 53: e39–e41.

37. Martinez-Orozco FJ, Vicario JL, Villalibre-Valderrey I, De Andres C, Fernandez-Arquero M, Peraita-Adrados R. Narcolepsy with cataplexy and comorbid immunopathological diseases. *J Sleep Res* 2014; 23: 414–419.

38. Jennum P, Ibsen R, Kjellberg J. Morbidity and mortality in children with obstructive sleep apnoea: A controlled national study. *Thorax* 2013; 68: 949–954.

39. Jennum P, Ibsen R, Kjellberg J. Morbidity prior to a diagnosis of sleep-disordered breathing: A controlled national study. *J Clin Sleep Med* 2013; 9: 103–108.

40. Ismailogullari S, Ferahbas A, Aksu M, Baydemir R, Utas S. Effects of isotretinoin treatment on sleep in patients with severe acne: A pilot study. *J Eur Acad Dermatol Venereol* 2012; 26: 778–781.

41. Gupta MA, Gupta AK. Isotretinoin use and reports of sustained dreaming. *Br J Dermatol* 2001; 144: 919–920.

42. Sarnes E, Crofford L, Watson M, Dennis G, Kan H, Bass D. Incidence and US costs of corticosteroid-associated adverse events: A systematic literature review. *Clin Ther* 2011; 33: 1413–1432.

43. Dafny N, Yang PB. Interferon and the central nervous system. *Eur J Pharmacol* 2005; 523: 1–15.

44. Kenedi CA, Goforth HW. A systematic review of the psychiatric side-effects of efavirenz. *AIDS Behav* 2011; 15: 1803–1818.

45. Vgontzas AN, Zoumakis E, Lin HM, Bixler EO, Trakada G, Chrousos GP. Marked decrease in sleepiness in patients with sleep apnea by etanercept, a tumor necrosis factor-alpha antagonist. *J Clin Endocrinol Metab* 2004; 89: 4409–4413.

46. Zamarron C, Maceiras F, Mera A, Gomez-Reino JJ. Effect of the first infliximab infusion on sleep and alertness in patients with active rheumatoid arthritis. *Ann Rheum Dis* 2004; 63: 88–90.

47. Thaci D, Galimberti R, Amaya-Guerra M et al. Improvement in aspects of sleep with etanercept and optional adjunctive topical therapy in patients with moderate-to-severe psoriasis: Results from the PRISTINE trial. *J Eur Acad Dermatol Venereol* 2014; 28: 900–906.

48. Strober BE, Sobell JM, Duffin KC et al. Sleep quality and other patient-reported outcomes improve after patients with psoriasis with suboptimal response to other systemic therapies are switched to adalimumab: Results from PROGRESS, an open-label Phase IIIB trial. *Br J Dermatol* 2012; 167: 1374–1381.

49. Al-Ahmad M. Omalizumab therapy in three patients with chronic autoimmune urticaria. *Ann Saudi Med* 2010; 30: 478–481.

50. Nonaka K, Nakazawa Y, Kotorii T. Effects of antibiotics, minocycline and ampicillin, on human sleep. *Brain Res* 1983; 288: 253–259.

51. Ahmad Z, Venus M, Kisku W, Rayatt SS. A case series of skin necrosis following use of non invasive ventilation pressure masks. *Int Wound J* 2013; 10: 87–90.

52. Egesi A, Davis MD. Irritant contact dermatitis due to the use of a continuous positive airway pressure nasal mask: 2 case reports and review of the literature. *Cutis* 2012; 90: 125–128.

53. Altemus M, Rao B, Dhabhar FS, Ding W, Granstein RD. Stress-induced changes in skin barrier function in healthy women. *J Invest Dermatol* 2001; 117: 309–317.

54. Mullington JM, Simpson NS, Meier-Ewert HK, Haack M. Sleep loss and inflammation. *Best Pract Res Clin Endocrinol Metab* 2010; 24: 775–784.

55. Callis Duffin K, Wong B, Horn EJ, Krueger GG. Psoriatic arthritis is a strong predictor of sleep interference in patients with psoriasis. *J Am Acad Dermatol* 2009; 60: 604–608.

56. Hirotsu C, Rydlewski M, Araujo MS, Tufik S, Andersen ML. Sleep loss and cytokines levels in an experimental model of psoriasis. *PLoS One* 2012; 7: e51183.

57. Matin A, Bliwise DL, Wellman JJ, Ewing HA, Rasmuson P. Resolution of dyshidrotic dermatitis of the hand after treatment with continuous positive airway pressure for obstructive sleep apnea. *South Med J* 2002; 95: 253–254.

58. Kapsimalis F, Basta M, Varouchakis G, Gourgoulianis K, Vgontzas A, Kryger M. Cytokines and pathological sleep. *Sleep Med* 2008; 9: 603–614.

59. Maari C, Bolduc C, Nigen S, Marchessault P, Bissonnette R. Effect of adalimumab on sleep parameters in patients with psoriasis and obstructive sleep apnea: A randomized controlled trial. *J Dermatolog Treat* 2014; 25: 57–60.

60. Sandu C, Dumas M, Malan A et al. Human skin keratinocytes, melanocytes, and fibroblasts contain distinct circadian clock machineries. *Cell Mol Life Sci* 2012; 69: 3329–3339.

61. Yosipovitch G, Sackett-Lundeen L, Goon A, Yiong Huak C, Leok Goh C, Haus E. Circadian and ultradian (12 h) variations of skin blood flow and barrier function in non-irritated and irritated skin-effect of topical corticosteroids. *J Invest Dermatol* 2004; 122: 824–829.

62. Yosipovitch G, Xiong GL, Haus E, Sackett-Lundeen L, Ashkenazi I, Maibach HI. Time-dependent variations of the skin barrier function in humans: Transepidermal water loss, stratum corneum hydration, skin surface pH, and skin temperature. *J Invest Dermatol* 1998; 110: 20–23.

63. Lee CH, Chuang HY, Shih CC, Jong SB, Chang CH, Yu HS. Transepidermal water loss, serum IgE and beta-endorphin as important and independent biological markers for development of itch intensity in atopic dermatitis. *Br J Dermatol* 2006; 154: 1100–1107.

64. Lopez-Jornet P, Molino Pagan D, Andujar Mateos P, Rodriguez Agudo C, Pons-Faster A. Circadian rhythms variation of pain in burning mouth syndrome. *Geriatr Gerontol Int* 2015; 15: 490–495.

65. Li WQ, Qureshi AA, Schernhammer ES, Han J. Rotating night-shift work and risk of psoriasis in US women. *J Invest Dermatol* 2014; 133: 565–567.

66. Mozzanica N, Tadini G, Radaelli A et al. Plasma melatonin levels in psoriasis. *Acta Derm Venereol* 1988; 68: 312–316.

67. Schwarz W, Birau N, Hornstein OP et al. Alterations of melatonin secretion in atopic eczema. *Acta Derm Venereol* 1988; 68: 224–229.

68. Gonzalez R, Sanchez A, Ferguson JA et al. Melatonin therapy of advanced human malignant melanoma. *Melanoma Res* 1991; 1: 237–243.

69. Buja A, Mastrangelo G, Perissinotto E. et al. Cancer incidence among female flight attendants: A meta-analysis of published data. *J Womens Health (Larchmt)* 2006; 15: 98–105.

70. Pukkala E, Aspholm R, Auvinen A. et al. Cancer incidence among 10,211 airline pilots: A Nordic study. *Aviat Space Environ Med* 2003; 74: 699–706.

71. Schwartzbaum J, Ahlbom A, Feychting M. Cohort study of cancer risk among male and female shift workers. *Scand J Work Environ Health* 2007; 33: 336–343.

72. Schernhammer ES, Razavi P, Li TY, Qureshi AA, Han J. Rotating night shifts and risk of skin cancer in the Nurses' Health Study. *J Natl Cancer Inst* 2011; 103: 602–606.

73. Ghaderi R, Sehatbakhsh S, Bakhshaee M, Sharifzadeh GR. Urinary melatonin levels and skin malignancy. *Iran J Med Sci* 2014; 39: 64–67.

74. Fischer TW, Kleszczynski K, Hardkop LH, Kruse N, Zillikens D. Melatonin enhances antioxidative enzyme gene expression (*CAT*, *GPx*, *SOD*), prevents their UVR-induced depletion, and protects against the formation of DNA damage (8-hydroxy-2′-deoxyguanosine) in *ex vivo* human skin. *J Pineal Res* 2013; 54: 303–312.

75. Fischer TW, Burmeister G, Schmidt HW, Elsner P. Melatonin increases anagen hair rate in women with androgenetic alopecia or diffuse alopecia: Results of a pilot randomized controlled trial. *Br J Dermatol* 2004; 150: 341–345.

76. Scheuer C, Pommergaard HC, Rosenberg J, Gogenur I. Melatonin's protective effect against UV radiation: A systematic review of clinical and experimental studies. *Photodermatol Photoimmunol Photomed* 2014; 30: 180–188.

77. Gilbert SS, van den Heuvel CJ, Ferguson SA, Dawson D. Thermoregulation as a sleep signalling system. *Sleep Med Rev* 2004; 8: 81–93.

78. Liguori R, Donadio V, Foschini E et al. Sleep stage-related changes in sympathetic sudomotor and vasomotor skin responses in man. *Clin Neurophysiol* 2000; 111: 434–439.

79. Liao WC, Wang L, Kuo CP, Lo C, Chiu MJ, Ting H. Effect of a warm footbath before bedtime on body temperature and sleep in older adults with good and poor sleep: An experimental crossover trial. *Int J Nurs Stud* 2013; 50: 1607–1616.

80. Fronczek R, Overeem S, Lammers GJ, van Dijk JG, Van Someren EJ. Altered skin-temperature regulation in narcolepsy relates to sleep propensity. *Sleep* 2006; 29: 1444–1449.

81. Kahn A, Van de Merckt C, Dramaix M et al. Transepidermal water loss during sleep in infants at risk for sudden death. *Pediatrics* 1987; 80: 245–250.

82. Savin JA, Paterson WD, Oswald I. Scratching during sleep. *Lancet* 1973; 2: 296–297.

83. Savin JA, Paterson WD, Oswald I, Adam K. Further studies of scratching during sleep. *Br J Dermatol* 1975; 93: 297–302.

84. Aoki T, Kushimoto H, Hishikawa Y, Savin JA. Nocturnal scratching and its relationship to the disturbed sleep of itchy subjects. *Clin Exp Dermatol* 1991; 16: 268–272.

85. Bender BG, Ballard R, Canono B, Murphy JR, Leung DY. Disease severity, scratching, and sleep quality in patients with atopic dermatitis. *J Am Acad Dermatol* 2008; 58: 415–420.

86. Brown DG, Kalucy RS. Correlation of neurophysiological and personality data in sleep scratching. *Proc R Soc Med* 1975; 68: 530–532.

87. Jenney ME, Childs C, Mabin D, Beswick MV, David TJ. Oxygen consumption during sleep in atopic dermatitis. *Arch Dis Child* 1995; 72: 144–146.

88. Koca R, Altin R, Konuk N, Altinyazar HC, Kart L. Sleep disturbance in patients with lichen simplex chronicus and its relationship to nocturnal scratching: A case control study. *South Med J* 2006; 99: 482–485.

89. Monti JM, Vignale R, Monti D. Sleep and nighttime pruritus in children with atopic dermatitis. *Sleep* 1989; 12: 309–314.

90. Bringhurst C, Waterston K, Schofield O, Benjamin K, Rees JL. Measurement of itch using actigraphy in pediatric and adult populations. *J Am Acad Dermatol* 2004; 51: 893–898.

91. Benjamin K, Waterston K, Russell M, Schofield O, Diffey B., Rees JL. The development of an objective method for measuring scratch in children with atopic dermatitis suitable for clinical use. *J Am Acad Dermatol* 2004; 50: 33–40.

92. Hon KL, Lam MC, Leung TF et al. Nocturnal wrist movements are correlated with objective clinical scores and plasma chemokine levels in children with atopic dermatitis. *Br J Dermatol* 2006; 154: 629–635.

93. Bender BG, Leung SB, Leung DY. Actigraphy assessment of sleep disturbance in patients with atopic dermatitis: An objective life quality measure. *J Allergy Clin Immunol* 2003; 111: 598–602.

94. Gupta MA, Gupta AK. Cutaneous sensory disorder. *Semin Cutan Med Surg* 2013; 32: 110–118.

95. Gupta MA, Gupta AK. Current concepts in psychodermatology. *Curr Psychiatry Rep* 2014; 16: 449.

96. Singareddy R, Moin A, Spurlock L, Merritt-Davis O, Uhde TW. Skin picking and sleep disturbances: Relationship to anxiety and need for research. *Depress Anxiety* 2003; 18: 228–232.

97. Lopez-Jornet P, Lucero-Berdugo M, Castillo-Felipe C, Zamora Lavella C, Ferrandez-Pujante A, Pons-Fuster A. Assessment of self-reported sleep disturbance and psychological status in patients with burning mouth syndrome. *J Eur Acad Dermatol Venereol* 2015; 29: 1285–1290.

98. Adamo D, Schiavone V, Aria M et al. Sleep disturbance in patients with burning mouth syndrome: A case–control study. *J Orofac Pain* 2013; 27: 304–313.

99. Chainani-Wu N, Madden E, Silverman S Jr. A case–control study of burning mouth syndrome and sleep dysfunction. *Oral Surg Oral Med Oral Pathol Oral Radiol Endod* 2011; 112: 203–208.

100. Lee CF, Lin KY, Lin MC, Lin CL, Chang SN, Kao CH. Sleep disorders increase the risk of burning mouth syndrome: A retrospective population-based cohort study. *Sleep Med* 2014; 15: 1405–1410.

101. Bieber T, Bussman C. Atopic dermatitis. In: Bolognia J, Jorizzo J, Schaffer J (Eds). *Dermatology* (3rd ed.). China: Elsevier Saunders, Vol 1, 2012, pp. 203–217.

102. Chng SY, Goh DY, Wang XS, Tan TN, Ong NB. Snoring and atopic disease: A strong association. *Pediatr Pulmonol* 2004; 38: 210–216.

103. Hon KL, Leung TF, Ma KC et al. Resting energy expenditure, oxygen consumption and carbon dioxide production during sleep in children with atopic dermatitis. *J Dermatolog Treat* 2005; 16: 22–25.

104. Reuveni H, Chapnick G, Tal A, Tarasiuk A. Sleep fragmentation in children with atopic dermatitis. *Arch Pediatr Adolesc Med* 1999; 153: 249–253.

105. Stores G, Burrows A, Crawford C. Physiological sleep disturbance in children with atopic dermatitis: A case control study. *Pediatr Dermatol* 1998; 15: 264–268.

106. Chang YS, Chou YT, Lee JH et al. Atopic dermatitis, melatonin, and sleep disturbance. *Pediatrics* 2014; 134: e397–405.

107. Hon KL, Leung TF, Wong Y, Fok TF. Lesson from performing SCORADs in children with atopic dermatitis: Subjective symptoms do not correlate well with disease extent or intensity. *Int J Dermatol* 2006; 45: 728–730.

108. Sandoval LF, Huang K, O'Neill JL et al. Measure of atopic dermatitis disease severity using actigraphy. *J Cutan Med Surg* 2014; 18: 49–55.

109. Wootton CI, Koller K, Lawton S, O'Leary C, Thomas KS. Are accelerometers a useful tool for measuring disease activity in children with eczema? Validity, responsiveness

to change, and acceptability of use in a clinical trial setting. *Br J Dermatol* 2012; 167: 1131–1137.

110. Murray CS, Rees JL. Are subjective accounts of itch to be relied on? The lack of relation between visual analogue itch scores and actigraphic measures of scratch. *Acta Derm Venereol* 2011; 91: 18–23.

111. van de Kerkhof P, Nestle F. Psoriasis. In: Bolognia J, Jorizzo J, Schaffer J (Eds). *Dermatology* (3rd ed.). China: Elsevier Saunders, Vol 1. 2012, pp. 135–156.

112. Gowda S, Goldblum OM, McCall WV, Feldman SR. Factors affecting sleep quality in patients with psoriasis. *J Am Acad Dermatol* 2010; 63: 114–123.

113. Gupta MA, Gupta AK, Kirkby S et al. Pruritus in psoriasis. A prospective study of some psychiatric and dermatologic correlates. *Arch Dermatol* 1988; 124: 1052–1057.

114. Tsai TF, Wang TS, Hung ST et al. Epidemiology and comorbidities of psoriasis patients in a national database in Taiwan. *J Dermatol Sci* 2011; 63: 40–46.

115. Yosipovitch G, Goon A, Wee J, Chan YH, Goh CL. The prevalence and clinical characteristics of pruritus among patients with extensive psoriasis. *Br J Dermatol* 2000; 143: 969–973.

116. Buslau M, Benotmane K. Cardiovascular complications of psoriasis: Does obstructive sleep apnoea play a role? *Acta Derm Venereol* 1999; 79: 234.

117. Karaca S, Fidan F, Erkan F et al. Might psoriasis be a risk factor for obstructive sleep apnea syndrome? *Sleep Breath* 2013; 17: 275–280.

118. Papadavid E, Vlami K, Dalamaga M et al. Sleep apnea as a comorbidity in obese psoriasis patients: A cross-sectional study. Do psoriasis characteristics and metabolic parameters play a role? *J Eur Acad Dermatol Venereol* 2013; 27: 820–826.

119. Grattan C. Urticaria and angioedema. In: Bolognia J, Jorizzo J, Schaffer J (Eds). *Dermatology* (3rd ed.) China: Elsevier Saunders, Vol 1, 2012, pp. 291–317.

120. Thakkar MM. Histamine in the regulation of wakefulness. *Sleep Med Rev* 2011; 15: 65–74.

121. Yosipovitch G, Ansari N, Goon A, Chan YH, Goh CL. Clinical characteristics of pruritus in chronic idiopathic urticaria. *Br J Dermatol* 2002; 147: 32–36.

122. Staevska M, Gugutkova M, Lazarova C et al. Night-time sedating H1-antihistamine increases daytime somnolence but not treatment efficacy in chronic spontaneous urticaria: A randomized controlled trial. *Br J Dermatol* 2014; 171: 148–154.

14

Neuroimaging of sleep and depression

BENJAMIN HATCH AND THIEN THANH DANG-VU

INTRODUCTION

The use of neuroimaging has contributed significantly to our understanding of both the anatomy and function of sleep in humans. Primarily, this has been done through the use of positron emission tomography (PET), in which compounds labeled with radioactive isotopes are injected into a subject. The distribution of these compounds is then assessed through the detection of positron emissions due to radioactive decay. The two compounds mainly used in sleep research are [^{18}F] fluorodeoxyglucose (FDG), a marker of glucose metabolism, and oxygen-15-labeled water (H$_2$^{15}O), an indirect marker of blood flow.

Due to the long half-life of FDG (around 110 min), subjects are injected while sleeping at the onset of the sleep phase of interest and then awoken usually around 20 min later and transferred to the PET scanner for assessment. This has the advantage of allowing subjects to sleep naturally unconstrained outside of the scanner, but as a result has more limited temporal resolution and does not allow several scans to be performed per night. Conversely, H$_2$^{15}O has a much shorter half-life (around 120 s), which allows for multiple

scans to be performed within a single session with greater temporal resolution, but does require the subject to sleep with their head restrained in the scanner.

More recently, functional magnetic resonance imaging (fMRI) has been used to assess various aspects of brain activation during sleep. This method indirectly measures cerebral blood flow through the detection of the blood oxygen level-dependent signal, which is based on relative changes in deoxyhemoglobin related to blood flow. The main advantages of this technique reside in its fine temporal resolution on the scale of seconds compared to PET, which is generally several minutes or more at best, as well as its superior spatial resolution.

Overall, there are different ways to approach the use of neuroimaging in order to study sleep, and as such each of these neuroimaging techniques has been used differentially. One approach is to examine the differences in global patterns of activation during each stage of sleep, generally compared to waking as a baseline. A large body of research has been devoted to this, and PET has been primarily used. Another approach is to examine the neural correlates of phasic sleep events within particular stages of sleep, such as spindles and slow waves. These events are identified through electroencephalography (EEG) and have been studied primarily through the use of fMRI, as its greater temporal resolution allows for precise matching of EEG recordings with neuroimaging scans. These two approaches and their results are reviewed below, and the use of neuroimaging of sleep in order to study depression is also assessed.

GLOBAL ACTIVATION PATTERNS

Overall, neuroimaging studies have consistently shown a global decrease in brain activity during synchronized non-rapid eye movement (NREM) sleep compared with waking, while levels of brain activity during desynchronized rapid eye movement (REM) sleep are comparable to those of waking.[1] These two broad categories of sleep also show regionally specific deactivation/activation patterns, which are discussed below.

NREM sleep

An early study using FDG PET imaging showed that global glucose metabolism in the brain decreases by around 40% in NREM sleep compared to waking, with particular decreases in the

two thalami.[2] Subsequent studies have primarily used the $H_2^{15}O$ method to identify particular brain regions whose patterns of activation change in NREM sleep. Overall, these studies indicate significant decreases in cerebral blood flow relative to waking in several areas, including subcortical structures such as the brainstem, thalami, basal ganglia, cerebellum, and basal forebrain, as well as areas of the prefrontal (dorsolateral and orbital), parietal, cingulate, and mesio-temporal cortices (see Figure 14.1).[3-7] Similar results were also observed in a more recent study employing fMRI.[8] These results are in accordance with the involvement of several of these subcortical structures in arousal, as well as the observation that many of the areas that show this decrease in activation are particularly active during waking.

Interestingly, a FDG study by Nofzinger and colleagues[9] demonstrated that after correcting for the global decrease in glucose metabolism seen in NREM sleep, several areas showed relative increases compared to waking. These included areas of the primary and association sensory motor cortices, hippocampus, amygdala, hypothalamus, basal forebrain, ventral striatum, anterior cingulate cortex, and pontine reticular formation. The authors suggest that these activations may be related to homeostatic regulation and the processing of memory traces.

Global activation patterns: NREM sleep

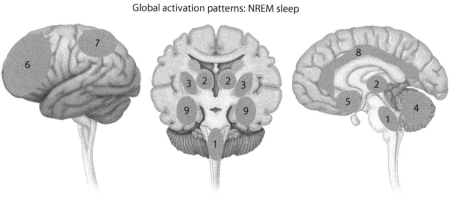

▨ Areas of decreased cerebral blood flow

1. Brainstem
2. Thalamus
3. Basal ganglia
4. Cerebellum
5. Basal forebrain
6. Prefrontal cortex
7. Parietal cortex
8. Cingulate cortex
9. Mesio-temporal cortex

Figure 14.1 Global activation patterns: NREM sleep. (Adapted from illustrations by Patrick J. Lynch and C. Carl Jaffe. http://creativecommons.org/licenses/by/2.5.)

REM sleep

In contrast to NREM sleep, whole-brain metabolism in REM sleep is comparable to that of waking.[2] However, at the regional level, several brain areas showed increased activation during REM sleep compared to wakefulness, as observed with FDG PET[10] as well as $H_2^{15}O$ PET.[3,11,12] These areas include subcortical and limbic structures such as the brainstem, thalami, basal forebrain, hippocampus, and amygdala, as well as cortical areas in the anterior cingulate and temporo-occipital areas. In addition, some of these studies have also shown areas of decreased activation when compared to waking and NREM sleep, particularly areas in the prefrontal, parietal, posterior cingulate, and primary visual cortices (see Figure 14.2).[3,11,12]

The activation of extrastriate and limbic structures, combined with a deactivation of primary visual and prefrontal cortices, appears to indicate the existence of a closed processing loop during REM sleep. The activation of these areas without primary visual input and higher-level processing from the prefrontal cortex has been suggested to underlie some of the aspects of dreaming (vivid/surreal/bizarre imagery, emotional content, and lack of awareness), which is commonly associated with REM sleep.[12]

PHASIC EVENTS

Another approach to the use of neuroimaging in order to study sleep that has been explored more recently is the examination of the neural correlates of phasic events during sleep. Instead of comparing global activity between stages of sleep and waking, this process aims to correlate regional activity within a particular stage to the occurrence of specific neural oscillations or events. Much of this work has been done with fMRI due to its better temporal resolution allowing hemodynamic responses to be time locked to these events, but it is possible to use PET as well by correlating overall activity in a period of sleep to the number of the events being studied. Of interest are sleep spindles and slow waves during NREM sleep and rapid eye movements and ponto-geniculo-occipital (PGO) waves during REM sleep.

Sleep spindles

Sleep spindles as seen on an EEG recording consist of transient bursts of around 12–14 Hz activity (sigma band) that are most prominent in stage 2 of NREM sleep.[13] An early $H_2^{15}O$ PET study found a negative correlation between cerebral blood flow in the medial thalamus and activity in the sigma band after

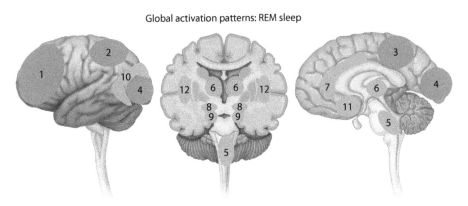

Global activation patterns: REM sleep

■ Areas of decreased cerebral blood flow
▨ Areas of increased cerebral blood flow

1. Prefrontal cortex
2. Parietal cortex
3. Posterior cingulate cortex/precuneus
4. Primary visual cortex
5. Brainstem
6. Thalamus
7. Anterior cingulate cortex
8. Amygdala
9. Hippocampus
10. Extrastriate cortex
11. Basal forebrain
12. Insula-temporal cortex

Figure 14.2 Global activation patterns: REM sleep. (Adapted from illustrations by Patrick J. Lynch and C. Carl Jaffe. http://creativecommons.org/licenses/by/2.5.)

Phasic events

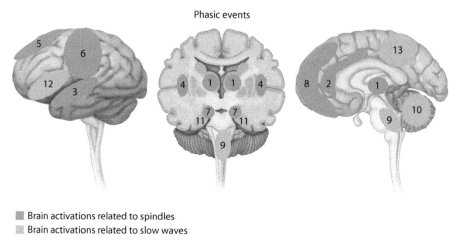

▨ Brain activations related to spindles
▨ Brain activations related to slow waves

1. Thalamus
2. Anterior cingulate cortex
3. Superior temporal gyrus
4. Insula
5. Superior frontal gyrus
6. Sensorimotor cortices
7. Hippocampus

8. Medial prefrontal cortex
9. Brainstem
10. Cerebellum
11. Parahippocampal gyrus
12. Inferior frontal gyrus
13. Posterior cingulate/precuneus

Figure 14.3 Phasic events. (Adapted from illustrations by Patrick J. Lynch and C. Carl Jaffe. http://creativecommons.org/licenses/by/2.5.)

controlling for blood flow changes due to slow-wave oscillations.[4] A later fMRI study found somewhat contrasting results in that spindles were correlated with increased activity in the thalami as well as anterior cingulate cortex, insula, and superior temporal gyrus (see Figure 14.3).[14] This difference is likely due to differing methodologies, in that PET requires the averaging of activity over a period of time of allowing hyperpolarization to dominate, while fMRI allows transient increases in activity to be time locked to specific events. In addition, this study subdivided spindles into fast (13–15 Hz) and slow (11–13 Hz) categories and found differences in activation patterns between the two. Fast spindles were associated with increased activity in sensorimotor areas, the hippocampus, and the mesial frontal cortex, while slow spindles were associated with activation in the superior frontal gyrus.[14] Spindles are of interest to scientists due to their implications in learning and memory[15–17] and in sensory gating during sleep.[18–20]

Slow waves

Slow waves are prominent characteristics of the later stages of NREM sleep and consist of high-amplitude delta waves at 1–4 Hz and a slower

oscillatory rhythm of less than 1 Hz.[13] In the same PET study as mentioned above, activity in the delta band was correlated with cerebral blood flow, and negative associations were found in the thalami, reticular formation, cerebellum, anterior cingulate, and orbitofrontal cortex.[4] A later study reanalyzing the data collected in several previous $H_2{}^{15}O$ PET studies found a negative correlation between delta power and blood flow in the ventromedial prefrontal cortex, basal forebrain, striatum, anterior insula, and precuneus.[21] In contrast, no association was found in the thalami as in the previous study. The authors suggest that these results may reflect non-thalamic cortex-based delta rhythms. Lastly, a recent fMRI study found increased activity in relation to slow waves in the brainstem, cerebellum, parahippocampal gyrus, inferior frontal gyrus, precuneus, and posterior cingulate cortex (see Figure 14.3).[22] When slow waves were subdivided into delta wave (1–4 Hz) and slow oscillation (<1 Hz) rhythms and then compared to baseline, delta waves were associated with activity in the prefrontal cortex, while slow oscillations were associated with activity in the brainstem, cerebellum, and parahippocampal gyrus. These cortical activations being related to slow oscillations is in accordance with evidence of their relation to memory processes.[16,17,23]

Collectively, these fMRI results on spindles and slow waves show that NREM sleep may not simply be a state of mere brain quiescence, but rather a dynamic state with transient neural activity against a baseline state of deactivation.

Rapid eye movements and PGO waves

Rapid eye movements are some of the most prominent hallmarks of REM sleep and are believed to be related to the PGO waves identified through cellular recordings in animals. These waves are produced in the transition to and throughout REM sleep, and consist of short, high-amplitude electrical potentials, prominently recorded from the pontine brainstem, lateral geniculate nuclei, and occipital cortex.[24] Due to their discovery through cellular recording in animals, their existence in humans has not been determined, but deep-brain recordings have pointed to their existence in humans,[25] as well as there being inferential evidence from neuroimaging. In a $H_2^{15}O$ PET study, the number of rapid eye movements during REM sleep compared to waking was found to be related to increased cerebral blood flow in the primary visual cortex and right geniculate nuclei, corroborating some of the recording sites for PGO waves in animals.[26] Another study employing fMRI found activations in the pontine tegmentum, ventoposterior thalamus, and primary visual cortex preceding the rapid eye movements and activations in the putamen and limbic areas (amygdala, parahippocampal gyrus, and anterior cingulate cortex) accompanying the movements (see Figure 14.3).[27] Due to the association between dreaming and REM sleep, it has been theorized that PGO waves are involved in this process, which is plausible given their prominence within the visual system of the brain and the visual nature of dreaming.[28]

DEPRESSION AND SLEEP

Sleep disturbances are a common facet of depression and include problems in sleep continuity, reductions of slow-wave sleep, decreased REM latency, and increased REM density.[29] Several studies have used neuroimaging in order to investigate the relationship between sleep and depression. Studies of NREM sleep using FDG PET have observed that depressed individuals show a smaller decrease in glucose metabolism relative to controls in broad cortical areas, particularly in the prefrontal region and the parietal and temporal cortices.[30,31] The authors suggest that this lack of reduction is likely due to hypofrontality during waking in depressed individuals. This is consistent with research showing lowered waking prefrontal glucose metabolism in depressed individuals, which improved with medication.[32] This hypofrontality is likely related to the broad cognitive deficits seen in depression, particularly in executive functions.[33] Furthermore, also note that there is hypermetabolism during both NREM sleep and waking in depressed individuals in the brainstem reticular formation, basal forebrain, left amygdala, anterior cingulate cortex, cerebellum, parahippocampal cortex, fusiform gyrus, and occipital cortex.[31] In REM sleep, a FDG PET study found increased glucose metabolism in depressed individuals compared to controls in the brainstem reticular formation and limbic and anterior paralimbic areas (including the hippocampus, basal forebrain, anterior cingulate, and medial prefrontal cortex), as well as cortical areas such as the dorsolateral prefrontal, left premotor, primary sensorimotor, and left parietal cortices (see Figure 14.4 for summary).[34] Hypermetabolism in REM sleep was further characterized by a recent PET study looking at both individuals with posttraumatic stress disorder and depression.[35] Both groups showed comparable hypermetabolism in limbic and paralimbic areas during REM sleep, with the depressed group showing this during waking as well.

Hyperarousal

These studies seem to indicate an overall hyperarousal during sleep in depressed individuals, which may underlie some of the aforementioned sleep disturbances. A common disturbance in depressed individuals is insomnia, and there is much evidence linking the two disorders, although the exact nature of this link and causal relationship remains unknown.[36] Hyperarousal has been heavily implicated in sufferers of insomnia,[37] and functional neuroimaging has shown increased cerebral glucose metabolism during both wake and sleep in insomniacs.[38] Further evidence of

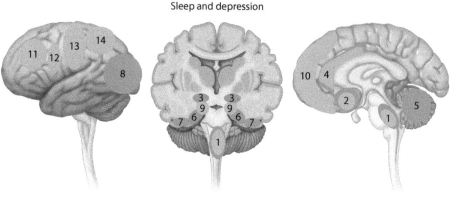

Sleep and depression

Areas of increased brain metabolism during NREM sleep
Areas of increased brain metabolism during REM sleep

1. Brainstem
2. Basal forebrain
3. Amygdala
4. Anterior cingulate cortex
5. Cerebellum
6. Parahippocampal gyrus
7. Fusiform gyrus
8. Occipital cortex
9. Hippocampus
10. Medial prefrontal cortex
11. Dorsolateral prefrontal cortex
12. Premotor cortex
13. Primary sensorimotor cortex
14. Parietal cortex

Figure 14.4 Sleep and depression. (Adapted from illustrations by Patrick J. Lynch and C. Carl Jaffe. http://creativecommons.org/licenses/by/2.5.)

hyperarousal in depressed individuals comes from EEG evidence showing a trend towards increased power in the beta band, a frequency band that is associated with the cortical arousal seen in waking and REM sleep, which was correlated with poorer subjective sleep quality.[39] Additionally, a PET study using a radioactive ligand for the histamine H_1 receptor showed decreased binding in depressed individuals, which correlated with self-reported measures of depression.[40] These results could indicate higher levels of endogenous histamine release, leading to more competition for binding, and given that the histaminergic system is a wake-promoting system, this would further support a model of hyperarousal.

Broadly, the cortical hyperarousal may be in part due to the increased activity in the brainstem noted in both NREM[31] and REM sleep,[34] as the brainstem sends many projections throughout the brain and is involved in waking and arousal. More particularly, studies seem to point to hyperactivity in limbic and paralimbic areas, which are commonly associated with the processing of affective states. Research indicates that sleep is important for the processing of emotional memory and modulating affective systems,[41] so it is plausible that hyperarousal in these areas during sleep could be indicative of abnormal or overly intense processing of affect, possibly underlying some of the emotional problems seen in waking, such as cognitive biases toward negative stimuli/experiences in attention and memory.[42]

Effects of treatment

Neuroimaging can also be used to assess the effects of the treatment of depression. Some of this research has looked at the influence of antidepressant drugs on brain activation in sleep. For instance, a FDG PET study looked at the effects of treatment with bupropion on REM sleep.[43] Although a reversal in activation deficits between waking and REM sleep was found in the anterior cingulate, medial prefrontal cortex, and right anterior insula after treatment, this was attributed to a reduction is waking metabolism in these areas, and not in REM sleep. A much more common method that has been used to study the effects of treatment on depression with neuroimaging is actually sleep deprivation. This has robustly been shown to have antidepressant effects, although these effects generally do not last for long, and early studies noted hyperactivity in the ventral anterior cingulate and orbital medial prefrontal cortex, which became normalized after sleep deprivation.[44] Recent studies have looked at the combined effects of sleep deprivation and antidepressant medication. A study in which depressed patients were given sertraline for 1 week and then scanned before and after sleep deprivation showed decreased Hamilton Depression ratings related to decreased glucose metabolism in several areas, particularly the ventral frontal lobe, as well as parts of the temporal and occipital cortices and the left insula.[45] They also noted increases in metabolism related to decreased Hamilton Depression ratings in a number of areas, particularly the dorsal frontal lobe, along with regions of the parietal, temporal, and occipital cortices and cerebellum. However, a placebo-controlled study of geriatric patients with depression treated with paroxetine did not reveal a synergistic effect of sleep deprivation combined with medication on either depressive symptoms or glucose metabolism in the brain areas that are generally associated with clinical improvement.[46]

CONCLUSION

Neuroimaging is an important tool for investigating the neural correlates of sleep, both in healthy individuals and those who suffer from depression. In particular, it shows that the brain is not merely "turned off" during sleep, but rather has specific and dynamic patterns of activation and deactivation. In NREM sleep, the brain is mostly deactivated, with transient fluctuations in neural activity related to phasic events like sleep spindles and slow waves. In REM sleep, the brain shows global levels of activity similar to waking, but with regional increases in limbic and visual association areas that may, along with rapid eye movements and PGO waves, be related to dreaming. Individuals with depression show abnormal brain activation patterns in sleep with a general tendency for increased brain metabolism, in line with a hyperarousal state, which may underlie some of the sleep problems related to depression, as well as cognitive/emotional dysfunctions in waking.

Future research should seek to clarify these findings through improved techniques and replication and further our understanding of sleep in both healthy and pathologic populations. Processes like sleep-related memory consolidation and dreaming have already been explored with neuroimaging, but still remain poorly understood, leaving much room for future research. Our understanding of neural oscillations during sleep also has much room for new discoveries, particularly in terms of understanding the functional purpose of phasic events. Lastly, sleep disturbances are widely known to be related to many psychiatric and neurologic disorders, but the exact relationship is unclear. Are sleep problems predictors or symptoms of these? How are changes in brain activation and/or morphology involved in this relationship? Neuroimaging provides a powerful means for investigating all of these questions, and much remains to be uncovered about the sleeping brain.

REFERENCES

1. Maquet P. Functional neuroimaging of normal human sleep by positron emission tomography. *J Sleep Res* 2000; 9(3): 207–231.
2. Maquet P, Dive D, Salmon E et al. Cerebral glucose utilization during sleep–wake cycle in man determined by positron emission tomography and [^{18}F]2-fluoro-2-deoxy-D-glucose method. *Brain Res* 1990; 513(1): 136–143.
3. Braun AR, Balkin TJ, Wesenten NJ et al. Regional cerebral blood flow throughout the sleep–wake cycle. An H$_2$^{15}O PET study. *Brain* 1997; 120(Pt 7): 1173–1197.
4. Hofle N, Paus T, Reutens D et al. Regional cerebral blood flow changes as a function of delta and spindle activity during slow wave sleep in humans. *J Neurosci* 1997; 17(12): 4800–4808.
5. Maquet P, Degueldre C, Delfiore G et al. Functional neuroanatomy of human slow wave sleep. *J Neurosci* 1997; 17(8): 2807–2812.
6. Andersson JL, Onoe H, Hetta J et al. Brain networks affected by synchronized sleep visualized by positron emission tomography. *J Cereb Blood Flow Metab* 1998; 18(7): 701–715.
7. Kajimura N, Uchiyama M, Takayama Y et al. Activity of midbrain reticular formation and neocortex during the progression of human non-rapid eye movement sleep. *J Neurosci* 1999; 19(22): 10065–10073.

8. Kaufmann C, Wehrle R, Wetter TC et al. Brain activation and hypothalamic functional connectivity during human non-rapid eye movement sleep: An EEG/fMRI study. *Brain* 2006; 129(Pt 3): 655–667.

9. Nofzinger EA, Buysse DJ, Miewald JM et al. Human regional cerebral glucose metabolism during non-rapid eye movement sleep in relation to waking. *Brain* 2002; 125(Pt 5): 1105–1115.

10. Nofzinger EA, Mintun MA, Wiseman M, Kupfer DJ, Moore RY. Forebrain activation in REM sleep: An FDG PET study. *Brain Res* 1997; 770(1–2): 192–201.

11. Maquet P, Peters J, Aerts J et al. Functional neuroanatomy of human rapid-eye-movement sleep and dreaming. *Nature* 1996; 383(6596): 163–166.

12. Braun AR, Balkin, TJ, Wesensten NJ et al. Dissociated pattern of activity in visual cortices and their projections during human rapid eye movement sleep. *Science* 1998; 279(5347): 91–95.

13. Pace-Schott EF. Sleep architecture. In: Matthew W, Stickgold R (Eds). *The Neuroscience of Sleep*. San Diego, CA: Academic Press, 2009, pp. 11–17.

14. Schabus M, Dang-Vu TT, Albouy G et al. Hemodynamic cerebral correlates of sleep spindles during human non-rapid eye movement sleep. *Proc Natl Acad Sci USA* 2007; 104(32): 13164–13169.

15. Schabus M, Hoedlmoser K, Pecherstorfer T et al. Interindividual sleep spindle differences and their relation to learning-related enhancements. *Brain Res* 2008; 1191: 127–135.

16. Tamminen J, Lambon Ralph MA, Lewis PA. The role of sleep spindles and slow-wave activity in integrating new information in semantic memory. *J Neurosci* 2013; 33(39): 15376–15381.

17. Astill RG, Piantoni G, Raymann RJ et al. Sleep spindle and slow wave frequency reflect motor skill performance in primary school-age children. *Front Hum Neurosci* 2014; 8: 910.

18. Dang-Vu TT, McKinney SM, Buxton OM, Solet JM, Ellenbogen JM. Spontaneous brain rhythms predict sleep stability in the face of noise. *Curr Biol* 2010; 20(15): R626–R627.

19. Dang-Vu TT, Bonjean M, Schabus M et al. Interplay between spontaneous and induced brain activity during human non-rapid eye movement sleep. *Proc Natl Acad Sci USA* 2011; 108(37): 15438–15443.

20. Schabus M, Dang-Vu TT, Heib DP et al. The fate of incoming stimuli during NREM sleep is determined by spindles and the phase of the slow oscillation. *Front Neurol* 2012; 3: 40.

21. Dang-Vu TT, Desseilles M, Laureys S et al. Cerebral correlates of delta waves during non-REM sleep revisited. *Neuroimage* 2005; 28(1): 14–21.

22. Dang-Vu TT, Schabus M, Desseilles M et al. Spontaneous neural activity during human slow wave sleep. *Proc Natl Acad Sci USA* 2008; 105(39): 15160–15165.

23. Marshall L, Helgadottir H, Molle M, Born J. Boosting slow oscillations during sleep potentiates memory. *Nature* 2006; 444(7119): 610–613.

24. Callaway CW, Lydic R, Baghdoyan HA, Hobson, JA. Pontogeniculooccipital waves: Spontaneous visual system activity during rapid eye movement sleep. *Cell Mol Neurobiol* 1987; 7(2): 105–149.

25. Lim AS, Lozano AM, Moro E et al. Characterization of REM-sleep associated ponto-geniculo-occipital waves in the human pons. *Sleep* 2007; 30(7): 823–827.

26. Peigneux P, Laureys S, Fuchs S et al. Generation of rapid eye movements during paradoxical sleep in humans. *Neuroimage* 2001; 14(3): 701–708.

27. Miyauchi S, Misaki M, Kan S, Fukunaga T, Koike T. Human brain activity time-locked to rapid eye movements during REM sleep. *Exp Brain Res* 2009; 192(4): 657–667.

28. Hobson JA, Friston KJ. Waking and dreaming consciousness: Neurobiological and functional considerations. *Prog Neurobiol* 2012; 98(1): 82–98.

29. Riemann D, Berger M, Voderholzer U. Sleep and depression—Results from psychobiological studies: An overview. *Biol Psychol* 2001; 57(1–3): 67–103.

30. Germain A, Nofzinger EA, Kupfer DJ, Buysse DJ. Neurobiology of non-REM sleep in depression: Further evidence for hypofrontality and thalamic dysregulation. *Am J Psychiatry* 2004; 161(10): 1856–1863.

31. Nofzinger EA, Buysse DJ, Germain A et al. Alterations in regional cerebral glucose metabolism across waking and non-rapid eye movement sleep in depression. *Arch Gen Psychiatry* 2005; 62(4): 387–396.

32. Baxter LR Jr, Schwartz JM, Phelps ME et al. Reduction of prefrontal cortex glucose metabolism common to three types of depression. *Arch Gen Psychiatry* 1989; 46(3): 243–250.

33. Levin RL, Heller W, Mohanty A, Herrington JD, Miller GA. Cognitive deficits in depression and functional specificity of regional brain activity. *Cogn Ther Res* 2007; 31(2): 211–233.

34. Nofzinger EA, Buysse DJ, Germain A et al. Increased activation of anterior paralimbic and executive cortex from waking to rapid eye movement sleep in depression. *Arch Gen Psychiatry* 2004; 61(7): 695–702.

35. Ebdlahad S, Nofzinger EA, James JA, Buysse DJ, Price JC, Germain A. Comparing neural correlates of REM sleep in post-traumatic stress disorder and depression: A neuroimaging study. *Psychiatry Res* 2013, 214(3): 422–428.

36. Staner L. Comorbidity of insomnia and depression. *Sleep Med Rev* 2010; 14(1): 35–46.

37. Riemann D, Spiegelhalder K, Feige B et al. The hyperarousal model of insomnia: A review of the concept and its evidence. *Sleep Med Rev* 2010; 14(1): 19–31.

38. Nofzinger EA, Buysse DJ, Germain A, Price JC, Miewald JM, Kupfer DJ. Functional neuroimaging evidence for hyperarousal in insomnia. *Am J Psychiatry* 2004, 161(11): 2126–2128.

39. Nofzinger EA, Price JC, Meltzer CC et al. Towards a neurobiology of dysfunctional arousal in depression: The relationship between beta EEG power and regional cerebral glucose metabolism during NREM sleep. *Psychiatry Res* 2000; 98(2): 71–91.

40. Kano M, Fukudo S, Tashiro A et al. Decreased histamine H1 receptor binding in the brain of depressed patients. *Eur J Neurosci* 2004; 20(3): 803–810.

41. Walker MP, van der Helm E. Overnight therapy? The role of sleep in emotional brain processing. *Psychol Bull* 2009; 135(5): 731–748.

42. Everaert J, Koster, EH, Derakshan N. The combined cognitive bias hypothesis in depression. *Clin Psychol Rev* 2012; 32(5): 413–424.

43. Nofzinger EA, Berman S, Fasiczka A et al. Effects of bupropion SR on anterior paralimbic function during waking and REM sleep in depression: Preliminary findings using [18F]-FDG PET. *Psychiatry Res* 2001; 106(2): 95–111.

44. Gillin JC, Buchsbaum M, Wu J, Clark C, Bunney W Jr. Sleep deprivation as a model experimental antidepressant treatment: Findings from functional brain imaging. *Depress Anxiety* 2001; 14(1): 37–49.

45. Wu JC, Gillin JC, Buchsbaum MS et al. Sleep deprivation PET correlations of Hamilton symptom improvement ratings with changes in relative glucose metabolism in patients with depression. *J Affect Disord* 2008; 107(1–3): 181–186.

46. Smith GS, Reynolds CF 3rd, Houck PR et al. Cerebral glucose metabolic response to combined total sleep deprivation and antidepressant treatment in geriatric depression: A randomized, placebo-controlled study. *Psychiatry Res* 2009; 171(1): 1–9.

15

Sleep alterations in schizophrenia

MATCHERI S. KESHAVAN AND RIPU D. JINDAL

INTRODUCTION

Schizophrenia is one of the most disabling of all mental illnesses. It is characterized by disordered thinking, disordered behavior, delusional beliefs, and perceptual disturbances such as hallucinations (positive symptoms), as well as deficits in motivation, socialization, and affect (negative symptoms). The illness typically begins in adolescence or early adulthood and leads to a marked decline in occupational and interpersonal function.

It has long been held that impaired sleep reflects a troubled mind. This view and the phenomenological similarity between hallucinations and dreams have led to an enduring interest in sleep studies in schizophrenia. A vast literature has accumulated in regard to sleep abnormalities in schizophrenic illness. Impairments are seen in subjective quality of sleep,[1] as well as in objective measures of sleep architecture in sleep studies. Notable findings, which will be summarized in this chapter, include reductions in total sleep, sleep continuity, rapid eye movement (REM) sleep latency, and amounts of slow wave sleep (SWS). The amounts of REM sleep are variably reduced in schizophrenia. In addition to changes in sleep architecture, there is evidence that suggests that there are alterations in the sleep–wake cycle.

Sleep disturbances seem clinically important in schizophrenia, predicting coping and perceived quality of life,[2] as well as symptomatic relapse following antipsychotic discontinuation.[3] Furthermore, a model of altered power in theta and delta and alpha peak frequencies ranges seem to predict conversion to psychosis in those who are deemed clinically at risk of psychosis.[4,5] Despite considerable investigational effort, the nature of the polysomnographic abnormalities and their relationships to the neurobiological underpinnings of schizophrenia have remained poorly understood. In this chapter, we will briefly review the nature of SWS, REM sleep, and sleep spindles, the current state of knowledge in regard to the alterations in sleep architecture, the relationship between antipsychotic medications and sleep, and the neurobiology of such impairments in schizophrenia.

NATURE OF SWS, REM SLEEP, AND SLEEP SPINDLES

In recent years, there has been an increasing understanding of the possible functions of SWS, REM sleep, and sleep spindles. SWS is characterized by large-amplitude, low-frequency electroencephalographic (EEG) rhythms, mainly occurring during the early part of sleep. The slow waves in SWS are associated with large-scale spatiotemporal synchrony across the neocortex, and are thought to be generated predominantly in the prefrontal cortex.[6,7] The SWS has been seen as reflecting overall synaptic density in the human cerebral cortex, in particular that of the prefrontal cortex,[8] although this has been recently debated[9]; deficits in SWS have therefore been thought to reflect parallel prefrontal cortical dysfunction.[7] Sleep deprivation in healthy subjects appears to cause impairments in sustained attention that closely resemble prefrontal cortical dysfunction.[10] SWS may be involved in restorative activity, reversing the "wear and tear" caused during wakefulness; there is evidence for increased protein synthesis during this sleep phase.[11,12]

REM sleep is characterized by low-amplitude, relatively fast EEG rhythms, saccadic eye movements, and decreased muscle tone. In contrast with SWS, REM sleep is associated with an increased activity in phylogenetically old limbic and paralimbic regions, such as the amygdala, hippocampus, and cingulate and entorhinal cortices.[13] Conversely, the dorsolateral prefrontal cortex, parietal cortex, and posterior cingulate cortex and precuneus are the least active brain regions during REM sleep. These observations are significant in view of the cognitive functions mediated by the prefrontal and limbic brain regions, as will be discussed later.

Sleep spindles are fast (11–16 Hz) EEG oscillations that seem to be unique to non-REM (NREM) sleep. Although for a long time they have been seen as an integral part of NREM sleep, especially of the stage 2 of NREM sleep, their functional role in memory consolidation and neuroplasticity was only recognized more recently.[14] In fact, converging evidence supports spindles as an index of an individual's intelligence and learning ability.[15] There is considerable evidence that spindles are generated in the thalamic reticular nucleus (TRN),[16] which comprises GABAergic neurons.[8,17] The TRN neurons project to glutamatergic thalamic neurons, which in turn project to the cortex. Cortical neurons send glutamatergic input back to the N-methyl-D-aspartic acid (NMDA) receptors on TRN neurons. Thus, spindles reflect a thalamocortical feedback loop regulated by both GABAergic and NMDA receptor-mediated glutamatergic transmission.[18] A two-stage model of long-term memory formation posits that newly encoded memories mediated by the hippocampal ripples/sharp waves traverse the thalamus in a feed-forward manner, leading to spindle oscillations, and then on to the neocortex to be consolidated via slow-wave oscillations (Figure 15.1).[14]

ALTERATIONS IN SLEEP ARCHITECTURE IN SCHIZOPHRENIA

Alterations of REM sleep

Many early studies of sleep architecture tested a hypothesis that schizophrenia results from a "spillover" of the dream state into wakefulness. While no evidence thus far directly supports this prediction, subtle alterations in the architecture of REM sleep have been described.[19] REM latency was found to be decreased in several early studies; however, this may result from either a deficit in SWS in the first NREM period, leading to a passive advance, the early onset of the first REM period, or from "REM pressure." Amounts of REM sleep have been reported to variably increase, decrease, or not change in different studies.[20] Since studies in treatment-naïve patients (with schizophrenia) show no increase in REM sleep,[21,22] the increases in REM sleep observed in previously treated subjects may reflect the effects of medication withdrawal and/or persistent changes related to the effects of medication.[22] Similarly, it is unlikely that the observed decreases in REM latency in some schizophrenia patients result from primary abnormalities in REM sleep.

Sleep deprivation provides a naturalistic, physiologic challenge for the dynamic manipulation of sleep processes, and can help clarify the primary nature of sleep abnormalities. An intriguing reduction in REM rebound following REM sleep deprivation has been described in several studies in acute schizophrenia, but there is a normal or exaggerated REM rebound in remitted schizophrenic patients.[10,19] This rebound failure in acute schizophrenia has been attributed to a possible

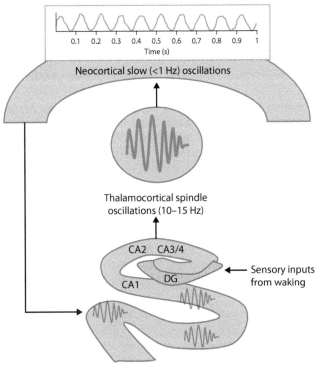

Figure 15.1 Thalamocortically generated spindles may work with hippocampal ripples and neocortical slow oscillations in order to mediate the consolidation of new memories during sleep. Cornu ammonis areas 1–4 (CA1–4) and the dentate gyrus (DG) are hippocampal subregions. (Adapted from Molle M, Born J. *Prog Brain Res* 2011; 193: 93–110.)

"leakage" of phasic REM events from REM sleep into NREM sleep, although no systematic investigation has supported this hypothesis.[19]

Lucid dreams, in which the individual has clear insight into the state of dreaming, offer an opportunity to understand the nature of impaired insight in psychosis. Recent neuroimaging studies suggest that lucid dreamers have more prefrontal gray matter volume, as well as frontoparietal and temporal activation during dreams, than those with non-lucid dreams.[23,24] These are the very same regions that show impaired structure and function in schizophrenia patients with poor insight. Thus, it is possible that similar brain circuits might underlie insight into the dream state, as well as into psychotic experiences. Consistent with this view, we have recently proposed that progressive frontotemporal and parietal gray matter reductions during the prodromal phase around late adolescence may underlie the loss of self-monitoring that leads to the emergence of psychosis.[25]

Alterations of SWS

SWS is of particular interest to schizophrenia researchers for two reasons: the evidence implicating the prefrontal cortex in the generation of SWS; and the evidence supporting a link between SWS and cognition.[26,27] Several, but not all, studies show reduced SWS in patients with schizophrenia; SWS deficits have been documented in acute, chronic, and remitted states.[10] Similarly, SWS deficits have been seen in studies of never-medicated as well as previously medicated patients. Studies that failed to find differences in SWS have generally used conventional visual scoring. On the other hand, studies that included quantified sleep EEG parameters have revealed reductions in SWS more consistently. Notably, Ganguli and associates observed no change in visually scored SWS, but detected reduced delta wave counts in treatment-naïve patients,[21] suggesting that visually scored SWS may not be sensitive enough, and automated counts may be a better marker of SWS deficiency in schizophrenia. Other groups have described similar reductions in delta counts. SWS deficits have been demonstrated in early-course schizophrenia using sensitive approaches such as spectral analysis.[28]

Reductions in low-frequency power may also be associated with alterations in high-frequency EEG activity (HFA; > 20Hz). HFA is associated with feature binding and attention. Tekell and associates reported that schizophrenic patients showed significantly greater HFA than healthy controls in all sleep stages.[29] Elevated HFA during sleep in unmedicated patients is associated with positive symptoms of illness.

Sleep deprivation studies can also help clarify whether SWS abnormalities in schizophrenia are secondary to pathology in neuronal circuits in this disorder, or whether they reflect primary homeostatic disruption in sleep processes. There is evidence that following total sleep deprivation, recovery of stage 4 sleep is diminished in schizophrenia.[30] SWS deprivation is known to consistently cause impaired attention, reaction times, verbal learning, and vigilance, similarly to what is seen in frontal lobe dysfunction and schizophrenia.[31] A defect in SWS recovery might be consistent with impairments in critical cognitive processes, such as the psychomotor vigilance observed in schizophrenia. Such a defect might also suggest impairment in restorative processes in schizophrenia.

Alterations in sleep spindles

Decreases in the amplitude and duration of spindles have been demonstrated in patients with schizophrenia.[32] By including a comparison group of normal controls who took antipsychotic medication, the investigators showed that their finding was more of a disease effect, rather than a medication effect.[32] Reductions in spindle activity were also seen amongst newly diagnosed patients with no prior exposure to antipsychotic medications, as well as non-psychotic relatives of patients with schizophrenia.[33] Even among the healthy population, spindle density seems to correlate inversely with magical ideation.[34]

The available studies suggest that spindle deficits that predate the onset of schizophrenia persist despite clinical improvement with antipsychotic

medication, and may well be an endophenotype of cognitive impairment in schizophrenia. Evidence of dysfunction in spindle generation is consistent with current models of schizophrenia that posit impaired thalamocortical circuitry.[35–38]

Circadian sleep abnormalities

In addition to alterations in sleep architecture, disturbances in rest–activity timing seem very common in schizophrenia.[4] In one study,[4] half of the participants with schizophrenia showed severe circadian misalignment, ranging from phase advance/delay to non-24-hour sleep–wake cycles and melatonin cycles, and the other half showed patterns ranging from excessive sleep to highly irregular and fragmented sleep periods, but with normally timed melatonin cycles. Notably, the circadian disruptions existed despite stability in mood and psychotic state. However, the question of whether these disturbances are secondary to some or all antipsychotic medications, or whether they are primary to the illness, remains to be clarified.

Circadian alterations appear to correlate with poor cognitive functioning in schizophrenia.[39] In one study,[40] patients with relatively intact sleep–wake cycles did better on frontal lobe function tasks, whereas the severity of positive or negative symptoms did not correlate with cognitive performance or the quality of sleep–wake cycles, which led to a conclusion that consolidated circadian rhythms may be a prerequisite for adequate cognitive functioning in schizophrenia.

RELATIONSHIP BETWEEN SLEEP ABNORMALITIES AND CLINICAL MEASURES IN SCHIZOPHRENIA

During the past two decades, researchers have increasingly made a distinction between the positive and negative symptoms of schizophrenia. Several studies have investigated the relationship between REM sleep parameters and clinical parameters. Tandon and colleagues reported an inverse association between REM latency and negative symptoms.[22] While no association was detected between sleep abnormalities and depressive symptoms,[22] there is evidence that increased REM sleep correlates with suicidal behavior in schizophrenia.[41,42]

In order to recognize the significance of sleep abnormalities for pathophysiology, it is important to understand the enduring nature of these abnormalities in schizophrenia; stage 4 sleep does not improve, while other sleep stages change following 3–4 weeks of conventional antipsychotic treatment.[43] In a longitudinal polysomnographic study, SWS deficits persisted at 1 year, but reduction in REM sleep appeared to improve. These observations suggest the possible trait-related nature of SWS deficits in schizophrenia.[44] Consistent with this view, SWS abnormalities correlate with negative symptoms[21] and with impaired outcome at 1 and at 2 years.[45]

Attentional impairment appears to correlate with SWS deficits in early studies of schizophrenia.[46] The thalamus—the main "switchboard" for information processing pathways in the brain—plays a crucial role in attention and gating of information because it is the major relay station receiving input from the reticular activating system and limbic and cortical association areas. A defect in the thalamus, therefore, could explain alterations in SWS, sleep spindles, and the psychopathology of schizophrenia. Impairments in visuospatial memory have been found to correlate with reductions in the amount of SWS and in sleep efficiency.[47] These results point to a functional interrelationship between regulation of SWS and performance in visuospatial memory in schizophrenia.

Reduced spindle activity in schizophrenia seems functionally significant. For instance, there is evidence that spindle density is associated with measures of attention and reasoning among newly diagnosed patients with schizophrenia who had no prior exposure to antipsychotic medications.[48] Furthermore, patients with schizophrenia do not show improvement on a performance on a motor task typically seen after a night's sleep, and lesser number and density of spindles seems to predict this poorer performance (after a night's sleep).[49] This is an important finding that indicates that dysfunction in spindle activity may account for the well-documented impairment in procedural memory in schizophrenia, and that spindle activity may be a therapeutic target in patients suffering from this disorder. Indeed, two separate studies have shown that the addition of a non-benzodiazepine hypnotic, eszopiclone, enhanced the number and density of spindles[50] and working memory[51] in patients with schizophrenia.

Taken together, studies of sleep and cognition suggest that sleep abnormalities may at least in part contribute to the cognitive impairments in schizophrenia.

RELATIONSHIP BETWEEN SLEEP FINDINGS AND NEUROBIOLOGY IN SCHIZOPHRENIA

Studies of the ontogeny of sleep during normal adolescence are of significance to our understanding of the pathophysiology of SWS deficits in the context of a neurodevelopmental framework for schizophrenia. Adolescence is characterized by a substantial reorganization of human brain function; a marked decline in synaptic density in the prefrontal cortex, pronounced reductions in cortical gray matter volume, and regional cerebral metabolism are seen during adolescence. In parallel, polysomnographic studies show robust SWS decreases across the age span from childhood to late adolescence.[52] The time courses for maturational changes in SWS, cortical metabolic rate, and synaptic density are strikingly similar, at least in humans. It has therefore been suggested that the maturational processes in sleep EEG, cortical synaptic density, and regional cerebral metabolism might reflect a common underlying biological change (i.e., a large-scale programmed synaptic elimination).[8]

Do the polysomnographic abnormalities in schizophrenia relate to the brain maturational changes discussed above? In addition to SWS deficits, consistent alterations in the structure and function of cortical and subcortical brain regions have been observed in schizophrenia. Studies of the correlations between such alterations and sleep can help us better understand the pathophysiologic substrate underlying schizophrenia.

Altered brain structure

Schizophrenia is associated with widespread reductions in cortical gray matter, notably in the frontal and temporal cortex, as well as in the thalamus.[53] The relationship between alterations in these brain structures and SWS is interesting since SWS is generated by a complex neural system involving the anterior brain regions and the thalamus. SWS is inversely correlated with the anterior horn ratio, a measure of frontal lobe size,[54] and positively correlated with lateral ventricular size.[55] These correlations may result from reductions in subcortical structures such as the thalamus, which forms a substantial part of the ventricular boundaries. A recent study collected both structural and sleep spindle-related cortical currents in schizophrenia and demonstrated reduced mediodorsal thalamic volumes that strongly correlated with the number of scalp-recorded anterior frontal spindles.[35]

Altered brain metabolism

SWS may result from several of the neurochemical processes involved in neural inhibition, excitation, and EEG synchrony. Activation of the cholinergic system facilitates arousal and enhances REM sleep. Therefore, cholinergic hyperfunction—postulated to underlie schizophrenia—could account for SWS and REM latency reductions in schizophrenic sleep.[56] Interestingly, schizophrenia is associated with supersensitive REM sleep induction with the cholinergic agonist R5 86,[57] suggesting cholinergic hyperfunction. Nicotinic receptors may also be involved; sensory gating deficits as evidenced by P50 event-related potentials, which are possibly related to central nicotinergic system alterations, are reversed following sleep in schizophrenia.[58] Disturbances in catecholaminergic mechanisms may also underlie the SWS deficits in schizophrenia. Serotonergic abnormalities may also be involved; an inverse correlation is seen between serotonin metabolites in the cerebrospinal fluid and SWS in schizophrenia.[59] Norepinephrinergic and serotoninergic neurotransmission, which are presumed to be abnormal in schizophrenia, are inhibitory to REM; therefore, it is plausible that cholinergic and monoaminergic abnormalities could mediate the constellation of reduced REM latency and SWS deficits without increases in REM sleep amounts in schizophrenia.[56]

Hormonal substances may also be related to delta sleep alterations. Adenosine, an amino acid neuromodulator, has drawn increasing interest in recent years as a possible endogenous sleep-promoting agent, as it tends to accumulate during waking hours.[60] Adenosine agonists such as dipyridamole, which increase delta sleep, have been suggested as having possible therapeutic benefits in schizophrenia.[61]

Altered physiology

There is evidence for decreased frontal lobe metabolism ("hypofrontality") in schizophrenia, as assessed by a variety of physiological imaging techniques. It may be instructive to examine SWS deficits in the context of such physiologic alterations. An association has been demonstrated between SWS deficits and reduced frontal lobe membrane phospholipid metabolism in schizophrenia, as examined by ^{31}P magnetic resonance spectroscopy.[45] It has been suggested that membrane phospholipid alterations are related to loss of synaptic neuropil (i.e. decreased synaptic density, postulated to underlie schizophrenia). Conceivably, this could result in reduced SWS by decreasing the membrane surface (fewer dendrites per neuron), causing a smaller-voltage response to the synchronizing stimulus, thereby leading to decreased SWS.

Single-cell recordings in cats have shown that slower (<1 Hz) synchronized oscillations originate mainly in the neocortex,[62] whereas delta waves (1–4 Hz) arise primarily from the activity of thalamocortical neurons. A finer analysis of these oscillations may clarify the nature of the pathophysiology in schizophrenia. Preliminary analysis of this question using period amplitude analyses suggested more prominent deficits in the <1 Hz range in schizophrenia, pointing to a thalamocortical dysfunction.[28] This finding deserves further study and application.

EFFECTS OF ANTIPSYCHOTIC DRUGS ON SLEEP

Studies of the acute effects of neuroleptics have consistently shown improvements in sleep continuity, as measured by reduced sleep latencies, improved sleep time, greater sleep efficiency, and prolongation of REM latency[10]; however, changes in SWS have been less consistent. Studies that have examined the sedative effect of conventional neuroleptics have reported either no effects or modest increases in SWS.

Recent studies have begun to examine the effect of atypical antipsychotics on sleep. There is evidence for increases in SWS but decreases in sleep spindles with olanzapine following acute administration.[63,64] Furthermore, olanzapine-induced increases in delta sleep may predict better treatment response.[65] On the other hand, clozapine increases stage 2 sleep, but may actually decrease stage 4 sleep.[66] In normal subjects, quetiapine increases total sleep time, sleep efficiency, percentage sleep stage 2, and subjective sleep quality.[67] These studies have frequently used small sample sizes; few studies have examined sleep variables in relation to acute versus long-term treatment with neuroleptics in a longitudinal design.

Investigations into the polysomnographic characteristics of schizophrenia have to account for the potential effects of neuroleptic discontinuation on sleep EEG. Neylan and colleagues reported significant worsening of REM and NREM sleep in a series of schizophrenia patients undergoing controlled neuroleptic discontinuation.[68] Patients experiencing relapse have larger impairments in sleep. The effects of neuroleptic discontinuation continued to worsen from 2 to 4 weeks of a neuroleptic-free condition, and did not correlate with clinical change.[69] These findings suggest that it is important to control for medication state in investigations of EEG sleep in schizophrenia.

CONCLUSIONS AND FUTURE DIRECTIONS

In summary, sleep disturbances are pervasive and cause substantial subjective distress as well as disability in schizophrenia. The emerging literature pointing to the relationship between sleep alterations and neurobiological changes in this illness suggests that sleep abnormalities may represent a window into the pathophysiology of schizophrenia. New knowledge regarding the brain mechanisms of sleep is likely to open new avenues for exploring such relationships. First, functional brain imaging studies suggest distinct patterns of regional brain activation in SWS and REM sleep; such studies could provide clues to the pathophysiology of schizophrenia, especially when used in conjunction with physiologic perturbation paradigms such as sleep deprivation.[70] Second, brain imaging studies that also examine different domains of cognition are providing important insights into the cognitive dysfunction in schizophrenia. Third, sleep architecture changes dramatically during development; sleep studies during development in health and disease could shed considerable light on developmentally mediated neuropsychiatric disorders.[71] Finally,

sleep changes are often the earliest signs of disturbance, and may even represent trait-related vulnerability markers for psychiatric disorders; sleep studies of individuals who are at risk of schizophrenia are likely to be fruitful.[5,72]

REFERENCES

1. Ritsner M et al. Perceived quality of life in schizophrenia: Relationships to sleep quality. *Qual Life Res* 2004; 13(4): 783–791.
2. Hofstetter JR, Lysaker PH, Mayeda AR. Quality of sleep in patients with schizophrenia is associated with quality of life and coping. *BMC Psychiatry* 2005; 5: 13.
3. Chemerinski E et al. Insomnia as a predictor for symptom worsening following antipsychotic withdrawal in schizophrenia. *Compr Psychiatry* 2002; 43(5): 393–396.
4. Wulff K et al. Sleep and circadian rhythm disruption in schizophrenia. *Br J Psychiatry* 2012; 200(4): 308–316.
5. van Tricht MJ et al. Can quantitative EEG measures predict clinical outcome in subjects at clinical high risk for psychosis? A prospective multicenter study. *Schizophr Res* 2014; 153(1–3): 42–47.
6. Horne J. Human slow-wave sleep and the cerebral cortex. *J Sleep Res* 1992; 1(2): 122–124.
7. Horne JA. Human sleep, sleep loss and behaviour. Implications for the prefrontal cortex and psychiatric disorder. *Br J Psychiatry* 1993; 162: 413–419.
8. Feinberg I. Schizophrenia: Caused by a fault in programmed synaptic elimination during adolescence? *J Psychiatr Res* 1982; 17(4): 319–334.
9. de Vivo L et al. Developmental patterns of sleep slow wave activity and synaptic density in adolescent mice. *Sleep* 2014; 37(4): 689–700, 700A–700B.
10. Zarcone VP Jr, Benson KL. Sleep and schizophrenia. In: Kryger MH, Roth T, Dement WC (Eds). *Principles and Practice of Sleep Medicine*. Philadelphia, PA: WB Saunders, 1994, pp. 105–214.
11. Ramm P, Smith CT. Rates of cerebral protein synthesis are linked to slow wave sleep in the rat. *Physiol Behav* 1990; 48(5): 749–753.

12. Nakanishi H et al. Positive correlations between cerebral protein synthesis rates and deep sleep in *Macaca mulatta*. *Eur J Neurosci* 1997; 9(2): 271–279.

13. Maquet P et al. Experience-dependent changes in cerebral activation during human REM sleep. *Nat Neurosci* 2000; 3(8): 831–836.

14. Molle M, Born J. Slow oscillations orchestrating fast oscillations and memory consolidation. *Prog Brain Res* 2011; 193: 93–110.

15. Fogel SM, Smith CT. The function of the sleep spindle: A physiological index of intelligence and a mechanism for sleep-dependent memory consolidation. *Neurosci Biobehav Rev* 2011; 35(5): 1154–1165.

16. Guillery RW, Harting JK. Structure and connections of the thalamic reticular nucleus: Advancing views over half a century. *J Comp Neurol* 2003; 463(4): 360–371.

17. Houser CR et al. GABA neurons are the major cell type of the nucleus reticularis thalami. *Brain Res* 1980; 200(2): 341–354.

18. Jacobsen RB, Ulrich D, Huguenard JR. GABA$_B$ and NMDA receptors contribute to spindle-like oscillations in rat thalamus *in vitro*. *J Neurophysiol* 2001; 86(3): 1365–1375.

19. Benson KL, Zarcone VP Jr. Testing the REM sleep phasic event intrusion hypothesis of schizophrenia. *Psychiatry Res* 1985; 15(3): 163–173.

20. Benca RM et al. Sleep and psychiatric disorders. A meta-analysis. *Arch Gen Psychiatry* 1992; 49(8): 651–668; discussion 669–670.

21. Ganguli R, Reynolds CF 3rd, Kupfer DJ. Electroencephalographic sleep in young, never-medicated schizophrenics. A comparison with delusional and nondelusional depressives and with healthy controls. *Arch Gen Psychiatry* 1987; 44(1): 36–44.

22. Tandon R et al. Electroencephalographic sleep abnormalities in schizophrenia. Relationship to positive/negative symptoms and prior neuroleptic treatment. *Arch Gen Psychiatry* 1992; 49(3): 185–194.

23. Dresler M et al. Neural correlates of insight in dreaming and psychosis. *Sleep Med Rev* 2015; 20C: 92–99.

24. Dresler M et al. Neural correlates of dream lucidity obtained from contrasting lucid versus non-lucid REM sleep: A combined EEG/fMRI case study. *Sleep* 2012; 35(7): 1017–1020.

25. Brent BK et al. Self-disturbances as a possible premorbid indicator of schizophrenia risk: A neurodevelopmental perspective. *Schizophr Res* 2014; 152(1): 73–80.

26. Keshavan MS, Anderson S, Pettegrew JW. Is schizophrenia due to excessive synaptic pruning in the prefrontal cortex? The Feinberg hypothesis revisited. *J Psychiatr Res* 1994; 28(3): 239–265.

27. Werth E, Achermann P, Borbely AA. Fronto-occipital EEG power gradients in human sleep. *J Sleep Res* 1997; 6(2): 102–112.

28. Keshavan MS et al. Delta sleep deficits in schizophrenia: Evidence from automated analyses of sleep data. *Arch Gen Psychiatry* 1998; 55(5): 443–448.

29. Tekell JL et al. High frequency EEG activity during sleep: Characteristics in schizophrenia and depression. *Clin EEG Neurosci* 2005; 36(1): 25–35.

30. Benson KL et al. The effect of total sleep deprivation on slow wave recovery in schizophrenia. *Sleep Res* 1993; 22: 143.

31. Horne J. Neuroscience. Images of lost sleep. *Nature* 2000; 403(6770): 605–606.

32. Ferrarelli F et al. Thalamic dysfunction in schizophrenia suggested by whole-night deficits in slow and fast spindles. *Am J Psychiatry* 2010; 167(11): 1339–1348.

33. Manoach DS et al. Sleep spindle deficits in antipsychotic-naive early course schizophrenia and in non-psychotic first-degree relatives. *Front Hum Neurosci* 2014; 8: 762.

34. Lustenberger C et al. Sleep spindles are related to schizotypal personality traits and thalamic glutamine/glutamate in healthy subjects. *Schizophr Bull* 2015; 41(2): 522–531.

35. Buchmann A et al. Reduced mediodorsal thalamic volume and prefrontal cortical spindle activity in schizophrenia. *Neuroimage* 2014; 102(Pt 2): 540–547.

36. Ferrarelli F, Tononi G. The thalamic reticular nucleus and schizophrenia. *Schizophr Bull* 2011; 37(2): 306–315.

37. Oh JS et al. Thalamo-frontal white matter alterations in chronic schizophrenia: A quantitative diffusion tractography study. *Hum Brain Mapp* 2009; 30(11): 3812–3825.

38. Smith RE et al. Expression of excitatory amino acid transporter transcripts in the thalamus of subjects with schizophrenia. *Am J Psychiatry* 2001; 158(9): 1393–1399.

39. Martin J et al. Actigraphic estimates of circadian rhythms and sleep/wake in older schizophrenia patients. *Schizophr Res* 2001; 47(1): 77–86.

40. Bromundt V et al. Sleep–wake cycles and cognitive functioning in schizophrenia. *Br J Psychiatry* 2011; 198(4): 269–276.

41. Keshavan MS et al. Sleep and suicidality in psychotic patients. *Acta Psychiatr Scand* 1994; 89(2): 122–125.

42. Lewis CF et al. Biological predictors of suicidality in schizophrenia. *Acta Psychiatr Scand* 1996; 94(6): 416–420.

43. Maixner S et al. Effects of antipsychotic treatment on polysomnographic measures in schizophrenia: A replication and extension. *Am J Psychiatry* 1998: 155(11): 1600–1602.

44. Keshavan MS et al. A longitudinal study of EEG sleep in schizophrenia. *Psychiatry Res* 1996; 59(3): 203–11.

45. Keshavan MS et al. Biological correlates of slow wave sleep deficits in functional psychoses: ^{31}P-magnetic resonance spectroscopy. *Psychiatry Res* 1995; 57(2): 91–100.

46. Orzack MH, Hartmann EL, Kornetsky C. The relationship between attention and slow-wave sleep in chronic schizophrenia [proceedings]. *Psychopharmacol Bull* 1977; 13(2): 59–61.

47. Goder R et al. Impairment of visuospatial memory is associated with decreased slow wave sleep in schizophrenia. *J Psychiatr Res* 2004; 38(6): 591–599.

48. Keshavan MS et al. Sleep correlates of cognition in early course psychotic disorders. *Schizophr Res* 2011; 131(1–3): 231–234.

49. Wamsley EJ et al. Reduced sleep spindles and spindle coherence in schizophrenia: Mechanisms of impaired memory consolidation? *Biol Psychiatry* 2012; 71(2): 154–161.

50. Wamsley EJ et al. The effects of eszopiclone on sleep spindles and memory consolidation in schizophrenia: A randomized placebo-controlled trial. *Sleep* 2013; 36(9): 1369–1376.

51. Tek C et al. The impact of eszopiclone on sleep and cognition in patients with schizophrenia and insomnia: A double-blind, randomized, placebo-controlled trial. *Schizophr Res* 2014; 160(1–3): 180–185.

52. Smith JR, Karacan I, Yang M. Ontogeny of delta activity during human sleep. *Electroencephalogr Clin Neurophysiol* 1977; 43(2): 229–237.

53. Andreasen NC et al. Thalamic abnormalities in schizophrenia visualized through magnetic resonance image averaging. *Science* 1994; 266(5183): 294–298.

54. Keshavan MS et al. Electroencephalographic sleep and cerebral morphology in functional psychoses: A preliminary study with computed tomography. *Psychiatry Res* 1991; 39(3): 293–301.

55. van Kammen DP et al. Decreased slow-wave sleep and enlarged lateral ventricles in schizophrenia. *Neuropsychopharmacology* 1988; 1(4): 265–271.

56. Keshavan MS, Tandon R. Sleep abnormalities in schizophrenia: Pathophysiological significance. *Psychol Med* 1993; 23(4): 831–835.

57. Riemann D et al. Cholinergic REM induction test: Muscarinic supersensitivity underlies polysomnographic findings in both depression and schizophrenia. *J Psychiatr Res* 1994; 28(3): 195–210.

58. Griffith JM et al. Nicotinic receptor desensitization and sensory gating deficits in schizophrenia. *Biol Psychiatry* 1998; 44(2): 98–106.

59. Benson KL, Faull KF, Zarcone VP Jr. Evidence for the role of serotonin in the regulation of slow wave sleep in schizophrenia. *Sleep* 1991; 14(2): 133–139.

60. Porkka-Heiskanen T. et al. Adenosine: A mediator of the sleep-inducing effects of prolonged wakefulness. *Science* 1997; 276(5316): 1265–1268.

61. Ferre S. Adenosine–dopamine interactions in the ventral striatum. Implications for the treatment of schizophrenia. *Psychopharmacology (Berl)* 1997; 133(2): 107–120.

62. Steriade M. Brain electrical activity and sensory processing during wake and sleep states. In: Kryger MH, Roth T, Dement WC (Eds). *Principles and Practice of Sleep Medicine*. Philadelphia, PA: WB Saunders 1994: 105–124.

63. Goder R et al. Effects of olanzapine on slow wave sleep, sleep spindles and sleep-related memory consolidation in schizophrenia. *Pharmacopsychiatry* 2008; 41(3): 92–99.

64. Salin-Pascual RJ et al. Olanzapine acute administration in schizophrenic patients increases delta sleep and sleep efficiency. *Biol Psychiatry* 1999; 46(1): 141–143.

65. Salin-Pascual RJ et al. Low delta sleep predicted a good clinical response to olanzapine administration in schizophrenic patients. *Rev Invest Clin* 2004; 56(3): 345–350.

66. Hinze-Selch D. et al. Effects of clozapine on sleep: A longitudinal study. *Biol Psychiatry* 1997; 42(4): 260–266.

67. Cohrs S et al. Sleep-promoting properties of quetiapine in healthy subjects. *Psychopharmacology (Berl)* 2004; 174(3): 421–429.

68. Neylan TC et al. Sleep in schizophrenic patients on and off haloperidol therapy. Clinically stable vs relapsed patients. *Arch Gen Psychiatry* 1992; 49(8): 643–649.

69. Nofzinger EA et al. Electroencephalographic sleep in clinically stable schizophrenic patients: Two-weeks versus six-weeks neuroleptic-free. *Biol Psychiatry* 1993; 33(11–12): 829–835.

70. Drummond SP et al. Altered brain response to verbal learning following sleep deprivation. *Nature* 2000; 403(6770): 655–657.

71. Dahl RE. The development and disorders of sleep. *Adv Pediatr* 1998; 45: 73–90.

72. Lauer CJ et al. In quest of identifying vulnerability markers for psychiatric disorders by all-night polysomnography. *Arch Gen Psychiatry* 1995; 52(2): 145–153.

Sleep dysfunction after traumatic brain injury

TATYANA MOLLAYEVA AND COLIN MICHAEL SHAPIRO

INTRODUCTION

Traumatic brain injury (TBI), defined as "an alteration in brain function, or other evidence of brain pathology, caused by an external force,"[1] is among the most serious and disabling neurological disorders affecting adults and children in all societies.[2-4] Recently endorsed is the view of TBI as a chronic disease process encompassing clinical, pathological, and cellular changes starting at the time of the head injury event.[5,6] Consequently, the classification systems of TBI can be clinical, pathological, and mechanistic.[7] The most commonly used clinical systems include the Glasgow Coma Scale (GCS)[8] and the American Academy of Neurology grading system[9] based on the duration of altered mental state. Pathological classifications are related to the anatomical (i.e., localization of injury—focal or diffuse)[10] and pathophysiological (i.e., primary impact injury or secondary pathological processes).[11] Mechanistic classifications are related to impact, force-loading, penetrating, and blast injuries.[12] While no single classification encompasses all of the features of the complex process of TBI,[7] a comprehensive model of TBI is becoming widely endorsed. Such models take into account accelerating/decelerating forces that cause mechanical strain; the direction of force that can determine the severity of injury[13]; rapid head rotation generating shear forces throughout the brain, causing shear-induced tissue damage, with lateral plane acceleration having the greatest likelihood for producing damage within the deep internal structures of the brain[14]; and the complex neuro-metabolic and neurochemical cascades occurring in neurons following stretch and compression.[15]

SLEEP AFTER TBI

There is considerable interest in the role of various sleep markers for use as diagnostic and prognostic measures in TBI. In 1949, researchers first introduced evidence of high and low arousal of the damaged brain, accompanied with changes in the sleep electroencephalogram (EEG), autonomic system, and motor activity.[16] Since then, the study of sleep abnormalities in persons with TBI has grown, resulting in the conceptualization of the pathophysiological processes occurring in sleep and their relationships with presenting daytime symptoms. In recent years, study has progressed towards more diverse clinical and nonclinical fields. As depicted in Figure 16.1, the number of papers published with Medical Subject Headings (MeSH) terms "sleep" and "TBI" saw an increase at the start of the new century. This trend parallels the interest in TBI by itself, and suggests that sleep is an important area of research in relation to TBI.

In reviewing the pathophysiology of posttraumatic sleep dysfunction, no single mechanism has been found to be responsible for disordered sleep and wakefulness after injury to the brain of any severity. Consequently, the goal of this chapter is to emphasize the importance of a multifaceted approach to dealing with sleep dysfunction in this population. In due course, sleep dysfunction in persons with TBI must not be viewed as a singular diagnosis, but as a manifestation of a complex interaction of sociodemographic, medical, psychiatric, behavioral, and environmental factors.

NEUROANATOMY OF SLEEP AND WAKEFULNESS

The reticular formation (RF), a diffuse network of cells deep in the tegmentum of the brainstem, has a central role in the complex neural regulation of the consciously aroused state.[17] This network is situated throughout the brainstem and spinal cord, and most of its neurons are interneurons with multiple efferent projections—ascending and descending—processing information from the ipsilateral and contralateral sides.[18] All networks

of the RF[19] receive input from other brain areas and can, in turn, influence the functioning of these other brain areas. The projections from the RF that ascend to the thalamus and cortex and play a role in the modulation of consciousness make up the reticular activating system (RAS).[19] It follows that the principal cause of coma, a deep sleep-like state from which the patient cannot be aroused, is bilateral damage to the RAS, or damage to both cerebral hemispheres that disrupts the RAS.[20] Unilateral damage may not be noticeable, or may involve only some alterations to the state of consciousness.[20]

Regulation of sleep is mediated by opposing actions of the anterior and posterior/lateral regions of the hypothalamus. The suprachiasmatic nucleus (SCN) serves as a central nervous system (CNS) pacemaker,[21] driving both physiological and behavioral circadian rhythms, including sleep-wake cycle, hormonal secretion, and thermoregulation. The CNS neurons have an approximately 24-hour rhythm of electrical activity—an intrinsic rhythm—even in the absence of environmental cues.[22] Environmental signals override this intrinsic rhythm through light input from the retina (reticulohypothalamic tract) and the secretion of melatonin, largely by the pineal gland, in a circadian pattern.[23] Secretion of melatonin is controlled by the CNS through projections to the visceral nervous systems and, in turn, a sympathetic projection from the superior cervical ganglion to the pineal gland. Thus, the circadian activities of endocrine secretion, visceral function, feeding, temperature, and behavior are interrelated.[24] Damage to the hypothalamus is expected to result in disturbances of neuroendocrine, autonomic, homeostatic, sleep–wake, and emotional functions. On the other hand, increased intracranial pressure as a result of head trauma can exert pressure on hypothalamic nuclei and alter sleep–wake function in various ways.[22]

EPIDEMIOLOGY

It is very difficult to determine the prevalence of sleep–wake dysfunction in TBI. This is not only because TBI is a complex disorder with multiple interdependent risk factors (i.e., age, sex, recurrent injury, comorbidities, prescription/illicit drugs, etc.) that can affect sleep after the injury, but also because TBI encompasses primary (i.e., focal, diffuse, or both) and secondary injuries (i.e., complex

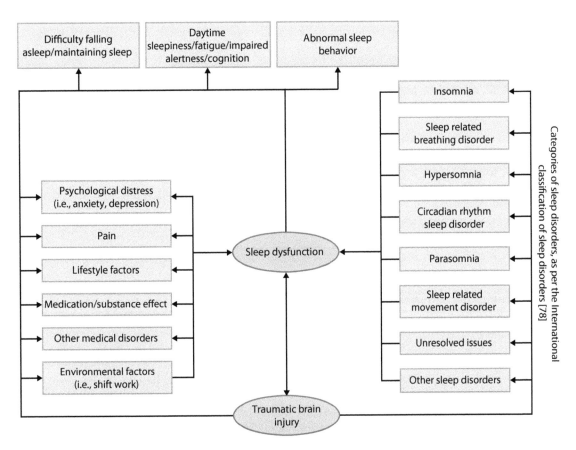

Figure 16.1 Construct of sleep dysfunction in traumatic brain injury. Unidirectional arrows from constructs (i.e., circles) to items (i.e., rectangles) represent reflective models, and from items to constructs, they represent formative models. Bidirectional arrows represent a combination of reflective and formative elements. (Based on Fayers PM, Hand DJ. *Qual Life Res* 1997; 6: 139–150.)

biochemical cascades, including disruption in cerebral blood flow, increases in intracranial pressure [brain swelling], cellular necrosis, and apoptosis), each of which impact sleep function differently.

In the acute phases post-injury, especially in moderate to severe TBI, medical care addresses the ABCs (airway, breathing, and circulation) of the injured person, as it is well established that hypoxemia and hypotension are associated with worse clinical outcomes, and a single episode of hypotension doubles mortality risk.[23] While research has provided some evidence on sleep pattern abnormalities in patients with TBI utilizing EEG, in the first 24 hours post-injury, the amplitude, frequency, and shape of wave potentials are not stable and of lesser prognostic significance than those taken between 24 and 48 hours post-injury.[24,25] Findings range from increased slow wave in the delta frequency band (<4 Hz) to amplitude suppression,

rises in slow focal or diffuse theta activity (4–8 Hz), and immediate decreases in the mean frequency of alpha waves (8–13 Hz).[24] Typical features of sleep (i.e., rapid eye movement [REM] and non-REM sleep) are reported to be more common among patients who show good recovery.[25–28] Although sleep EEG has value when assessing injury severity and prognosis, it does not provide great spatial resolution and is impractical for long-term monitoring.[29] In patients with mild TBI, the utility of the EEG is even more limited, as not all patients sustaining these injuries will seek medical care. It is important to note that studies that utilized conventional EEG in mild TBI did not report any early EEG abnormalities, even in cases where structural magnetic resonance imaging abnormalities were present.[30] However, given the recent reports that an index of brain electrical activity derived from 10 minutes of monitoring was found to be a highly

Table 16.1 Prevalence of the most common sleep disorders, by International Classification of Sleep Disorders category

Sleep disorder	Traumatic brain injury population (%)	General population (%)
Insomnia	30[82]	<10[113]
Sleep-related breathing disorders	25[73]–35[62]	6[114]
Hypersomnia, not due to sleep-related breathing disorders: narcolepsy	3[112]	0.05[115]
Circadian rhythm sleep disorders	36[35]	6.6[116]
Sleep-related movement disorders: periodic limb movement disorder	17[80]	4[117]–11[118]
Sleep-related movement disorders: rapid eye movement behavior disorder	13[58]	0.5[119]

sensitive measure for the detection of potentially life-threatening traumatic intracranial hematomas following closed head injury,[31] the study of brain activity in sleep early after mild TBI may prove useful.

Investigation of sleep function usually starts in the post-acute phase post-injury, when patients report or seek a physician's help because of one or more of the following: (1) inability to sleep at night; (2) daytime symptoms of fatigue, tiredness, or excessive daytime sleepiness (EDS); (3) behavioral manifestations in sleep; or a combination of these complaints. There have been several studies aiming to determine the prevalence of sleep dysfunction in chronic TBI (Table 16.1).[32–39] Because of the wide variation in definitions and methodologies utilized to study it, prevalence values ranged significantly. In a meta-analysis of 21 studies featuring a total of 1706 participants with TBI, Mathias and Alvaro reported that, overall, 50% of people suffered from some form of sleep disturbance after a TBI, and 25%–29% had a diagnosed sleep disorder.[40] These numbers are much higher than those observed in the general population.[41]

MOST COMMON SLEEP DISORDERS IN TBI

Insomnia

Insomnia is characterized as difficulty falling asleep, maintaining sleep, awakening in early morning, or non-restorative sleep. It is one of the most endorsed symptoms in persons with TBI. A study performed by Cantor and colleagues at five National Institute of Disability and Rehabilitation Research TBI Model Systems involving 334 individuals with TBI who completed 1-year (n = 213) or 2-year (n = 121) follow-up interviews between 2008 and 2012 reported insomnia incidence of 11%–24%.[42] A meta-analysis of Mathias and Alvaro reported insomnia to be one of the most commonly occurring disorders, being self-reported by 941 TBI participants of varying injury severities (50%).[40] When the standardized criteria of the *Diagnostic and Statistical Manual of Mental Disorders, Fourth Edition, Text Revision* (DSM-IV-TR)[43] was applied, Fichtenberg and colleagues reported insomnia syndrome as occurring in 30% of persons with mild to moderate TBI[44]; similar findings were reported by Ouellet and colleagues (29.4%).[45] In the general population, based on a review of over 50 epidemiologic studies published worldwide, a prevalence rate of approximately 6% was obtained for insomnia diagnosis based on DSM-IV-TR criteria.[46]

The higher prevalence of insomnia in the TBI population has been related to the injury itself in a recent study of 204 patients with TBI (mean age 33 years), in which 40.2% of participants were found to have insomnia, as measured by the Insomnia Severity Index, and the only variables that were associated with insomnia were severity and duration of TBI.[47] TBI patients with moderate injury severity (70.7%) had a significantly higher occurrence of insomnia than those with mild injury severity (19.7%; p < 0.0001). Over half (63.4%) of the TBI patients reporting insomnia did so within the first 3 months after injury. Neuroanatomical localization was also correlated with insomnia.

Cerebral contusion was the most common (40.2%) site of impact. Almost half (42.4%) of the patients with insomnia had multiple contusions. Similar findings were reported in a military TBI clinic located in Iraq, where 150 male military patients completed standardized self-report measures and clinical interviews.[48] Patients were categorized into three groups according to history of TBI: zero TBIs (n = 18), single TBI (n = 54), and multiple TBIs (n = 78). Rates of clinical insomnia, as measured by ISI, were significantly increased across these TBI groups (p < 0.001): 5.6% for no TBIs, 20.4% for single TBI, and 50.0% for multiple TBIs, a pattern that remained even after controlling for depression, posttraumatic stress disorder, and concussion symptoms.[48] Based on these results, it is reasonable to assume that there is a risk for the development of insomnia as a result of TBI. The current data are limited, however, making it difficult to determine the exact level of risk posed, the type of injury to the brain that is responsible, and whether insomnia in TBI is an independent problem or one that is related to multiple medical comorbidities and/or psychopathologies.

Excessive daytime sleepiness

Daytime impairment due to dysfunction of nocturnal sleep should be distinguished from hypersomnia due to a neurologic disorder as a result of brain injury. In particular, this may be important when selecting a treatment intervention. For example, in a TBI patient, daytime sleepiness can occur due to a primary sleep disorder (i.e., sleep-related breathing disorder, periodic limb movement disorder [PLMD], etc.) as a result of nocturnal sleep fragmentation,[49] have the etiology of a circadian disorder (i.e., EDS during the afternoon hours due to advance sleep phase syndrome),[50] neurologic hypersomnia (i.e., with long sleep time but poor sleep efficiency),[51] be a manifestation of narcolepsy as a result of cerebrospinal fluid hypocretin-1 level deficiency,[52] psychiatric disorder,[53] medication effects,[54] and/or a combination of these. Moreover, while the definition of sleepiness implies the risk of falling asleep, the description of this impairment is sometimes used in reference to physical tiredness, reduced mental alertness, or fatigue.[55] Finally, prolonged symptom duration, as well as brain injury, may affect judgment and ability to report any difficulty with

insight.[56] Therefore, a comprehensive differential diagnosis is of high importance.

Studies utilizing self-report measures reported EDS in TBI in a frequency range of between 14% and 55%.[57–59] In a sample of 184 patients with post-traumatic hypersomnia, Guilleminault and colleagues reported that a majority of participants were involved in litigation, and in addition to daytime sleepiness, had memory difficulty, poor concentration, depressive symptoms, and were unable to work.[57] Only 28% of the patients with EDS by self-report were found to be so following the multiple sleep latency test (MSLT).[57] Verma and colleagues reported that over 50% of persons with EDS by self-report in his sample had mean onset sleep latencies of <5 minutes on the MSLT.[58] A recent case–control study of TBI patients with increased sleep need following their injury assessed sleep through sleep logs, actigraphy, polysomnography (PSG), and the MSLT. Actigraphy recordings revealed that TBI patients had substantially longer estimated sleep durations than controls (10.8 hours per 24 hours, compared to 7.3 hours for controls). When using sleep logs, TBI patients underestimated their sleep need.[59] During nocturnal sleep, patients had higher amounts of slow-wave sleep than controls (20% versus 13.8%). The MSLTs revealed EDS in 15 patients (42%), and ten of them had indicators of chronic sleep deprivation.[59] It should be noted that the MSLT requires a criterion of average time to fall asleep in four or five tests. However, the same average can be derived in a variety of different ways (e.g., consistent times to fall asleep over all sessions, or very long times in some and abrupt in others), and therefore will not have the same implications.

Severe head trauma can affect the hypothalamic system to such an extent as to alter levels of the neurotransmitter hypocretin, transiently or permanently. Prevalence of narcolepsy, a rare disorder, is reported to be 60-times higher in persons with TBI than in the general population.[41] Baumann and colleagues reported that extensive loss of the hypothalamic neurons that produce the wake-promoting neuropeptide hypocretin (orexin) causes the severe sleepiness that is characteristic of narcolepsy, and partial loss of these cells may contribute to the sleepiness in patients with severe TBI.[60] These findings highlight the often-overlooked hypothalamic injury in TBI and provide new insights into the causes of chronic sleepiness in patients with TBI (i.e., deficits in orexin neurotransmission in the lateral hypothalamus). Nardone and colleagues applied transcranial magnetic stimulation to study posttraumatic sleep–wake disturbances in persons with mild to moderate TBI.[61] The researchers reported changes in excitability of the cerebral cortex. Resting motor threshold was higher in the patients with EDS, similar to that reported in patients with narcolepsy. The researchers proposed that this cortical hypoexcitability might reflect a deficit in the excitatory hypocretin/orexin neurotransmitter system.[61]

Behavioral manifestations in sleep

Sleep-related movement disorders are commonly reported in persons with TBI. Castriotta and colleagues reported PLMD to affect 17% of the TBI patients in their study.[62] This disorder is nonspecific to TBI and occurs in a wide range of medical and sleep disorders, including narcolepsy, sleep apnea, REM behavior disorder (RBD), and various forms of insomnia.[63,64] The prevalence of PLMD in population-based studies ranges between 4% and 11% of adults, with incidence reported to increase with age.[41] The etiology of this disorder is poorly understood; however, animal models with high spinal transections and continued PLMD support the possibility of there being a spinal origin of the disorder.[65]

RBD is characterized by dramatic REM motor activation resulting in dream enactment, often with violent or injurious results. A population-based survey indicated an overall 2% prevalence of violent behavior during sleep, a quarter of which was likely to be RBD, putting the prevalence of RBD at 0.5%.[65] Verma and colleagues examined the spectrum of sleep disorders in chronic TBI patients and reported complaints of parasomnia in 25% of participants, with RBD being the most frequently reported parasomnia (13%).[58] It has been proposed that the increased RBD incidence in TBI relative to that of the general population is attributed to damage to brainstem mechanisms mediating descending motor inhibition during REM sleep.[58] Animal studies with bilateral lesions of the pontine tegmentum in areas controlling REM sleep motor inhibition support this hypothesis.[66] It has also been reported that individuals taking

serotonergic antidepressant medications may be at an increased risk of developing RBD, particularly with increasing age.[67,68]

Sleep-related breathing disorders manifesting as disturbed nocturnal sleep with EDS

Respiratory dysfunction in sleep is common in persons with TBI, occurring at a significantly higher rate than in the general population (25%–35% vs. 4%–9%).[40,41] The pathogenesis of obstructive sleep apnea (OSA) is occlusion of the upper airways in sleep.[69] The main factor leading to airway collapse is the generation of pressure during inspiration that exceeds the ability of airway dilator and abductor muscles to maintain stability within the airway. Various factors can increase the risk of collapsing the upper airway after TBI, including, but not limited to: disturbed coordination of the upper-airway muscles due to damage to the brainstem (i.e., hypoglossal nucleus); sleep-related withdrawal of noradrenergic and serotonergic excitatory drive to upper-airway muscles; muscle relaxants commonly prescribed after the injury; and alcohol, an important cofactor because of its selective depressant influence on upper-airway muscles and on arousal response.[70–72] In many TBI patients, the patency of airways can also be compromised structurally, or be compromised due to obesity as a result of compressing of the pharynx by superficial fatty tissue.

Central sleep apneas in patients with TBI

Webster and colleagues reported on the nature of sleep apnea after TBI, with a majority of the apneic episodes (36%) in their sample of 29 TBI patients being central sleep apnea (CSA) events rather than OSA events (11%), concluding that the brain injury may be part of the underlying cause of this observation.[73] Although CNS-active medications could potentially play a role in this occurrence, the researchers found there to be no difference between the rate of CNS-active medication use among subjects with respiratory disturbance index (RDI) values over 5 and those with RDI values of less than 5. The etiology of CSA is still unknown; however, when the ventilatory control system is unstable, large fluctuations in ventilation can produce central events. Furthermore, because distinguishing between central and obstructive hypopneas cannot be reliably accomplished without noninvasive techniques (i.e., esophageal balloons or respiratory-induced wall motion strain), for a definite diagnosis, only apneic events should be documented for a CSA diagnosis in persons with TBI.

Circadian rhythm sleep disorders manifesting as inability to fall asleep or EDS

Circadian rhythm sleep disorders are important to consider when dealing with insomnia and/or EDS due to their common occurrence in patients with TBI.[74] Numerous research studies to date have reported a delayed sleep phase syndrome, which can potentially be confused with insomnia if the person with TBI goes to bed at a socially acceptable time. Ayalon and colleagues investigated complaints of insomnia following mild TBI and reported 36% of people to be diagnosed with circadian rhythm sleep disturbances (CRSDs).[75] Clinical presentation (i.e., insomnia, excessive sleepiness, or both) is expected to depend on whether an individual schedules sleep according to their internal circadian clock or in accordance with a socially accepted schedule. Llompart-Pou and colleagues applied cerebral microdialysis techniques in order to study brain interstitial cortisol levels at 08:00, 16:00, and 24:00 hours in the acute phase of TBI in ten patients (median GCS score after resuscitation was 5 [range 3–10]). Intra-individual analysis showed that circadian variability was lost in all patients, both in serum and brain interstitial cortisol samples.[76] The study of the circadian rhythm in TBI patients warrants further investigation.

Shift-work sleep disorders

The negative effects of shift work include sleepiness and insomnia, reduced alertness, and greater risk of re-injury, overall poor health, low work productivity, and poor quality of life.[77] Although the cause–effect relationship between pre-morbid shift work and insomnia after TBI remains unclear, study of the pre-morbid misalignment of the output of the endogenous circadian pacemaker in association with post-morbid insomnia is timely.

INVESTIGATING SLEEP DYSFUNCTION IN PATIENTS WITH TBI

Sleep in TBI represents something of a challenge to investigate because of the broad array of possible diagnoses for the manifested problems. One approach would be to use a carefully constructed interview or screening algorithm that incorporates the recommendations of the American Association of Sleep Medicine pertaining to clinical sleep dysfunction evaluation.[78] Such a theoretical framework can provide a context for examining a problem and serve as a guide for the systematic study of a precisely defined relationship between factors that can constitute or contribute to sleep dysfunction in TBI. Based on a comprehensive literature review,[41] we developed a construct of sleep dysfunction for patients with TBI (Figure 16.2).

Self-report questionnaires

A recent systematic review of the various sleep-related self-report measures used in TBI research and clinical practice to date resulted in the identification of 16 measures from more than 100 currently available in the field of sleep medicine.[79] These measures were comprehensively described.[79] Two of the measures—the Pittsburgh Sleep Quality Index (PSQI) and the Epworth Sleepiness Scale (ESS)—have been partially validated in TBI samples of varying injury severity. Fichtenberg and colleagues reported on the concurrent validity of the PSQI in a sub-acute TBI sample of 50 consecutive patients who were distinguished with respect to insomnia by the DSM-IV criteria.[44] The overall agreement of the PSQI with the DSM-IV diagnosis of insomnia was 94%, with a sensitivity of 100% and a specificity of 96%. The proposed global PSQI cutoff score of 8 was found to be appropriate for discriminating 96% of insomnia cases correctly, and a cutoff score of 9 accurately established the condition in 98% of cases. The concurrent validity of the ESS was reported by Masel and colleagues.[80] The study investigated the relationship of ESS scores to MSLT scores in 71 TBI patients with time since injury ranging from 3 to 27 years. The authors found no significant correlation between the ESS score and mean sleep latency by the MSLT.[80]

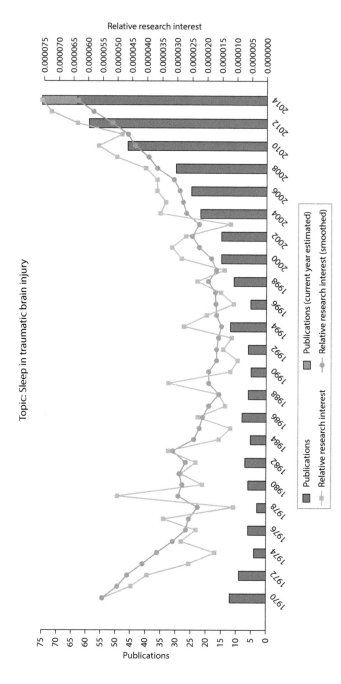

Figure 16.2 Publications including the MeSH term "sleep traumatic brain injury" from 1970 to the present. Retrieved from the GoPubMed website, http://gopubmed.org/web/gopubmed/.

Sleep diary

The sleep diary has been used in the TBI population for assessing sleep quality and quantity, and for identifying factors that may affect sleep and wakefulness. Results were generally applied towards a clinical diagnosis of insomnia and quantification of sleep and waking behavior over a specific period (usually weeks). The strength of the instrument lies in its ability to prospectively depict sleep and daytime functioning. However, the low return rates of diaries, response burden, possible impaired judgment, and common sleep state misperception in the TBI population requires careful interpretation of results.[79] While self-report is important and evaluation of sleep by self-report is cost-effective, researchers have described under-/over-reporting of sleep disturbances by patients and recommended the utilization of self-report measures in conjunction with objective methods for the assessment of sleep post-TBI.[79–82]

Polysomnography

There are numerous qualities that allow for an informed clinical decision with regards to sleep quality to be made based on PSG findings in patients with TBI. This technique provides objective evidence of sleep continuity, efficiency, architecture, and physiological measures (e.g., heart rate, breathing, oxygen saturation, and muscle movements), and allows for extensive study of EEG activity and arousability.[83] Continuous video/audio is particularly relevant for differentiation between sleep disorders and epileptic phenomena post-TBI, and there is the potential to capture infrequent events. Table 16.2 presents a list of sleep disorders that can be diagnosed or ruled out based

Table 16.2 Sleep disorders diagnosed/ruled out with polysomnography, and frequency of relevant publications based on the traumatic brain injury population (as of July 2014)

Sleep disorder	Value of polysomnography	# PubMed MeSH term hits
Circadian rhythm sleep disorder	• Diagnostic accuracy, differentiation from other disorders	"shift work sleep disorder" (MeSH): 5
Insomnia	• Observation of long sleep-onset latency, frequent nocturnal awakenings, increased light-stage sleep • Allows diagnosis and exclusion of other disorders given multiple possible causes of insomnia	"sleep initiation and maintenance disorders" (MeSH): 22
Narcolepsy	• Multiple sleep latency test for verification of daytime somnolence, observation of number of sleep-onset rapid eye movements	"narcolepsy" (MeSH): 14
Nocturnal seizures	• Observation of episodes by video recording • Diagnostic accuracy, differentiation from other disorders	"epilepsy" (MeSH): 13
Parasomnia	• Observation of confusional arousals, sleep walking/talking, enuresis, bruxism, etc. • Diagnostic accuracy, differentiation from other disorders	"sleepwalking" (MeSH): 3 "night terrors" (MeSH): 1
Sleep-related breathing disorder	• Respiratory parameters, arterial oxygen saturation, heart rate, partial pressure of carbon dioxide • Identification and classification of apneas and hypopneas	"sleep apnea syndromes" (MeSH): 37
Sleep-related movement disorder	• Observation of muscle activity, movement by electromyography and video • Effect of sleep movement disorder based on sleep architecture	"movement disorders" (MeSH): 11 "restless legs syndrome" (MeSH): 4

on PSG, as well as data on the use of the technique in research addressing specific sleep disorders in the TBI population. PSG is currently the gold standard for the diagnosis of sleep disorders such as sleep-related breathing disorders, narcolepsy, PLMD, and nocturnal seizures, among others.

MSLT and maintenance of wakefulness test

The MSLT is a validated measure that is used to quantify daytime sleepiness and has been applied to the TBI population.[84] It was reported to aid in the differential diagnosis of pathological sleep abnormalities from self-report sleepiness and post-TBI fatigue. To diagnose narcolepsy, a patient must fall asleep within 10 minutes and present with at least two sleep-onset REMs across four to five nap opportunities at 2-hour intervals. The maintenance of wakefulness test evaluates one's ability to remain alert, which is highly relevant to TBI, as test results can be linked to variations in daytime vigilance and/or daytime functioning.[85]

Actigraphy

Actigraphy records sleep–wake parameters based on the absence or presence of motor limb or head activity over a period (typically 2 weeks), and it has promising value for people whose self-reports are in question. For individuals with brain injury, it is reported to aid in the study of CRSDs. Zollman and colleagues utilized actigraphy in a TBI sample and raised caution regarding its application for patients with spasticity, paresis, agitation, and impulsivity.[86] Duclos and colleagues applied actigraphy in order to study rest/activity cycles and the association with injury severity and outcome in acute moderate to severe TBI. The researchers found that the patients with a more rapid return to a consolidated rest/activity cycle acutely after injury were more likely to have cleared posttraumatic amnesia and have lower disability at hospital discharge.[87] Sinclair and colleagues reported on the association between sleep diary and actigraphic assessments of sleep disturbance (i.e., wake after sleep onset [WASO] and sleep-onset latency), finding weaker agreement between the methods for WASO measures, supporting the view that actigraphy may prove to be useful as a supplement to self-report measures of sleep following TBI, being an overall more naturalistic and cost-effective approach compared to PSG.[88]

TREATMENT OF SLEEP DISORDERS IN PATIENTS WITH TBI

The effectiveness of pharmacological and non-pharmacological treatments of various sleep disorders has been demonstrated in TBI samples. The evidence for their efficacy and specific considerations for their use are discussed below.

Non-pharmacological interventions

Non-pharmacological treatments for insomnia in TBI include cognitive–behavioral therapy (CBT) and acupuncture. Eight weeks of CBT with stimulus control, sleep restriction, cognitive restructuring, and sleep hygiene education was proven to be successful for 8 of 11 TBI participants in one study.[89] Progress was maintained at the 1- and 3-month follow-ups. Sleep improvements were accompanied by a reduction in reported fatigue.

Acupuncture has been applied to the treatment of posttraumatic insomnia in 24 adults with TBI at up to 5 years post-injury.[90] A pilot randomized control trial supported the contention that acupuncture has a beneficial effect on perception of sleep or sleep quality and on cognition in patients with TBI, and resulted in the tapering of sleep medication use.

Sleep apnea in 13 TBI patients was managed by continuous positive airway pressure (CPAP), eliminating respiratory events and their associated outcomes.[91] While researchers reported significantly decreased Apnea–Hypopnea Index scores and improved sleep quality 3 months after treatment administration, EDS was not resolved. The compliance of this treatment option in TBI is still under study.

Pharmacological interventions

Li Pi Shan and colleagues studied the effects of benzodiazepine and non-benzodiazepine hypnotics on sleep restfulness and the effects of these medications on cognition. The researchers reported no difference between the two drug classes in terms of their effects on sleep duration and sleep continuity, as well as cognition in individuals with TBI.[92]

Kemp and colleagues conducted a randomized double-blind trial with melatonin and amitriptyline (1 month/drug, 2-week washout period between treatments) involving seven TBI patients who complained of difficulty initiating and maintaining sleep 3 years after injury.[93] The researchers reported no improvement in sleep duration or reduced sleep onset with either drug; the effect size, however, revealed positive changes: patients on melatonin reported improved daytime alertness, and those on amitriptyline had improved sleep duration compared to baseline. The researchers reported no adverse drug effects. Nagtegaal and colleagues described a case of delayed sleep phase syndrome developed following TBI that was successfully treated with melatonin in a 15-year-old girl.[94] Several physiological markers of the sleep–wake rhythm, including plasma melatonin, body temperature, wrist activity, and sleep architecture (EEG), were delayed by almost half a day, returning to normal after treatment with 5 mg of melatonin.

TBI patients with narcolepsy and EDS have been reported to respond well to neurostimulant agents such as methylphenidate. Kaiser and colleagues reported the positive effect of modafinil in alleviating EDS after TBI in ten patients who received 100–200 mg of modafinil every morning compared to ten patients who received placebo.[95] No clinically relevant residual effects were found.

MEDICATION EFFECTS: A WORD OF CAUTION

Persons with TBI are more likely to have or develop co-occurring conditions related to pain, mood, adjustment and anxiety disorders, among others, which may require treatment with a number of different medications. Two decades ago, Goldstein reviewed the medical records of 100 patients with head trauma admitted to a university hospital over 1 year and reported on the frequencies of medication prescriptions during their hospital stays.[96] While only 14% of patients with head injury were taking medications at the time of injury, during the hospital stay, all received pharmacological interventions. A total of 72% of the patients received one or a combination of drugs that, in animal studies, have been implicated in impaired sleep and recovery (i.e., neuroleptics and other central dopamine receptor antagonists, benzodiazepines, and

the anticonvulsants phenytoin and phenobarbitone). Goldstein warned that until the true impact of these classes of drugs on the recovery process is better understood, care should be exercised in their use. A recent retrospective cohort study of 306 persons with moderate to severe TBI who were discharged from a large rehabilitation hospital reported that the prevalence of prescription medication was 58.9% in the sample, and this was greater in females (65.6%) than in males (56.1%).[97] The most prescribed medication types were anticonvulsants (25.8%) followed by antidepressants (8.2%), painkillers (8.2%), and antianxiety medications (5.9%). On average, persons with TBI were prescribed 2.6 (standard deviation = 2.1) medications, with a range of 1–12.

Another study reported that veterans with TBI being treated with nervous system and muscular skeletal drug classes of medications can develop addiction and various adverse effects, including those related to sleep.[98] These effects might be difficult to disentangle from the effects of the brain injury itself, and also more difficult to treat.

Opioid-induced CSA is very difficult to manage. While not studied in TBI, a recent systematic review of the various modalities of positive airway pressure (PAP) in a total of 127 patients who had been on opioids (10–450 mg daily of morphine-equivalent dose) for at least 6 months reported CPAP as being mostly ineffective for reducing central apneic events; however, bi-level PAP with and without supplemental oxygen achieved elimination of central apneas in 62% of patients, while adaptive servo-ventilation yielded conflicting results, with 58% of participants attaining a central apnea index of less than 10 per hour.[99] With the increasing use of opioids for managing pain in patients with TBI (59.7%),[100] the threat of untreated CSA is real.

Hypersomnia in TBI is often treated with sodium oxybate, a neurotransmitter that is found in the human brain and exerts sedative effects. A recent case report described de novo CSA induced by sodium oxybate in a patient without pre-existing sleep-disordered breathing.[101] Previous publications have also reported a deterioration of sleep-disordered breathing by this drug in patients with narcolepsy and comorbid OSA.[102] The safety of sodium oxybate in TBI patients with concurrent breathing disorders should be revisited.

Drug-induced Parkinsonism caused by the concurrent use of donepezil and risperidone in a patient with TBI has been recently reported. When drugs were stopped, these symptoms rapidly disappeared in several days.[103]

While currently there is no strong evidence directly relating sleep difficulties in persons with TBI to the side effects of neurotropic medications, data on their safety for chronic use are also lacking. While our discussion of medication effects and sleep and wakefulness in TBI is limited, given the complexity of sleep topics, future research should consider medication effects, as the potential for medications to cross the blood–brain barrier and mimic neurological deficits and cause or exacerbate sleep dysfunction is great. Moreover, continuous medication exposure produces changes in neural functions, including those that regulate sleep and wakefulness. The proposed mechanisms include medication effect changes in receptors and ion channel functions, signal transduction, synaptic reorganization, and gene expression, among others.[104–106] Such long-term use of sleep-affecting medications could lead to alterations in the brain pathways underlying sleep dysfunction in TBI. Given the vulnerability of the already-compromised cerebral function from the brain injury itself, careful consideration should be given to the medicines taken by TBI patients, as this will provide invaluable insights into the proper differential diagnosis and treatment of sleep dysfunction post-injury.

PRACTICAL CONSIDERATIONS

Sleep complaints are common in TBI patients. Increasing evidence of the significant consequences of disturbed sleep necessitates comprehensive assessment and treatment. Treatment should be highly specific and follow general principles based on one's medical and sleep dysfunction profile, ability to tolerate side effects, effects of treatment on patients' neural recovery, rehabilitation, safety, and quality of life.[107] We propose the following steps for consideration in establishing the treatment plan of a person with TBI and their sleep dysfunction, abbreviated to "DEDEF":

- *Diagnosis.* Differential diagnosis is extremely important, and the identification of numerous factors that contribute to a patient's insomnia, hypersomnia, or behavioral sleep disturbances is key to targeted therapy.

- *Education.* While generally the patient is considered as the primary target of treatment, in TBI, the involvement of family members/caregivers is of key importance. Therefore, they should be educated and informed about sleep and its roles in recovery, specifically in situations where self-awareness and motivation to follow through with the treatment plan are reduced.

- *Dosage.* It is important to focus on treatment interventions with minimal side effects. Pharmacological studies in TBI report variable efficacy. The dosing schedule requires considerable attention. It is recommended to begin with a lower dose than that which is recommended for the healthy persons, and then to titrate the medication and dosage to the individual patient.[108,109]

- *Emphasis* should be put on sleep hygiene, behavioral and environmental strategies with non-pharmacological interventions for pain, depression, anxiety, and insomnia and hypersomnia.[98]

- *Following* the patient with TBI closely, monitoring compliance with treatment planning should be conducted.

SUMMARY

Although the effects of a multidisciplinary approach to the treatment and rehabilitation of persons with TBI are well documented, the rates of poor rehabilitation outcomes post-injury are still high.[110] This fact highlights the significance of primary and secondary injury prevention[110]; it is easier to prevent than to cure. Regular check-ups of one's health and activity demands (productive duties timing /duration and duty difficulty/intensity) have to be in alliance with circadian timing and amount of time available for sleep, which are primary determinants of daytime performance, alertness, and therefore safety.

TBIs, although largely preventable, nevertheless do occur. A thorough investigation of sleep function in persons who have sustained a TBI must begin as early as possible with an investigation of what constitutes "having difficulty falling asleep and/or maintaining sleep" and/or "sleeping too much." Next, the cause of poor sleep and/or excessive sleepiness must be determined and a diagnosis established. Although the study of the pathophysiology of sleep dysfunction after TBI is still evolving, the goal is to determine whether the problem is caused by a correctable factor (i.e., primary sleep disorder, medication/substance effects, depression, endocrine dysfunction, poor sleep hygiene, etc.) so that interventions are applied in a timely fashion and appropriately. Integrating screening for sleep disorders into post-acute and chronic TBI care may reduce healthcare and disability costs.

To summarize, sleep is disturbed in patients with TBI. The construction of this disturbance is complex. While the question of whether sleep dysfunction is the cause, the consequence, or develops on its own after the injury as the person ages and more comorbid conditions accumulate remains to be answered, the implications of sleep dysfunction for the injured person and the clinician call for timely and proper differential diagnosis, followed by highly specific treatment.

REFERENCES

1. Brain Injury Association of America. *About Brain Injury*. 2011. Retrieved June 10, 2014, from http://www.biausa.org/about-brain-injury.htm.
2. World Health Organization. *Neurological Disorders: Public Health Challenges*. 2010. Retrieved October 19, 2014, from http://www.who.int/mental_health/publications/neurological_disorders_ph_challenges/en/.
3. World Health Organization. *Projection of Mortality and Burden of Disease to 2030*. Geneva: Death by Income Group, 2002.
4. Canadian Institute for Health Information. *Head Injuries in Canada: A Decade of Change (1994–1995–2003–2004)*. Toronto: Canadian Institute for Health Information, 2006.
5. Masel BE. *Conceptualizing Brain Injury as a Chronic Disease*. Vienna, VA: Brain Injury Association of America, 2009.
6. Masel BE, DeWitt DS. Traumatic brain injury: A disease process, not an event. *J Neurotrauma* 2010; 27: 1529–1540.
7. Saatman KE, Duhaime AC, Bullock R, Maas AI, Valadka A, Manley GT; Workshop Scientific Team and Advisory Panel Members. Classification of traumatic brain injury for targeted therapies. *J Neurotrauma* 2008; 25(7): 719–738.
8. Balestreri M, Czosnyka M, Chatfield DA et al. Predictive value of Glasgow Coma Scale after brain trauma: Change in trend over the past ten years. *J Neurol Neurosurg Psychiatry* 2004; 75: 161–162.
9. Giza CC, Kutcher JS, Ashwal S et al. Summary of evidence-based guideline update: Evaluation and management of concussion in sports: Report of the Guideline Development Subcommittee of the American Academy of Neurology. *Neurology* 2013; 80(24): 2250–2257.
10. Andriessen TM, Jacobs B, Vos PE. Clinical characteristics and pathophysiological mechanisms of focal and diffuse traumatic brain injury. *J Cell Mol Med* 2010; 14(10): 2381–2392.
11. Hardman JM, Manoukian A. Pathology of head trauma. *Neuroimaging Clin N Am* 2002; 12(2): 175–187.
12. Davis AE. Mechanisms of traumatic brain injury: Biomechanical, structural and cellular considerations. *Crit Care Nurs Q* 2000; 23(3): 1–13.
13. Ommaya AK, Gennarelli TA. Cerebral concussion and traumatic unconsciousness. Correlation and clinical observations of blunt head injuries. *Brain* 1974; 97: 633–654.
14. Crisco JJ, Chu JJ, Greenwald RM. An algorithm for estimating acceleration magnitude and impact location using multiple nonorthogonal single-axis accelerometers. *J Biomech Eng* 2004; 126: 849–854.
15. Barkoudarian G, Hovda DA, Giza CC. The molecular pathophysiology of concussive brain injury. *Clin Sport Med* 2011; 30: 33–48.
16. Moruzzi G, Magoun HW. Brain stem reticular formation and activation of the EEG. *Electroencephalogr Clin Neurophysiol* 1949; 1(4): 455–473.
17. Sprague JM. The effects of chronic brainstem lesions on wakefulness, sleep and behavior. *Res Publ Assoc Res Nerv Ment Dis* 1967; 45: 148–194.
18. Fuller PM, Sherman D, Pedersen NP, Saper CB, Lu J. Reassessment of the structural basis of the ascending arousal system. *J Comp Neurol* 2011; 519(5): 933–956.

19. Jones BE. Modulation of cortical activation and behavioral arousal by cholinergic and orexinergic systems. *Ann N Y Acad Sci* 2008; 1129: 26–34.
20. Krebs C, Weinberg J, Akesson E. *Lippincott's Illustrated Review of Neuroscience*. Baltimore, MD: Lippincott Williams & Wilkins, 2012.
21. DelRosso LM, Hoque R, James S, Gonzalez-Toledo E, Chesson AL Jr. Sleep–wake pattern following gunshot suprachiasmatic damage. *J Clin Sleep Med* 2014; 10(4): 443–445.
22. Hastings MH, Brancaccio M, Maywood ES. Circadian pacemaking in cells and circuits of the suprachiasmatic nucleus. *J Neuroendocrinol* 2014; 26(1): 2–10.
23. Pandi-Perumal SR, BaHammam AS, Brown GM et al. Melatonin antioxidative defense: Therapeutical implications for aging and neurodegenerative processes. *Neurotox Res* 2013; 23(3): 267–300.
24. Bratton SL, Chestnut RM, Chajar J et al. Guidelines for management of severe traumatic brain injury. Blood pressure and oxygenation. *J Neurotrauma* 2007; 24(Suppl 1): S7–S13.
25. Thatcher RW, North DM, Curtin RT. An EEG severity index of traumatic brain injury. *J Neuropsychiatry Clin Neuriosc* 2001; 13(1): 77–88.
26. Broicolo A, Turella G. Electroencephalographic patterns of acute traumatic coma: Diagnostic and prognostic value. *J Neurosurg Sci* 1973; 17: 278–285.
27. Synek VM. Revised EEG coma scale in diffuse acute head injuries in adults. *Clin Exp Neurol* 1990; 27: 99–111.
28. Urakami Y. Relationship between sleep spindles and clinical recovery in patients with traumatic brain injury: A simultaneous EEG and MEG study. *Clin EEG Neurosci* 2012; 43(1): 39–47.
29. Arciniegas BD, Anderson AC, Rojas DC. Electrophysiological assessment. In: Silver JM, McAllister TW, Yudofsky SC (Eds). *Textbook of Traumatic Brain Injury* (2nd ed.). Arlington, VA: American Psychiatric Publishing, Inc., 2011, pp. 115–126.
30. Jacome DE, Risko M. EEG features of post-traumatic syndrome. *Clin Electroencephalogr* 1984; 15(4): 214–221.
31. Prichep LS, Naunheim R, Bazarian J, Mould WA, Hanley D. Identification of hematomas in mild traumatic brain injury using an index of quantitative brain electrical activity. *J Neurotrauma* 2015; 32(1): 17–22.
32. Ouellet MC, Beaulieu-Bonneau S, Morin CM. Insomnia in patients with traumatic brain injury: Frequency, characteristics, and risk factors. *J Head Trauma Rehabil* 2006; 21(3): 199–212.
33. Seyone C, Kara B. Head injuries and sleep. In: Lader MH, Cardinali DP, Pandi-Perumal SR (Eds). *Sleep and Sleep Disorders: A Neuropsychopharmacological Approach*. New York, NY: Landes Bioscience/Eurekah. com & Springer Science + Business Media, 2006, pp. 210–215.
34. Lankford DA, Wellman JJ, O'Hara C. Posttraumatic narcolepsy in mild to moderate close head injury. *Sleep* 1994; 17(Suppl 8): S25–S28.
35. Ayalon L, Borodkin K, Dishon L, Kanety H, Dagan Y. Circadian rhythm sleep disorders following mild traumatic brain injury. *Neurology* 2007; 68(14): 1136–1140.
36. Shekleton JA, Parcell DL, Redman JR, Phipps-Nelson J, Ponsford JL, Rajaratnam SM. Sleep disturbance and melatonin levels following traumatic brain injury. *Neurology* 2010; 74(21): 1732–1738.
37. Steele D, Rajaratnam S, Redman J, Ponsford J. The effect of traumatic brain injury on timing of sleep. *Chronobiol Int* 2005; 22(1): 89–105.
38. Castriotta RJ, Wilde MC, Lai JM et al. Prevalence and consequences of sleep disorders in traumatic brain injury. *J Clin Sleep Med* 2007; 3(4): 349–356.
39. Rao V, Spiro J, Vaishnavi S et al. Prevalence and types of sleep disturbances acutely after traumatic brain injury. *Brain Inj* 2008; 22(5): 381–386.
40. Mathias JL, Alvaro PK. Prevalence of sleep disturbances, disorders, and problems following traumatic brain injury: A meta-analysis. *Sleep Med* 2012; 13(7): 898–905.
41. Mollayeva T, Colantonio A, Mollayeva S, Shapiro CM. Screening for sleep dysfunction after traumatic brain injury. *Sleep Med* 2013; 14(12): 1235–1246.

42. Cantor JB, Bushnik T, Cicerone K et al. Insomnia, fatigue, and sleepiness in the first 2 years after traumatic brain injury: An NIDRR TBI model system module study. *J Head Trauma Rehabil* 2012; 27(6): E1–E14.

43. American Psychiatric Association. *Diagnostic and Statistical Manual of Mental Disorder* (4th ed.) (Text Revision). Washington, DC: American Psychiatric Association, 2000.

44. Fichtenberg NL, Putnam SH, Mann NR, Zafonte RD, Millard AE. Insomnia screening in postacute traumatic brain injury: Utility and validity of the Pittsburgh Sleep Quality Index. *Am J Phys Med Rehabil* 2001; 80(5): 339–345.

45. Ouellet M, Morin CM. Subjective and objective measures of insomnia in the context of traumatic brain injury: A preliminary study. *Sleep Med* 2006; 7(6): 486–497.

46. Ohayon MM. Epidemiology of insomnia: What we know and what we still need to learn. *Sleep Med Rev* 2002; 6: 97–111.

47. Jain A, Mittal RS, Sharma A, Sharma A, Gupta ID. Study of insomnia and associated factors in traumatic brain injury. *Asian J Psychiatr* 2014; 8: 99–103.

48. Bryan CJ. Repetitive traumatic brain injury (or concussion) increases severity of sleep disturbance among deployed military personnel. *Sleep* 2013; 36(6): 941–946.

49. Bartlett DJ, Marshall NS, Williams A, Grunstein RR. Predictors of primary medical care consultation for sleep disorders. *Sleep Med* 2008; 9(8): 857–864.

50. Flygare J, Parthasarathy S. Narcolepsy: Let the patient's voice awaken us! *Am J Med* 2015; 128(1): 10–13.

51. Drakatos P, Leschziner GD. Update on hypersomnias of central origin. *Curr Opin Pulm Med* 2014; 20(6): 572–580.

52. Byrd K, Gelaye B, Tadessea MG, Williams MA, Lemma S, Berhanec Y. Sleep disturbances and common mental disorders in college students. *Health Behav Policy Rev* 2014; 1(3): 229–237.

53. Cabrera MA, Dellaroza MS, Trelha CS et al. Psychoactive drugs as risk factors for functional decline among noninstitutionalized dependent elderly people. *J Am Med Dir Assoc* 2010; 11(7): 519–522.

54. Belmont A, Agar N, Azouvi P. Subjective fatigue, mental effort, and attention deficits after severe traumatic brain injury. *Neurorehabil Neural Repair* 2009; 23 (9): 939–944.

55. Siebern AT, Guilleminault C. Sleepiness and fatigue following traumatic brain injury: A clear relationship? *Sleep Med* 2012; 13(6): 559–560.

56. Port A, Willmott C, Charlton J. Self-awareness following traumatic brain injury and implications for rehabilitation. *Brain Inj* 2002; 16(4): 277–289.

57. Guilleminault C, Yuen KM, Gulevich MG, Karadeniz D, Leger D, Philip P. Hypersomnia after head-neck trauma: A medicolegal dilemma. *Neurology* 2000; 54(3): 653–659.

58. Verma A, Anand V, Verma NP. Sleep disorders in chronic traumatic brain injury. *J Clin Sleep Med* 2007; 3(4): 357–362.

59. Sommerauer M, Valko PO, Werth E, Baumann CR. Excessive sleep need following traumatic brain injury: A case–control study of 36 patients. *J Sleep Res* 2013; 22(6): 634–639.

60. Baumann CR, Bassetti CL, Valko PO et al. Loss of hypocretin(orexin) neurons with traumatic brain injury. *Ann Neurol* 2009; 66(4): 555–559.

61. Nardone R, Bergmann J, Kunz A et al. Cortical excitability changes in patients with sleep–wake disturbances after traumatic brain injury. *J Neurotrauma* 2011; 28(7): 1165–1171.

62. Castriotta RJ, Wilde MC, Lai JM, Atanasov S, Masel BE, Kuna ST. Prevalence and consequences of sleep disorders in traumatic brain injury. *J Clin Sleep Med* 2007; 15, 3(4): 349–356.

63. Dikeos D, Georgantopoulos G. Medical comorbidity of sleep disorders. *Curr Opin Psychiatry* 2011; 24(4): 346–354.

64. Sasai T, Inoue Y, Matsuura M. Clinical significance of periodic leg movements during sleep in rapid eye movement sleep behavior disorder. *J Neurol* 2011; 258(11): 1971–1978.

65. Esteves AM, de Mello MT, Lancellotti CL, Natal CL, Tufik S. Occurrence of limb movement during sleep in rats with spinal cord injury. *Brain Res* 2004; 1017(1–2): 32–38.

66. Lai YY, Hsieh KC, Nguyen D, Peever J, Siegel JM. Neurotoxic lesions at the ventral mesopontine junction change sleep time and muscle activity during sleep: An animal model of motor disorders in sleep. *Neuroscience* 2008; 154(2): 431–443.

67. Morrison AR. Paradoxical sleep without atonia. *Arch Ital Biol* 1988; 126: 275–289.

68. Winkelman JW, James L. Serotonergic antidepressants are associated with REM sleep without atonia. *Sleep* 2004; 27(2): 317–321.

69. Van Holsbeke C, De Backer J, Vos W et al. Anatomical and functional changes in the upper airways of sleep apnea patients due to mandibular repositioning: A large scale study. *J Biomech* 2011; 44(3): 442–449.

70. Jin H, Wang S, Hou L et al. Clinical treatment of traumatic brain injury complicated by cranial nerve injury. *Injury* 2010; 41(9): 918–923.

71. Pappius HM. Cortical hypometabolism in injured brain: New correlations with the noradrenergic and serotonergic systems and with behavioral deficits. *Neurochem Res* 1995; 20(11): 1311–1321.

72. Zafonte R, Elovic EP, Lombard L. Acute care management of post-TBI spasticity. *J Head Trauma Rehabil* 2004; 19(2): 89–100.

73. Webster JB, Bell KR, Hussey JD, Natale TK, Lakshminarayan S. Sleep apnea in adults with traumatic brain injury: A preliminary investigation. *Arch Phys Med Rehabil* 2001; 82(3): 316–321.

74. Boone DR, Sell SL, Micci MA et al. Traumatic brain injury-induced dysregulation of the circadian clock. *PLoS One* 2012; 7(10): e46204.

75. Ayalon L, Borodkin K, Dishon L, Kanety H, Dagan Y. Circadian rhythm sleep disorders following mild traumatic brain injury. *Neurology* 2007; 68(14): 1136–1140.

76. Llompart-Pou JA, Pérez G, Raurich JM et al. Loss of cortisol circadian rhythm in patients with traumatic brain injury: A microdialysis evaluation. *Neurocrit Care* 2010; 13(2): 211–216.

77. Vallières A, Azaiez A, Moreau V, LeBlanc M, Morin CM. Insomnia in shift work. *Sleep Med.* 2014; 15(12): 1440–1448.

78. American Academy of Sleep Medicine. *The International Classification of Sleep Disorders* (2nd ed.). Westchester, IL: American Academy of Sleep Medicine, 2005.

79. Mollayeva T, Kendzerska T, Colantonio A. Self-report instruments for assessing sleep dysfunction in an adult traumatic brain injury population: A systematic review. *Sleep Med Rev* 2013; 17: 411–423.

80. Masel BE, Scheibel RS, Kimbark T, Kuna ST. Excessive daytime sleepiness in adults with brain injuries. *Arch Phys Med Rehabil* 2001; 82: 1526–1532.

81. Parcell DL, Ponsford JL, Redman JR, Rajaratnam SM. Poor sleep quality and changes in objectively recorded sleep after traumatic brain injury: A preliminary study. *Arch Phys Med Rehabil* 2008; 89(5): 843–850.

82. Fichtenberg NL, Zafonte RD, Putnam S, Mann NR and Millard AE. Insomnia in a post-acute brain injury sample. *Brain Inj* 2002; 16(3): 197–206.

83. Lu W, Cantor J, Aurora RN et al. Variability of respiration and sleep during polysomnography in individuals with TBI. *Neurorehabilitation* 2014; 35(2): 245–251.

84. Mahowald MW. Sleep in traumatic brain injury and other acquired CNS conditions. In: Culebras A (Ed.). *Sleep Disorders and Neurological Disease.* New York, NY: Dekker, 2000, pp. 365–385.

85. Castriotta RJ, Atanasov S, Wilde MC, Masel BE, Lai JM, Kuna ST. Treatment of sleep disorders after traumatic brain injury. *J Clin Sleep Med* 2009; 5(2): 137–44.

86. Zollman FS, Cyborski C, Duraski SA. Actigraphy for assessment of sleep in traumatic brain injury: Case series, review of the literature and proposed criteria for use. *Brain Inj* 2010; 24(5): 748–754.

87. Duclos C, Dumont M, Blais H et al. Rest–activity cycle disturbances in the acute phase of moderate to severe traumatic brain injury. *Neurorehabil Neural Repair* 2013; 28(5): 472–482.

88. Sinclair KL, Ponsford J, Rajaratnam SM. Actigraphic assessment of sleep disturbances following traumatic brain injury. *Behav Sleep Med* 2014; 12(1): 13–27.

89. Ouellet M, Morin CM. Efficacy of cognitive–behavioral therapy for insomnia associated with traumatic brain injury: A single-case experimental design. *Arch Phys Med Rehabil* 2007; 88(12): 1581–1592.

90. Zollman FS, Larson EB, Wasek-Throm LK, Cyborski CM, Bode RK. Acupuncture for treatment of insomnia in patients with traumatic brain injury: A pilot intervention study. *J Head Trauma Rehabil* 2012; 27(2): 135–142.

91. Castriotta RJ, Atanasov S, Wilde MC, Masel BE, Lai JM, Kuna ST. Treatment of sleep disorders after traumatic brain injury. *J Clin Sleep Med* 2009; 5(2): 137–144.

92. Li Pi Shan RS, Ashworth NL. Comparison of lorazepam and zopiclone for insomnia in patients with stroke and brain injury: A randomized, crossover, double-blinded trial. *Am J Phys Med Rehabil* 2004; 83(6): 421–427.

93. Kemp S, Biswas R, Nuemann V, Coughlan A. The value of melatonin for sleep disorders occurring post-head injury: A pilot RCT. *Brain Inj* 2004; 18(9): 911–919.

94. Nagtegaal JE, Kerkhof GA, Smits MG, Swart AC, van der Meer YG. Traumatic brain injury-associated delayed sleep phase syndrome. *Funct Neurol* 1997; 12(6): 345–348.

95. Kaiser PR, Valko PO, Werth E et al. Modafinil ameliorates excessive daytime sleepiness after traumatic brain injury. *Neurology* 2010; 75(20): 1780–1785.

96. Goldstein LB. Prescribing of potentially harmful drugs to patients admitted to hospital after head injury. *J Neurol Neurosurg Psychiatry* 1995; 58(6): 753–755.

97. Yasseen B, Colantonio A, Ratcliff G. Presrption medication use in persons many years following traumatic brain injury. *Brain Inj* 2008; 22(10): 752–757.

98. Farinde A. An examination of co-occurring conditions and management of psychotropic medication use in soldiers with traumatic brain injury. *J Trauma Nurs* 2014; 21(4): 153–157.

99. Reddy R, Adamo D, Kufel T, Porhomayon J, El-Solh AA. Treatment of opioid-related central sleep apnea with positive airway pressure: A systematic review. *J Opioid Manag.* 2014; 10(1): 57–62.

100. Meares S, Shores EA, Taylor AJ, Batchelor J. The prospective course of postconcussion syndrome: The role of mild traumatic brain injury. *Neuropsychology* 2011; 25(4): 454–465.

101. Frase L, Schupp J, Sorichter S, Randelshofer W, Riemann D, Nissen C. Sodium oxybate-induced central sleep apnea. *Sleep Med* 2013; 14(9): 922–924.

102. Zvosec DL, Smith SW, Mahowald MW. Further research on Xyrem®/sodium oxybate treatment of patients with obstructive sleep apnea is needed. *Sleep Breath* 2011; 15(4): 619–620.

103. Kang SH, Kim DK. Drug induced Parkinsonism caused by the concurrent dose of donepezil and risperidone in a patient with traumatic brain injuries. *Ann Rehabil Med* 2013; 37(1): 147–150.

104. Amara SG, Sonders MS. Neurotransmitter transporters as molecular targets for addictive drugs. *Drug Alcohol Depend* 1998; 51(1–2): 87–96.

105. Jaffe JH, Sharpless SK. Pharmacological denervation supersensitivity in the central nervous system: A theory of physical dependence. In: Wikler AH (Ed.). *The Addictive States*. Baltimore, MD: Williams and Wilkins, 1968, pp. 226–246.

106. Wise RA, Koob GF. The development and maintenance of drug addiction. *Neuropsychopharmacology* 2014; 39(2): 254–262.

107. De La Rue-Evans L, Nesbitt K, Oka RK. Sleep hygiene program implementation in patients with traumatic brain injury. *Rehabil Nurs* 2013; 38(1): 2–10.

108. Wortzel HS, Arciniegas DB. Treatment of post-traumatic cognitive impairments. *Curr Treat Options Neurol* 2012; 14(5): 493–508.

109. Larson EB, Zollman FS. The effect of sleep medications on cognitive recovery from traumatic brain injury. *J Head Trauma Rehabil* 2010; 25(1): 61–67.

110. Dams-O'Connor K, Cuthbert JP, Whyte J, Corrigan JD, Faul M, Harrison-Felix C. Traumatic brain injury among older adults at level I and II trauma centers. *J Neurotrauma* 2013; 30(24): 2001–2013.

111. Fayers PM, Hand DJ. Factor analysis, causal indicators, and quality of life. *Qual Life Res* 1997; 6: 139–150.

112. Baumann CR, Werth E, Stocker R, Ludwig S, Bassetti CL. Sleep–wake disturbances 6 months after traumatic brain injury: A prospective study. *Brain* 2007; 130(7): 1973–1983.

113. Lichstein KL, Durrence HH, Riedel BL. *Epidemiology of Sleep: Age, Gender, and Ethnicity*. Mahwah, NJ: Erlbaum, 2004.

114. Young T, Peppard PE, Taheri S. Excess weight and sleep-disordered breathing. *J Appl Physiol*. 2005; 99(4): 1592–1599.

115. Longstreth WT Jr, Koepsell TD, Ton TG, Hendrickson AF, van Belle G. The epidemiology of narcolepsy. *Sleep* 2007; 30: 13–26.

116. Weitzman ED, Czeisler CA, Zimmerman JC, Moore-Ede MC. Biological rhythms in man: Relationship of sleep–wake, cortisol, growth hormone, and temperature during temporal isolation. *Adv Biochem Psychopharmacol* 1981; 28: 475–499.

117. Hornyak M, Feige B, Voderholzer U. Periodic leg movements in sleep and periodic limb movement disorder: Prevalence, clinical significance and treatment. *Sleep Med Rev* 2006; 10(3): 169–177.

118. Scofield H, Roth T, Drake C. Period limb movements during sleep: Population prevalence, clinical correlates, and racial differences. *Sleep* 2008; 31(9): 1221–1227.

119. Ohayon MM, Caulet M, Priest RG. Violent behavior during sleep. *J Clin Psychiatry* 1997; 58(8): 369–376.

17

Dreaming in psychiatric patients

MILTON KRAMER AND ZVJEZDAN NUHIC

INTRODUCTION FOR 2014

We have extended our search of Medline for dreams and the seven mental illness categories from 2005 to 2008, which we reported in 2010.[162] We found 90 articles in our search: 5 for schizophrenia, 24 for depression, 48 for posttraumatic stress disorder (PTSD), 5 for eating disorders, 1 for brain damage, none for mental retardation, and 7 for alcoholism and drug abuse. We have now extended our Medline search from 2009 to 2014.[163] In our most recent search from 2009 to 2014, we found 137 articles that came from crossing dreams with one of the seven illness-specific categories: 15 for schizophrenia, 28 for depression, 81 for PTSD, 3 for eating disorders, 3 for brain damage, 1 for mental retardation, and 6 for alcoholism and drug abuse. None of the 227 (90 + 137) studies met the illness and content criteria, so our descriptions of dream content described in the first edition are unchanged.

The interest in dreams has not diminished over the past 10 years, as the average number of studies over the two time periods is essentially the same (22.5/year compared to 23.0/year). The most frequent questions asked by investigators are comparative and treatment oriented rather than content focused for an illness category and do not contribute to what the dream content may be for a patient with a particular illness.

In my seven articles on dream content,[20–24,162,163] I have focused on the second interest of Freud's, "modifications to which dream-life is subject in cases of mental disease."[5] The other two areas—etiology and analogies to psychosis—were seen occasionally as we explored the dreams for content reported from 1969 to 2004 and should be systematically studied.

INTRODUCTION

There has long been the assumption that there is an intimate relationship between dreams and mental disorders. Epigrammatic statements that "the madman is a waking dreamer,"[1] that "dreams [are] a brief madness and madness a long dream,"[2] that if we "let the dreamer walk about and act like a person awake…, we [would] have the clinical picture of dementia praecox [schizophrenia],"[3] and that if "we could find out about dreams, we would find out about insanity,"[4] reflect the conviction about the close relationship between dreams and profound emotional disturbance. This view enlivened efforts to study dreaming in order to gain insights into the problems of the mentally ill.

In the literature review that introduces *The Interpretation of Dreams*,[5] Freud has a section on "The Relations between Dreams and Mental Diseases." He points out that when he "speaks of the relationship of dreams to mental disorders [he] has three things in mind: (1) etiological and clinical connections, as when a dream represents a psychotic state, or introduces it, or is left over from it; (2) modifications to which dream-life is subject in cases of mental disease; and (3) intrinsic connections between dreams and psychosis, analogies pointing to their being essentially akin."

The published work on dreams and psychopathologic states touches on all three areas of Freud's concern.[6] There are reports of psychotic states appearing to begin with a dream or a series of dreams, and there is certainly literature that continues to pursue analogies between dreams and psychosis. However, the vast majority of the work that has been done on dreams and psychopathologic states devotes itself to the "modifications to which dream life is subject in cases of mental disease" and will be the focus of this report. Freud was of the opinion that as we better understand dreams, this will enhance our understanding of psychosis. Hartmann is of similar opinion.[7]

There is a potential confusion between a psychopathology of dreams and dreams in a psychopathologic state. The former refers to alternations in the dreaming process that may be seen as abnormal, whereas the latter refers to the dreams that are the concomitants of a mental disorder. A dream that awakens the dreamer in a terrified state generally with accompanying frightening dream content—a nightmare—would be a psychopathologic dream.[7] A dream report from a patient suffering from schizophrenia would be a dream from a person in a psychopathologic state. The dream may or may not be unique, either pathognomonically or statistically, to that state. Strangers occurring more frequently in the dreams of schizophrenics than in normal or depressed individuals is a statistical change in dream content in a psychopathologic condition.[8]

A psychological examination of the dream is a study of the manifest content of the dream. Jones[9] has pointed out that a psychology of dreams must rest on a study of the elements of which it is composed—the manifest dream images. Even Freud has pointed out "…that in some cases the façade of the dream directly reveals the dream's actual nucleus."[10] However, his almost exclusive focus on the latent dream content and his dismissal of the reported manifest dream retarded the study of dream reports. In the modern era, Hall and Van de Castle[11] have presented quantitative methods for assessing dream content and have encouraged a scientific approach to the examination of dreams.

The study of the dream is an undertaking that is fraught with many difficulties. The dream experience cannot be directly observed and its study is still dependent on the dream report. The dream is experienced during one state—sleep—and reported during another state—wakefulness. The problems of examining verbal reports of inner experiences are compounded by the change in state necessary to obtain the dream report. The study of lucid dreaming[12] opens the possibility of examining the dream experience while it is occurring, but the work so far on lucid dreaming has been more directed at demonstrating its occurrence than utilizing it as a method for studying the dream experience.

The verbal nature of the dream report needs to be addressed. Does the form of the dream obtained in the dream report reflect the dream as experienced or is it a result of the verbal style of the dreamer? An appropriate report of a waking experience becomes a necessary control if the study of the form of the dream is undertaken. For example, is the "apparently" nonlinear description of an experience the same for a subject in describing a waking experience as when describing a dream experience? If it is, then the finding of nonlinearity cannot be considered a property of the dream experience, but rather it is an aspect of the dreamer's verbal style.

There are those who see the dream as an ineffable experience whose essence is destroyed by scientific study, by quantification.[13] The present survey is of quantitative reports. Quantification does not have to damage the essence of the dream experience.

There are significant methodological issues that influence the dream content found in dream reports. These issues relate to the collection and measurement of the dream report.[14] They apply generally to dreams obtained either from nighttime awakenings or from morning reports. There are seven collection factors that influence dream content: (1) the place in which the dream is experienced and collected; (2) the method of awakening the dreamer; (3) the context of the interpersonal situation in which the dream report is given; (4) the style of the collection interview; (5) the time of night and stage of sleep from which the sleeper is awakened; (6) the method of recording the dream report; and (7) the type of subject from whom the dream is collected. In addition, there are five problems that are related to the quantification of dream content that need to be considered: (1) the verbal nature of the dream report; (2) the definition of the scoreable protocol; (3) the effect of dream length on the type of measurement made; (4) the methods of quantifying dream content; and (5) the validity and reliability of the measurement. These factors need to be considered in studying the reports of dreams both separately and in interaction, as they affect the results obtained.

Interest in the dream has been kept alive by depth psychologists'[15] and by the man in the street,[16] while the scientific study of the dream has been significantly stimulated by the amount of the dream experience that can be recovered relatively close to the time of occurrence in rapid eye movement (REM) sleep,[17,18] and this has opened the possibility for manipulative (experimental) studies of dreaming.

In 1953, Ramsey[19] published a review of the studies of dreaming. These were all from the pre-REM literature. Overall, he cites some 121 articles and books, of which 20 at most were studies of the dreams of six patient groups. The amount of information available from 20 publications would be woefully inadequate for characterizing the dreams of psychopathologic groups. Ramsey concludes that the research was scientifically inadequate. Very few of the studies were so designed and reported that they could be replicated in order to validate their findings. He found that the dream studies were weak in not adequately: (1) describing the population under study (i.e., their gender, age, intelligence, health, economic status, and education); (2) limiting the group of subjects under

study; (3) using control groups; (4) defining more adequately the characteristics of the dreams; (5) treating the data statistically; and (6) controlling for interviewer bias.

The literature dealing with the nature of the relationship between dreaming and mental illness has been reviewed on several occasions,[20-24] with the last detailed review published in 1979.[23] That review focused on 75 reports in 71 articles in six patient groups (i.e., schizophrenia, depression, disturbing dreams, alcoholism, chronic brain syndrome, and mental retardation). Seventy-one articles were covered, four of which referred to more than one diagnostic group of interest. The scientific adequacy of the publications covered in that review[23] was quite problematic, but a picture of dream content in some psychopathologic states began to emerge.

Since that review, only one other review article on dream content in psychiatric conditions has appeared.[25] The report was an extension of the previous work[23] and used as its database relevant studies and cases reports found in *Psychological Abstracts* from 1977 to 1990. It was of interest in that many of the findings of the previous review[23] were supported and some additional groups were examined. Unfortunately, as only 35 articles were cited, covering nine diagnostic groups, the scope of the review was limited. Meanwhile, a review of sleep physiology in psychiatric disorders has been published by Benca.[26]

METHOD

For the present report, an extensive title search of the English language periodical literature on dreams was undertaken. The basic source of bibliographic information was Medline (1966–2005). Dream was used as a descriptor to generate a basic 5484-item list (see Table 17.1). This list was then searched for the six psychopathologic categories of interest (i.e., schizophrenia, depression, PTSD, eating disorders, organic brain disorder, and alcoholism and drug abuse). Some 496 articles published since 1976 were selected.

Of those 496 articles, 493 were obtained and examined. Ninety-four articles with dream content were included in the review. In the case of PTSD, all of the appropriate articles available from 1966 onwards were reviewed. Two of the publications had three studies each and one had two studies.

Table 17.1 Extent of periodical literature search for dreams in psychiatric disorders

	1975[23]	2005
Citations found	2503	5484
Citations requested	1410	496[a]
Citations obtained	1359	493
Citations unavailable	51	3
Citations reviewed	71[b]	94[c]
Citations not reviewed	17	0

[a] Limited to schizophrenia, depression, posttraumatic stress disorder, eating disorders, organic brain disorders, and alcoholism and drug abuse.
[b] Two articles had three studies each.
[c] Two articles had three studies each and one had two studies.

Table 17.2 Psychiatric disorder studies with dream content reported

	Number of studies		References
	1975[23]	2005	2005
Schizophrenia	30	8	35–42
Depression	14	20[a]	45–65
Posttraumatic stress disorder	17[b]	46[a]	66–129
Eating disorders	0	11	130–140
Organic brain disorder	5	6	141–146
Alcoholism and drug abuse	5	7	147–153
	71	98	

[a] Articles 55 and 67 each had three studies.
[b] Included nightmares and anxiety dreams.

Each study was reviewed as a separate article, giving a total of 98 studies (see Table 17.2).

The studies utilized for this review were categorized along 53 parameters covering the areas of: (1) the type, nature and site of the studies; (2) the description of the patient sample; (3) the description of the control sample; (4) the method of dream content collection; (5) the method of scoring the dream content; (6) the nature of the statistical analysis; and (7) the dream content results obtained. The review of the various parameters established only the presence or absence of the category, not its adequacy, with the exception of the statistical category.

RESULTS

The current update (1976–2005) yielded 94 articles covering 98 studies in six psychiatric conditions of concern (see Table 17.2). There were eight articles about schizophrenia, 18 articles with 20 studies about depression, 44 articles with 46 studies about PTSD, 11 articles on eating disorders, six articles on organic brain disorder and seven articles on alcoholism and drug abuse. In the earlier review, there had been 30 studies on schizophrenia, 14 on depression, 17 on disturbing dreams (PTSD and nightmares), five on organic brain disorder, and five on alcoholism. The findings from the 98 studies are presented under six content headings and are compared with the findings from the earlier review.[23]

The type, nature, and site of the studies

In the current update (1976–2005), 67% of the reports on the six conditions are of studies (see Table 17.3). This is encouraging, as single case reports are less likely to provide leads about the fundamental aspects of the dream life of a particular patient group. Interestingly, there is a 10% increase in case reports compared to our 1975 review, which showed 23% case reports and 77% studies.

Table 17.3 Type, nature, and site of studies

	Percentage	
	1975[23]	2005
A. Type of report		
1. Study	77	67
2. Case report	23	33
B. Nature of report		
1. Descriptive	37	69
2. Separate groups	51	27
3. Repeated measures	12	4
C. Site of data collection		
1. Sleep laboratory	32	27
2. Non-laboratory	68	73
D. Subject–control in separate groups		
1. Sick-sick	53	46
2. Sick-well	29	29
3. Both	18	25

The increase in case reports may reflect an interest in trying to capture some aspect of the reported dream life of a patient group that has not been adequately described in the literature (e.g., effects of physical illness, medication, or residence).[22]

The trend towards increased description is highlighted in that currently 69% of the reports were descriptive in nature, compared to 37% reported earlier. Comparison of the target patient group to another (control) group has fallen from 51% to 27%, while following an aspect of the manifest dream content across time or condition change—ill to well or vice versa—has declined from 12% to 4%.

It appears that the concern voiced by Ramsey[19] in reviewing the pre-REM era of dream studies, and in the earlier review[23] about the neglect of appropriate research designs for building a scientifically sound literature about dream studies, remains an issue to this day.

The sleep laboratory has not become an increased source for collecting dream reports, as only 27% of current studies and 32% in the earlier review were in the dream laboratory. This may reflect the fact that researchers are not convinced of the value of waking subjects at night to collect a larger, more complete sample of dreams, or that those who are interested in dream content do not have such a facility available and therefore base their work instead on reports of spontaneously recalled dreams from home studies. We know that the site in which the dream is experienced and collected influences the content.[14]

It is heartening to see that the value of comparing the index group to another ill group or to both an ill and well (normal) group has been maintained. In both reviews, 71% of the studies had at least a sick–sick comparison.

It is unfortunate that the available designs have not been systematically applied. Based on the earlier review,[23] it had been suggested that a content or theme generated from a descriptive report, case study, or literature review be appropriately compared in a separate group study to an ill and well (normal) group and then be studied in a repeated-measure design to establish if the finding was limited to the illness (state) or was linked to a predisposition to the condition (trait). Our knowledge of the dream life of psychopathological groups will remain limited without more systematic study.

The adequacy of the description of the patient sample

SAMPLE SIZE

The overall sample sizes for both sleep laboratory and non-laboratory studies appear to be adequate (see Tables 17.4 and 17.5). The mean laboratory sample size has almost doubled from 12 to 22 when comparing the earlier review to the current one. The mean number of subjects in the non-laboratory studies, although still adequate, has decreased from 246 to 97.

PATIENT SELECTION

There has been greater attention paid to describing the basis for selecting patients in the various reports. In 56% of the studies, the selection basis is provided, whereas in earlier studies, only 31%

Table 17.4 Patient sample size

	Total		Original	
	1975[23]	2005	1975[23]	2005
1. Number of studies				
a. Laboratory	24	27	17	22
b. Non-laboratory	51	71	46	69
2. Total number of subjects				
a. Laboratory	297	594	234	496
b. Non-laboratory	12,528	6693	5966	5034
3. Mean number of subjects per study				
a. Laboratory	12	22	14	22
b. Non-laboratory	246	97	130	73

Table 17.5 Description of patient sample

	Percentage of studies			
	1975[23]		2005	
	Yes	No	Yes	No
A. Basis of selection	31	69	56	44
B. Basis of diagnosis	45	55	48	52
C. Specificity of diagnosis	48	52	23	77
D. Drugs or physical treatment	24	76	36	64
E. Demography				
1. Sex	85	15	88	12
2. Age	69	31	67	33
3. Race	27	73	26	74
4. Education	29	71	22	78
5. Marital status	29	71	30	70
6. Socioeconomic class	25	75	12	88
F. General health	5	95	29	71
G. Original sample	81	19	89	11
H. Site of patient residence				
1. In hospital	57		28	
2. Out of hospital	28		62	
3. Both	11		10	
4. Not given	4		0	

provided this information. This information is essential to understanding the nature of the group studied, as well as providing the minimal information needed for replication.

BASIS OF DIAGNOSIS

There is a slight increase from 45% to 48% in providing the basis for the diagnosis. Without this information, no judgment can be made about whether the classification of the patient was made on reasonable and reproducible grounds. Was the patient classified as schizophrenic because he or she was delusional or because he or she met some criteria, such as those in the *Diagnostic and Statistical Manual of Mental Disorders, 4th Edition* (DSM-IV)?[27]

SPECIFICITY OF DIAGNOSIS

The problem of categorizing the patient as to "subtype" has been considerably less well attended to in the current than in the earlier review. Only 23% of studies provided the operational basis for the subtyping of the patient (e.g., major depressive disorder or dysthymia), while 48% included this information in the earlier review.[23]

DRUGS OR PHYSICAL TREATMENT

In slightly over a third of the patient studies (36%), there is mention made of the treatment status of the patient. This is up from the 24% in the earlier review. In neither case is this adequate, as we have reason to believe that medication and physical treatments can affect the psychology of dreaming, as well as the physiology of sleep.[22] The problem is that if one obtains a positive finding in a study, one would not know whether to attribute it to the psychopathological state, the treatment, or an interaction between the two.

DEMOGRAPHY

There is an awareness in the current and prior review[23] that the gender and age of subjects are parameters that do indeed influence the content of dreams. Gender is reported in 88% of the current studies, and age is reported in 67%. This is compared to 85% and 69% in the earlier review. The other demographic variables—race, education, marital status, and social class—are reported less often (12%–30%) and, indeed, they have less influence on dream content.[28]

GENERAL HEALTH

There has been a significant increase in reporting the general health status of the patient population from 5% in the 1975 review to 29% of studies in the current review. The physical health status of a patient group is a potentially confounding variable, as it may either independently or in interaction affect the dream report content.[22]

ORIGINAL SAMPLE

In 89% of studies, the sample used was an original one. This is slightly larger than the 81% of studies in the earlier review.

SITE OF PATIENT RESIDENCE

There has been a major shift in dream studies of patients with psychopathology from studying those in the hospital (57%) to those out of the hospital (62%). This may reflect the current practice of treating patients outside of the hospital. Setting impacts the contents of the dream[29] and is universally reported (96% in the 1975 review and 100% in the present review).

The adequacy of the description of the control sample (separate-group studies)

The descriptions of the control group in the 27 separate-group studies are more complete than those found for all of the other studies and are significantly improved over those reported in the earlier review (see Table 17.6).[23] To illustrate, 56% of the current overall sample reported the basis for

selecting their target group. A total of 37% of the separate-group studies reported the basis for selection of the control sample in the earlier review,[23] while in the current review, 83% of the studies provided the basis for selecting the control sample. For 12 descriptors (Table 17.7), the studies in the current review presented more complete data for 8 of 12 parameters compared to what was reported in the 1975 review.

The method of dream content collection

The major interest in dream content studies, whether of psychopathological or normal individuals, is what is contained in the dream report. How the report is obtained, scored or categorized, counted and statistically analyzed is basic to establishing what the subject is dreaming about.

The nature and extent of the sampling of the dream life of a psychopathological population is reflected in the number of nights (laboratory studies) or days (non-laboratory studies) of dream collection that was attempted, the number of dreams collected, and the percentage of dream recall achieved. The sampling process in the current review reflects less attention to the number of days or nights of collection or to the percentage recall than in the earlier studies (see Table 17.8). The number of days or nights was reported in 52% in 1975 versus 38% in the current study. In addition, the percentage dream recall was reported in 35% of the earlier studies versus 22% in the current study.

Table 17.6 Control patient sample sizes

	Total		Original	
	1975[23]	2005	1975[23]	1997
1. Number of studies				
a. Laboratory	10	10	8	9
b. Non-laboratory	28	17	23	17
2. Total number of subjects				
a. Laboratory	166	167	143	127
b. Non-laboratory	1843	1007	1703	1007
3. Mean number of subjects per study				
a. Laboratory	17	17	18	14
b. Non-laboratory	66	72	74	72

Table 17.7 Description of control sample

| | Percentage of studies | | | |
| | 1975[23] | | 2005 | |
	Yes	No	Yes	No
A. Basis of selection	37	63	83	17
B. Basis of diagnosis	26	74	71	29
C. Specificity of diagnosis	24	76	50	50
D. Drugs or physical treatment	5	76	54	46
E. Demography				
1. Sex	76	24	92	8
2. Age	68	32	62	38
3. Race	18	82	25	75
4. Education	33	67	21	79
5. Marital status	24	76	12	88
6. Socioeconomic class	26	74	12	88
F. General health	11	89	33	67
G. Original sample	80	20	88	12
H. Site of patient residence				
1. In hospital	42		25	
2. Out of hospital	34		75	
3. Both	24		0	

Table 17.8 Dream content collection variables

| | Percentage of studies | | | |
| | 1975[23] | | 2005 | |
	Yes	No	Yes	No
A. Number of days or nights	52	48	38	62
B. Number of dreams	57	43	58	42
C. Percent dream recall	35	65	22	78
D. Who collected dreams	67	33	46	54
E. When dreams collected	55	45	51	49
F. Mode of awakening[a]	38	62	22	78
G. Protocol for obtaining dreams	15	85	19	81
H. Mode of recording dreams	87	13	27	73
I. Associations obtained	37	63	34	66

[a] Only applicable to laboratory studies.

The interpersonal setting[30] and the mode of dream inquiry[13] may influence the nature of any content obtained. In the 1975 review, 67% of the articles reported who and 55% reported when the dreams were collected, compared respectively to 46% and 51% currently. There has been a decrease in reporting of both variables that is more significant in terms of who collected the dream, indicating a failure to appreciate the effect of the interpersonal situations on what is reported.

Although it is known to influence content, the mode of awakening[31] the subject in laboratory studies remains infrequently reported at 22%. The protection against interviewer bias provided by a fixed protocol for obtaining a dream report also remains low at 19%. The mode of recording the report and whether associations are obtained are reported in a third or fewer of the current reports.

Unfortunately, the major methodological problems in dream collection studies remain unchanged. The basic sampling procedures are often not reported. The failure in 81% of studies to use a fixed protocol in order to protect against interviewer bias is of great concern. Moreover, the low rate of reporting of the mode of awakening in laboratory studies (22%) also contributes to the problems in characterizing the dream lives of patient groups. The neglect of these crucial parameters contributes to our difficulties in assessing the adequacy of a study, in comparing the results from one study to another and in being able to resolve discrepancies between studies.

The method of scoring dream content

Little or no attention is paid in the articles under review to reporting on protocol preparation (6%), on whether the raters were "blind" (18%), or on the reliability of the scorers (11%) (see Table 17.9). In 33% of the studies, only one rater performed the rating. The type of scale and source of the scale is given in only 38% of the studies. These issues are reported at a lower frequency than in the earlier review, in which they were also given inadequate attention. The failure to use more than one "blind" rater whose reliability has been established

and checked periodically attributes a remarkable faith in the objectivity and consistency of human performance.

The limited use of standard rating scales (18% currently and 22% in the prior review) highlights a core problem in the field of dream research. Failure to describe the basic elements in the manifest dream report with a standard device, and then to build special or inferential scoring from these identifiable parameters, severely limits the development of a body of knowledge in which one study builds on another and in which any given study is potentially relatable to another.[32]

The nature of the statistical analysis

The percentage of studies in which any statistical analyses was reported is only 33% currently compared to 41% earlier (see Table 17.10).[23] In the 30 studies that reported statistical results, the number of tests is reported in almost all studies (90%), the statistics are by and large appropriate to the design and the data (80%), and the comparisons are generally preplanned (77%).

The content of the dream in psychopathologic states can be developed only from studies with acceptable statistical treatment of the data collected. This limits the core data pool to 30 of the

Table 17.9 Dream content scoring variables

| | Percentage of studies | | | |
| | 1975[23] | | 2005 | |
	Yes	No	Yes	No
A. Protocol preparation	28	72	6	94
B. "Blind" raters	28	72	18	82
C. Reliability reported	27	73	11	89
D. Number of scorers used				
1. One	68		33	
2. Two	27		6	
3. Three	5		0	
4. Not given	0		61	
E. Type of scale				
1. Item	20[a]		32	
2. Thematic	77[a]		6	
3. Not given	3[a]		62	
F. Scale source				
1. Standard	22[a]		18	
2. Ad hoc	78[a]		20	
3. Not given	0[a]		62	

[a] Modified.

Table 17.10 Nature of the statistical analysis

| | Percentage of studies | | | |
| | 1975[23] | | 2005 | |
	Yes	No	Yes	No
A. Are statistical tests reported?	41	59	33	67
B. Are the numbers of tests reported?[a]	81	19	90	10
C. Are the statistics appropriate?[a]				
1. To the design?	81	19	80	20
2. To the data?	55	45	80	20
D. What is the nature of the comparison?[a]				
1. Preplanned	48		77	
2. Post hoc	52		23	
E. Significant results[a]	81		73	

[a] Studies that reported statistical tests.

90 studies. The nonstatistical articles—either case reports or studies—provide leads for further, more systematic study. However, only a small group of studies remain for characterizing a large area of interest.

DREAM CONTENT RESULTS

In 1976, Frosch[6] wrote a critical review of the "The Psychoanalytic Contributions to the Relationship between Dreams and Psychosis" that provided some partial answers to Freud's questions about the relationship. Frosch concluded: (1) "...that although there are many apparent similarities between dreams and psychosis, they do differ in some basic respects; certainly insofar as the factors are concerned which play a role in their production"; (2) "...that there is no consensus as to whether the manifest dream was of itself a meaningful guide to the presence of psychosis...[some] felt it was the latent content that was most telling [while others] seemed to feel that there might be features about the manifest form and content which could be of significance, indicating the presence of a psychosis. It [was] felt by some investigators that the patient's attitude toward the dream, difficulties in differentiating the dream from reality, and the persistence of dreamlike states invading the waking life [that] may offer clues to the possibly psychotic nature of dreams"; and, (3) "...[in regard] to whether there are dreams which presage psychosis, there was some suggestive evidence that this was the case."

Schizophrenia

Schizophrenics[23] are less interested in their dreams and their dreams are more primitive (i.e., less complex and more direct, sexual, anxious, and hostile, and showed evidence of their thought disorder in being more bizarre and implausible). In a mixed patient population that included schizophrenics, Lesse[33] reported that mounting anxiety could increase or decrease dream reporting. With increasing anxiety, motion and affect in the manifest dream are increased. In addition, a decrease in affect in the manifest dream is the first change that is seen during successful phenothiazine therapy.

Strangers were the most frequent dream characters in schizophrenics. Hallucinations and dream content were relatable and the degree of paranoia—awake and in dreaming—was similar, contrary to Freud's compensatory view of waking and dreaming in paranoia.[34] An updated literature review yielded only eight articles on dream content in schizophrenic states,[35–42] a surprisingly small number that added little to our understanding.

Lobotomized schizophrenics[43] had a lower dream recall rate in the laboratory (10.4%) than non-schizophrenics (46.7%), but both were lower than in another study.[44]

Depression

The depressed[23] patient was found to dream as frequently as the nondepressed, but the dreams were shorter and had a paucity of traumatic or depressive content, even after the depression had lifted. Family members were more frequent in their dreams. When hostility was present, it could be directed at or away from the dreamer, while in schizophrenia it was directed at the dreamer. In their dreams, the depressed had more friendly and fewer aggressive interactions than schizophrenics, but more failure and misfortune. With clinical improvement, hostility decreased while intimacy, motility, and heterosexuality increased.

The view that begins to emerge more clearly from the updated review[32,45–65] is that, in depression, there is a decrease in the frequency[47,49–52,55,64,65] and length[51,52,55,56,64] of the dream reports. Their dreams are often commonplace, but at times have content characteristics[45,50,52,57] of high interest.

There is an increase in the dreaming of death themes in depressed suicidal patients and in bipolar patients before becoming manic.[45,52] An increase in family roles in the dreams of the depressed may also be the case.[46,51,55]

Masochism in the dreams of the depressed[32,53,54,57–59] appears more clearly in women than in men and is more likely a trait than a state characteristic. It was evident that a past focus[49–53,56,58] was not universal in the dreams of the depressed, nor was it unique to the depressed state. Affects such as anxiety and hostility were not prominent in the dreams of the depressed.[45,53,57,61] The content of their dreams may have prognostic significance for the response of the depressed patient to treatment or the spontaneous outcome of the depression.[49,54]

A most striking implication of these findings about dreaming in the depressed is that the affective state of the dreamer covaries with the content of the dream.[45,48,50,55-57] In addition, changes in dreams across the night may contribute to the dreamer's coping capacity,[56,57,60,63,65] as was suggested by Kramer.[62] Changes in dream content across the night alter the affective condition of the dreamer and contribute to the adaptive state of the dreamer during the next day. Mood regulation processes may have implications for the treatment of depression. Untreated depressed subjects reporting more negative dreams at the beginning of the night than at the end of the night were more likely to be in remission after 1 year.[63] In contrast, a failure to self-regulate mood was associated with a suicidal tendency.[65]

Posttraumatic stress disorder

A widespread interest in PTSD has developed since the Vietnam War, including the dreams of such patients.[68-128] PTSD was only included in the official nomenclature of the American Psychiatric Association in 1980, although it had been described in the psychiatric literature for over 100 years.[66] In a review article, Ross and his colleagues[68] attempt to demonstrate that a sleep disturbance is the hallmark of PTSD. They base their hypothesis on the mentation difference between REM and non-REM sleep. They characterize the dreams of PTSD patients as vivid, affect laden, disturbing, outside the realm of current waking experience (although representative of an earlier life experience), repetitive, stereotyped, and easy to recall. They are of the opinion that the dream disturbance is relatively specific to the disorder[69] and that PTSD may fundamentally be a disorder of the REM sleep mechanism. However, as the nightmare in REM sleep occurs early in the night when there is less REM and is associated with gross body movements, the authors[68] acknowledge that abnormal non-REM sleep mechanisms may be involved as well, and speculate that the neural circuitry involved in PTSD may be similar to that in accentuated startle behavior. Ross et al.[68] take exception to Reynold's[69] suggestion that the dream in PTSD is the same that occurs in traumatized depressives, pointing out that the dreams of traumatized depressives are not dreamlike and do not incorporate the trauma.[56]

Ross's group[68] sees the dream in PTSD as repetitive and, more importantly, stereotyped.

In contrast, Kramer[71] views disturbing dreams as the hallmark of PTSD, rather than sleep disturbance. Green and collaborators[72] have suggested that the unique aspect of PTSD is indeed the intrusive symptoms, including intrusive images and recurrent dreams and nightmares. They point out that not all dreams are direct recapitulations of the trauma. For them, these intrusive images may be the hallmark of PTSD. The view of Green's group[72] was based on the suggestion by Brett and Ostroff[73] that there has been a neglect of posttraumatic imagery, which they postulate is the core of PTSD. They lament the lack of research into the range, content, and patterning of this imagery. Interestingly, Fisher and coworkers[74] point out that trauma sufferers may have disturbing arousals that can come out of both REM and non-REM (stage 4 and stage 2) sleep. This is a view that Schlosburg and Benjamin,[75] Kramer and Kinney,[76] and Dagan et al.[124] confirm. Questions arise as to whether the sleep disturbance in PTSD involves more than REM sleep mechanisms and whether the imagery and dreams reported by PTSD patients are (1) stereotyped and (2) REM bound, as Ross's group[68] postulate.

There has been a relative lack of attention to the range, content, and patterning of the nightmares in PTSD. It was found that there can be different types of nightmares,[84,85,88] that themes may be unrelated to the trauma,[67,90,128] with one study finding a strong association between recalled dream content and a war experience,[127] and that the traumatic dreams can change across time.[100] The traumatic nightmare is seen to reflect classical Freudian dream work mechanisms,[77,79,80,84-86] and not to be a meaningless reenactment of the trauma.

An adequate characterization of the phenomenology of the disturbing dream in PTSD remains to be done. The dream experience is disturbing, but this may be more of a reaction to the dream than the dream itself.[120] The affect-laden nature of the disturbing dream cannot be confirmed, and expectations may influence the perception of what the dream should be like in PTSD.[129] The content of the disturbing dream may be outside of the realm of current waking experience, but it is linked to earlier childhood experiences[77,79,80,85-87] and can be reactivated later in life.[82-84,91-93] The vividness of the dream has not been adequately addressed. The dream in PTSD is not easily recalled. Patients with active PTSD have a lower dream recall rate[67,104,106] than normal subjects, but higher than well-adjusted former PTSD patients.[113]

A consensus has begun to emerge from the PTSD dream literature suggesting that the hallmark of PTSD is a disturbance in psychological dreaming and possibly of non-REM sleep early in the night. Disturbed dreaming covaries with combat exposure[117] and being tortured,[118] not with the complaint of a sleep disturbance. The disturbing dream tends to occur early during sleep,[108] as do increases in movement,[67,119] spontaneous awakenings,[120,121] autonomic discharge,[122] elevated arousal threshold,[123–125] and a heightened startle response.[125,126] The disturbing dream is not sleep stage bound and may emerge out of REM or non-REM sleep.[74–76] Stereotypical dream content is not the sine qua non of the dream in PTSD.

The failure or avoidance of dream recall may be an adaptational strategy in PTSD.[113,114]

Eating disorders

Anorexics and bulimics[130–140] both report dreams, and their dreams are seen as useful in therapy. For eating disorder patients, the rate of dream recall is low on self-report questionnaires, but normal in the sleep laboratory. Dreaming of food is high in eating disorder patients, and is higher in bulimics than anorexics. Aggressive dreams are less common in eating disorder patients than normal subjects.

Brain damage

The previous review[23] found five articles on brain-damaged patients. These studies reported that there was a decrease in dream reporting with age and dementia.[154,155] The more recent studies[141–146] of brain-injured patients report the value of dream exploration in psychotherapy with these patients. A questionnaire study of aged individuals found no relationship between dream report frequency and the degree of brain atrophy on computed tomography scan. Repetitive visual imagery in brain-damaged patients was not REM bound. Focal brain damage studies suggest the anatomical substrate for dream formation. The dream content of right hemispherectomized patients was similar to the content of control subjects, which suggests that the left hemisphere has a critical role in dream generation.[146]

Alcoholism and drug abuse

In the previous review,[23] the dreams of alcoholics could be distinguished from those of non-alcoholics. The alcoholic had more oral references in his or her dreams, was more often the object of aggression, and had fewer sexual interactions. Those detoxifying alcoholics who dreamed about drinking maintained sobriety longer. The implication of dreaming about drinking or drug use as a predictor of abstinence remains unclear.[147–153] However, it raises the possibility that what one dreams about may have adaptive significance. Drug dreams in patients with cocaine dependence and bipolar disorder are similar to those in patients with pure substance dependence.[153]

CONCLUSION

It is apparent that the mysteries of psychosis have not been revealed through the study of dreams. The paucity of studies in some conditions and the relative lack of scientific rigor throughout continue to plague the study of dreams in psychiatric conditions. However, in some areas, such as depression and PTSD, we do know more about dreaming than we did previously.

Detailed phenomenological descriptions are needed of the dream experience in normal subjects and in the various psychiatric illnesses—both in and out of the laboratory—utilizing quantitative techniques in order to capture various aspects of the experience. These results can then be statistically compared in between-group and repeated-measure experiments, as was recommended in a previous review.[23] Further, study of the dream construction process, of the dream as a dependent measure and applying the manipulations suggested by Tart[156] and the analytic techniques described by Kramer et al.,[157] Montangero,[158] and Cipolli[159] would enhance our understanding of the cognitive process in dreaming.

The most intriguing insight that emerges from this review is that what one does or does not dream about may contribute to the waking adaptational process.[57,62,113,114,150,151,160] Manipulating the dream by the controlled incorporation of characters or events into the dream[161] and assessing the daytime consequences would treat the dream as an independent variable and contribute to our understanding of the functional significance of dreaming.

ACKNOWLEDGMENT

The assistance of Mike Douglas, Valerie Ratchford, and Linda Kittrell, the library staff at Bethesda Oak Hospital, Cincinnati, OH, is gratefully acknowledged, as well as the support of Lydia Friedman, James Verlander, and the library staff of the Maimonides Medical Center, Brooklyn, New York.

REFERENCES

1. Kant I. Quoted in Freud S. *The Interpretation of Dreams* (Standard ed.). London: Hogarth Press, 1953, Vol. IV and V, p. 90.

2. Schopenhauer A. Quoted in Freud S. *The Interpretation of Dreams* (Standard ed.). London: Hogarth Press, 1953, Vol. IV and V, p. 90.

3. Jung C. *The Psychology of Dementia Praecox*. London: Princeton University Press, 1960, p. 86.

4. Taylor J, Holmes G, Walshe FMR (EDs). *Selected Writings of John Hughlings Jackson* London: Hodder & Stoughtonon, 1932, Vol. 2, 510 pp.

5. Freud S. *The Interpretation of Dreams* (Standard ed.). London: Hogarth Press, 1953, Vol. IV and V.

6. Frosch J. Psychoanalytic contributions to the relationship between dreams and psychosis—A critical survey. *Int J Psychoanal Psychother* 1976; 5: 39–63.

7. Hartmann E. *The Nightmare: The Psychology and Biology of Terrifying Dreams*. London: Harper Row, 1981.

8. Kramer M, Baldridge B, Whitman R et al. An exploration of the manifest dream in schizophrenia and depressed patients. *Dis Nerv Syst* 1969; 30: 126–136.

9. Jones R. Dream interpretation and the psychology of dreaming. *J Am Psychoanal Assoc* 1965; 13: 304–319.

10. Freud S. *On Dreams* (Standard ed.). London: Hogarth Press, 1958, Vol. V, p. 667.

11. Hall C, Van de Castle R. *The Content Analysis of Dreams*. New York, NY: Appleton, 1966.

12. LaBerge S, Rheingold H. *Exploring the World of Lucid Dreaming*. New York, NY: Ballentine, 1990.

13. Boss M. *The Analysis of Dreams*. New York, NY: Philosophical Library, 1958.

14. Kramer M, Winget C, Roth T. *Problems in the Definition of the REM Dream: Sleep, 1974*. Basel: S. Karger, 1975.

15. Kramer M (Ed.). *Dream Psychology and the New Biology of Dreaming*. Springfield, IL: Charles C. Thomas Publishers, 1969.

16. Weiss H. Oneirocritica americana. *Bull New York Public Library* 1944; 48: 519–541.

17. Aserinsky E, Kleitman N. Regularly occurring periods of eye motility and concomitant phenomena during sleep. *Science* 1953; 118: 273–274.

18. Aserinsky E, Kleitman N. Two types of ocular motility in sleep. *J Appl Physiol* 1955; 8: 1–10.

19. Ramsey G. Studies of dreaming. *Psychol Bull* 1953; 50: 432–455.

20. Kramer M. Manifest dream content in psychopathological states. In: Kramer M (Ed.). *Dream Psychology and the New Biology of Dreaming*. Springfield, IL: Charles C. Thomas Publishers, 1969, pp. 377–396.

21. Kramer M. Manifest dream content in normal and psychopathological states. *Arch Gen Psychiatry* 1970; 22: 149–159.

22. Kramer M, Roth T. Dreams in psychopathological groups: A critical review. In: Williams R, Karacan I (Eds). *Sleep Disorders: Diagnosis and Treatment*. New York, NY: John Wiley & Sons, 1978, pp. 323–349.

23. Kramer M, Roth T. Dreams in psychopathology. In: Wolman B (Ed.). *Handbook of Dreams: Research, Theories and Applications*. New York, NY: Von Norstrand Reinhold, Co., 1979, pp. 361–387.

24. Kramer M. Dream content in psychiatric conditions: An overview of sleep laboratory studies. In: Perris C, Struwe G, Jansson B (Eds). *Biological Psychiatry*. New York, NY: Elsevier/North Holland Biomed/-Cal Press, 1981, pp. 306–309.

25. Mellen R, Duffey T, Craig S. Manifest content in the dreams of clinical population. *J Mental Health Counseling* 1993; 15: 170–183.

26. Benca R. Sleep in psychiatric disorders. *Neur Clinics* 1996; 14: 739–764.
27. American Psychiatric Association. *Diagnostic and Statistical Manual of Mental Disorders* (4th ed.). Washington, DC: American Psychiatric Association, 1994.
28. Kramer M, Winget C, Whitman R. A city dreams: A survey approach to normative dream content. *Am J Psychiatry* 1971; 127: 1350–1356.
29. Piccione P, Thomas S, Roth T, Kramer M. Incorporation of the laboratory situation in dreams. *Sleep Res* 1976; 5: 120.
30. Whitman R, Kramer M, Baldridge B. Which dream does the patient tell? *Arch Gen Psychiatry* 1963; 8: 277–282.
31. Goodenough D, Lewis H, Shapiro A et al. Dream reporting following abrupt and gradual awakenings from different types of sleep. *J Pers Soc Psychol* 1965; 2: 170–179.
32. Clark J, Trinder J, Kramer M et al. An approach to the content analysis of dream content scales. *Sleep Res* 1972; 1: 118.
33. Lesse S. Psychiatric symptoms in relationship to the intensity of anxiety. *Psychother Psychosom* 1974; 23: 94–102.
34. Freud S. *Some Neurotic Mechanisms in Jealousy, Paranoia and Homosexuality* (Standard ed.). London: Hogarth Press, 1955, Vol. XVIII.
35. Ushijima S. On recovery from the post psychotic collapse in schizophrenia. *Jpn J Psychiatry Neurol* 1988; 42: 199–207.
36. Deutsch H. A case that throws light on the mechanism of regression in schizophrenia. *Psychoanal Rev* 1985; 72: 1–8.
37. Meloy J. Thought organization and primary process in the parents of schizophrenics. *Br J Med Psychol* 1984; 57: 279–281.
38. Wilmer H. Dream seminar for chronic schizophrenic patients. *Psychiatry* 1982; 45: 351–360.
39. Ohira K, Kato N, Namura I et al. A psychopathology of schizophrenic dreaming: A feeling of passivity. *Sleep Res* 1979; 8: 170.
40. Van de Castle R. Manifest content of schizophrenic dreams. *Sleep Res* 1974; 3: 126.
41. Hadjez J, Stein D, Gabbay U et al. Dream content of schizophrenic, nonschizophrenic mentally ill, and community control adolescent. *Adolescence* 2003; 38: 331–342.
42. Stompe T, Ritter K, Ortwein-Swoboda G et al. Anxiety and hostility in the manifest dreams of schizophrenic patients. *J Nerv Ment Dis* 2003; 191: 806–812.
43. Jus A, Jus K, Villeneuve A et al. Studies on dream recall in chronic schizophrenic patients after prefrontal lobotomy. *Biol Psychiatry* 1973; 6: 275–293.
44. Solms M. *The Neuropsychology of Dreams: A Clinico-anatomical Study*. Mahwah, NJ: Lawrence Erlbaum Associates, 1997.
45. Beauchemin K, Hays P. Prevailing mood, mood changes and dreams in bipolar disorder. *J Affect Disord* 1995; 35: 41–49.
46. Brenman E. Separation: A clinical problem. *Int J Psychoanal* 1982; 63: 303–310.
47. Mathew R, Largen J, Cleghorn J. Biological symptoms of depression. *Psychosom Med* 1979; 41: 439–443.
48. Levitan H. The relationship between mania and the memory of pain: A hypothesis. *Bull Menninger Clin* 1977; 41: 145–161.
49. Greenberg R, Pearlman C, Blacher R et al. Depression: Variability of intrapsychic and sleep parameters. *J Am Acad Psychoanal* 1990; 18: 233–246.
50. Beauchemin K, Hays P. Dreaming away depression: The role of REM sleep and dreaming in affective disorders. *J Affect Disord* 1996; 41: 125–133.
51. Barrett D, Loeffler M. Comparison of dream content of depressed versus non-depressed dreamers. *Psychol Rep* 1992; 70: 403–406.
52. Firth S, Blouin J, Natarajan C et al. A comparison of the manifest content in dreams of suicidal, depressed and violent patients. *Can J Psychiatry* 1986; 31: 48–51.
53. Dow B, Kelsoe J, Gillen J. Sleep and dreams in Vietnam PTSD and depression. *Biol Psychiatry* 1996; 39: 42–50.
54. Cartwright R, Wood E. The contribution of dream masochism to the sex ratio difference in major depression. *Psychiatry Res* 1993; 46: 165–173.
55. Riemann D, Low H, Schredl M et al. Investigations of morning and laboratory dream recall and content in depressive patients during baseline conditions and under antidepressive treatment with trimipramine. *Psychiatr J Univ Ott* 1990; 15: 93–99.

56. Cartwright R, Lloyd S, Knight S et al. Broken dreams. A study of the effects of divorce and depression on dream content. *Psychiatry* 1984; 47: 251–259.

57. Trenholme I, Cartwright R, Greenberg G. Dream dimension differences during a life change. *Psychiatry Res* 1984; 12: 35–45.

58. Hauri P. Dreams of patients remitted from reactive depression. *J Abnorm Psychol* 1976; 85: 1–10.

59. Beck A. *Depression: Clinical, Experimental and Theoretical Aspects.* New York, NY: Harper and Row, 1967.

60. Cartwright R. Dreams that work: The relation of dream incorporation to adaptation to stressful events. *Dreaming* 1991; 1: 3–9.

61. Strauch I, Meier B. *In Search of Dreams: Results of Experimental Dream Research.* Albany, NY: State University of New York Press, 1966, p. 234.

62. Kramer M. The selective mood regulatory function of dreaming: An update and revision. In: Moffitt A, Kramer M, Hofmann R (Eds). *The Functions of Dreaming.* Albany, NY: State University of New York Press, 1993, pp. 139–195.

63. Cartwright R, Young AM, Mercer P et al. Role of REM sleep and dream variables in the prediction of remission from depression. *Psychiatry Res* 1998; 80: 249–255.

64. Armitage R, Rochlen A, Fitch T et al. Dream recall and major depression: A preliminary report. *Dreaming* 1995; 5: 189–197.

65. Agargun YM, Cartwright R. REM sleep, dream variables and suicidality in depressed patients. *Psychiatry Res* 2003; 119: 33–39.

66. Erichson J. *On Concussion of the Spine.* New York, NY: Bermingham, 1882. Cited in: Modlin H. Is there an assault syndrome. *Bull Am Acad Psychiatry Law* 1985; 13: 139–145.

67. Mellman T, Kulick-Bell R, Ashlock L et al. Sleep events among veterans with combat-related post traumatic stress disorder. *Am J Psychiatry* 1995; 152: 110–115.

68. Ross R, Ball W, Sullivan K et al. Sleep disturbances as the hallmark of post traumatic stress disorder. *Am J Psychiatry* 1989; 146: 697–707.

69. Reynold C. Sleep disturbance in post traumatic stress disorder: Pathogenic or epiphenomenal? *Am J Psychiatry* 1989; 146: 695–696.

70. Ross R, Ball W, Sullivan K et al. Sleep disturbance in post traumatic stress disorder. *Am J Psychiat* 1990; 147: 374.

71. Kramer M. Dream disturbances. *Psychiatry Ann* 1979; 9: 366–376.

72. Green B, Lindy J, Grace M. Post traumatic stress disorder: Toward DSM IV. *J Nerv Ment Dis* 1985; 173: 406–411.

73. Brett E, Ostroff R. Imagery and post traumatic stress disorder: An overview. *Am J Psychiatry* 1985; 142: 417–424.

74. Fisher C, Kahn E, Edwards A et al. A physiological study of nightmares and night terrors. *J Nerv Ment Dis* 1973; 2: 275–298.

75. Schlosberg A, Benjamin M. Sleep patterns in three acute combat fatigue cases. *J Clin Psychiatry* 1978; 39: 546–549.

76. Kramer M, Kinney L. Sleep patterns in trauma victims with disturbed dreaming. *Psychiatr J Univ Ott* 1988; 13: 12–16.

77. Lansky M. Nightmares of a hospitalized rape victim. *Bull Menninger Clin* 1995; 59: 4–14.

78. Straker G. Integrating African and western healing practices in South Africa. *Am J Psychother* 1994; 48: 455–467.

79. Lansky M. The transformation of affect in post traumatic nightmares. *Bull Menninger Clin* 1991; 55: 470–490.

80. Silvan-Adams A, Silvan M. "A dream is the fulfillment of a wish": Traumatic dream, repetition compulsion and the pleasure principle. *Int J Psychoanal* 1990; 71: 513–522.

81. Modlin H. Is there an assault syndrome? *Bull Am Acad Psychiatry Law* 1985; 13: 139–145.

82. Van Dyke C, Zilberg N, McKinnon J. Post traumatic stress disorder: A thirty year delay in a World War II veterans. *Am J Psychiatry* 1985; 142: 1070–1073.

83. Wells B, Chu C, Johnson R et al. Buspirone in the treatment of post traumatic stress disorder. *Pharmacotherapy* 1991; 11: 340–343.

84. Siegel L. Holocaust survivors in Hasidic and ultra-orthodox Jewish populations. *J Contemp Psychother* 1980; 11: 5–31.

85. Dowling S. Dreams and dreaming in relation to trauma in childhood. *Int J Psychoanal* 1983; 63: 157–166.

86. de Saussure J. Dreams and dreaming in relation to trauma in childhood. *Int J Psychoanal* 1982; 63: 167–175.

87. Puk G. Treating traumatic memories: A case report on the eye movement desensitization procedure. *J Behav Ther Exp Psychiat* 1991; 22: 149–151.

88. Schreuder J. Post traumatic re-experiencing in older people: Working through or covering up? *Am J Psychother* 1996; 50: 231–242.

89. Helzer J, Robins L, McEvoy M. Post traumatic stress disorder in the general population: Findings of the epidemiologic catchment area survey. *N Engl J Med* 1987; 317: 1630–1634.

90. Watson I. Post traumatic stress disorder in Australian Prisoners of the Japanese: A clinical study. *Aust N Z J Psychiatry* 1993; 27: 20–29.

91. Kuch K, Cox B. Symptoms of PTSD in 124 survivors of the holocaust. *Am J Psychiatry* 1992; 149: 337–340.

92. Mollica R, Wyshak G, Lavelle J. The psychosocial impact of the war trauma and torture on Southeast Asian refugees. *Am J Psychiatry* 1987; 144: 1567–1572.

93. Goldstein G, van Kammen W, Shelly C et al. Survivors of imprisonment in the Pacific theater during World War II. *Am J Psychiatry* 1987; 144: 1210–1213.

94. Burstein A. Dream disturbances and flashbacks. *J Clin Psychiatry* 1984; 45: 46.

95. Woodward S, Arsenault E, Bliwise D et al. The temporal distribution of combat nightmares in Vietnam combat veterans. *Sleep Res* 1991; 20: 152.

96. Woodward S, Arsenault E, Bliwise D et al. Physical symptoms accompanying dream reports in combat veterans. *Sleep Res* 1991; 20: 153.

97. Horowitz M, Wilner N, Kaltreider N et al. Signs and symptoms of post traumatic stress disorder. *Arch Gen Psychiatry* 1980; 37: 85–92.

98. Wilkinson C. Aftermath of a disaster: The collapse of the Hyatt Regency Hotel skywalks. *Am J Psychiatry* 1983; 140: 1134–1139.

99. Terr L. Children of chowchilla. In: *The Psychoanalytic Study of the Child*. New Haven, CT: Yale University Press, 1979, pp. 547–623.

100. Titchener J, Kapp F. Family and character change at Buffalo Creek. *Am J Psychiatry* 1976, 133: 295–299.

101. Brockway S. Group treatment of combat nightmares in post-traumatic stress disorder. *J Contemp Psychother* 1987; 17: 270–284.

102. Kramer M, Schoen L, Kinney L. Nightmares in Vietnam veterans. *J Am Acad Psychoanal* 1987; 15: 67–81.

103. Kinzie J, Sack R, Riley C. The polysomnographic effects of clonidine on sleep disorders in post traumatic stress disorder: A pilot study with Cambodian patients. *J Nerv Ment Dis* 1994; 182: 585–587.

104. Hefez A, Metz L, Lavie P. Long-term effects of extreme situational stress on sleep and dreaming. *Am J Psychiat* 1987; 144: 344–347.

105. Fisher C, Byrne J, Edwards A et al. A physiological study of nightmares. *J Am Psychoanal Assoc* 1970; 18: 747–782.

106. Dagan Y, Lavie P. Subjective and objective characteristics of sleep and dreaming in war related PTSD patients: Lack of relationships. *Sleep Res* 1991; 20A: 270.

107. Deekin M, Bridenbaugh R. Depression and nightmares among Vietnam veterans in a military psychiatry outpatient clinic. *Mil Med* 1987; 152: 590–591.

108. van der Kolk B, Blitz R, Burr W et al. Nightmares and trauma: A comparison of nightmares after combat with lifelong nightmares in veterans. *Am J Psychiatry* 1984; 14: 187–190.

109. Terr L. Life attitudes, dreams and psychic trauma in a group of "normal" children. *J Am Acad Child Psychiatry* 1983; 22: 221–230.

110. Defazio V, Rustn S, Diamond A. Symptom development in Vietnam era veterans. *Am J Orthopsychiatry* 1975; 45: 158–163.

111. Archibald H, Long D, Miller C et al. Gross stress reaction in combat—A 15 year follow-up. *Am J Psychiatry* 1962; 119: 317–322.

112. Ross R, Ball W, Dinges D et al. Rapid eye movement sleep disturbance in post traumatic stress disorder. *Biol Psychiatry* 1994; 35: 195–202.

113. Lavie P, Kaminer H. Dreams that poison sleep: Dreaming in Holocaust survivors. *Dreaming* 1991; 1: 11–21.

114. Kramer M, Schoen L, Kinney L. The dream experience in dream-disturbed Vietnam veterans. In: van der Kolk B (Ed.). *Post Traumatic Stress Disorders: Psychological*

and Biologic Sequelae. Washington, DC: American Psychiatric Association Press, 1984, pp. 81–95.

115. Kellett S, Beail N. The treatment of chronic post-traumatic nightmares using psychodynamic–interpersonal psychotherapy: A single case study. *Br J Med Psychol* 1997; 70: 35–49.

116. Terr L. Chowchilla revisited: The effects of psychic trauma four years after a school-bus kidnapping. *Am J Psychiatry* 1983; 140: 1543–1550.

117. Neylan T, Marmar C, Metzler M et al. Sleep disturbances in the Vietnam generation: Findings from a nationally representative sample of male veterans. *Am J Psychiatry* 1998; 155: 929–933.

118. Shrestha N, Sharma B, Van Ommeren M et al. Impact of torture on refugees within the development world: Symptomatology among Bhutanese refugees in Nepal. *JAMA* 1998; 280: 443–448.

119. Lavie P, Hertz G. Increased sleep motility and respiration rates in combat neurotic patients. *Biol Psychiatry* 1979; 14: 983–987.

120. Kramer M, Schoen L, Kinney L. Psychological and behavioral features of disturbed dreamers. *Psychiatr J Univ Ott* 1984; 9: 102–106.

121. Schoen L, Kramer M, Kinney L. Arousal patterns in non-REM dream disturbed veterans. *Sleep Res* 1983; 12: 315.

122. Wilmer H. The healing nightmare: War dreams of Vietnam veterans. In: Barrett D (Ed.). *Trauma and Dreams.* Cambridge, MA: Harvard University Press, 1996, p. 92.

123. Schoen L, Kramer M, Kinney L. Auditory thresholds in the dream disturbed. *Sleep Res* 1984; 13: 102.

124. Dagan Y, Lavie P, Bleich A. Elevated awakening threshold in sleep stage 3–4 in war related post-traumatic stress disorder. *Biol Psychiatry* 1991; 30: 618–622.

125. Kramer M, Kinney L. Vigilance and avoidance during sleep in U.S. Vietnam War veterans with post traumatic stress disorder. *J Nerv Ment Dis* 2003; 191: 1–3.

126. Kinney L, Schoen L, Kramer M. Responsivity of night terror patients in sleep. *Sleep Res* 1983; 12: 193.

127. Schreuder BJ, van Egmond M, Klein WC et al. Daily reports of posttraumatic nightmares and anxiety dreams in Dutch war victims. *J Anxiety Disord* 1998; 12: 511–524.

128. Esposito K, Benitez A, Barza L et al. Evaluation of dream content in combat-related PTSD. *J Trauma Stress* 1999; 12: 681–687.

129. Taub J, Kramer M, Arand D et al. Nightmare dreams and nightmare confabulations. *Compr Psychiatry* 1978; 19: 285–291.

130. Jackson C, Tabin J, Russell J et al. Themes of death: Helmut Thoma's "Anorexia nervosa" (1967)—A research note. *Int J Eat Disord* 1993; 14: 433–437.

131. Jackson C, Beumont P, Thornton C et al. Dreams of death: Von Weizsacker's dreams in so-called endogenic anorexia—A research note. *Int J Eat Disord* 1993; 13: 329–332.

132. Wilson C. Dream interpretation. In: Wilson C (Ed.). *Fear of Being Fat: The Treatment of Anorexia Nervosa and Bulimia.* New York, NY: Jason Aronson, 1983, pp. 245–254.

133. Wilson C. The fear of being fat and anorexia nervosa. *Int J Psychoanal Psychother* 1982–1983; 9: 233–255.

134. Wells L. Anorexia nervosa: An illness of young adults. *Psychiatr Q* 1980; 52: 270–282.

135. Sprince M. Early psychic disturbances in anorexic and bulimic patients as reflected in the psychoanalytic process. *J Child Psychother* 1984; 10: 199–215.

136. Levitan H. Implications of certain dreams reported by patients in a bulimic phase of anorexia nervosa. *Can J Psychiatry* 1981; 26: 228–231.

137. Hudson J, Bruch H, DeTrinis J et al. Content analysis of dreams of anorexia nervosa patients. *Sleep Res* 1978; 7: 176.

138. Brink S, Allan J. Dreams of anorexic and bulimic women: A research study. *J Anal Psychol* 1992; 37: 275–297.

139. Frayn D. The incidence and significance of perceptual qualities in the reported dreams of patients with anorexia nervosa. *Can J Psychiatry* 1991; 36: 517–520.

140. Dippel B, Lauer C, Riemann D, Majer-Trendel K, Krieg JC, Berger M. Sleep and dreams in eating disorders. *Psychother Psychosom* 1987; 48: 165–169.

141. Stern B, Stern J. On the use of dreams as a means of diagnosis of brain injured patients. *Scand J Rehabil Med Suppl* 1985; 12: 44–46.

142. Stern M, Stern B. Psychotherapy in cases of brain damage: A possible mission. *Brain Inj* 1990; 4: 297–304.

143. Benyakar M, Tadir M, Groswasser Z et al. Dreams in head-injured patients. *Brain Inj* 1988; 2: 351–356.

144. Nathan R, Rose-Itkoff C, Lord G. Dreams, first memories and brain atrophy in the elderly. *Hillside J Clin Psychiatry* 1981; 3: 139–148.

145. Askenasy J, Gruskiewicz J, Braun J et al. Repetitive visual images in severe war head injuries. *Resuscitation* 1986; 13: 191–201.

146. McCormik L, Nielsen M, Ptito M et al. REM sleep dream mentation in right hemispher-ectomized patients. *Neuropsychologia* 1997; 35: 695–701.

147. Cernovsky Z. MMPI and nightmares in male alcoholics. *Percept Mot Skills* 1985; 61: 841–842.

148. Cernovsky Z. MMPI and nightmare reports in women addicted to alcohol and other drugs. *Percept Mot Skills* 1986; 62: 717–718.

149. Fiss H. Dream content and response to with-drawal from alcohol. *Sleep Res* 1980; 9: 152.

150. Christo G, Franex C. Addicts' drug related dreams: Their frequency and relationship to six-month outcomes. *Subst Use Misuse* 1996; 31: 1–15.

151. Denizen N. Alcoholic dreams. *Alcohol Treat Q* 1988; 5: 133–139.

152. Reid SD, Simeon DT. Progression of dreams of crack cocaine abusers as a predictor of treatment outcome: A preliminary report. *J Nerv Ment Dis* 2001; 189: 854–857.

153. Yee T, Perantie DC, Dhanani N et al. Drug dreams in outpatients with bipolar disorder and cocaine dependence. *J Nerv Ment Dis* 2004; 192: 238–242.

154. Turner J, Graffam J. Deceased loved ones in the dreams of mentally retarded adults. *Am J Ment Retard* 1987; 92: 282–289.

155. Voelm C, Kossor M, Duran E. Dream work with the mentally retarded. *Psychiatr J Univ Ott* 1988; 13: 85–90.

156. Tart C. From spontaneous event to lucidity: A review of attempts to consciously control nocturnal dreaming. In: Wolman B (Ed.). *Handbook of Dreams: Research, Theories and Applications*. New York, NY: Von Nostrand Reinhold, Co., 1979, pp. 226–268.

157. Kramer M, Whitman R, Baldridge B et al. Patterns of dreaming: The interrelationship of the dreams of a night. *J Nerv Ment Dis* 1964; 139: 426–439.

158. Montangero J. Dream, problem solving and creativity. In: Corrado C, Foulkes D (Eds). *Dreaming as Cognition*. New York, NY: Harvester-Wheatsheaf, 1993, pp. 93–113.

159. Cipolli C. The narrative structure of dreams: Linguistic tools of analysis. In: Horne J (Ed.). *Sleep '90*. Bochum: Pontenagel Press, 1990, pp. 281–84.

160. Koulack D. *To Catch a Dream: Exploration of Dreaming*. Albany, NY: State University of New York Press, 1991.

161. Kramer M, Kinney L, Scharf M. Dream incorporation and dream function. In: Koella W (Ed.). *Sleep 1982*. Basel: S. Karger, 1983, pp. 369–371.

162. Kramer M. Dream differences in psychiatric patients. In: Pandi-Perumal SR, Kramer M (Eds). *Sleep and Mental Illness*. New York, NY: Cambridge University Press, 2010, pp. 375–382.

163. Kramer M, Nuhic Z. A review of dream-ing by psychiatric patients: An update. In: Pandi-Perumal SR, Ruoti RP, Kramer M (Eds). *Sleep and Psychosomatic Medicine*. New York, NY: Informa Healthcare, 2007, pp. 137–159.

Medication effects on sleep

JAMES F. PAGEL

Sleep is defined behaviorally—a reversible state of perceptual isolation. Most categories of medication affect this state in which we spend at least a third of our lives. Historically, the effects that medications exert on sleep and alertness were viewed as global and nonspecific; however, in the last 25 years, the selective neuromodulating effects of most sedative–hypnotic drugs on primary neurotransmitters have been defined. This improvement in understanding has, however, occurred as the field of neuropharmacology has changed. We now understand that psychoaffective medications exert effects and side effects not just on the classic synaptic transmission systems, but also on intracerebral neuroendocrine, electrophysiological, and generalized arousal systems that are potentially as complex and important as the interconnecting systems described by classic neuroanatomy.[1]

The field of sleep disorders medicine has matured into an increasingly complex field involved in the diagnosis and treatment of more than 90 diagnoses each with clear diagnostic criteria, many treated with specific pharmacological therapies.

An even larger group of medical and psychiatric diseases produce mental or physical discomfort that can adversely affect sleep.

Sleep disorders can be generally divided into three large groups: (1) those producing insomnia (the complaint of difficulty falling asleep, staying asleep, or non-restorative sleep); (2) those with a primary complaint of daytime sleepiness; and (3) those associated with disruptive behaviors during sleep—the disorders of arousal.[2,3] A wide spectrum of medications can be used to treat these disorders, each with particular benefits as well as potential for harm. The medications that have been used to treat sleep disorders have a long and checkered history that included limited efficacy, misuse, serious side effects, addiction, and lethal toxicity in overdose. One of the significant advances in the development of sleep medicine as a medical specialty has been the development and use of efficacious medications to treat these disorders; medications with minimal side effects, low addiction potential, and limited toxicity in overdose.

MEDICATIONS INDUCING DISORDERED SLEEP

Drug-induced sleepiness is perhaps the most commonly reported side effect of central nervous system (CNS)-active pharmacological agents (the 1990 Drug Interactions and Side Effect Index of the Physicians' Desk Reference lists drowsiness as a side effect of 584 prescription or over-the-counter preparations). Unfortunately, the terminology describing daytime sleepiness, generally considered to be "the subjective state of sleep need," is poorly defined, interchangeably including such contextual terminology as sleepiness, drowsiness, languor, inertness, fatigue, and sluggishness. The results of questionnaire, cognitive, and performance tests for daytime sleepiness correlate only loosely with the actual effects of sleepiness on complex tasks such as the operation of a motor vehicle.[4]

Most medications affecting CNS functioning induce insomnia in some patients. The neuroanatomical systems modulating waking and sleep are contained within the isodendritic core of the brain extending from the medulla, through to the brainstem and hypothalamus up to the basal forebrain. Neurochemically, multiple factors and systems are involved. Most neuromodulators affecting sleep exert effects on GABA, the primary negative neurotransmitter utilized in the CNS; however, no single neurochemical has been identified as necessary or sufficient for modulating sleep and wakefulness. Medications affecting the neuromodulators norepinephrine, serotonin, acetylcholine, orexin, muscarine, and dopamine, often induce insomnia and/or sleepiness. However, agents such as antibiotics, antihypertensives, antivirals, oral contraceptives, and thyroid replacements can induce insomnia in susceptible individuals (Table 18.1).[5] Over-the-counter medications can induce insomnia, including decongestants (including nose sprays), weight loss agents, ginseng preparations, and high-dose vitamins, notably vitamin B1 (niacin). Finally, chronic and long-term sedative/hypnotics and sedating medications used to induce sleep may develop tolerance to the sedative effect, contributing to chronic insomnia.[3,6]

Diagnoses that lead to alterations in sleep and alertness are quite common. Obstructive sleep apnea (OSA), with its well-described effect of

Table 18.1 Medication types known to cause insomnia

Adrenocorticotropin and cortisone
Antibiotics—quinalones
Anticonvulsants
Antihypertensives (alpha-agonists, beta-blockers, and central acting agents)
Antidepressants (selective serotonin reuptake inhibitors)
Antineoplastic agents
Appetite suppressants
Beta-agonists
Caffeine
Decongestants
Diuretics
Dopamine agonists
Ephedrine and pseudoephedrine
Ethanol
Ginseng
Lipid- and cholesterol-lowering agents
Niacin
Oral contraceptives
Psychostimulants and amphetamines
Sedative/hypnotics
Theophylline
Thyroid preparations

daytime somnolence, affects 5%–10% of the population. Many of these patients with OSA also have disrupted sleep, yet treatment of these patients with sedative/hypnotic medications can, in some patients, cause respiratory depression, increased apnea, and worsened sleep. Patients with narcolepsy often paradoxically report improved sleep with daytime amphetamine use. Periodic limb movement disorder (PLMD) and its symptomatic co-diagnosis restless legs syndrome (RLS) often respond to treatment with low-dose dopamine agonists, yet increase in intensity with the use of some antidepressants. Increased daytime arousal typifies a spectrum of common diagnoses, including chronic insomnia, anxiety disorder, and posttraumatic stress disorder. Such patients may demonstrate altered responses to medications, inducing alertness and/or sleepiness, with unexpected results. Stimulants may induce sleepiness in some patients, while hypnotics may induce agitation and insomnia even when used in anesthetic settings and dosages.[7]

MEDICATIONS FOR THE TREATMENT OF INSOMNIA

Sedative/hypnotics

Insomnia is an extremely common complaint. Transient insomnia (<2 weeks in duration) affects up to 80% of the population on a yearly basis.[8] Depending on criteria, chronic insomnia affects 7%–15% of the population.[9,10]

Historically, sedative/hypnotics have been some of the most commonly prescribed drugs. Many sedatives were initially utilized as anesthetics. Chloral hydrate was the original "Mickey Finn" that was slipped into the drinks of unsuspecting marks for the purposes of criminal activity. Unfortunately, the potentially fatal dose for chloral hydrate is quite close to the therapeutic dose, and murders rather than robberies were often the result. In the years leading up to the discovery of benzodiazepines, barbiturates were commonly utilized for their sedative effects. Unfortunately, these medications can be drugs of abuse and have a significant danger of overdose. Marilyn Monroe, Elvis Presley, and Jim Morrison, among others, were celebrities who died during this era from overdoses that included sleeping pills. Barbiturates and barbiturate-like medications (methaqualone [Quaalude, Sopor], glutethimide [Doriden], ethchlorvynol [Placidyl], and methyprylon [Noludar]) are still available, but are rarely used because of their limited efficacy, cognitive effects, potential for abuse, and lethal toxicity associated with overdose.[5]

In the 1970s, benzodiazepines became available for the treatment of insomnia. These drugs are non-specific GABA agonists and have far less overdose danger and abuse potential than barbiturate-like medications. The many drugs in this class are best viewed therapeutically based on their pharmacodynamics (Table 18.2).[11] Rapid onset of action is characteristic of flurazepam (Dalmane) and triazolam (Halcion), indicating that both of these agents have excellent sleep-inducing effects. Flurazepam, like diazepam (Valium) and clorazepate (Tranzene), has the characteristic of having active breakdown products. This results in an extraordinarily long active half-life, which can approach 11 days. This prolonged effect in the elderly has been associated with increased automobile accidents and falls with hip fractures.[12,13] Withdrawal from long-acting

agents can be difficult, with an initial syndrome of insomnia followed by persistent anxiety that may extend beyond the half-life of the agent.

Benzodiazepines are rapid eye movement (REM) sleep–suppressant medications, and withdrawal often results in episodes of increased REM sleep (REM rebound). REM sleep is known to play a role in learning and memory consolidation. For short-acting agents such as triazolam (Halcion), this rebound occurs during the same night in which the medication was taken, and has been associated with daytime memory impairment, particularly at higher dosages.[14] Temazepam (Restoril) and estazolam (Prosom) have half-lives that are compatible with an 8-hour night of sleep. Temazepam, because of its slower onset of action, is less efficacious as a sleep-inducing agent than other drugs used as hypnotics in this class. All benzodiazepines can result in respiratory depression in patients with pulmonary disease and tend to lose sleep-inducing efficacy with prolonged use.[15] Chronic hypnotic (particularly benzodiazepine) use has been associated with the development of mood disorders (depression) and hypnotic-dependent disorders of sleep.[16] Therefore, the underlying reasons and the diseases resulting in chronic insomnia should be addressed.

The newer hypnotics zolpidem (Ambien), zaleplon (Sonata), eszopiclone (Lunesta), and indiplon are agents that exert specific effects on the same GABA receptors (Figure 18.1). Benzodiazepine withdrawal is not blocked by these agents. These agents demonstrate excellent efficacy with minimal side effects. Although any agent used to induce sleep can result in a dependence on that agent for inducing sleep, the abuse potential of these agents is minimal. Idiosyncratic reactions of persistent daytime somnolence and/or memory loss have been reported in some patients. Tachyphylaxis is unusual, and these agents can be used on a long-term basis. Sleep is altered minimally, and REM rebound is not associated with these agents (Table 18.2).[17,18] Zolpidem and eszopiclone have a 6–8-hour half-life, while zaleplon and indiplon are shorter acting (3–4 hours). Clinical comparison of these agents suggests that zolpidem and eszopiclone may have greater sleep-inducing efficacy, and zaleplon may have fewer side effects. Insomnia often occurs as a comorbid symptom in psychiatric and medical diagnoses associated nocturnal discomfort. While the importance of treating the

Table 18.2 Sedative/hypnotics

Class	Drug	Sleep stage effects	Significant side effects	Indications
Benzodiazepine		Decreased amplitude of stages 3 and 4 Increased stage 2 (all)	Loss of effect with chronic use Dependence	
Short onset, short half-life— <4 hours	Triazolam (Halcion)	Shortened sleep latency In-night rapid eye movement (REM) sleep rebound	Anterograde amnesia	Transient insomnia
Short onset, medium half-life—8.5 hours	Estazolam (Prosom)	Shortened sleep latency Decreased REM sleep	Daytime sleepiness	Transient insomnia
Short onset, long half-life— 50–110 hours	Flurazepam (Dalmane)	Shortened sleep latency Decreased REM sleep Withdrawal REM sleep rebound	Daytime sleepiness, chronic buildup (car accidents, hip fractures)	Transient insomnia, anxiety
Medium onset, medium half-life—7–10 hours	Temazepam (Restoril) Clonazepam (Klonopin)	Decreased REM sleep	Daytime sleepiness, poor sleep induction	Transient insomnia, anxiety (parasom-nias)

GABA receptor agents

Class	Drug	Sleep stage effects	Significant side effects	Indications
(a) Short onset, medium half-life (b) Short onset, short half-life	(a) Zolpidem (Ambien) (b) Zaleplon (Sonata)	Shortened sleep latency, Benzodiaz-epine effects with dose above that normally prescribed	Idiosyncratic daytime sleepi-ness or antero-grade amnesia	Transient insomnia, chronic insomnia

Other agents

Class	Drug	Sleep stage effects	Significant side effects	Indications
Chloral hydrate Barbiturates and barbiturate-like agents Sedating antihista-mines H1 blockers	Chloral hydrate Phenobarbital Methaqualone Glutethimide Ethchlorvynol Methyprylon Diphenhydr-amine	1) Short sleep latency, decreased REM sleep, withdrawal REM sleep rebound 2) REM sleep suppres-sion, short sleep latency, decreased REM sleep, withdrawal REM sleep rebound 3) Decreased sleep latency in some patients	Low lethal dose, loss of effect with chronic use Addiction, low lethal dose, loss of effect with chronic use Daytime sedation, anti-cholinergic	Transient insomnia in controlled settings No sleep indications Over-the-counter insomnia
Melatonin agonists	Ramelteon	Shortened sleep latency	Neurohormonal interactions	Sleep-onset insomnia

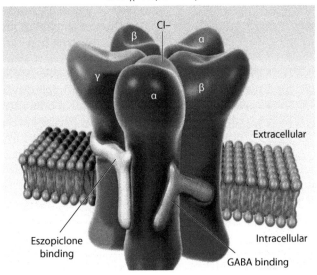

GABA$_A$ receptor complex

Cl–

β

α

γ

α

β

Extracellular

Eszopiclone
binding

Intracellular

GABA binding

Figure 18.1 A model of the GABA receptor.

underlying disorder cannot be overemphasized, these newer hypnotics can, in most cases, be safely utilized on a short-term basis for the treatment of such secondary/comorbid insomnia.

Melatonin agonists

Melatonin is a neural hormone that is effective at resetting circadian rhythms of sleep and body core temperature through its actions on the suprachiasmatic nucleus. For individuals with insomnia secondary to disruptions in circadian rhythms, melatonin can act as a hypnotic and is a useful adjunct to treatment that often includes cognitive therapies, light exposure, and other hypnotics. Ramelteon (Rozerem) is a melatonin receptor agonist that is indicated for the treatment of transient and chronic sleep-onset insomnia. Tasimelteon is a new agent that is designed to treat 24-hour sleep–wake disorder in the blind.

The side effect profiles of melatonin agonists are neurohormonal. These differ significantly from the side effects associated with the hypnotics affecting GABA.[19]

HYPNOTICS: OVERVIEW

Most of the newer hypnotic medications, in general, can be safely utilized on a short-term basis for the treatment of transient insomnia. For persistent chronic insomnia due to anxiety disorders, idiopathic insomnia (persistent life-long insomnia without other sleep-associated diagnoses), post-menopausal insomnia, agitated depression, and other less common diagnoses associated with persistent insomnia, chronic hypnotic use with the newer non-benzodiazepine hypnotics can be justified and is indicated if medication use leads to improvement in waking performance.[3,6] Eszopiclone (Lunesta) has been approved by the Food and Drug Administration (FDA) for extended use in patients with chronic insomnia.[17] For zolpidem in a sustained-release preparation (Ambien-CR) and ramelteon (Rozarem), long-term use is no longer contraindicated. These newer hypnotic agents are less likely to have deleterious side effects than most over-the-counter treatments for insomnia.[20]

Treating arousal

Individuals with chronic insomnia are hyperaroused with increased corticotropin secretion throughout the sleep–wake cycle and have associated greater whole-brain metabolism.[21] The primary neurotransmitters and neuromodulators affecting sleep can be divided into positive activators (affecting arousal) and negative agents

Table 18.3 Sleep/wake neurotransmitters and modulators

Wakefulness	Sleep
• Norepinephrine	• Adenosine
• Serotonin	• γ-aminobutyric acid (GABA)
• Acetylcholine	
• Histamine	• Galanin
• Orexin/hypocretin	• Melatonin

(affecting sleep tendency) (Table 18.3). Many recent approaches to treating insomnia have been directed towards addressing this higher level of arousal in individuals complaining of insomnia.

HISTAMINE

Almost all over-the-counter sleeping pills contain sedating antihistamines, usually diphenhydramine. These agents are varyingly effective, but may result in daytime sleepiness, cognitive impairment, and anticholinergic effects that persist into the day after use, affecting driving performance.[22] These agents are not recommended for use in the elderly.[23] Seizure thresholds can be lowered by their use in epileptic patients. Sedating antihistamines are associated with decreased performance on daytime driving tests and an increase in automobile accidents. Sedation is infrequent with H2 antagonists (e.g., cimetidine, ranitidine, famotidine, and nizatidine), but somnolence as a side effect is evidently reproducible in susceptible individuals. The side effect profiles of the newer sedative/hypnotics are generally more benign that those of the sedating antihistamines.[22,24]

Doxepin, a tricyclic antidepressant with high H1 selectivity, has been rereleased recently for the treatment of sleep-maintenance insomnia at lower doses (3–6 mg) than previously utilized clinically. At this dosage, next-day sedation and anticholinergic effects are minimized.[25]

OREXIN

Recent studies indicate that the orexin (hypocretin) system that is abnormal in individuals with a diagnosis of narcolepsy is a critical regulator of sleep and wake. Recently, the orexin antagonist, surorexant has been released for the treatment of insomnia associated with daytime arousal.[26] Side effects in these clinical trials include sleepiness during the next day and issues with driving, as well as unusual dreams and thoughts of suicide. These agents may turn out to be excellent drugs for treating hyperarousal in patients with insomnia; however, it should be noted that most of the primary side effects associated with the use of sedative/hypnotics were not discovered in clinical trials, but only after extensive clinical utilization. Such recent concerns as to the potential effects of insomnia and hypnotics use on falls in the elderly were raised only after decades of the widespread use of generic and easily available medication.[27]

NONPRESCRIPTION SEDATING AGENTS

Ethanol is probably the most widely used hypnotic medication. In patients with chronic insomnia, 22% report using ethanol as a hypnotic.[17] Unfortunately, chronic use of ethanol to induce sleep can result in tolerance, dependence, and diminished sleep efficiency and quality. When used in excess with other sedative/hypnotic agents, overdose can be fatal. Among drugs of abuse, marijuana has significant hypnotic effects. The central sedative effects of barbiturates, barbiturate-like agents, benzodiazepines, and opioids can induce fatal respiratory suppression at higher doses, particularly when abuse is coupled with ethanol.

In addition to over-the-counter melatonin supplements and the sedating antihistamines, a variety of non-prescription and herbal agents are marketed as hypnotics. Tryptophan has a history of known efficacy in the treatment of chronic insomnia. In the late 1980s, use of this agent was associated with severe eosinophilia that was lethally toxic in some cases. This agent was removed from the market, despite speculation that the toxicity was not secondary to the drug itself, but to deficiencies in the preparation process. Kava, which is considered to be a drug of abuse in some cultures, has been used for insomnia, but has shown potential for hepatic toxicity in some patients. The best data supporting the sedative effect of an herbal agent are for Valerian.[28] Evidence supporting the hypnotic efficacy of other herbal agents including chamomile, passionflower, and skullcap is limited.

ANTIDEPRESSANTS

Sedating antidepressants are sometimes used to treat insomnia. A significant percentage of individuals with chronic insomnia and/or daytime sleepiness also have depressive symptoms. Chronic insomnia itself can predispose patients to developing depression.[16] Depression associated with insomnia is likely to be a different diagnostic entity from depression without insomnia, and treatment of the former with nonsedating antidepressants may produce no improvement in sleep, even when the underlying depression resolves.[29] Use of antidepressants is limited by side effects (anticholinergic effects, daytime hangover, etc.) and danger of overdose (particularly with the tricyclics).[30] Sedating antidepressants include the tricyclics (amitriptyline, imipramine, nortriptyline, etc.), Trazadone (Deseryl), and the newer agents mirtazapine (Remeron) and nefazodone (Serzone) (Table 18.4). The selective serotonin reuptake inhibitors (SSRIs) have a tendency to induce insomnia; however, in some patients, paroxetine (Paxil) may induce mild sedation. Depression-related insomnia responds to sedating antidepressants more rapidly and at lower doses compared to other symptoms of depression.[31] In patients with insomnia and concomitant depression, antidepressants are often used in combination with the newer sedative/hypnotic medications.[32] Use of sedating

Table 18.4 Antidepressants

Class	Drug	Sleep stage effects	Indications
Tricyclics	**Trimipramine** **Nortriptyline** **Doxepin** **Amoxapine** **Amitriptyline** **Imipramine** **Amoxapine** **Protriptyline**[a]	Increased rapid eye movement (REM) sleep latency Decreased REM sleep,[b] short-wave sleep (SWS) latency Deep sleep, sleep latency	Depression with insomnia, REM sleep and SWS suppression, chronic pain, fibromyalgia, enuresis, etc.
Non-tricyclic sedating	**Desimprinine** **Maprotiline** **Mirtazapine**	Increased REM sleep latency Decreased SWS latency, REM sleep,[b] sleep latency	Depression, depression with insomnia, REM sleep suppression
Monoamine oxidase inhibitors	**Phenelzine** **Tranylcypromine**	Increased stage 4 sleep Decreased REM sleep latency, REM sleep[b]	Depression, REM sleep suppression
Selective serotonin reuptake inhibitors (SSRIs)	*Fluoxetine*[a] *Paroxetine* *Sertraline* Fluvoxamine Citalopram HBr	Increased REM sleep latency, sleep latency, stage 1 sleep Decreased REM sleep	Depression, posttraumatic stress disorder, obsessive–compulsive disorder, phobias, cataplexy, etc.
SSRIs + tricyclics	Venlafaxine	Increased REM sleep latency Decreased sleep latency, REM sleep	Depression
Da-na-SSRI	Bupropion	Increased REM sleep latency and increased sleep latency	Depression, nicotine withdrawal
Non-tricyclic Non-SSRI	**Nefazodone**	Increased REM sleep Decreased sleep latency	Depression, depression with insomnia and anxiety
SSNRI	***Duloxetine HCl***	Increased REM sleep and sleep latency	Depression
5HT$_{1A}$ agonist	Buspirone	Increased REM sleep latency Decreased REM sleep	Anxiety

Note: Sedating agents are in bold and insomnia-inducing agents are in italics.
[a] Documented as a respiratory stimulant.
[b] Higher levels of effect.

antidepressants has been associated with declines in daytime performance, driving test performance, and an increased potential for involvement in motor vehicular accidents.[33]

Daytime sleepiness

Excessive daytime sleepiness (EDS), which is varyingly defined, is present in 5%–15% of the population.[4,12,31] Many patients with EDS, particularly those who also complain of snoring, will require overnight sleep evaluation (polysomnography) because of the potential diagnosis of OSA. OSA is usually treated with continuous positive airway pressure (CPAP; a system that utilizes positive nasal pressure to maintain airway patency during sleep). Other treatment approaches for OSA include ear, nose, and throat (ENT) surgery and dental mouthpieces. Symptoms of mood disorders (e.g., depression), which also common causes of daytime sleepiness, can be difficult to distinguish from the symptoms of OSA.[23,33] Chronic sleep deprivation as a basis for daytime sleepiness is particularly common in the adolescent and young adult populations and in individuals involved in occupations requiring nocturnal shift work. Less common causes of EDS are neurological diseases inducing sleepiness: narcolepsy and idiopathic hypersomnolence. Daytime sleepiness is probably the most common side effect of CNS-active medications (Table 18.5). A major concern in such sleepy patients is the potential danger to self and others while working and/or driving motor vehicles.[34,35]

ALERTING MEDICATIONS

Medications that are used in somnolent patients to induce alertness include the amphetamines (dextroamphetamine [Dexedrine], methylphenidate [Ritalin]) and pemoline (Cylert). The tendency of pemoline to cause acute hepatic failure has limited its usefulness. The amphetamines are considered to have high abuse potential and are Schedule II prescription drugs. Side effects of these drugs include personality changes, tremor, hypertension (Dexedrine and Ritalin), headaches, and gastrointestinal reflux.[36]

The newer alerting agents modafinil (Provigil) and armodafinil (Nuvigil) are pharmacologically distinct from the amphetamines and have much lower potential for abuse (Schedule IV). These agents are indicated for the treatment of narcolepsy as well as the persistent sleepiness associated with OSA in patients who are already being treated with CPAP. Modafinil is also indicated for the treatment of fatigue and daytime sleepiness in patients with shift-work disorder.[37]

MEDICATION-INDUCED ALTERATIONS IN SLEEP STAGES AND SLEEP ELECTROENCEPHALOGRAM

Many psychoactive mediations alter the recorded physiologic parameters of sleep. CNS-active medications often alter the occurrence, latency, and electroencephalogram (EEG) characteristics of specific sleep/dream states, either with therapeutic intent or as side effects. Even some non-pharmacologic therapies, such as oxygen, CPAP, and electroconvulsive therapy, can alter REM sleep and deep sleep.[38] Typically, psychoactive medications alter background EEG frequencies and the occurrence, frequency, and latency of the various defined stages of sleep and waking consciousness.[39] Since drug-induced EEG changes are associated with characteristic behavioral effects, this relationship can be utilized in order to suggest therapeutic possibilities for medications producing characteristic EEG effects.[40]

Sleep state-specific diagnoses and symptoms

Parasomnias are sleep disorders occurring during arousal, partial arousal, or sleep state transition.[41] The arousal disorders (sleep terrors, somnambulism [sleepwalking], and confusional arousals) are associated with arousals from deep sleep, confused dream reports, autonomic behaviors, and, sometimes, extreme fright.

REM sleep parasomnias include sleep paralysis, nightmare disorder, and REM behavior disorder. Nightmares are the most common of these parasomnias. The most common symptom of posttraumatic stress disorder are stereotypic frightening nightmares that can occur either at sleep onset or during REM sleep. REM sleep alters many physiological processes, and therefore it is not surprising that a variety of physical illnesses become symptomatic during REM sleep. Respiratory muscle atonia associated with REM sleep can result in increased sleep apnea, particularly in patients with chronic

Table 18.5 Medications reported in clinical trials and case reports to have sleepiness as a side effect

Medication class	Neurochemical basis for sleepiness	Specific medications	
Antihistamines	Histamine receptor blockade	Azatadine (Optimine) Chlorpheniramine (Chlor-Trimeton) Dexbrompheniramine (Polaramine) Clemastine (Tavist) Cyproheptadine (Periactin)	Diphenhydramine (Benadryl) Doxylamine (Unisom) Promethazine (Phenergan) Triprolidine (Actifed)
Anti-Parkinso-nian agents	Dopamine receptor agonists	Benztropine (Cogentin) Biperiden (Akineton)	Procyclidine (Kemadrin) Trihexiphendyl (Artane)
Anti-muscarinic/ anti-spasmodic	Varied effects	Atropine Belladonna Dicyclomine (Bentyl) Glycopyrrolate (Robinul) Hyoscyamine	Ipratropium bromide (Atrovent) Mepenzolate bromide (Cantil) Methscopolamine bromide (Pamine) Scopolamide
Skeletal muscle relaxants	Varied effects	Baclofen (Lioresal) Carisoprodol (Soma) Chlorzoxazone (Parafon Forte) Cyclobenzaprine (Flexeril)	Dantroline (Dantrium) Metaxalone (Skelaxin) Methocarbamol (Robaxin) Orphenadrine (Norflex)
Alpha-adrenergic blocking agents	Alpha-1-adrener-gic antagonists	Doxazosin (Cardura) Prazosin (Minipress)	Terazosin (Hytrin)
Beta-adrenergic blocking agents	Beta-adrenergic antagonists	Acebutolol (Sectral) Atenolol (Tenormin) Betaxolol (Kerlone) Bisoprolol (Zobeta) Carvedilol (Coreg) Esmolol (Brevibloc)	Labetalol (Normodyne) Metoprolol (Lopressor) Nadolol (Corgard) Pindolol (Visken) Propranolol (Inderal) Sotalol (Betapace) Timolol (blocadren)
Opiate agonists	Opioid receptor agonists (general central nervous system [CNS] depression)	Codeine Fentanyl (Sublimaze) Hydrocodone Hydromorphine (Dilaudid) Levomethadyl (Orlamm) Levorphanol (Levo-Dromoran) Meperidine (Demerol)	Methadone Morphine Opium Oxycodone Oxymorphone (Numorphan) Propoxyphene (Darvon) Sufentanil (Sufenta) Tramadol (Ultram)
Opiate partial agonists	Opioid receptor agonists (general CNS depression)	Buprenorphine (Buprenex) Butorphanol (Stadol)	Nalbuphine (Stadol) Pentazocine (Talwin)

(Continued)

Table 18.5 (*Continued*) Medications reported in clinical trials and case reports to have sleepiness as a side effect

Medication class	Neurochemical basis for sleepiness	Specific medications	
Anticonvulsants			
Barbiturates	GABA receptor agonists	Mephobarbital (Mebaral)	Phenobarbital (Luminal) Primidone (Mysoline)
Benzodiazepines	GABA receptor agonists	Elonazepam (Klonopin)	Fosphenytoin (Cerebyx)
Hydantoins	General or poorly defined effects	Ethotoin (Peganone) Phenytoin (Dilantin)	
Succinimides	General or poorly defined effects	Ethosuximide (Zarontin)	Methsuximide (Celontin)
Other	Varied effects including GABA potentiation	Carbamazepine (Tegretol) Felbamate (Felbatol) Gabapentin (Neurontin) Lamotrigine (Lamictal) Levetiracetam (Keppra)	Oxcarbazepine (Trileptal) Tigabine (Gabitril) Topiramate (Topamax) Valproic acid (Depakene) Zonisamide (Zonegran)
Antidepressants			
Monoamine oxidase inhibitors	Norepinephrine, 5HT, and dopamine effects	Phenelzine (Nardil)	Tranylcypromine (Parnate)
Tricyclics	Acetylcholine blockade, norepinephrine, and 5HT uptake inhibition	Amitryptyline (eEavil) Clomipramine (Anafranil) Desipramine (Norpramin) Doxepin (Sinequan) Imipramine (Tofranil)	Maprotiline (Ludiomil) Nortriptyline (Pamelor) Protriptyline (Vivactil) Tripramine (Surmontil)
Selective serotonin reuptake inhibitors	5HT uptake inhibition	Citalopram (Celexa) Escitalopram (Lexapro) Fluoxetine (Prozac)	Fluvoxamine (Luvox) Paroxetine (Paxil) Sertraline (Zoloft)
Others	5HT, dopamine, and norepinephrine effects	Bupropion (Wellbutrin) Mirtazapine (Remeron) Nefazodone (Serazone)	Trazadone (Deseryl) Venlafaxine (Effexor)
Antipsychotics	Dopamine receptor blockade, varied effects on histaminic, cholinergic, and alpha-adrenergic receptors	Fluphenazine (Prolixin) Mesoridazine (Serentil) Perphenazine (Trilafon) Prochlorperazine (Compazine) Thioridazine (Mellaril) Trifluperazine (Stelazine) Aripiprazole (Abilify) Clozapine (Clozaril)	Haloperidol (Haldol) Loxapine (Loxitane) Molidone (Moban) Olanzapine (Zyprexia) Pimozide (Orap) Quetiapine (Seroquel) Risperidone (Risperdal) Thiothixene (Navane) Ziprasidone (Geodone)

(Continued)

Table 18.5 (*Continued*) Medications reported in clinical trials and case reports to have sleepiness as a side effect

Medication class	Neurochemical basis for sleepiness	Specific medications	
Barbiturates	GABA agonists	Amobarbital (Amytal)	Phenobarbital (Luminal)
		Butabarbital (Butisol)	Secobarbital (Seconal)
		Mephobarbital (Mebaral)	Secobarbital/amobarbital
		Pentobarbital (Nembutal)	(Tuinal)
Benzodiazepines	GABA agonists	Alprazolam (Xanax)	Lorazepam (Ativan)
		Chlordiazepoxide (Librium)	Midazolam (Versed)
		Clorazepate (Tranxene)	Oxazepam (Serax)
		Diazepam (Valium)	Quazepam (Doral)
		Estazolam (Prosom)	Emezepam (Restoril)
		Flurazepam (Dalmane)	Triazolam (Halcion)
Anxiolytics, miscellaneous sedative/ hypnotics	GABA agonists, varied effects	Buspiirone (Buspar)	Meprobamate (Equanil, Miltown)
		Chloral hydrate	Promethazine (Phenergan)
		Dexmedetomidine (Precedex)	Zaleplon (Sonata)
		Droperidol (Inapsine)	Zolpidem (Ambien)
		Hydroxyzine (Vistaril, Atarax)	Eszopiclone (Estorra)
Anti-tussives	General or poorly defined	Benzonatate (Tessalon)	Dextromethorphan (robitussin)
Anti-diarrhea agents	Opioid, general or poorly defined	Diphenoxylate (Lomotile)	Loperamide (imodium)
Anti-emetics	Antihistamine and varied effects	Dimenhydrinate (Dramamine)	Thiethylperazine (Torecan)
		Diphenidol (Vontrol)	Trimethobenzamide (Tigan)
		Meclizine (Antivert)	Metoclopramide (Reglan)
		Prochlorperazine (Compazine)	
Genitourinary smooth muscle relaxants	General or poorly defined	Flavoxate (Urispas)	Tolterodine (Detrol)
		Oxybutynin (Ditropan)	

obstructive pulmonary disease (COPD). Lower esophageal pressure, which is also characteristic, of REM sleep, can result in symptomatic gastrointestinal reflux. Chronic diseases that manifest symptoms during REM sleep include angina, migraines, and cluster headaches. Nocturnal seizures, asthma, and panic attacks are more likely to occur in light sleep (stage 2).

Medication effects on sleep stages and EEG

Medication-induced changes in sleep and EEG activity can lead to an increase in symptoms occurring during specific sleep/dream states. For example, insomnia and nightmares are associated with the REM sleep rebound that occurs after the discontinuation of REM suppressive drugs (i.e., ethanol, barbiturates, and benzodiazepines). Medications such as lithium, opiates, and gamma-hydroxybutyrate (sodium oxybate), which can cause an increase in deep sleep, can induce the occurrence of arousal disorders such as somnambulism.[41]

The influence of psychoactive medications on sleep states has a positive side as well. For example, REM sleep-suppressive medications can be useful adjuncts in the treatment of REM sleep parasomnias and sleep stage-specific symptoms. Both benzodiazepines and antidepressants can be used to decrease REM sleep. Similarly, the arousal disorders can be treated with medications that affect deep sleep (benzodiazepines and others).[42] Clonazepam (Klonopin) is the medication that is most commonly utilized in the treatment of parasomnias, particularly in REM behavior disorder.

SLEEP DIAGNOSIS-SPECIFIC MEDICATION EFFECTS

Respiratory effects

Most sedative medications depress respiratory drive with increasing dosage. Benzodiazepines, barbiturates, and opiates can exacerbate respiratory failure in patients with COPD, central sleep apnea, and restrictive lung disease. These medications can also negatively affect OSA and may increase the potential for symptomatic sleep apnea in some population groups, such as patients being treated for chronic pain. The newer non-benzodiazepine hypnotics have demonstrated lower potential for respiratory depression. Methylprogesterone (Provera), protriptyline (Vivactil), and fluoxetine (Prozac) have been documented to have respiratory-stimulant effects that may be clinically useful in some patients.[38]

Enuresis

Persistent bedwetting is present in up to 15% of 5-year-olds. Medication has been shown to be symptomatically useful. Tricyclic antidepressants have been used for decades in this disorder, but there has been concern about cardiac effects and long-term safety in children. The current treatment of choice is desmopressin (DDAVP), which corrects the lack of cyclic antidiuretic hormone increase during sleep that is typically seen in these patients. Symptoms can be controlled until neurophysiological maturity occurs, with a resolution of nocturnal enuresis.[43]

RLS and PLMD

Symptoms of RLS include uncomfortable limb sensations at sleep onset and motor restlessness relieved by exercise and exacerbated by relaxation. PLMD is characterized by repetitive, stereotypic limb movements occurring in 15–40 cycles in non-REM sleep and often leading to recurrent arousals from sleep.[44] These associated disorders are quite common, occurring in up to 15% of the population and increasing in frequency with age. PLMD/RLS may develop in patients during pregnancy, OSA, renal failure, low serum ferritin levels, and in patients taking antidepressants, particularly the SSRIs. Both PLMD and RLS are most commonly treated with low dosages of dopamine precursors and dopamine receptor agonists at bedtime. Possible side effects of these medications, which include carbidopa/levodopa (Sinemet), pramipexole (Mirapex), ropinirole (Requip), and rotigotine, are nausea, headache, tachyphylaxis and augmentation of symptoms.[45]

Circadian rhythm disturbance

A number of sleep disorders are linked to abnormally timed sleep–wake cycles. These include delayed and advance sleep phase syndromes in which the sleep period is markedly later or earlier than what is socially accepted, jet lag, shift work, and certain sleep abnormalities associated with aging. Low doses of melatonin—the photoneuroendocrine transducer that conveys information controlling sleep–wake cycles and circadian rhythms in the central nervous system—may be useful in treating these disorders. Because it is marketed as a dietary supplement, there is minimal data on safety, side effects, and drug interactions for this compound.[46] Jet lag and shift work disorders can also be effectively treated with bright light therapy and the repetitive short-term use of sedative/hypnotics.[47] Ramelteon and tasimelteon are melatonin receptor agonists with high affinity for melatonin MT_1 and MT_2 receptors in the suprachiasmatic nucleus. Ramelteon reduces sleep latency in most patients and is not a controlled substance since there is no potential for abuse. The most frequent adverse events leading to discontinuation were somnolence, dizziness, nausea, fatigue, headache, and insomnia.[19]

Cataplexy

The primary symptom of narcolepsy is daytime sleepiness and sleep attacks. A significant number of narcolepsy patients, however, also have hypnagogic hallucinations, sleep paralysis and cataplexy—skeletal motor weakness associated with emotion when awake. For a subgroup of narcolepsy patients, cataplexy can be incapacitating. Sodium oxybate (also called gamma-hydroxybutyrate) has shown excellent efficacy in the treatment of cataplexy. Sodium oxybate, available in liquid form, is given as two doses in the first 2 hours of the night. This agent has a history of recreational misuse and has limited pharmacological distribution, being available as a class 3 agent for narcolepsy patients.[48]

CONCLUSION

Many medications are used therapeutically for specific sleep disorders. The effects of most sedative/hypnotic drugs on sedation and arousal are secondary to selective neurotransmitter effects, rather than through nonspecific CNS depression. The newer non-benzodiazepine sedative/hypnotic agents have lower addictive potential and toxicity than older agents and can be utilized on a long-term basis in patients with chronic insomnia. Benzodiazepines, antidepressants, and other agents are utilized for their sedative side effects in anxious and insomniac patients. Conversely, medications may induce insomnia, disturb sleep, or exacerbate the effects of chronic illnesses on sleep. Drugs known to induce daytime sleepiness are associated with declines in daytime performance and increased rates of automobile accidents. One of the significant advances in sleep medicine has been the development and use of efficacious medications to treat sleep disorders—medications with minimal side effects, low addiction potential, and limited toxicity in overdose. The fields of sleep and neuropharmacology are changing rapidly, leading to new understanding and comprehension of the complexity of the state of sleep disorders and the medications used clinically to address the complaints and disorders of sleep.

REFERENCES

1. Wiedemann K. Biomarkers in development of psychotropic drugs. *Dialogues Clin Neurosci* 2014; 13: 225–234.
2. Lee-Chiong T, Pagel JF. Medication effects on sleep. In: Waldman SA, Terzic A (Eds). *Pharmacology and Therapeutics—Principles to Practice*. Philadelphia, PA: Saunders/Elsevier. 2009, pp. 849–856.
3. Pagel JF. Pharmachological treatment of insomnia. In: Pagel JF, Pandi-Perumal SR (Eds). *Primary Care Sleep Medicine: A Practical Guide* (2nd ed.). New York, NY: Springer Press, 2014, pp. 91–98.
4. Buysse DJ. Drugs affecting sleep, sleepiness and performance. In: Monk TM (Ed.). *Sleep, Sleepiness, and Performance*. West Sussex: Wiley, 1991, pp. 4–31.
5. Pagel JF. The treatment of insomnia. *Am Fam Physician* 1994; 49: 1417–1422.
6. Richardson GS. Managing insomnia in the primary care setting: Raising the issues. *Sleep* 2000; 23(1): S9–S12.
7. Wilens TE, Biederman J. The stimulants. *Psychiatr Clin North Am* 1992; 41: 191–222.
8. Walsh JK, Coulouvrat C, Hajak G et al. Nighttime insomnia symptoms and perceived health in the America Insomnia Survey (AIS). *Sleep* 2011; 34(8): 997–1011.
9. Morin CM, LeBlanc M, Belanger L et al. Prevalence of insomnia and its treatment in Canada. *Can J Psychiatry* 2011; 56(9): 540–548.
10. Roth T, Coulouvrat C, Hajak G et al. Prevalence and perceived health associated with insomnia based on DSM-IV-TR. International Statistical Classification of Diseases and Related Health Problems, Tenth Revision; and Research Diagnostic Criteria/International Classification of Sleep Disorders, Second Edition Criteria: Results from the America Insomnia Survey. *Biol Psychiatry* 2011; 69: 592–600.
11. Mitler MM. Nonselective and selective benzodiazepine receptor agonists—Where are we today? *Sleep* 2000; 23(S1): S39–S47.
12. Quera Salva MA, Barbot F, Hartley S et al. Sleep disorders, sleepiness, and near-miss accidents among long-distance highway drivers in the summertime. *Sleep Med* 2014; 15: 23–26.
13. Ray WA, Griffen MR, Downey W. Benzodiazepines of long and short elimination half-life and the risk of hip fracture. *JAMA* 1989; 262: 3303–3307.
14. Greenblatt DJ. Benzodiazepine hypnotics: Sorting the pharmacodynamic facts. *J Clin Psychiatry* 1991; 52: 4–10.

15. George CFP. Perspectives in the management of insomnia in patients with chronic respiratory disorders. *Sleep* 2000; 23(S1): S31–S35.

16. Kessler RC, McGonagle KC, Zhao S. Epidemiology of psychiatric disorders. *Arch Gen Psychiatry* 1994; 51: 8–19.

17. Sateia MJ, Doghramji K, Hauri PJ, Morin CM. Evaluation of chronic insomnia. *Sleep* 2000; 23(2): 243–314.

18. Krystal A, Walsh J, Laska E et al. Sustained efficacy of eszopiclone over 6 months of nightly treatment: Results of randomized, double blind, placebo-controlled study in adults with chronic insomnia. *Sleep* 2003; 26(7): 793–799.

19. Abbott S, Soca R, Zee P. Circadian rhythm sleep disorders. In: Pagel JF, Pandi-Perumal SR (Eds). *Primary Care Sleep Medicine: A Practical Guide* (2nd ed.). New York, NY: Springer Press, 2014, pp. 297–310.

20. Pagel JF, Parnes BL. Medications for the treatment of sleep disorders: An overview. *Prim Care Companion J Clin Psychiatry* 2001; 3: 118–125.

21. Vgontzas A, Fernandez-Mendoza J. Objective measures are useful in subtyping chronic insomnia. *Sleep* 2014; 36: 1125–1126.

22. O'Hanlon JF, Ramaekers JG. Antihistamine effects on actual driving performance in a standard driving test: A summary of Dutch experience, 1989–94. *Allergy* 1995; 50: 234–242.

23. Breslau N, Roth T, Rosenthal L et al. Sleep disturbance and psychiatric disorder: A longitudinal epidemiological study of young adults. *Biol Psychiatry* 1996; 39: 411–418.

24. Weiler JM, Bloomfield JR, Woodworth GG et al. Effects of fexofenadine, diphenhydramine, and alcohol on driving performance—A randomized, placebo controlled trial in the Iowa driving stimulator. *Ann Intern Med* 2000; 132: 354–363.

25. Krystal AD, Lankford A, Durrence HH et al. Efficacy and safety of doxepin 3 and 6 mg. in a 35-day sleep laboratory trial in adults with chronic primary insomnia. *Sleep* 2011; 34: 1433–1442.

26. Mieda M, Sakurai T. Orexin (hypocretin) receptor agonists and antagonists for treatment of sleep disorders. *CNS Drugs* 2013; 27: 83–90.

27. Ancoli-Israel S. Insomnia in the elderly: A review for the primary care practitioner. *Sleep* 2000; 23(S1): S23–S30.

28. Stevenson C, Ernst E. Valerian for insomnia: A systematic review of randomized clinical trials. *Sleep Med* 2000; 1(2): 91–99.

29. Vermeeren A, Danjou PE, O'Hanlon JF. Residual effects of evening and middle-of-the-night administration of zaleplon 10 and 20 mg on memory and actual driving performance. *Hum Psychopharacol Clin Exp* 1998; 13: S98–S107.

30. Beitinger ME, Fulda S. Long term effects of antidepressants on sleep. In: Pandi-Perumal SR, Kramer M (Eds). *Sleep and Mental Illness*. Cambridge: Cambridge University Press, 2010, pp. 183–201.

31. Pagel JF. Disease, psychoactive medication, and sleep states. *Prim Psychiatry* 1996; 3(3): 47–51.

32. Rickles K, Schweizer E, Clary C et al. Nefazodone and imipramine in major depression: A placebo controlled trial. *Br J Psychiatry* 1994; 164: 802–805.

33. Volz HP, Sturm Y. Antidepressant drugs and psychomotor performance. *Neuropsychobiology* 1995; 31: 146–155.

34. Chervin RD. Sleepiness, fatigue, tiredness and lack of energy in obstructive sleep apnea. *Chest* 2000; 118: 372–379.

35. Pagel JF. Medications that induce sleepiness. In: Lee-Chiong T (Ed.). *Sleep: A Comprehensive Handbook*. Hoboken, NJ: Wiley-Blackwell, 2006, pp. 175–182.

36. Nishino S, Kotorii N. Overview of the management of narcolepsy. In: Goswami M, Pandi-Perumal SR, Thorpy M (Eds). *Narcolepsy—A Clinical Guide*. New York, NY: Springer/Humana Press, 2010, pp. 251–266.

37. McClellan KJ, Spencer CM. Modafinil: A review of its pharmacology and clinical efficacy in the management of narcolepsy. *CNS Drugs* 1998; 9: 311–324.

38. Hart LL, Middleton RK, Schott WJ. Drug treatment for sleep apnea. *DICP Ann Pharmcother* 1989; 23: 308–315.

39. Pagel JF. The synchronous electrophysiology of conscious states. *Dreaming* 2012; 22: 173–191.

40. Itil TM. The discovery of psychotropic drugs by computer-analyzed cerebral bioelectrical potentials (CEEG). *Drug Dev Res* 1981; 1: 373–407.

41. Pagel JF. Drugs, dreams and nightmares. In: Pagel J (Ed). *Dreaming and Nightmares, Sleep Medicine Clinics*. Philadelphia, PA: Saunders/Elsevier, 2010, pp. 277–288.

42. Schenck CH, Mahowald MW. Long-term, nightly benzodiazepine treatment of injurious parasomnias and other disorders of disrupted nocturnal sleep in 170 adults. *Am J Med* 1996: 100(3): 333–337.

43. Klauber GT. Clinical efficacy and safety of desmopressin in the treatment of nocturnal enuresis. *J Pediatr* 1989; 114(4 pt 2): 719–722.

44. Walters AS. Toward a better definition of the restless leg syndrome. The International Restless Legs Syndrome Study Group. *Mov Disord* 1995; 10(5): 634–642.

45. Verma N, Kushida C. Restless leg and PLMD. In: Pagel JF, Pandi-Perumal SR (Eds). *Primary Care Sleep Medicine: A Practical Guide* (2nd ed.). New York, NY: Springer Press, 2014, pp. 339–344.

46. Stone BM, Turner C, Mills SL, Nicholson AN. Hypnotic activity of melatonin. *Sleep* 2000; 23: 663–670.

47. Buxton OM, Copinschi G, Van Onderbergen A et al. A benzodiazepine hypnotic facilitates adaptation of circadian rhythms and sleep–wake homeostasis to an eight hour delay shift simulating westward jet lag. *Sleep* 2000; 23: 915–928.

48. Mamelak M, Black J, Montplasier J et al. A pilot study on the effects of oxybate on sleep architecture and daytime alertness in narcolepsy. *Sleep* 2004; 27(7): 1327–1334.

19

Behavioral intervention for sleep disorders

CHIEN-MING YANG AND ARTHUR J. SPIELMAN

INTRODUCTION

Sleep disturbance is one of the most common health complaints in the general population. It is a symptom of various conditions of different pathogeneses. Some sleep disorders have clearly identified physiologically etiologies. Examples include breathing-related sleep disorders, narcolepsy, and sleep-related movement disorders.[1] Others are primarily caused by psychosocial or behavioral factors, such as stress-related transient insomnia (adjustment sleep disorder) or sleep disturbance associated with poor sleep hygiene. Moreover, sleep complaints in some other cases are associated with psychiatric or medical conditions. It was estimated that about 40% of patients with insomnia and 47% of patients with hypersomnia carried one or more comorbid psychiatric diagnoses, with depressive/bipolar disorders, anxiety disorders, and substance-related and addictive disorders among the most prevalent ones.[2–4] In addition, 16%–82% of patients with chronic medical

conditions, such as musculoskeletal and other painful disorders, heart disease, airway disease, and diabetes, reported difficulty sleeping.[5–9] The sleep complaints in these cases are often assumed to be symptoms that are secondary to the primary conditions or caused by the discomforts associated with the primary conditions. Ideally, the different aspects of sleep pathologies in a patient should be differentiated and treatment should be instituted accordingly. However, in clinical practice, the pathogeneses are usually multifaceted and the causal inferences among the factors are usually difficult to tease apart. For example, a patient's sleep may be initially disrupted by the pain caused by a medical condition. The resulting worries over the consequences of sleeplessness may further worsen sleep and exacerbate the sleep problems. A longitudinal study that examined the association between chronic illnesses and insomnia reported that newly developed medical conditions might precipitate or worsen insomnia; resolution of comorbid medical conditions, however, had no significant impacts on preexisting insomnia.[5] Often, psychological and behavioral factors can develop after the onset of insomnia and further disrupt the neurophysiological systems that regulate sleep and wake.[10,11] Even in a physiologically based sleep disorder, such as mild sleep-related breathing disorder, anxiety over poor sleep and daytime functioning, or inappropriate sleep practices can further disrupt the already fragile sleep. Therefore, behavioral intervention is beneficial not only in the treatment of sleep disorders of psychosocial origins, but also in the management of physiologically based sleep problems.

Many behavioral and psychological interventions are effective in the treatment of sleep disorders, especially in the management of chronic insomnia.[12-14] Some of the behavioral techniques have also been applied to insomnia associated with medical or psychiatric disorders with positive results.[12,15-18] In addition to improvements in sleep, psychiatric symptoms decreased.[19,20] Widely applicable behavioral interventions have been used to enhance the compliance with medical treatment in sleep-related breathing disorders.[21,22]

In this chapter, we will describe how psychological and behavioral factors influence the neurophysiological mechanisms of normal sleep regulation, and will provide a framework for the clinician to conceptualize the development of insomnia in individual patients. Further, we will review the behavioral techniques that are effective in the treatment of insomnia. Since there are some recent review articles on the treatment efficacy of cognitive–behavioral therapy for insomnia,[23-25] our chapter will focus more on the conceptual rationales and practical aspects of the cognitive–behavioral interventions. Lastly, we will briefly review the behavioral interventions for specific sleep disorders other than insomnia.

PSYCHOSOCIAL AND BEHAVIORAL FACTORS AFFECTING SLEEP–WAKE REGULATION

Sleep disturbances can be conceptualized as disruptions of the mechanisms that regulate the normal processes of sleep and wakefulness. Thus, understanding the mechanisms of normal sleep control is important for the evaluation and treatment of sleep disorders. It is now recognized that human sleep is regulated by the interactions of two major systems: a homeostatic system that regulates the optimal level of sleep drive in order to maintain the internal balance between sleep and wakefulness; and a circadian process that generates a biological rhythm of sleep and wake tendency over a day.[26-28] In addition, the neural mechanism that promotes wakefulness and/or arousal can counteract sleep drive and disrupt normal sleep regulation.[26]

The homeostatic sleep drive is determined by the amount of sleep acquired previously and the duration of prior wakefulness. Like other homeostatic systems, such as those that regulate body temperature, the system seeks to maintain a set amount of sleep. Therefore, sleep deprivation is followed by enhanced sleep drive and extra recovery sleep. Satisfaction of sleep drive is, in contrast, followed by decreased sleep propensity and shorter or lighter sleep. In addition, increased physical activities may enhance homeostatic sleep drive.[29]

The circadian system, on the other hand, generates a near 24-hour rhythm of sleep and wake tendency that is independent of prior amount of sleep. The tendency for sleeping and waking is determined by temporal factors associated with the phase of the endogenous circadian cycle. Studies with both animals and humans have identified the genetic determinants of the circadian sleep tendency.[30,31] In addition, exposure to environmental time cues, especially bright light, can stabilize or shift the endogenous circadian rhythm. It has been recognized that the endogenous circadian cycle in human beings is slightly over 24 hours[32,33]; therefore, there is a natural tendency for the endogenous rhythm to drift later in time. Morning light exposure can advance the circadian sleep phase and maintain a 24-hour circadian period.[34,35] With the interactions of the homeostatic and circadian systems, the sleep propensity and amount of sleep obtained at a given time is determined by the accumulated sleep debt and the circadian sleep tendency at that time.

Behavioral practices and environmental factors can optimize or disrupt the operations of these systems. Maladaptive behavioral practices, such as napping or sleeping late on weekends, may trigger and perpetuate sleep problems. These sleep episodes intrude into the typical waking period of the day and lead to insufficient sleep drive being accumulated by bedtime. Furthermore, sleeping-in will limit the morning light cue and thus will permit a delay in the individual's endogenous circadian sleep tendency, making it difficult to fall asleep at night.

Irregular sleep–wake habits may destabilize the endogenous circadian rhythm and produce fluctuations in the homeostatic sleep drive, both of which can interfere with a regular sleep–wake rhythm.

In addition to the homeostatic and circadian processes, the wakefulness and arousal-promoting process can be conceived of as being reciprocal to the sleep processes. This process can be boosted by stress and emotional and environmental stimuli, leading to disruption of the normal sleep processes. It can be considered to be a protective system that is intended to arouse the organism when it is at risk. However, when it is activated at bedtime, sleep disturbance may result. Several etiological models addressing the role of arousal as a cause of insomnia have been proposed.[36,37] Characteristics of over-activation of the arousal system in insomniacs from either psychosocial or physiological perspectives have been reported in previous research. Individuals with insomnia have been shown to have elevated indices of autonomic activity, such as higher metabolic rate, body temperature, heart rate, urinary cortisol, adrenaline excretion, skin conduction, and muscle tension.[38,39] In addition, increased cognitive processing around sleep onset or during sleep was reflected by disturbing mental activity, faster electroencephalography (EEG) frequencies,[40–42] and reduced inhibitory event-related potentials.[43,44] The elevated high-frequency EEG power spectral findings have been shown to be reduced after successful treatment with cognitive–behavioral therapy for insomnia (CBT-I), but were still maintained at a level that is higher than in good sleepers.[45] These results imply that hyperarousal might in part be a predisposing trait and in part aggravated in the course of chronic insomnia.

Similar to homeostatic and circadian regulation, the arousal system is also highly susceptible to the effects of behavioral practices and psychological status. Vigorous presleep activities and worries about sleep (e.g., watching the clock, counting the number of hours left for sleep, and the consequences of poor sleep) may activate the arousal system; emotional disturbances, such as feeling depressed over a loss or guilt and anticipatory anxiety of challenges to come, also lead to over-activation of the arousal system and further disruption of sleep processes. Both the behavioral practices and emotional arousal may be influenced by attitudes and beliefs concerning sleep. For example, a belief that one should obtain 8 hours

of sleep can lead an insomnia sufferer to spend extra time in bed during the weekend, which further results in decreased homeostatic sleep drive at bedtime on the next night and may delay the phase of endogenous circadian rhythms. In addition, the failure to obtain 8 hours of sleep may lead to excessive worry, which further aggravates the individual's sleep condition. Therefore, sleep cognition plays a major role in the pathogeneses of insomnia.[46–49] Changes in sleep cognition after treatment have also been found to be associated with sleep improvement.[47,50] Figure 19.1 shows a conceptual model of the interactions among the psychological/behavioral factors and the neurophysiological systems that regulate sleep. With this model in mind, the clinician can try to evaluate a patient's sleep problems by examining the functioning of the three neurophysiological systems and the various factors that may have impacts on these systems.

As discussed above, the factors contributing to sleep disturbance in an individual are usually multidimensional. The 3-P model proposed by Spielman can serve as a conceptual framework to organize these contributing factors along the timeline of the development of chronic insomnia.[10] The model categorizes the factors into predisposing, precipitating, or perpetuating factors. Predisposing factors are the individual traits that set the stage for the development of insomnia, but do not necessarily lead to insomnia. For example, vulnerability to stress-related sleep disturbance[51] has been shown to predict the onset of insomnia.[52] Other examples include anxiety-prone personality styles, hypersensitive physiological arousal system, a rigid circadian system, and vulnerable sleep systems. The clinician may need to assess the patient's premorbid sleep pattern, sleep history at younger age, and his/her general personality style in order to identify these factors. Understanding the risk factors can help a patient to establish a realistic expectation of his/her sleep and to avoid getting into situations that may trigger a sleep problem. Precipitating factors are usually readily identified as the events or conditions that initiate the sleep disturbance. The most common precipitating factor is major life stressors, such as loss, having a baby, change of working schedule, major illness, retiring, or any event that heightens arousal or leads to significant changes in daily life routines. In some cases of transient insomnia, resolving the triggering event is the only treatment that is needed in order to address

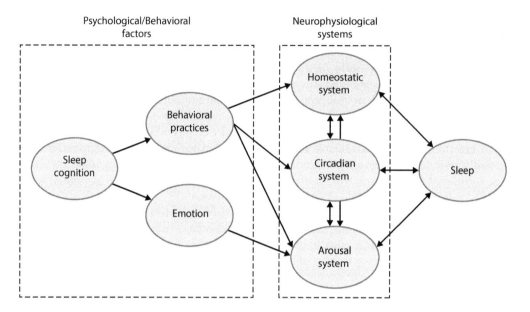

Figure 19.1 A conceptual model showing that psychological/behavioral factors influence sleep mediated by the neurophysiological systems for sleep regulation.

the sleeping problem. However, in many cases, the insomnia persists and becomes chronic over time. Common perpetuating factors include the conditioned association between arousal and bedtime cues, maladaptive sleep–wake habits intended to gain more sleep, strategies to minimize daytime deficits, and excessive worries over sleeplessness. In these cases, the perpetuating factors should then become the focus of the treatment. Figure 19.2 illustrates the relative contribution of the factors along the development of chronic insomnia. Thus, the model can be helpful for understanding the formation of the sleep problem and the formulation of a treatment plan.

To illustrate this model, take a man with a high anxiety trait as an example. The anxiety tendency predisposes him psychologically to worry about trivial matters and physiologically to be easily aroused. The trait alone may not ordinarily be sufficient to produce a persistent sleep disturbance, but may lead to occasional poor sleep. However, a life stressor (e.g., a breakup with his girlfriend) may boost the arousal system and precipitate the onset of insomnia. He starts tossing and turning in bed and ruminating over the conflicts in the relationship. Stress management to help him cope with the stressor by direct actions, or temporizing and allowing time to dampen these problems may be the first therapeutic plan. However, in reacting

to his insomnia, he may develop behavioral adaptations and thoughts that are counterproductive. For example, he may start to spend too much time in bed to "get rest" or to "make up for lost sleep." These behaviors will lead to more chances of being awake in bed, and may strengthen the association between the bed and fretful tossing and turning, leading to a conditioned insomnia. The patient may

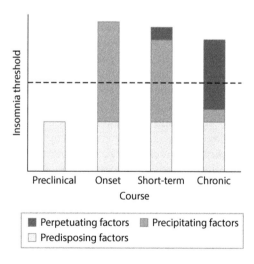

Figure 19.2 The 3-P model of the development of insomnia. (Modified from Spielman AJ, Caruso LS, Glovinsky PB. *Psychiatr Clin North Am.* 1987; 10(4): 541–553. Review. PubMed PMID: 3332317.)

Table 19.1 Common contributing factors associated with the development of insomnia

Predisposing factors

Homeostatic process

- Abnormality or weakness of the neurophysiological system that generates sleep

Circadian process

- Extreme circadian-type functioning better during late evening or early morning as an individual trait
- Less flexible circadian system

Arousal system

- Anxiety-prone and depressive personality traits and tendencies towards neuroticism and somatization lead to a higher level of emotional and physiological arousal
- Personality traits associated with sustained level of arousal, such as perfectionism and excessive need for control
- Heightened or more sensitive physiological arousal system

Precipitating factors

Homeostatic process

- Lack of or decrease of daytime activities, such as retirement

Circadian process

- Change of sleep–wake schedule, such as jet lag or start a nightshift job

Arousal system

- Life stressors or events leading to emotional and physiological distress

Perpetuating factors

Homeostatic process

- Increased resting in bed
- Discharge of the sleep drive by sleeping outside of the nocturnal sleep period, such as increased daytime naps or frequent dozing-offs
- Reduced daytime activities

Circadian process

- Sleeping-in during weekends to catch up on sleep

Arousal system

- Dysfunctional beliefs and attitudes about sleep that lead to increased emotional arousal and worries over sleep loss
- Conditioning between bedtime cues and arousal

start to worry about not being able to fall asleep or about job performance on the next day. These worries further activate his arousal system and exacerbate the sleep difficulties. At this point in time, treatment should target the reduction of arousal, the changed view of sleep, and the elimination of maladaptive sleep practices. Without proper treatment, the individual may start to use hypnotic medication regularly and have difficulty stopping because of the rebound effects when withdrawing from the drug. At this point in the course of the insomnia, education about the effects of hypnotic use and withdrawal, as well as strategies to taper off

medication, should be included as part of the treatment. Therefore, this model provides a framework for the clinician to organize the multiple determinants of insomnia and to formulate a comprehensive treatment plan. Table 19.1 shows the common contributing factors that are associated with the development of insomnia.

BEHAVIORAL INTERVENTIONS FOR INSOMNIA

Various behavioral techniques have been applied for the treatment of insomnia. Many of these techniques are effectively combined as a multi-component approach that is usually referred to as CBT-I. CBT-I can be conducted in groups with a structured program or conducted individually with techniques being selected that are tailored for an individual. Meta-analyses and review articles on the efficacy of individual behavior techniques as well as multicomponent CBT-I programs showed moderate to large effect sizes on the major sleep parameters.[12,13,53] Furthermore, the treatment effects were shown to be equal to or better than pharmacological treatments at the end of treatment sessions and with long-term follow-up of up to 2 years.[14,23,54]

Before the initiation of a cognitive and behavioral intervention for insomnia, several issues needed to be addressed: first, it is important to conduct a thorough evaluation in order to identify the contributing factors and to share the case formulation with the patient. Patients with sleep disturbances are often puzzled by their condition. The feeling of being out of control and the mystery of what is causing the problem often generates worries and anxiety that may further disrupt their sleep. Thus, sharing the formulation with the patient may be therapeutic by restoring some modicum of control and reducing unnecessary worry. Furthermore, the patient's grasp of the formulation will facilitate the understanding of treatment rationales and may motivate the patient to practice the behavioral techniques. In addition to the evaluation and etiological formulation, providing the rationale and describing the treatment procedures comprise a comprehensive first treatment session.

Second, it is useful for patients to know what to expect from treatment. For example, providing information that the behavioral techniques require a few weeks to have significant impact may prevent premature demoralization when improvement is not immediate. The patient needs to understand that reliably carrying out behavioral practices is crucial for the treatment to be effective. Changes in sleep habits and daily life routines and the practice of some techniques are all likely to be required. A series of office visits or group sessions is then scheduled on a weekly or biweekly base, with sleep log recording and/or phone contacts in between sessions. For both individual and group CBT-I, the whole program usually takes 4–8 weeks. Lately, abbreviated, two-session CBT-I was also reported to be effective in the primary care setting.[55]

As mentioned above, CBT-I comprises various cognitive and behavioral techniques. The major ones include sleep hygiene education, cognitive therapy, stimulus control instruction, sleep restriction, and relaxation training. The following are the rationales and practical guidelines of the major CBT-I techniques for insomnia.

Sleep hygiene education

Sleep hygiene refers to practices of everyday living and sleep-related activities that promote good-quality sleep or that make sleep more resistant to disruption. The objectives of sleep hygiene education are to improve basic knowledge about sleep and to modify counterproductive sleep practices.[56] Sleep hygiene education usually includes both a knowledge part and a practice part. Firstly, sleep hygiene education provides basic knowledge about sleep and sleep disorders, including information about the homeostatic process and function of normal sleep, the influence of circadian rhythms on sleep, the influences of stress and emotion on sleep, the variability in night-to-night sleep, the developmental changes of sleep, the effect of daily activities on sleep, and the effect of sleep disturbances on daytime function. Understanding empowers the patient and eliminates unnecessary worry about the consequences of sleep loss. It also provides a rationale for sleep-promoting behavioral practices. Secondly, good sleep hygiene requires the patient to modify daily living practices that are counterproductive to sleep. The clinician reviews the lifestyle and sleep–wake habits with the patient and identifies a set of practices that are not consistent with good sleep hygiene. The patient is asked to refrain from maladaptive activities and, in some cases, engage in sleep-promoting behaviors. Common behavioral practices that are incompatible with good sleep are listed in Table 19.2.

Sleep hygiene education is usually part of a more comprehensive treatment program. Sleep hygiene education alone has been shown to be less effective than the other behavioral treatments.[53] Insomnia patients were found to engage in poorer sleep hygiene practices in some studies,[57–59] but not in other studies.[60,61] One study categorized sleep hygiene practices into different categories and found that the behavioral practices that might increase the level of arousal correlated most significantly with insomnia symptoms and sleep quality in patients with insomnia.[62]

Table 19.2 Sleep-related habits and daily life practices that may interfere with sleep

Practices that reduce homeostatic drive at bedtime

Daily life behaviors
- Insufficient activities during the day
- Lying down to get rest during the day

Sleep-related habits
- Napping, nodding-off, and dozing-off during the day or evening
- In a trance or semi-awake in the evening
- Spending too much time in bed
- Extra sleep during the weekends

Practices that disrupt circadian regularity

Daily life behaviors
- Insufficient morning light exposure, leading to a phase delay in circadian rhythm
- Early-morning light exposure, producing early-morning awakening due to a phase advance in circadian rhythm

Sleep-related habits
- Irregular sleep–wake schedule
- Sleeping-in in the morning during weekends

Practices that enhance the level of arousal

Daily life behaviors
- Excess caffeine consumption or caffeine later in the day
- Smoking in the evening
- Alcohol consumption in the evening
- Exercising in the late evening
- Late evening meal or fluid (may cause frequent urination)
- Getting home late or not enough time to wind down

Sleep-related habits
- Evening apprehension of sleep
- Preparations for bed are arousing
- No regular presleep ritual
- Distressing pillow talk
- Watching TV, reading, or engaging in other sleep-incompatible behaviors in bed before lights out, or falling asleep with the TV or radio left on
- Trying too hard to sleep
- Clock-watching during the night
- Staying in bed during awakenings, or lingering in bed awake in the morning
- Poor sleep environment, such as bed-partner snoring, noises, direct morning sunlight, or pets in the bedroom

Many patients are aware of sleep hygiene practices, but do not believe that the practices will produce significant changes in their sleep. It is important to convey to such individuals that insomnia is the result of the interaction of a number of factors, and that an effective treatment should address multiple factors at the same time.

Eliminating these habits may not solve the problem. However, a successful treatment result may be prevented or delayed due to poor sleep hygiene practices. In addition, patients with chronic sleep problems may engage in counterproductive sleep practice, such as daytime napping, going to bed earlier, or staying in bed in the morning to catch up on sleep, as ways of coping with the consequences of their sleep problem. It is important to convey to the patient that these coping strategies may help in the short term, but they sacrifice robust sleep in the long term.

Cognitive restructuring

Worrying about sleeplessness may promote arousal and become a self-fulfilling prophecy that further exacerbates sleep difficulties. Misconceptions about sleep may also lead to sleep-disruptive behavioral practices. Faulty beliefs and attitudes about sleep have been shown to be associated with the symptoms of insomnia.[47,49,50] Common dysfunctional cognitions about sleep can be classified into five categories: misconceptions of the causes of insomnia, misattributions or amplifications of its consequences, unrealistic sleep expectations, diminished perceptions of control, and predictability of sleep.[63] Changes of dysfunctional thoughts can reduce the worries and therefore break the vicious cycle that leads to arousal. Disruptive cognitions about sleep may be corrected with sleep hygiene education. Relaxation training can also be helpful in distracting the patient from excessive worry and reducing physical arousal. Cognitive restructuring, on the other hand, addresses the sleep-disturbing cognitions directly, and replaces these thoughts with more realistic thoughts and positive ideas.[46,63]

To institute cognitive restructuring, the clinician should discuss with the patients their beliefs regarding sleep in general and regarding their sleep problems. Beliefs that lead to maladaptive behavioral practices or enhance worries about sleep should be identified. The Dysfunctional Beliefs and Attitudes Scale[63] can also be utilized to help with the evaluation of a patient's sleep cognitions. The clinician can provide correct information about sleep and the consequences of sleep loss in order to help the patient develop more positive and less disruptive ideas about their sleep problems. For example, a patient with a belief that "without an adequate night's sleep, I can hardly function the next day" will become anxious prior to bedtime and make a lot of effort to try to fall asleep. The patient should be guided to understand that the worries over poor sleep and efforts to gain more sleep tend to increase presleep arousal, which may further disrupt the natural process of sleep. This dysfunctional belief should be replaced by a more adaptive belief, such as "a poor night of sleep might have a minor influence on my functioning the next day; what really ruins my sleep and my day is my over-concern about the negative consequences of poor sleep." The change in belief could decrease the bedtime worries and allow the natural sleep process to occur around bedtime. Alternatively, a strategy of behavioral experiment may be employed by enlisting the patient as a coinvestigator to gather data that will address the validity of specific beliefs. For example, the patient who believes that poor sleep will ruin their day may be asked to rate job performance on daily sleep logs. In this way, data could be collected to disconfirm dysfunctional thoughts and beliefs regarding sleep.

Stimulus control therapy

Stimulus control therapy is designed to break the maladaptive association of bedtime cues with wakefulness and worrying. Patients are instructed to get out of bed if they are unable to fall asleep and return to bed when they feel ready to fall asleep. Over time, the repeated association of bedroom cues with rapid sleep onset brings sleep under the stimulus control of the bedroom environment.[64] Specific instructions are listed in Table 19.3.

While the instructions are very straightforward, carrying them out regularly is difficult and needs encouragement from the therapist. Initially, patients will spend considerable time out of bed and thus will suffer some sleep loss. It is important to motivate the patients by letting them fully understand the rationale of this treatment

Table 19.3 Instructions for stimulus control therapy

1. Go to sleep only when feeling sleepy.
2. Do not use the bed or bedroom for other activities except sleep (sexual activity is the only exception).
3. If you do not fall asleep within about 20 minutes, get up and go into another room to do something relaxing.
4. Go back to bed only when feeling sleepy again.
5. Repeat the procedure of getting out of bed if you still cannot fall asleep rapidly.
6. Get up at the same time each morning, regardless of how much sleep you obtain.

procedure. Although daytime functioning and mood may be impaired temporarily, the partial sleep deprivation will foster both rapid sleep onset and increased sleep. The patient needs to be educated that short-term sacrifice will produce long-term gains. In addition, some patients may worry that they do not know what to do when they are out of bed during the night. The clinician may need to prescribe activities to perform during the night that are not too taxing or activating. Stimulus control is one of the CBT-I techniques that shows the most consistent effects when administered alone.[4]

Sleep restriction therapy

Sleep restriction therapy was developed with the assumption that the homeostatic sleep process can self-correct when sleep disturbance leads to sleep loss. Sleep restriction therapy promotes sleep by inducing a mild sleep loss initially and gradually increasing sleep time after sleep is stabilized. Like stimulus control therapy, this procedure can also prevent or break the maladaptive association between anxiety and bedtime cues by decreasing the chance of lying awake in bed.[65,66] Specific procedures are listed in Table 19.4.

Some modified procedures to adjust time in bed have been proposed and utilized. One way is to increase time in bed progressively by 15 or 30 minutes each week until the patient is spending 7 hours in bed. Further changes of time in bed may be made based on daytime functioning, fatigue, and sleepiness.[67] In addition, a sleep compression procedure has been utilized in order to gradually reduce time in bed instead of curtailing it abruptly. As in the original procedure, mean total sleep time and time in bed were determined from the 2 weeks of the baseline sleep log. The difference between mean total sleep time and mean time in bed was then divided by 5, and the prescribed time in bed was compressed by this amount weekly. By the end of the fifth session, time in bed had been gradually compressed to match the initial total sleep time. Like in the original procedure, time in bed was increased by 15 minutes if sleep efficiency surpassed 90%. This procedure was found to be helpful for elderly patients with insomnia.[68]

As with stimulus control instructions, patients should be told to expect mild sleep loss and daytime deficits at the initiation of the treatment. It should be stressed that this temporary worsening

Table 19.4 Instructions for sleep restriction therapy

1. Patients complete a sleep log that records the daily sleep pattern over a 2-week period.
2. The average total sleep time (TST) per night of this 2-week period is then prescribed as the time in bed (TIB) for the following week. Time of arising from bed is set to the time when the patient is required to wake up or when the patient generally awakens, and time of retiring is calculated accordingly. To avoid the effects of severe sleep deprivation, the minimum TIB is never set below 4.5 hours. Lying down or napping outside of the scheduled bedtimes is not permitted.
3. Patients fill out a sleep daily log to record TIB and estimated TST. Sleep efficiency (SE = TST/TIB × 100%) is evaluated every week. Prescribed TIB for the next week is adjusted by three criteria:
   a. SE ≥ 90% (85% in older individuals): TIB is increased by 15 minutes by setting the retiring time earlier.
   b. SE < 85% (80% in seniors): TIB is decreased by 15 minutes by delaying bedtime.
   c. SE ≥ 85% (80% in seniors) and <90% (85% in seniors): TIB remains the same.

of daytime mood, performance, and energy will lead to deeper and more consolidated sleep after a few weeks of practicing this method. Patients may have difficulty resisting the temptation to spend more time in bed. Clinicians should encourage patients to continue to follow the instruction and to plan specific activities for their increased time in the evening before going to bed.

Relaxation training

As described earlier, thoughts and behaviors resulting in arousal may interfere with sleep. For the same reasons, activities that reduce arousal may facilitate sleep. Various relaxation techniques have been developed to assist in the reduction of tension and anxiety. Examples include progressive muscle relaxation that reduces muscle tension by sequential tensing and relaxing of the main muscle groups,[69,70] autogenic training that produces somatic relaxation by inducing sensations of warmth and heaviness of the body,[70] and guided imagery that aims to channel mental processes into a vivid storyline.[71] Biofeedback has also been utilized to assist with the mastering of relaxation techniques.[72,73] In general, relaxation training starts with a demonstration of how it is done by the clinician. The patients are then instructed to practice the technique at home in between sessions. The instructions of the relaxation procedures can be recorded during the session for the patients to practice at home. Commercial relaxation training audio or video media are also available to facilitate the practice at home. The patient is asked to practice the relaxation protocol once or twice a day. A patient's level of relaxation should be assessed before and after each practice session with a simple subjective rating scale in order to monitor the progress of the training. It may take weeks for some individuals to develop the skill to relax on cue. Only after mastering the procedure to a certain degree is the patient told to use it to facilitate falling asleep or returning to sleep following a nocturnal awakening, otherwise the patient may be frustrated when failing to use the skill to assist sleep. At the start of the training, patients are often distracted and deflected from the tasks, and find themselves thinking and worrying about other things. The patients should be told to anticipate that their mind will wander and to avoid self-criticism. It is crucial to help motivate the patients to continue practicing the technique and to help them deal with any obstacles that they encounter.

Mindfulness-based techniques

CBT with a mindfulness-based approach, known as mindfulness-based cognitive therapy (MBCT) and mindfulness-based stress reduction (MBSR), have gained great popularity in the fields of psychiatry and behavioral medicine. The main goal of these programs is to develop metacognitive awareness cultivated through meditation practices and mindfulness exercises in order to manage the negative emotional reactions associated with various distressful mental and physiological symptoms. Many empirical studies have demonstrated the effectiveness of MBCT and MBSR in improving mental and physical health. These approaches have also been applied in the treatment of sleep disturbance. They are often combined with the concepts and techniques of CBT-I and have generated positive results in both primary and comorbid insomnia.[74–83]

To date, there is no standard procedure for conducting mindfulness-based treatments for insomnia. Taking the mindfulness-based therapy for insomnia developed by Ong and colleagues[79,80,82] as an example, the treatment adopts the principles of acceptance and letting go and integrates the mindfulness training and exercises from the MBSR and the behavioral techniques from the CBT-I. It emphasizes not only the alleviation of the nocturnal insomnia symptoms, but also the decrease of emotional and physiological arousal both during the day and at nighttime. The intervention is usually delivered in groups of six to eight participants and consists of six to eight 2-hour weekly sessions. The sessions could be roughly divided into two parts. The first part begins with a practice of mindfulness meditation (body scan and sitting and walking meditation), followed by a discussion of the application of mindfulness principles (e.g., beginner's mind, non-striving, letting go, non-judging, acceptance, trust, and patience) in order to deal with reactions to disturbed sleep and real-world situations. Participants are instructed to engage in meditation practice between sessions for at least 30 minutes per day, 5 days per week. The second part of the session is spent implementing and discussing the behavioral techniques from CBT-I (e.g., sleep restriction, stimulus control, and sleep hygiene education).

BEHAVIORAL INTERVENTION FOR ADJUSTMENTS OF CIRCADIAN RHYTHM

Sleep disturbances can result from a mismatch between the endogenous circadian phase and environmental time. In these cases, a person's sleep per se may not be problematic when they are allowed to choose their preferred sleep schedule. Individuals with a delayed endogenous circadian sleep–wake system, for example, tend to have difficulty falling asleep in the evening, and also experience difficulty getting up in the morning. However, when allowed to sleep at their preferred schedule, such as during the weekend or on extended vacations, they usually choose to go to bed late and to sleep-in in the morning without experiencing sleep disruption. This condition is more frequently seen in adolescents and young adults.[84,85] In contrast, patients with an advanced endogenous circadian rhythm usually have no problem initiating sleep, but complain of sleepiness in the evening and early-morning awakenings. This sleep pattern is more commonly reported by elderly individuals.[86,87] In clinical cases, mild phase shifting of the circadian rhythm may interact with other sleep pathologies and make the evaluation and treatment more complicated.

Although the mechanisms that are responsible for circadian rhythms have been shown to include genetic operations, particular neuroanatomical loci, and environmental time cues, bright light has also been shown to be quite effective for phase shifting human circadian rhythms.[34,35] Light exposure can be used to treat sleep disturbances that are related to phase shifts of endogenous circadian rhythms. The influence of light on the phase of circadian rhythms depends on the timing, wavelength, intensity, and duration of light exposure. There is a relationship between the timing of light exposure and the size and direction of the phase shift induced. This relationship can be plotted as a phase response curve. Light exposure in late subjective night and early subjective day produces a phase advance in circadian rhythms; in contrast, light exposure in late subjective day and early subjective night induces a phase delay.[35] Therefore, for patients complaining of difficulty falling asleep in the evening and difficulty getting up in the morning, indicating possible delayed sleep phase, bright light administration upon awakening in the morning can advance the patients' circadian rhythms to the desired time.[88] Bright light exposure in the evening has also been used to treat patients with advanced sleep phase who complain of early-evening sleepiness and early-morning awakening.[89,90] Since in these individuals endogenous circadian oscillators are advanced in time, bright light exposure in the evening should be applied in order to delay their circadian rhythms. In addition, manipulation of light exposure during work hours and avoiding light exposure after work have also been shown to facilitate the adjustment to shift work.[91]

To institute light therapy for phase shifting, patients complete a sleep log that records their daily sleep patterns over a 1–2-week period in order to estimate their endogenous circadian phase. The time of the afternoon/evening dip in alertness may help estimate the endogenous circadian phase, with a later dip (evening flagging attention and energy) suggesting more of a delay in the phase. The patient should also identify an ideal sleep schedule as the goal of the phase shifting. In the beginning, patients are asked to set a sleep–wake schedule that is close to the estimated endogenous sleep phase. As the treatment proceeds, the wake-up time is progressively shifted to an earlier time for patients with a delayed circadian phase; in contrast, the retiring time is gradually shifted to a later time for patients with an advanced sleep phase. Most patients can shift their circadian phase by over 1 hour each week with little difficulty with the help of light exposure. In terms of the light exposure, bright light exposure should be administered in the morning as close to the patient's scheduled wake-up time as possible for patients with delayed sleep–wake patterns, and administered in the early evening for patients with advanced sleep–wake patterns. The source of bright light can be an artificial light device that is made for light therapy or natural outdoor sunlight. Illuminance of approximately 2500 lux or more at eye level is usually required to obtain successful results. The optimal duration of exposure must be determined individually. A 1–2-hour period of treatment each day usually achieves adequate effect. There is a precaution that patients with eye pathology should consult with an ophthalmologist before receiving light therapy.

Another behavioral procedure that is used for the treatment of delayed sleep phase is chronotherapy. Since it was recognized that the sleep–wake cycle in human beings assumes a period length of slightly greater than 24 hours,[32,33] there is therefore a natural tendency for humans to be able to delay their sleep–wake schedule more easily than to advance their phase. Based on this observation, chronotherapy was invented as a treatment for individuals with delayed sleep phase by gradually shifting the sleep–wake schedule later in time until it reaches the desired schedule.[92] Again, conducting this procedure requires an estimation of baseline circadian phase by sleep log and setting up the initial sleep–wake schedule accordingly. The patient is then instructed to delay retiring time and arising time by 2 or 3 hours each day until it matches the desired sleep schedule. Once the desired sleep schedule is achieved, the patient should be told to avoid sleeping late and to arrange for exposure to morning sunlight in order to prevent relapses of the condition. The primary problem in carrying out chronotherapy is that it requires the patients to follow unusual sleep–wake schedules for several days. There will be a number of days in a row in which the patient is sleeping during daylight hours and is awake during the night. Therefore, patients usually take several days off from work and must ensure that they are not disturbed while they sleep. Since the relapse rate for patients with delayed sleep phase syndrome is as high as 91.5%,[93] clinicians need to emphasize the importance of following a strict sleep–wake schedule after the patient reaches their desired schedule. Morning exposure to bright light can be instituted in order to assist in the maintenance of a stable circadian phase.

BEHAVIORAL INTERVENTIONS FOR OTHER SLEEP DISORDERS

Behavioral techniques have also been used to assist in the treatment of other sleep disorders, although evidence of their effectiveness is still preliminary. The behavioral intervention can be the primary treatment or a supplementary treatment combined with pharmacological therapy and/or other medical procedures.

Behavioral intervention has been applied as a supplementary treatment to facilitate compliance with the use of continuous positive airway pressure (CPAP) in patients with sleep-related breathing disorders. The treatment usually includes psychological support, education regarding sleep-related breathing disorders and the function of CPAP, modeling of CPAP use, and systematic desensitization to overcome the anxiety associated with CPAP use. Studies of the effects of education and support on CPAP compliance have generated mixed results. Some studies showed a facilitation of CPAP use,[22,94,95] while others did not.[96,97] Evidence supporting the effects of systematic desensitization is also limited. So far, there is only one case study reporting efficacy.[21]

Behavioral intervention has also been developed for the management of nightmares. Clinicians have the patient recall and rehearse the content of nightmares and change the dream narrative ending to a more positive outcome. The patient is then instructed to rehearse the revised version of the dream while awake each day for a few days. The dream rehearsal was found to be effective in reducing the frequency and intensity of nightmares, and also to improve their sense of control. This technique was also applied successfully to patients with posttraumatic stress disorder (PTSD). Patients showed improved sleep and PTSD symptoms with this technique.[20,98]

In addition to these specific techniques, the basic conceptual framework and behavioral intervention developed for insomnia may also be applied in order to optimize sleep in patients with other sleep disorders. Although their sleep pathologies may disrupt the normal processes of sleep regulation, maximizing the sleep drive at bedtime, maintaining a stable circadian rhythm and reducing the interference from cognitive and emotional arousal can strengthen the sleep cycle and may make the patients more resistant to disruption by the primary sleep pathology.

CONCLUSION

Classical western medicine favors a single explanation for a cluster of symptoms. However, in the case of sleep disorders, the sleep symptom is often a manifestation of the interactions between multiple contributing factors. Frequently, both physiological and psychological factors are involved in the production of the sleep disturbance. Clinicians should take a broad prospective and avoid viewing the problem from a single perspective. A careful evaluation of the patient's overall condition is often the key to a successful treatment. The basic rationale of behavioral interventions for sleep disorders is to make an individual's lifestyle and sleep pattern more compatible with the nature of the systems that regulate sleep. This often requires the patients to change daily life routines and to engage in specific behavioral practices. Patient motivation is another key for effective treatment.

Although effective behavioral techniques for sleep disorders are available, the percentage of patients seeking help remains relatively low. Survey studies have reported that only around 5% of patients with insomnia seek help specifically for their sleep problem,[99] and only a third of patients with sleep difficulties discussed their problems with a healthcare professional.[100] Public education is important for promoting awareness of the treatment options and promoting the wisdom to seek help at an early stage of the sleep problem. In addition, inclusion of sleep medicine training for professionals is important for providing physicians, psychologists, and other medical professionals with sufficient knowledge and techniques to help patients with sleep problems.

ACKNOWLEDGMENTS

The author would like to dedicate this chapter to the memory of late Professor Arthur J. Spielman, who passed away in 2014, for his mentorship and inspiration.

REFERENCES

1. American Academy of Sleep Medicine (AASM). *The International Classification of Sleep Disorders* (2nd ed.). Westchester, IL: American Academy of Sleep Medicine.

2. Ford D, Kamerow D. Epidemiologic study of sleep disturbances and psychiatric disorders: An opportunity for prevention? *JAMA* 1989; 262: 1479–1484.

3. Ohayon MM, Roth T. Place of chronic insomnia in the course of depressive and anxiety disorders. *J Psychiatr Res* 2003; 37(1): 9–15.

4. Breslau N, Roth T, Rosenthal L, Andreski P. Sleep disturbance and psychiatric disorders: A longitudinal epidemiological study of young adults. *Biol Psychiatry* 1996; 39: 411–418.

5. Katz DA, McHorney CA. Clinical correlates of insomnia in patients with chronic illness. *Arch Intern Med* 1998; 158: 1099–1107.

6. Smith MT, Perlis ML, Smith MS, Giles DE, Carmody TP. Sleep quality and presleep arousal in chronic pain. *J Behav Med* 2000; 23: 1–13.

7. Sridhar GR, Madhu K. Prevalence of sleep disturbances in diabetes mellitus. *Diabetes Res Clin Pract* 1994; 23: 183–186.

8. Klink M, Quan S, Kaltenborn W, Lobowitz M. Risk factors associated with complaints of insomnia in a general adult population. *Arch Intern Med* 1992; 152: 1634–1637.

9. Ohayon MM. Relationship between chronic painful condition and insomnia. *J Psychiatr Res* 2005; 39: 151–159.

10. Spielman AJ. Assessment of insomnia. *Clin Psychol Rev* 1986; 6: 11–25.

11. Yang C-M, Spielman AJ, Glovinsky P. Nonpharmacologic strategies in the management of insomnia. *Psychiatr Clin North Am* 2006; 29(4): 895–919.

12. Morin CM, Culbert JP, Schwartz SM. Nonpharmacological interventions for insomnia: A meta-analysis of treatment efficacy *Am J Psychiatry* 1994; 151: 1172–1180.

13. Murtagh DRR, Greenwood KM. Identifying effective psychological treatment for insomnia: A meta-analysis. *J Consult Clin Psychol* 1995; 63: 79–89.

14. Smith MT, Perlis ML, Park A et al. Comparative meta-analysis of pharmacotherapy and behavior therapy for persistent insomnia. *Am J Psychiatry* 2002; 159: 5–11.

15. Lichstein KL, Wilson NM, Johnson CT. Psychological treatment of secondary insomnia. *Psychol Aging* 2000; 15: 232–240.

16. Perlis ML, Sharpe MC, Smith MT, Greenblatt DW, Giles DE. Behavioral treatment of insomnia: Treatment outcome and the relevance of medical and psychiatric morbidity. *J Behav Med* 2001; 24: 281–296.

17. Rybarczyk B, Lopez M, Benson R, Alsten C, Stepanski E. Efficacy of two behavioral treatment programs for comorbid geriatric insomnia. *Psychol Aging* 2002; 17: 288–298.

18. Rybarczyk B, Stepanski E, Fogg L, Lopez M, Barry P, Davis A. A placebo-controlled test of cognitive–behavioral therapy for comorbid insomnia in older adults. *J Consult Clin Psychol* 2005; 73: 1164–1174.

19. Kuo T, Manber R, Loewy D. Insomniacs with comorbid conditions achieved comparable improvement in a cognitive behavioral group treatment program as insomniacs without comorbid depression. *Sleep* 2001; 14: A62.

20. Krakow B, Hollifield M, Johnston L et al. Imagery rehearsal therapy for chronic nightmares in sexual assault survivors with posttraumatic stress disorder. *JAMA* 2001; 286: 537–545.

21. Edinger JD, Radtke RA. Use of *in vivo* desensitization to treat a patient's claustrophobia response to nasal CPAP. *Sleep* 1993; 16: 678–680.

22. Likar LL, Panciera TM, Erickson AD, Rounds S. Group education sessions and compliance with nasal CPAP therapy. *Chest* 1997; 111: 1273–1277.

23. Morin CM, Colecchi C, Stone J et al. Behavioral and pharmacological therapies for late-life insomnia. *JAMA* 1999; 281: 991–999.

24. Edinger JD, Means MK. Cognitive–behavioral therapy for primary insomnia. *Clin Psychol Rev* 2005; 25: 539–558.

25. Smith MT, Huang MI, Manber R. Cognitive behavior therapy for chronic insomnia occurring within the context of medical and psychiatric disorders. *Clin Psychol Rev* 2005; 25: 559–592.

26. Mignot E, Taheri S, Nishino S. Sleeping with the hypothalamus: Emerging therapeutic targets for sleep disorders. *Nat Neurosci* 2002; 5: 1071–1075.

27. Saper CB, Scammell TE, Lu J. Hypothalamic regulation of sleep and circadian rhythms. *Nature* 2005; 437: 1257–1263.

28. Borbely AA. A two process model of sleep regulation. *Hum Neurobiol* 1982; 1: 195–204.

29. Youngstedt SD, O'Connor PJ, Dishman RK. The effects of acute exercise on sleep: A quantitative synthesis. *Sleep* 1997; 20: 203–214.

30. Wager-Smith K, Kay SA. Circadian rhythm generics: From flies to mice to humans. *Nat Genet* 2000; 26: 23–27.

31. Klei L, Reitz P, Miller M et al. Heritability of morningness–eveningness and self-report sleep measures in a family-based sample of 521 hutterites. *Chronobiol Int* 2005; 22: 1041–1054.

32. Dijk DJ, Czeisler CA. Paradoxical timing of the circadian rhythm of sleep propensity serves to consolidate sleep and wakefulness in humans. *Neurosci Lett* 1994; 166: 63–68.

33. Wyatt JK, Ritz-De Cecco A, Czeisler CA, Dijk DJ. Circadian temperature and melatonin rhythms, sleep, and neurobehavioral function in humans living on a 20-h day. *Am J Physiol* 1999; 277: R1152–R1163.

34. Czeisler CA, Allan JS, Strogatz SH et al. Bright light resets the human circadian pacemaker independent of the timing of the sleep–wake cycle. *Science* 1986; 233: 667–671.

35. Minors DS, Waterhouse JM, Wirz-Justice A. A human phase–response curve to light. *Neurosci Lett* 1991; 133: 354–361.

36. Bonnet MH, Arand DL. Hyperarousal and insomnia. *Sleep Med Rev* 1997; 2: 97–108.

37. Perlis ML, Giles DE, Mendelson WB et al. Psychophysiological insomnia: The behavioral model and a neurocognitive perspective. *J Sleep Res* 1997; 6: 179–188.

38. Vgontzas AN, Tsigos C, Bixler EO et al. Chronic insomnia and activity of the stress system: A preliminary study. *J Psychosom Res* 1998; 45: 21–31.

39. Bonnet MH, Arand DL. 24-hour metabolic rate in insomniacs and matched normal sleepers. *Sleep* 1995; 18: 581–588.

40. Freedman RR. EEG power spectra in sleep-onset insomnia. *Electroencephalogr Clin Neurophysiol* 1986; 63: 408–413.

41. Merica H, Blois R, Gaillard J-M. Spectral characteristics of sleep EEG in chronic insomnia. *Eur J Neurosci* 1998; 10: 1826–1834.

42. Perlis ML, Smith MT, Andrews PJ et al. Beta/gamma EEG activity in patients with primary and secondary insomnia and good sleeper controls. *Sleep* 2001; 24: 110–117.

43. Yang C-M, Lo HS. ERP evidence of enhanced excitatory and reduced inhibitory processes of auditory stimuli during sleep in patients with primary insomnia. *Sleep* 2007; 30(5): 585–592.

44. Bastien CH, St-Jean G, Morin CM, Turcotte I, Carrier J. Chronic psychophysiological insomnia: Hyperarousal and/or inhibition deficits? An ERPs investigation. *Sleep* 2008; 31(6): 887–898.

45. Jacobs GD, Benson H, Friedman R. Home-based central nervous system assessment of a multifactor behavioral intervention for chronic sleep-onset insomnia. *Behav Ther* 1993; 24: 159–174.

46. Harvey AG, Tang NK, Browning L. Cognitive approaches to insomnia. *Clin Psychol Rev* 2005; 25: 593–611.

47. Morin CM, Stone J, Trinkle D, Mercer J, Remsberg S. Dysfunctional beliefs and attitudes about sleep among older adults with and without insomnia complaints. *Psychol Aging* 1993; 8: 463–467.

48. Fichten CS, Creti L, Amsel R, Brender W, Weinstein N, Libman E. Poor sleepers who do not complain of insomnia: Myths and realities about psychological and lifestyle characteristics of older good and poor sleepers. *J Behav Med* 1995; 18: 189–223.

49. Edinger JD, Fins AI, Glenn DM et al. Insomnia and the eye of the beholder: Are there clinical markers of objective sleep

disturbances among adults with and without insomnia complaints? *J Consult Clin Psychol* 2000; 68: 586–593.

50. Harvey L, Inglis SJ, Espie CA. Insomniacs' reported use of CBT components and relationship to long-term clinical outcome. *Behav Res Ther* 2002; 40: 75–83.

51. Drake C, Richardson G, Roehrs T, Scofield H, Roth T. Vulnerability to stress-related sleep disturbance and hyperarousal. *Sleep* 2004; 27(2): 285–891.

52. Yang CM, Hung CY, Lee HC. Stress-related sleep vulnerability and maladaptive sleep beliefs predict insomnia at long-term follow-up. *J Clin Sleep Med* 2014; 10(9): 997–1001.

53. Lacks P, Morin CM. Recent advances in the assessment and treatment of insomnia. *J Consult Clin Psychol* 1992; 60: 586–594.

54. Jacobs GD, Pace-Schott EF, Stickgold R et al. Cognitive behavior therapy and pharmacotherapy for insomnia. *Arch Intern Med* 2004; 164: 1888–1896.

55. Edinger JD, Sampson WS. A primary care "friendly" cognitive behavioral insomnia therapy. *Sleep* 2003; 26: 177–182.

56. Hauri PJ. Sleep hygiene, relaxation therapy, and cognitive interventions. In: Hauri PJ (Ed.). *Case Studies in Insomnia*. New York, NY: Plenum, 1991, pp. 65–84.

57. Lacks P, Rotert M. Knowledge and practice of sleep hygiene techniques in insomniacs and good sleepers. *Behav Res Ther* 1986; 24: 365–368.

58. Kohn L, Espie CA. Sensitivity and specificity of measures of the insomnia experience: A comparative study of psychophysiologic insomnia, insomnia associated with mental disorder and good sleepers. *Sleep* 2005; 28: 104–112.

59. Jefferson CD, Drake CL, Scofield HM et al. Sleep hygiene practices in a population-based sample of insomniacs. *Sleep* 2005; 28: 611–615.

60. Harvey AG. Sleep hygiene and sleep-onset insomnia. *J Nerv Ment Dis* 2000; 188: 53–55.

61. McCrae CS, Rowe MA, Dautovich ND et al. Sleep hygiene practices in two community dwelling samples of older adults. *Sleep* 2006; 29: 1551–1560.

62. Yang C-M, Lin SC, Hsu SC, Cheng CP. Maladaptive sleep hygiene practices in good sleepers and patients with insomnia. *J Health Psychol* 2010; 15(1): 147–155.

63. Morin CM. *Insomnia: Psychological Assessment and Management*. New York, NY: Guilford, 1993.

64. Bootzin RR. Stimulus control treatment for insomnia. *Proc Am Psychol Assoc* 1972; 7: 395–396.

65. Spielman AJ, Saskin P, Thorpy MJ. Treatment of chronic insomnia by restriction of time in bed. *Sleep* 1987; 10: 45–56.

66. Friedman L, Bliwise DL, Yesavage JA et al. A preliminary study comparing sleep restriction and relaxation treatments for insomnia in older adults. *J Gerontol B Psychol Sci* 1991; 46: 1–8.

67. Rubenstein ML, Rothenberg SA, Maheswaran S et al. Modified sleep restriction therapy in middle-aged and elderly chronic insomniacs [abstract]. *Sleep Res* 1990; 19: 276.

68. Lichstein KL, Riedel BW, Wilson NM, Lester KW, Aguillard RN. Relaxation and sleep compression for late-life insomnia: A placebo-controlled trial. *J Consult Clin Psychol* 2001; 69: 227–239.

69. Borkovec TD, Grayson JB, O'Brien GT et al. Relaxation treatment of pseudoinsomnia and idiopathic insomnia: An electroencephalographic evaluation. *J Appl Behav Anal* 1979; 12: 37–54.

70. Nicassio P, Bootzin R. A comparison of progressive relaxation and autogenic training as treatment for insomnia. *J Abnorm Psychol* 1974; 83: 253–260.

71. Woolfolk RL, Carr-Kaffashan L, McNulty TF. Meditation training as a treatment for insomnia. *Behav Ther* 1976; 7: 359–365.

72. Haynes SN, Sides H, Lockwood G. Relaxation instructions and frontalis electromyographic feedback intervention with sleep-onset insomnia. *Behav Ther* 1977; 8: 644–652.

73. Hauri PJ. Treating psychophysiologic insomnia with biofeedback. *Arch Gen Psychiat* 1981; 38: 752–758.

74. Ong JC, Manber R, Segal Z, Xia Y, Shapiro S, Wyatt JK. A randomized controlled trial of mindfulness meditation for chronic insomnia. *Sleep* 2014; 37(9): 1553–1563.

75. Nicolau ZF, Bezerra AG, Andersen ML, Tufik S, Hachul H. Mindfulness-based intervention to treat insomnia in elderly people. *Contemp Clin Trials* 2014; 39(1): 166–167.

76. Garland SN, Carlson LE, Stephens AJ, Antle MC, Samuels C, Campbell TS. Mindfulness-based stress reduction compared with cognitive behavioral therapy for the treatment of insomnia comorbid with cancer: A randomized, partially blinded, noninferiority trial. *J Clin Oncol* 2014; 32: 449–457.

77. Britton WB, Haynes PL, Fridel KW, Bootzin RR. Mindfulness-based cognitive therapy improves polysomnographic and subjective sleep profiles in antidepressant users with sleep complaints. *Psychother Psychosom* 2012; 81(5): 296–304.

78. Gross CR, Kreitzer MJ, Reilly-Spong M et al. Mindfulness-based stress reduction versus pharmacotherapy for chronic primary insomnia: A randomized controlled clinical trial. *Explore (NY)* 2011; 7(2): 76–87.

79. Ong J, Sholtes D. A mindfulness-based approach to the treatment of insomnia. *J Clin Psychol* 2010; 66(11): 1175–1184.

80. Ong JC, Shapiro SL, Manber R. Mindfulness meditation and cognitive behavioral therapy for insomnia: A naturalistic 12-month follow-up. *Explore (NY)* 2009; 5(1): 30–36.

81. Yook K, Lee SH, Ryu M et al. Usefulness of mindfulness-based cognitive therapy for treating insomnia in patients with anxiety disorders: A pilot study. *J Nerv Ment Dis* 2008; 196(6): 501–503.

82. Ong JC, Shapiro SL, Manber R. Combining mindfulness meditation with cognitive–behavior therapy for insomnia: A treatment-development study. *Behav Ther* 2008; 39(2): 171–182.

83. Heidenreich T, Tuin I, Pflug B, Michal M, Michalak J. Mindfulness-based cognitive therapy for persistent insomnia: A pilot study. *Psychother Psychosom* 2006; 75(3): 188–189.

84. Weitzman ED, Czeisler CA, Coleman RM et al. Delayed sleep phase syndrome: A chronobiological disorder with sleep-onset insomnia. *Arch Gen Psychiatry* 1981; 38: 737–746.

85. Thorpy MJ, Korman E, Spielman AJ, Glovinsky PB. Delayed sleep phase syndrome in adolescents. *J Adolesc Health Care* 1988; 9: 22–27.

86. Czeisler CA, Dumont M, Duffy JF et al. Association of sleep–wake habits in older people with changes in output of circadian pacemaker. *Lancet* 1992; 340: 933–936.

87. Kramer CJ, Kerkhof GA, Hofman WF. Age differences in sleep-wake behavior under natural conditions. *Pers Indiv Differ* 1999; 27: 853–860.

88. Rosenthal NE, Joseph-Vanderpool JR, Levendosky AA et al. Phase-shifting effects of bright morning light as treatment for delayed sleep phase syndrome. *Sleep* 1990; 13: 354–361.

89. Campbell SS, Murphy PJ, van den Heuvel CJ, Roberts ML, Stauble TN. Etiology and treatment of intrinsic circadian rhythm sleep disorders. *Sleep Med Rev* 1999; 3: 179–200.

90. Lack L, Wright H, Kemp K, Gibbon S. The treatment of early-morning awakening insomnia with 2 evenings of bright light. *Sleep* 2005; 28: 616–623.

91. Eastman CI, Boulos Z, Terman M, Campbell SS, Dijk DJ, Lewy AJ. Light treatment for sleep disorders: Consensus report. VI. Shift work. *J Biol Rhythms* 1995; 10: 157–164.

92. Czeisler CA, Richardson GS, Coleman RM et al. Chronotherapy: Resetting the circadian clocks of patients with delayed sleep phase insomnia. *Sleep* 1981; 4: 1–21.

93. Dagan Y, Yovel I, Hallis D, Eisenstein M, Raichik I. Evaluating the role of melatonin in the long-term treatment of delayed sleep phase syndrome (DSPS). *Chronobiol Int* 1998; 15: 181–190.

94. Hoy CJ, Vennelle M, Kingshott RN, Engleman HM, Douglas, NJ. Can intensive support improve continuous positive airway pressure use in patients with the sleep apnea/hypopnea syndrome? *Am J Resp Crit Care Med* 1999; 159: 1096–1100.

95. Chervin RD, Theut S, Bassetti C, Aldrich MS. Compliance with nasal CPAP can be improved by simple interventions. *Sleep* 1997; 20: 284–289.

96. Hui DS, Chan JK, Choy DK et al. Effects of augmented continuous positive airway pressure education and support on compliance and outcome in a Chinese population. *Chest* 2000; 117: 1410–1416.

97. Fletcher EC, Luckett RA. The effect of positive reinforcement on hourly compliance in nasal continuous positive airway pressure users with obstructive sleep apnea. *Am Rev Respir Dis* 1991; 143: 936–941.

98. Krakow B, Johnston L, Melendrez D et al. An open-label trial of evidence-based cognitive behavior therapy for nightmares and insomnia in crime victims with PTSD. *Am J Psychiat* 2001; 158: 2043–2047.

99. Gallup Organization. *Sleep in America.* Princeton, NJ: Gallup Organization, 1991.

100. Gallup Organization. *Sleep in America.* Princeton, NJ: Gallup Organization, 1995.

101. Spielman AJ, Caruso LS, Glovinsky PB. A behavioral perspective on insomnia treatment. *Psychiatr Clin North Am.* 1987; 10(4): 541–553. Review. PubMed PMID: 3332317.

20

Sleep deprivation therapy:
A rapid-acting antidepressant

JOSEPH C. WU AND BLYNN G. BUNNEY

I never thought my brain could be normal. I now have hope.

—Quote from a treatment-resistant patient who improved after one night of sleep deprivation therapy

INTRODUCTION

Sleep deprivation therapy (SDT), also referred to as "wake therapy," is a robust, rapid-acting (within 24–48 hours), noninvasive antidepressant treatment. First proposed by Schulte[1] and Pflug and Tolle,[2] SDT has been administered to more than 2000 patients worldwide. One night of enforced sleep loss significantly decreases depressive symptoms in approximately 40%–60% of depressed patients and can be administered as an adjunctive treatment to ongoing medications.[3] Although SDT is relatively safe, there is a slight risk of switches into hypomania/mania that is comparable to or lower than that associated with conventional antidepressants. Mood stabilizers can quickly reverse symptoms in order to restore euthymia.[4–6]

The robustness of the antidepressant action of SDT is reported to be equivalent to that of slower-acting conventional antidepressants that usually require 2–8 weeks for a full clinical response.[7] It is efficacious in both major depressive disorder (MDD) and bipolar disorder (BPD), as well as in treatment-resistant depression, although rates in refractory patients may be lower.[8,9]

In Europe, SDT is considered to be a first-line treatment option.[6] The World Federation of Societies of Biological Psychiatry (WFSBP) includes guidelines for SDT for both MDD[10] and BPD.[11] Its use is recommended: (1) in unmedicated depressed patients; (2) to accelerate response to medication; and (3) to potentiate an ongoing anti-depressant drug therapy. The guidelines also note that SDT plus sleep phase advance is an effective treatment for MDD and BPD. To minimize risk of switches, it is recommended that mood stabilizers be initiated prior to SDT in BPD patients. (For further details on administering chronotherapy including sleep deprivation, Wirz-Justice and colleagues have published a second edition of a clinician's manual.[12]) There is also a website (www.cet.org/education/clinicians-forum-intro/) that acts as a forum for clinicians who are interested in SDT, which allows clinicians to ask questions and share their experiences with the use of sleep deprivation.

Variations in SDT

Total sleep deprivation (TSD) involves keeping patients awake for 36–40 continuous hours; partial sleep deprivation (PSD) refers to keeping patients awake for the first half of the night—"early PSD"—or patients being kept awake for the second part of the night—"late PSD." Cycles of sleep deprivation can be administered as a single treatment or may be repeated over the course of a week or more.

Strategies to block relapse and sustain improvement

SDT by itself rapidly but transiently decreases depressive symptoms. Relapse rates, however, can be as high as 80% following recovery sleep.[13] Naps, lasting only a few minutes, can precipitate symptoms in some,[3,14–17] but not all patients.[18] A study of 30 depressed SDT responders found that morning

naps and shorter naps were more detrimental than afternoon naps.[14]

Concomitant treatment with medications (e.g., antidepressants and lithium) and post-sleep deprivation chronotherapies such as bright light therapy and sleep phase advance can lengthen the maintenance of improvement from weeks to months. Bright light therapy typically involves exposing patients to 10,000 lux in the morning following SDT. Sleep phase advance involves setting patient bedtimes earlier and then advancing bedtimes over subsequent days. Table 20.1[9,19–28] and Table 20.2[4,8,29–38] summarize 22 long-term studies (>2 weeks) showing the feasibility of sustaining SDT antidepressant responses. Table 20.1 includes 11 studies using sleep deprivation plus adjunctive medications only. Table 20.2 summarizes 11 studies combining SDT with medications plus chronotherapies (light therapy and sleep phase advance). Follow-up times ranged from 2 weeks to 9 months. Protocol designs included: TSD (one night) in 15/22 studies; PSD (late; second half of the night) in 4/22 studies; and both partial and total SDT in 3/22 studies. One study administered PSD in the first half of the night.[38] Repeated SDTs were administered in the majority of protocols—14/22 studies (63%).

In our study,[31] we randomly assigned 49 medicated (sertraline and lithium) BPD depressed patients to either chronotherapy (n = 32; one night of TSD and bright light therapy plus sleep phase advance; see Figure 20.1) or to a medication-only (n = 17; treatment as usual) group. Significant decreases in depression ratings in the chronotherapy group were seen in the first 24 hours following a night of sleep deprivation. Improvement was immediate and significantly better than the treatment-as-usual group at every time-point over the 7-week trial with the exception of day 6. By the end of the study, 12/19 responders (63%) fulfilled the criteria for remission.[31]

SDT: Reducing suicidality

A major challenge in the treatment of depression is to rapidly decrease the risk for suicide. Two studies reported the efficacy of SDT plus chronotherapy. Benedetti et al.[4] observed that SDT plus lithium and extended bright light treatment produced an immediate (day 1) and

Table 20.1 Studies reporting efficacy in sustaining the antidepressant response of sleep deprivation therapy (SDT) with medication

Study	Subjects	Sleep deprivation method	Medication	Responders (n)	Finding
Caliyurt and Guducu[19]	24 MDD Recurrent (n = 14) Acute (n = 10)	6 PSD	Sertraline and late PSD (n = 13) Sertraline only (n = 11)	12/13 responders to sertraline + PSD 5/11 responders to sertraline only	4 weeks
Benedetti et al.[99]	40 BPD	3 TSD	Chronic lithium (>6 months) vs. lithium post-SDT	13/20 responders to chronic lithium 2/20 responders to lithium post-SDT	3 months
Smith et al.[20]	6 elderly depressed; 6 age-matched controls	TSD	Paroxetine	5/6 responders	12 weeks
Smeraldi et al.[21]	40 BPD Placebo-controlled study	3 TSD (7 a.m.–7 p.m.) on alternate days	TSD + pindolol (n = 20) TSD alone (n = 20) Initiated lithium post-SDT to maintain response	14/20 responders to TSD + pindolol 3/20 responders to TSD only	6 months
Kuhs et al.[22]	44 MDD Crossover design	PSD (late) group A: 2×/2 weeks; group b: 1×/ week for 2 weeks	Amitriptyline	Timing of PSD was not significant Group A: 9/20 responders Group B: 8/22 responders	35 days
Bump et al.[23]	13 elderly MDD 8 recurrent; 5 acute	1 night of TSD (open trial)	Paroxetine	8/13 responders within 2 weeks 11/13 responders by 12 weeks	3 months
Kuhs et al.[24]	51 MDE	6 PSD (late)	Amitriptyline (n = 27) Amitriptyline plus PSD (n = 24)	Significant improvement in amitriptyline + PSD on day 14 compared to amitriptyline alone Continued improvement only in the amitriptyline/PSD group at day 28	28 days

(Continued)

Table 20.1 (Continued) Studies reporting efficacy in sustaining the antidepressant response of sleep deprivation therapy (SDT) with medication

Study	Subjects	Sleep deprivation method	Medication	Responders (n)	Finding
Leibenluft et al.[9]	n = 29 4 BPD 25 MDD Medicated for 3 months but still chronically depressed	4 cycles of PSD early (slept from 3 a.m. to 7 a.m.) or late (slept from 10 p.m. to 3 a.m.) on days 5, 6, 12, and 13	Fluoxetine (n = 15) Lithium (n = 5) Sodium valproate (n = 1) Other antidepressants (n = 8)	No difference between late or early PSD 15/26 responders 1/3 BPD responders 1 BPD spontaneous responder	3 weeks
Shelton and Loosen[25]	20 MDD	TSD (36 hours)	Nortriptyline administered after TSD	11/20 responders TSD expedited the antidepressant response	2 weeks
Kasper et al.[26,27]	41 MDD	TSD: 2 nights (40 hours) 1 week apart	Fluvoxamine Maprotiline Started on evening after SDT	No relapse in either group after recovery sleep	4 weeks
Elsenga and van den Hoofdakker[28]	30 MDD	2 TSD/week for 2 weeks (double-blind)	n = 10 in each group Clomipramine TSD plus clomipramine TSD alone	Faster response in first week; no difference in second week between clomipramine and clomipramine + SDT	15 days

MDD: major depressive disorder; PSD: partial sleep deprivation; BPD: bipolar disorder; TSD: total sleep deprivation; SDT: sleep deprivation therapy; MDE: major depressive episode.

Table 20.2 Studies reporting efficacy in sustaining the antidepressant effects of sleep deprivation therapy[a] with additional chronotherapies (bright light therapy and sleep phase advance)

Study	Subjects	PSD or TSD	Medication	Additional chronotherapy	Findings	Follow-up (length of trial)
Benedetti et al.[4]	141 BPD 32 with history of suicidality 117 TRD	3 TSDs in a 1-week period	Lithium maintained or initiated at start of trial	BLT (10,000 lux) for 30 minutes at 3 a.m. post-SDT and for 1 week following acute response	99/141 responders; 78/99 achieved remission at the end of the trial Suicidality decreased after first SDT + BLT in responders and nonresponders, although improvement was more rapid in the responders 78/99 TRD patients responded	1 month
Martiny et al.[29,30]	64 MDD 30 SDT 34 exercise group	3 TSDs in a 1-week period	Duloxetine	BLT (10,000 lux for 30 minutes) + sleep time stabilization[b] (29 weeks)	Chronotherapy vs. exercise group Remission rate = 61.9% vs 37.9% in exercise group	29 weeks
Wu et al.[31]	49 BPD	TSD	n = 17 sertraline + lithium n = 32 sertaline + lithium + CAT	BLT (5000 lux/2 hours) for 3 consecutive days post-SDT SPA on recovery night for 3 days	CAT group significantly improved over treatment-as-usual group at every time point with the exception of day 6 12/19 CAT patients were in remission at 7 weeks	7 weeks
Moscovici and Kotler[32]	12 MDD	TSD 3 PSDs (late)	Medication-free	BLT (green light dawn simulation) SPA	7 remitters 2 responders 1 partial responder	4 weeks
Benedetti et al.[8]	60 BPD (TRD) and non-TRD	3 TSDs	n = 34 drug-free n = 26 on medication n = 16 lithium + medication	400 lux green light for first 30 minutes after awakening for 7 days	23/33 non-TRD responders 12/27 TRD responders At 9 months: 13/23 non-TRD responders 2/12 TRD responders	9 months

(Continued)

Table 20.2 (*Continued*) Studies reporting efficacy in sustaining the antidepressant effects of sleep deprivation therapy[a] with additional chronotherapies (bright light therapy and sleep phase advance)

Study	Subjects	PSD or TSD	Medication	Additional chronotherapy	Findings	Follow-up (length of trial)
Benedetti et al.[33]	30 TRD BPD	TSD	Lithium vs. no lithium Responders to SPA were maintained or initiated on lithium	SPA	9/16 responders on lithium 4/14 responders initiated on lithium	3 months
Fritzsche et al.[34]	33 MDD 7 SAD 14 males, 26 females	TSD	Medicated	BLT or dim red light (50 lux) from day 3 post-SDT until day 16 from 7 to 9 a.m. Dim red light (50 lux) or bright white light (2500 lux) was administered	20/40 responders to TSD (improved by 30%). It was concluded that only the sleep deprivation responders further improved with light treatment. (No difference between bright or dim light)	16 days
Szuba et al.[35]	10 MDD 6 BPD	—	Lithium or desipramine	PSD (late) or TSD	PSD had greater efficacy than TSD 0/4 TSD showed response at day 30. (2/4 responded by day 20) 4/6 PSD responders maintained response through to day 30 2/3 TSD on desipramine sustained improvement over 30 days	30 days

(*Continued*)

Table 20.2 (*Continued*) Studies reporting efficacy in sustaining the antidepressant effects of sleep deprivation therapy[a] with additional chronotherapies (bright light therapy and sleep phase advance)

Study	Subjects	PSD or TSD	Medication	Additional chronotherapy	Findings	Follow-up (length of trial)
Holsboer-Trachsler[36]	12 first-episode MDD 23 MDD 6 BPD 1 dysthymic	PSD (late) 3×/week for first week and then weekly for 3 weeks	Trimipramine	Medication alone Medication + PSD Medication + BLT (5000 lux for 1 week and last 3 days of week for 3 weeks)	Medication alone (11/14 responders Medication + PSD (6/14 responders) Medication + BLT (6/14 responders	6 weeks
van den Burg et al.[37]	21 MDD 1 BPD 1 atypical	2 TSD/2 weeks	Drug-free Medication	BLT (n = 11) Dim light (n = 12)	8/23 responders SDT + BLT sustained responses	2 weeks
Souetre et al.[38]	5 TSD responders (open study)	PSD (early)	Drug-free	SPA for 2 weeks	4/5 responders sustained improvement with SPA	2 weeks

[a] SDT is also referred to as "wake therapy." In this table, for simplification, the term "sleep deprivation therapy" is used.
[b] Sleep time stabilization: patients were instructed to keep a stable sleep–wake cycle, prevent oversleeping and encouraged to go to sleep earlier than midnight.
PSD: partial sleep deprivation; TSD: total sleep deprivation; BPD: bipolar disorder; TRD: treatment-resistance depression; BLT: bright light therapy; SDT: sleep deprivation therapy; MDD: major depressive disorder; SPA: sleep phase advance; CAT: chronotherapeutic augmentation therapy; SAD: seasonal affective disorder.

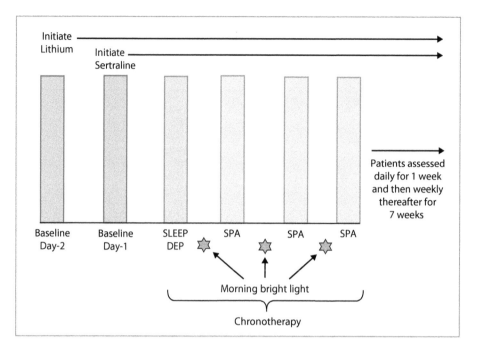

Figure 20.1 Protocol for triple chronotherapy. One night of sleep deprivation therapy is followed by three mornings of bright light therapy and three nights of sleep phase advance. (For more details, see Wu JC et al. *Biol Psychiatry* 2009; 66: 298–301.)

sustained decrease in suicidality (as measured by the Hamilton Depression Rating Scale, item #3). (Significant improvement was seen after the first of three nights of TSD.) Nonresponders also benefited from the sleep deprivation protocol, although their course of improvement was slower.

In an independent study using an open design, Sahlem et al.[39] treated 10 suicidal depressed patients with one night of TSD, three nights of sleep phase advance, and four 30-minute sessions of morning bright light (10,000 lux). Primary outcome measures included the Columbia Suicide Severity Rating Scale (CSSRS) and the Hamilton Depression Scale (HAM-17). The results showed a dramatic decrease in CSSRS ratings from 19.5 at baseline to 6.5 on the first day following SDT. Six of the 10 suicidal patients fulfilled remission criteria on both the CSSRS and the HAM-17 by day 4.

Two case reports of the use of chronotherapy (one night of sleep deprivation plus sleep phase advance) in a hospital setting in treatment-resistant patients with severe suicidality suggest that chronotherapy is a useful adjunct to medications. One patient was in remission for 1 month post-hospitalization after receiving SDT. A second chronotherapy patient also remitted and was doing well at a 10-month follow-up. Family members attributed the improvement to SDT.[40]

SDT: Treatment-resistant depression

SDT is beneficial in treatment-resistant patients, despite the failure of response to conventional medications. Two studies conducted by Benedetti's group in BPD patients involved three nights of TSD plus light therapy administered on alternating nights over a 1-week period. All patients were medicated with lithium. In the first study,[8] light therapy consisted of morning green light (400 lux) for 7 days and reported a response rate of 17%. In the second study,[4] light therapy consisted of exposure to morning bright light (10,000 lux) for 7 days. Response rates dramatically increased from 17% to 55% in the second study. The investigators attributed the improvement to prolonged bright light exposure and the initiation of lithium.[4] (As noted above, SDT was successfully used to reduce suicidality in treatment-resistant patients.)

Novel combination of SDT plus repetitive transcranial magnetic stimulation in treatment-resistant depression

Coadministration of repetitive transcranial magnetic stimulation (rTMS) with PSD produced a slower but prolonged response. Krstic et al.[41] treated 19 treatment-resistant MDD females with PSD (kept awake for part of the night) in combination with rTMS. All patients were treated with antidepressants. Patients were administered a 1-Hz rTMS or sham stimulation over the right dorsolateral prefrontal cortex for 10 daily sessions. PSD was administered once per week. Full remission was seen in 5 out of 11 patients at week 24 in the PSD plus rTMS groups.

PREDICTION OF SDT RESPONSE

Wiegand et al.,[17] conducted a series of six SDT trials in 12 medicated inpatients (each cycle consisted of 36 hours of SDT followed by a night of recovery sleep) over a period of 3 weeks. Subjects relapsed to baseline levels prior to each cycle so that the pre-sleep deprivation depression ratings were consistent throughout the protocol. The results showed that: (1) individuals who do not respond to one cycle of SDT may respond to another; (2) there is no predictable pattern of response for any individual from one cycle to the next; and (3) patients do not become habituated or sensitized to repeated SDT treatments.

Diurnal mood changes as predictors of response

It is reported that mood swings characterized by early-morning worsening and improvement by late afternoon may increase the likelihood of response to SDT in a subgroup of depressed patients with stabilized mood.[3,29,42-44] Wirz-Justice[45] proposes that this is not due to the directionality of the mood swings, but rather to the propensity for diurnal mood changes.

Diffusion tensor imaging data as predictors of response

Diffusion tensor imaging (DTI) data reveal white matter changes in the brains of a wide range of BPD patients (i.e., depressed, manic, drug-naïve, untreated first-episode psychosis, untreated BPD depressed, drug-naïve manic, and euthymic BPD patients). DTI pathology is also reported in high-risk unaffected relatives.[46-48] Bollettini[49] conducted a DTI study of 70 BPD inpatients in order to evaluate DTI as a biomarker of response to SDT. Patients were administered three consecutive SDT cycles on alternating days (1 week). Bright light therapy (10,000 lux for 30 minutes) was administered at 3:00 a.m. on the night of sleep deprivation and 30 minutes after awakening on recovery sleep mornings. All patients were medicated with lithium. The response rate to chronotherapy was 65.7% (46/70 patients). DTI results showing increased mean and radial diffusivity correlated with poor antidepressant response to total SDT plus light therapy.

GENETIC BIOMARKERS ASSOCIATED WITH SDT OUTCOME

Single-genomic polymorphisms may influence responses to antidepressants, including those associated with SDT.[50]

Serotonin transporter promoter genotype

Benedetti and colleagues conducted a study of depressed patients treated with one night of TSD.[51] The authors identified a functional polymorphism within the promoter of the serotonin transporter (5-HTTLPR) that may influence response to antidepressants as well as SDT. The data showed that homozygotes (*l/l*) for the long variant of 5-HTTLPR have higher response rates to SDT (plus bright light) compared to heterozygotes and homozygotes for the short variant.

We hypothesized that depressed patients that are homozygous for 5-HTTPLR (*l/l*) may have a greater capability for mobilizing the serotonin system under sleep deprivation conditions.[52]

In other work, the serotonergic receptor gene (*5HT2A*) polymorphism rs6313 was shown to play a role in sleep deprivation responses. Bipolar patients with the T/T variant of the polymorphism had a 36% higher response rate to three cycles of SDT compared to patients with the T/C or C/C variants.[53]

GSK3β

$GSK3\beta$ codes for an enzyme that is a target of lithium and valproic acid. In animal studies, inhibition of $GSK3\beta$ increases antidepressant behaviors, including the rapid-acting effects of low-dose ketamine.[54] Benedetti et al.[50] reported higher rates of antidepressant response to one night of SDT (TSD) in BPD patients who were homozygous for the mutant allele of the $GSK3\beta$ promoter (−50 T/C) single-nucleotide polymorphism, which inhibits $GSK3\beta$. A limitation to these findings is that only a few BPD patients carry the $GSK3\beta$ (C/C) allele.

FUNCTIONAL BRAIN IMAGING STUDIES

Functional brain imaging studies provide compelling evidence for predicting response to SDT. This evidence includes positron emission tomography, single-photon emission computerized tomography, and functional magnetic resonance imaging (fMRI) data. The findings suggest that increased activation at baseline is associated with response to TSD and PSD. With clinical improvement, activation levels usually return to normal. Regions associated with response include the anterior cingulate cortex, the fronto-orbital cortex,[55–61] the left superior and inferior temporal cortices,[59] and the amygdala.[55,62] In the largest MDD study to date,[59] we identified 12 responders and 24 nonresponders to TSD and demonstrated that improvement was associated with baseline activation in a central region of the medial prefrontal cortex (BA 32). Increased amygdalar perfusion at baseline was associated with therapeutic response and was reported in a fMRI study involving 17 unmedicated MDD patients compared to 8 controls using a PSD protocol. Sleep deprivation response produced a significant reduction in amygdalar perfusion, which correlated with clinical improvement.[62]

In healthy controls, sleep deprivation decreases metabolic activity, particularly in the thalamus,[63,64] as well as in the prefrontal cortex, basal ganglia, posterior parietal association cortices, and cerebellum.[63,64] Nofzinger et al.[65] suggests that depression may be associated with hypermetabolism (in a ventral emotional neural system), which may persist during waking and sleep (non-rapid eye movement) in a subgroup of depressed patients.

POSSIBLE MECHANISM OF ACTION OF SDT

Circadian abnormalities in depression

Accumulating data support depression as a circadian-related illness, as evidenced by abnormalities in sleep, mood, hormonal secretions, and/or temperature in a subgroup of depressed patients (for reviews, see Bunney and Bunney[66,67]). Biological rhythms are under the control of the circadian machinery (including core clock genes). When circadian rhythms are dysregulated, the risk for depression increases.[68,69] Moreover, the degree of circadian desynchronization is reported to correlate with the severity of the illness; as depressive symptoms subside, circadian rhythms often normalize.[70–73]

SLEEP

An estimated 70%–80% of depressed patients complain of sleep disturbances, including insomnia, early-morning awakening, and difficulty in initiating sleep.[74] Patients with severe insomnia are likely to have poorer treatment outcomes[75] and may be at increased risk of suicidality.[76] Sleep disturbances are considered to be risk factors in the onset of depressive disorders, even in nonpsychiatric patients. Normalization of sleep patterns can be an early predictor of treatment response.[70,77]

TEMPERATURE

Increases in minimal core body temperatures are reported to predict response to SDT.[78] Clinical findings suggest that there is a relationship between nocturnal core body temperature, severity of depression, and the timing of sleep, possibly reflecting a "misalignment" of circadian rhythms.[68] As depressive symptoms improve, temperature rhythms usually normalize.[72,73]

HORMONE SECRETION

Melatonin and cortisol are abnormally regulated in a subgroup of depressed patients. Melatonin is the major hormone of the pineal gland. Secretion of melatonin serves as a marker of circadian "phase" relative to the timing of its release under dim light conditions (dim light melatonin onset). In healthy subjects, melatonin is secreted 1–2 hours before sleep onset. Melatonin secretion is delayed and its levels are lower in depressed subjects.[79,80]

Cortisol is also abnormally regulated in a subgroup of patients, as characterized by increased nocturnal levels and earlier phase shifts in late night.[81,82] Cortisol-related corticotrophin-releasing factor levels are dysregulated, which relates to dysfunction of the hypothalamic–pituitary–adrenal axis.[83] Cortisol levels frequently normalize as depressive symptoms remit.[71]

First evidence that circadian clock gene expression is altered in the MDD brain

Indirect evidence supports circadian abnormalities in the pathophysiology of depression. However, it was not until recently that the expression of clock genes, which control circadian rhythms in the human brain, was investigated by our group.[84] Time of death of gene expression data of 24-hour cyclic patterns from six cortical and limbic regions (dorsolateral prefrontal cortex, anterior cingulate, hippocampus, nucleus accumbens, amygdala, and cerebellum postmortem tissue) compared findings from 55 nonpsychiatric controls and 34 MDD patients. Microarray analyses including 12,000 transcripts identified the genes with the most robust 24-hour sinusoidal rhythms. To discover the cyclic genes, the expression values for each of the genes were fit by a sinusoidal function of time using the method of least squares and fixing the period at 24 hours. High-quality postmortem brain tissue was collected (all cases had sudden death and high pH values >6.5). Additionally, a 141-item questionnaire, as part of a psychological autopsy, was completed by interviews with next-of-kin, and medical records were used to confirm diagnoses.

The results showed a significant and marked loss in rhythmicity in the top-ranked cyclic genes in MDD patients compared to controls and provided the first direct evidence of dysregulation in clock gene expression across six brain regions. The findings were independent of medication and cause of death. Dysregulated genes included the core circadian clock genes (i.e., *BMAL1*, *PER1*, *PER2*, *PER3*, *Rev-Erbα*, *DBP*, *DEC1*, and *DEC2*), which are essential to modulating virtually all rhythms throughout the body.[84]

SDT: POSSIBLE MECHANISMS OF ACTION ON THE CIRCADIAN SYSTEM

We hypothesize that a subgroup of depressed patients with abnormal circadian rhythms may have a state-related defect in clock gene machinery. With remission, many of the dysregulated rhythms normalize. Figure 20.2 illustrates that chronotherapeutic interventions such as SDT, bright light therapy, and sleep phase advance may act, in part, by resetting and stabilizing circadian function. Clinical relapse may reflect a desynchronization in circadian function associated with the reactivation of abnormal clock gene expression.

A body of data suggests that low-dose ketamine—a rapid-acting antidepressant—may alter circadian rhythmicity as reported in animal studies[85–87] and a neuronal cell culture study.[88] It is hypothesized that SDT and low-dose ketamine may have overlapping mechanisms of action that

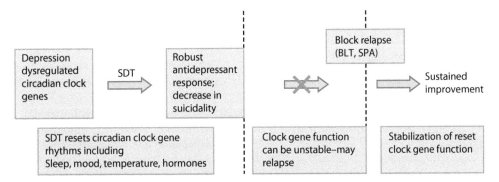

Figure 20.2 Hypothesized mechanism of action of sleep deprivation therapy (SDT) in robustly and rapidly (<24 hours) reducing depression and suicidality by "resetting" circadian rhythms and stabilizing response with chronotherapies (bright light therapy and sleep phase advance). A subgroup of "non"-SDT responders are reported to benefit from SDT, but at a slower rate.

modulate circadian rhythmicity and promote rapid antidepressant responses. However, in contrast to shorter-acting antidepressant effects of ketamine (4–5 days), SDT plus light and lithium has longer-lasting effects (up to 7 weeks).[31] As reviewed by Bunney et al.,[69] both interventions alter circadian-related systems involving NMDA, AMPA, mTOR, and GSK3β.

FUTURE DIRECTIONS

Sleep deprivation as a research tool

Studies of the mechanisms of action of sleep deprivation may provide potentially important clues as to the switch process. The rapid antidepressant actions of SDT and the equally rapid relapses into depressive episodes associated with recovery sleep provide a potentially critical framework for investigating the molecular underpinnings of the switch process.

Insomnia

Since 80% of MDD patients complain of insomnia,[89] it is somewhat counterintuitive that depriving depressed patients of sleep can rapidly improve symptoms. In otherwise-healthy individuals, insomnia (greater than 2 weeks) increases the risk of depressive illness within 1–3 years.[90] Relatively little is understood about the interaction between sleep and depression and why sleep patterns frequently normalize with improvement in symptoms.

Animal models of depression

Sleep deprivation in animals has a wide range of physiological and cognitive effects.[91] In animal models of depression, 12 hours of sleep deprivation has been shown to reverse depressive-like behavior in rodents.[92–94]

Convergence of antidepressant actions of SDT with conventional interventions

Future research may determine whether SDT (as well as low-dose ketamine) activates a "fast-track" pathway for antidepressant response, or whether it converges on a common neuronal circuitry in order to augment the actions of conventional antidepressants.

Technology advances

Emerging new technologies, including optogenetics, clustered regularly interspaced short palindromic repeats (CRISPR), induced pluripotent stem cells (iPSCs), and nanomedicine strategies, are expected to provide exciting data in the near future that will help us to understand the mechanisms of rapid-acting antidepressants. Optogenetics enables researchers to selectively manipulate individual neural circuits.[95] CRISPR enables the editing of DNA,[96] while iPSCs can differentiate into many cell subtypes, thus providing the potential ability to rebuild neuronal circuits.[97] Finally, a recent breakthrough in nanotechnology involves the introduction of novel DNA nanostructures in order to manipulate gene expression.[98] Using these tools to target the genes and their pathways that are relevant to the rapid-acting antidepressant effects of SDT (and fast-acting, low-dose ketamine) could help to revolutionize the treatment of depression.

REFERENCES

1. Schulte W. Sequelae of sleep deprivation. *Med Klin (Munich)* 1959; 54(20): 969–973.
2. Pflug B, Tolle R. Disturbance of the 24-hour rhythm in endogenous depression and the treatment of endogenous depression by sleep deprivation. *Int Pharmacopsychiatry* 1971; 6(3): 187–196.
3. Wu JC, Bunney WE. The biological basis of an antidepressant response to sleep deprivation and relapse: Review and hypothesis. *Am J Psychiatry* 1990; 147(1): 14–21.
4. Benedetti F, Riccaboni R, Locatelli C, Poletti S, Dallaspezia S, Colombo C. Rapid treatment response of suicidal symptoms to lithium, sleep deprivation, and light therapy (chronotherapeutics) in drug-resistant bipolar depression. *J Clin Psychiatry* 2014; 75(2): 133–140.
5. Colombo C, Benedetti F, Barbini B, Campori E, Smeraldi E. Rate of switch from depression into mania after therapeutic sleep deprivation in bipolar depression. *Psychiatry Res* 1999; 86(3): 267–270.
6. Benedetti F, Barbini B, Colombo C, Smeraldi E. Chronotherapeutics in a psychiatric ward. *Sleep Med Rev* 2007; 11(6): 509–522.

7. Gillin JC, Buchsbaum M, Wu J, Clark C, Bunney W Jr. Sleep deprivation as a model experimental antidepressant treatment: Findings from functional brain imaging. *Depress Anxiety* 2001; 14(1): 37–49.

8. Benedetti F, Barbini B, Fulgosi MC et al. Combined total sleep deprivation and light therapy in the treatment of drug-resistant bipolar depression: Acute response and long-term remission rates. *J Clin Psychiatry* 2005; 66(12): 1535–1540.

9. Leibenluft E, Moul DE, Schwartz PJ, Madden PA, Wehr TA. A clinical trial of sleep deprivation in combination with antidepressant medication. *Psychiatry Res* 1993; 46(3): 213–227.

10. Bauer M, Pfennig A, Severus E, Whybrow PC, Angst J, Moller HJ; Disorders World Federation of Societies of Biological Psychiatry; Task Force on Unipolar Depressive. World Federation of Societies of Biological Psychiatry (WFSBP) guidelines for biological treatment of unipolar depressive disorders, part 1: Update 2013 on the acute and continuation treatment of unipolar depressive disorders. *World J Biol Psychiatry* 2013; 14(5): 334–385.

11. Grunze H, Vieta E, Goodwin GM et al.; WFSBP Task Force on Treatment Guidelines for Bipolar Disorders. The World Federation of Societies of Biological Psychiatry (WFSBP) guidelines for the biological treatment of bipolar disorders: Update 2010 on the treatment of acute bipolar depression. *World J Biol Psychiatry* 2010; 11(2): 81–109.

12. Wirz-Justice A, Benedetti F, Terman M. *Chronotherapeutics for Affective Disorders: A Clinician's Manual for Light and Wake Therapy.* Basel: Karger, 2013.

13. Mendlewicz J. Disruption of the circadian timing systems: Molecular mechanisms in mood disorders. *CNS Drugs* 2009; 23(Suppl 2): 15–26.

14. Wiegand M, Riemann D, Schreiber W, Lauer CJ, Berger M. Effect of morning and afternoon naps on mood after total sleep deprivation in patients with major depression. *Biol Psychiatry* 1993; 33(6): 467–476.

15. Hemmeter U, Bischof R, Hatzinger M, Seifritz E, Holsboer-Trachsler E. Microsleep during partial sleep deprivation in depression. *Biol Psychiatry* 1998; 43(11): 829–839.

16. Reist C, Chen CC, Chhoeu A, Berry RB, Bunney WE Jr. Effects of sleep on the antidepressant response to sleep deprivation. *Biol Psychiatry* 1994; 35(10): 794–797.

17. Wiegand MH, Lauer CJ, Schreiber W. Patterns of response to repeated total sleep deprivations in depression. *J Affect Disord* 2001; 64(2–3): 257–260.

18. Gillin JC, Kripke DF, Janowsky DS, Risch SC. Effects of brief naps on mood and sleep in sleep-deprived depressed patients. *Psychiatry Res* 1989; 27(3): 253–265.

19. Caliyurt O, Guducu F. Partial sleep deprivation therapy combined with sertraline affects subjective sleep quality in major depressive disorder. *Sleep Med* 2005; 6(6): 555–559.

20. Smith GS, Reynolds CF 3rd, Pollock B et al. Cerebral glucose metabolic response to combined total sleep deprivation and antidepressant treatment in geriatric depression. *Am J Psychiatry* 1999; 156(5): 683–689.

21. Smeraldi E, Benedetti F, Barbini B, Campori E, Colombo C. Sustained antidepressant effect of sleep deprivation combined with pindolol in bipolar depression. A placebo-controlled trial. *Neuropsychopharmacology* 1999; 20(4): 380–385.

22. Kuhs H, Kemper B, Lippe-Neubauer U, Meyer-Dunker J, Tolle R. Repeated sleep deprivation once versus twice a week in combination with amitriptyline. *J Affect Disord* 1998; 47(1–3): 97–103.

23. Bump GM, Reynolds CF 3rd, Smith G et al. Accelerating response in geriatric depression: A pilot study combining sleep deprivation and paroxetine. *Depress Anxiety* 1997; 6(3): 113–118.

24. Kuhs H, Farber D, Borgstadt S, Mrosek S, Tolle R. Amitriptyline in combination with repeated late sleep deprivation versus amitriptyline alone in major depression. A randomised study. *J Affect Disord* 1996; 37(1): 31–41.

25. Shelton RC, Loosen PT. Sleep deprivation accelerates the response to nortriptyline. *Prog Neuropsychopharmacol Biol Psychiatry* 1993; 17(1): 113–123.

26. Kasper S, Kick H, Voll G, Vieira A. Therapeutic sleep deprivation and antidepressant medication in patients with major depression. *Eur Neuropsychopharmacol* 1991; 1(2): 107–111.

27. Kasper S, Voll G, Vieira A, Kick H. Response to total sleep deprivation before and during treatment with fluvoxamine or maprotiline in patients with major depression—Results of a double-blind study. *Pharmacopsychiatry* 1990; 23(3): 135–142.

28. Elsenga S, van den Hoofdakker RH. Clinical effects of sleep deprivation and clomipramine in endogenous depression. *J Psychiatr Res* 1982; 17(4): 361–374.

29. Martiny K, Refsgaard E, Lund V et al. The day-to-day acute effect of wake therapy in patients with major depression using the HAM-D6 as primary outcome measure: Results from a randomised controlled trial. *PLoS One* 2013; 8(6): e67264.

30. Martiny K, Refsgaard E, Lund V et al. Maintained superiority of chronotherapeutics vs. exercise in a 20-week randomized follow-up trial in major depression. *Acta Psychiatr Scand* 2015; 131(6): 446–457.

31. Wu JC, Kelsoe JR, Schachat C et al. Rapid and sustained antidepressant response with sleep deprivation and chronotherapy in bipolar disorder. *Biol Psychiatry* 2009; 66(3): 298–301.

32. Moscovici L, Kotler M. A multistage chronobiologic intervention for the treatment of depression: A pilot study. *J Affect Disord* 2009; 116(3): 201–207.

33. Benedetti F, Barbini B, Campori E, Fulgosi MC, Pontiggia A, Colombo C. Sleep phase advance and lithium to sustain the antidepressant effect of total sleep deprivation in bipolar depression: New findings supporting the internal coincidence model? *J Psychiatr Res* 2001; 35(6): 323–329.

34. Fritzsche M, Heller R, Hill H, Kick H. Sleep deprivation as a predictor of response to light therapy in major depression. *J Affect Disord* 2001; 62(3): 207–215.

35. Szuba MP, Baxter LR Jr, Altshuler LL et al. Lithium sustains the acute antidepressant effects of sleep deprivation: Preliminary findings from a controlled study. *Psychiatry Res* 1994; 51(3): 283–295.

36. Holsboer-Trachsler E. Monitoring the neurobiological and psychopathological course in therapy of depression. Trimipramine, sleep deprivation and light. *Bibl Psychiatr* 1994; 166: 1–138.

37. van den Burg W, Bouhuys AL, van den Hoofdakker RH, Beersma DG. Sleep deprivation in bright and dim light: Antidepressant effects on major depressive disorder. *J Affect Disord* 1990; 19(2): 109–117.

38. Souetre E, Salvati E, Pringuey D, Plasse Y, Savelli M, Darcourt G. Antidepressant effects of the sleep/wake cycle phase advance. Preliminary report. *J Affect Disord* 1987; 12(1): 41–46.

39. Sahlem GL, Kalivas B, Fox JB et al. Adjunctive triple chronotherapy (combined total sleep deprivation, sleep phase advance, and bright light therapy) rapidly improves mood and suicidality in suicidal depressed inpatients: An open label pilot study. *J Psychiatr Res* 2014; 59: 101–107.

40. Casher MI, Schuldt S, Haq A, Burkhead-Weiner D. Chronotherapy in treatment-resistant depression. *Psychiatr Ann* 2012; 42(5): 166–169.

41. Krstic J, Buzadzic I, Milanovic SD, Ilic NV, Pajic S, Ilic TV. Low-frequency repetitive transcranial magnetic stimulation in the right prefrontal cortex combined with partial sleep deprivation in treatment-resistant depression: A randomized sham-controlled trial. *J ECT* 2014; 30(4): 325–331.

42. Bunney WE, Bunney BG. Molecular clock genes in man and lower animals: Possible implications for circadian abnormalities in depression. *Neuropsychopharmacology* 2000; 22(4): 335–345.

43. Reinink E, Bouhuys N, Wirz-Justice A, van den Hoofdakker R. Prediction of the antidepressant response to total sleep deprivation by diurnal variation of mood. *Psychiatry Res* 1990; 32(2): 113–124.

44. Elsenga S, Van den Hoofdakker RH. Response to total sleep deprivation and clomipramine in endogenous depression. *J Psychiatr Res* 1987; 21(2): 151–161.

45. Wirz-Justice A. Diurnal variation of depressive symptoms. *Dialogues Clin Neurosci* 2008; 10(3): 337–343.

46. Benedetti F, Bollettini I, Poletti S et al. White matter microstructure in bipolar disorder is influenced by the serotonin transporter gene polymorphism 5-HTTLPR. *Genes Brain Behav* 2015; 14(3): 238–250.

47. Benedetti F, Yeh PH, Bellani M et al. Disruption of white matter integrity in bipolar depression as a possible structural marker of illness. *Biol Psychiatry* 2011; 69(4): 309–317.

48. Sprooten E, Fleming KM, Thomson PA et al. White matter integrity as an intermediate phenotype: Exploratory genome-wide association analysis in individuals at high risk of bipolar disorder. *Psychiatry Res* 2013; 206(2–3): 223–231.

49. Bollettini I, Poletti S, Locatelli C et al. Disruption of white matter integrity marks poor antidepressant response in bipolar disorder. *J Affect Disord* 2014; 174C: 233–240.

50. Benedetti F, Serretti A, Colombo C, Lorenzi C, Tubazio V, Smeraldi E. A glycogen synthase kinase 3-beta promoter gene single nucleotide polymorphism is associated with age at onset and response to total sleep deprivation in bipolar depression. *Neurosci Lett* 2004; 368(2): 123–126.

51. Benedetti F, Colombo C, Serretti A et al. Antidepressant effects of light therapy combined with sleep deprivation are influenced by a functional polymorphism within the promoter of the serotonin transporter gene. *Biol Psychiatry* 2003; 54(7): 687–692.

52. Wu JC, Buchsbaum M, Bunney WE Jr. Clinical neurochemical implications of sleep deprivation's effects on the anterior cingulate of depressed responders. *Neuropsychopharmacology* 2001; 25(5 Suppl): S74–S48.

53. Benedetti F, Barbini B, Bernasconi A et al. Serotonin 5-HT2A receptor gene variants influence antidepressant response to repeated total sleep deprivation in bipolar depression. *Prog Neuropsychopharmacol Biol Psychiatry* 2008; 32(8): 1863–1866.

54. Liu RJ, Fuchikami M, Dwyer JM, Lepack AE, Duman RS, Aghajanian GK. GSK-3 inhibition potentiates the synaptogenic and antidepressant-like effects of subthreshold doses of ketamine. *Neuropsychopharmacology* 2013; 38(11): 2268–2277.

55. Ebert D, Feistel H, Barocka A. Effects of sleep deprivation on the limbic system and the frontal lobes in affective disorders: A study with Tc-99m-HMPAO SPECT. *Psychiatry Res* 1991; 40(4): 247–251.

56. Wu JC, Gillin JC, Buchsbaum MS, Hershey T, Johnson JC, Bunney WE Jr. Effect of sleep deprivation on brain metabolism of depressed patients. *Am J Psychiatry* 1992; 149(4): 538–543.

57. Volk SA, Kaendler SH, Hertel A et al. Can response to partial sleep deprivation in depressed patients be predicted by regional changes of cerebral blood flow? *Psychiatry Res* 1997; 75(2): 67–74.

58. Holthoff VA, Beuthien-Baumann B, Pietrzyk U et al. Changes in regional cerebral perfusion in depression. SPECT monitoring of response to treatment. *Nervenarzt* 1999; 70(7): 620–626.

59. Wu J, Buchsbaum MS, Gillin JC et al. Prediction of antidepressant effects of sleep deprivation by metabolic rates in the ventral anterior cingulate and medial prefrontal cortex. *Am J Psychiatry* 1999; 156(8): 1149–1158.

60. Smith GS, Reynolds CF 3rd, Houck PR et al. Glucose metabolic response to total sleep deprivation, recovery sleep, and acute antidepressant treatment as functional neuroanatomic correlates of treatment outcome in geriatric depression. *Am J Geriatr Psychiatry* 2002; 10(5): 561–567.

61. Ebert D, Feistel H, Barocka A, Kaschka W. Increased limbic blood flow and total sleep deprivation in major depression with melancholia. *Psychiatry Res* 1994; 55(2): 101–109.

62. Clark CP, Brown GG, Archibald SL et al. Does amygdalar perfusion correlate with antidepressant response to partial sleep deprivation in major depression? *Psychiatry Res* 2006; 146(1): 43–51.

63. Wu JC, Gillin JC, Buchsbaum MS et al. The effect of sleep deprivation on cerebral glucose metabolic rate in normal humans assessed with positron emission tomography. *Sleep* 1991; 14(2): 155–162.

64. Thomas M, Sing H, Belenky G et al. Neural basis of alertness and cognitive performance impairments during sleepiness. I. Effects of 24 h of sleep deprivation on waking human regional brain activity. *J Sleep Res* 2000; 9(4): 335–352.

65. Nofzinger EA, Buysse DJ, Germain A et al. Alterations in regional cerebral glucose metabolism across waking and non-rapid eye movement sleep in depression. *Arch Gen Psychiatry* 2005; 62(4): 387–396.

66. Bunney BG, Bunney WE. Rapid-acting antidepressant strategies: Mechanisms of action. *Int J Neuropsychopharmacol* 2012; 15(5): 695–713.

67. Bunney BG, Bunney WE. Mechanisms of rapid antidepressant effects of sleep deprivation therapy: Clock genes and circadian rhythms. *Biol Psychiatry* 2013; 73(12): 1164–1171.

68. Hasler BP, Buysse DJ, Kupfer DJ, Germain A. Phase relationships between core body temperature, melatonin, and sleep are associated with depression severity: Further evidence for circadian misalignment in non-seasonal depression. *Psychiatry Res* 2010; 178(1): 205–207.

69. Bunney BG, Li JZ, Walsh DM et al. Circadian dysregulation of clock genes: Clues to rapid treatments in major depressive disorder. *Mol Psychiatry* 2015; 20(1): 48–55.

70. Gupta R, Lahan V. Insomnia associated with depressive disorder: Primary, secondary, or mixed? *Indian J Psychol Med* 2011; 33(2): 123–128.

71. Scharnholz B, Lederbogen F, Feuerhack A et al. Does night-time cortisol excretion normalize in the long-term course of depression? *Pharmacopsychiatry* 2010; 43(5): 161–165.

72. Souetre E, Salvati E, Wehr TA, Sack DA, Krebs B, Darcourt G. Twenty-four-hour profiles of body temperature and plasma TSH in bipolar patients during depression and during remission and in normal control subjects. *Am J Psychiatry* 1988; 145(9): 1133–1137.

73. Avery DH, Shah SH, Eder DN, Wildschiodtz G. Nocturnal sweating and temperature in depression. *Acta Psychiatr Scand* 1999; 100(4): 295–301.

74. Mendlewicz J. Sleep disturbances: Core symptoms of major depressive disorder rather than associated or comorbid disorders. *World J Biol Psychiatry* 2009; 10(4): 269–275.

75. O'Brien EM, Chelminski I, Young D, Dalrymple K, Hrabosky J, Zimmerman M. Severe insomnia is associated with more severe presentation and greater functional deficits in depression. *J Psychiatr Res* 2011; 45(8): 1101–1105.

76. Krakow B, Ribeiro JD, Ulibarri VA, Krakow J, Joiner TE Jr. Sleep disturbances and suicidal ideation in sleep medical center patients. *J Affect Disord* 2011; 131(1–3): 422–427.

77. Salo P, Sivertsen B, Oksanen T et al. Insomnia symptoms as a predictor of incident treatment for depression: Prospective cohort study of 40,791 men and women. *Sleep Med* 2012; 13(3): 278–284.

78. Elsenga S, Van den Hoofdakker RH. Body core temperature and depression during total sleep deprivation in depressives. *Biol Psychiatry* 1988; 24(5): 531–540.

79. Buckley TM, Schatzberg AF. A pilot study of the phase angle between cortisol and melatonin in major depression—A potential biomarker? *J Psychiatr Res* 2010; 44(2): 69–74.

80. Lewy AJ. Depressive disorders may more commonly be related to circadian phase delays rather than advances: Time will tell. *Sleep Med* 2010; 11(2): 117–118.

81. Carpenter WT Jr, Bunney WE Jr. Adrenal cortical activity in depressive illness. *Am J Psychiatry* 1971; 128(1): 31–40.

82. Koenigsberg HW, Teicher MH, Mitropoulou V et al. 24-h monitoring of plasma norepinephrine, MHPG, cortisol, growth hormone and prolactin in depression. *J Psychiatr Res* 2004; 38(5): 503–511.

83. Nemeroff CB. The corticotropin-releasing factor (CRF) hypothesis of depression: New findings and new directions. *Mol Psychiatry* 1996; 1(4): 336–342.

84. Li JZ, Bunney BG, Meng F et al. Circadian patterns of gene expression in the human brain and disruption in major depressive disorder. *Proc Natl Acad Sci USA* 2013; 110(24): 9950–9955.

85. Cao R, Li A, Cho HY, Lee B, Obrietan K. Mammalian target of rapamycin signaling modulates photic entrainment of the suprachiasmatic circadian clock. *J Neurosci* 2010; 30(18): 6302–6314.

86. Cao R, Lee B, Cho HY, Saklayen S, Obrietan K. Photic regulation of the mTOR signaling pathway in the suprachiasmatic circadian clock. *Mol Cell Neurosci* 2008; 38(3): 312–324.

87. Zunszain PA, Horowitz MA, Cattaneo A, Lupi MM, Pariante CM. Ketamine: Synaptogenesis, immunomodulation and glycogen synthase kinase-3 as underlying mechanisms of its antidepressant properties. *Mol Psychiatry* 2013; 18(12): 1236–1241.

88. Bellet MM, Vawter MP, Bunney BG, Bunney WE, Sassone-Corsi P. Ketamine influences CLOCK:BMAL1 function leading to altered circadian gene expression. *PLoS One* 2011; 6(8): e23982.

89. Ohayon MM. Epidemiology of insomnia: What we know and what we still need to learn. *Sleep Med Rev* 2002; 6(2): 97–111.

90. Riemann D, Voderholzer U. Primary insomnia: A risk factor to develop depression? *J Affect Disord* 2003; 76(1–3): 255–259.

91. Thompson CL, Wisor JP, Lee CK et al. Molecular and anatomical signatures of sleep deprivation in the mouse brain. *Front Neurosci* 2010; 4: 165.

92. Meerlo P, Overkamp GJ, Benning MA, Koolhaas JM, Van den Hoofdakker RH. Long-term changes in open field behaviour following a single social defeat in rats can be reversed by sleep deprivation. *Physiol Behav.* 1996; 60(1): 115–119.

93. Serchov T, Clement HW, Schwarz MK, Iasevoli F, Tosh DK, Idzko M et al. Increased signaling via adenosine A1 receptors, sleep deprivation, imipramine, and ketamine inhibit depressive-like behavior via induction of Homer1a. *Neuron* 2015; 87(3): 549–562.

94. Hines DJ, Schmitt LI, Hines RM, Moss SJ, Haydon PG. Antidepressant effects of sleep deprivation require astrocyte-dependent adenosine mediated signaling. *Transl Psychiatry* 2013; 3: e212.

95. Steinberg EE, Christoffel DJ, Deisseroth K, Malenka RC. Illuminating circuitry relevant to psychiatric disorders with optogenetics. *Curr Opin Neurobiol* 2015; 30: 9–16.

96. Hartenian E, Doench JG. Genetic screens and functional genomics with CRISPR/Cas9 technology. *FEBS J* 2015; 282(8): 1383–1393.

97. Jessberger S, Gage FH. Adult neurogenesis: Bridging the gap between mice and humans. *Trends Cell Biol* 2014; 24(10): 558–563.

98. Jones MR, Seeman NC, Mirkin CA. Nanomaterials. Programmable materials and the nature of the DNA bond. *Science* 2015; 347(6224): 1260901.

99. Benedetti F, Colombo C, Barbini B, Campori E, Smeraldi E. Ongoing lithium treatment prevents relapse after total sleep deprivation. *J Clin Psychopharm* 1999; 19(3): 240–245.

Non-pharmacological interventions for insomnia in cancer patients

GILLA K. SHAPIRO, SAMARA PEREZ, ZEEV ROSBERGER, AND JOSÉE SAVARD

BACKGROUND

Cancer is a leading cause of morbidity and mortality worldwide. The World Health Organization estimates that there are 14 million new cancer cases and 8.2 million cancer deaths annually.[1] Alarmingly, given demographic trends, this already colossal number of new annual cancer cases is expected to increase to approximately 22 million over the next two decades.[2,3] The picture is slightly different in developed and developing countries, with a greater global burden in the developed world.[4] For example, cancer is responsible for more than one in four deaths in Canada,[5] the United Kingdom,[6] and the United States.[7] Beneficially, there has been progress in cancer prevention, early detection, and treatment. However, a paradoxical result of advances in detection and treatments is that, overall, cancer patients are surviving longer and living with short- and long-term side effects and consequences that affect their quality of life and may require clinical attention.

Available evidence suggests that sleep–wake disturbances—including delayed sleep latency, waking episodes after sleep onset, non-refreshing sleep, reduced quality of sleep, and reduced sleep efficiency—are amongst the most common symptoms reported by patients with cancer.[8] Compared to the sleep disturbance rate of 10%–15% that is reported in the general population,[9,10] it is estimated that 30%–60% of cancer patients experience sleep disturbances.[11-15] The large range in prevalence rates is a result of multiple factors, including differences in assessment methods and criteria used to identify sleep disturbances, type and stage of cancer, socio-environmental stressors, adverse effects from medications and cancer treatments, medical comorbidities and age.[14,16] For example, contrary to what is found in the general population, younger cancer patients are more likely to report having a sleep problem.[11,17,18] Further, although sleep disturbances occur at diagnosis, during treatment and can continue into survivorship or palliative care, the underlying cause may differ at the different stages of cancer. Notably, sleep disturbances in cancer patients cluster with pain, fatigue, depression, anxiety, and vasomotor symptoms.[14,19]

The most common sleep disorder in the general population and among cancer patients is insomnia.[20] Insomnia is defined as difficulty initiating sleep, maintaining sleep or early-morning awakenings, or a combination of these difficulties.[21,22] A complaint of fatigue and daytime impairment is often the consequence of insomnia.[21] According to the *Diagnostic and Statistical Manual of Mental Disorders Fifth Edition* (DSM-5), the insomnia disorder is a combination of criteria regarding frequency (i.e., at least three nights per week), duration (i.e., symptoms present for at least 3 months to be considered persistent), and consequences on one's functioning (i.e., including impairment in social, occupational, educational, academic, behavioral, or other daytime functioning) (see Table 21.1).[20] While approximately 10% of individuals in the general population meet the criteria for insomnia disorder,[20] a population-based study conducted in cancer patients has found rates that are at least two- to three-times higher.[15,23]

The underlying etiological causes of insomnia are multifactorial. Figure 21.1 describes the predisposing, precipitating, and perpetuating factors of insomnia in cancer patients that have been discussed in the literature.[12,24] As mentioned, the complaint of insomnia can occur at any point during the cancer trajectory, including diagnosis, treatment, survivorship, and palliative care.[12] In cancer patients, insomnia is also related to physical, psychological, interpersonal, or treatment factors.[12,25–27] Evidently, cancer patients who have insomnia are not a homogeneous group. The different underlying causes of each case should be considered for beneficial patient-centered insomnia treatment.

Insomnia can profoundly affect every aspect of a cancer patient's life.[28] Disruption of sleep may interfere with physiological and behavioral functioning and can have multiple and wide-reaching repercussions. Sleep problems in cancer patients are associated with sexual disorders (e.g., loss of libido and inability to achieve orgasm) and increased physical symptoms (i.e., pain).[12,29] Insomnia can also decrease cognitive functioning (e.g., in memory and concentration) and work productivity, and increase safety risks, accidents, and medication misuse and abuse.[17,29–32] In addition, because of these complications and their possible repercussions on reduced productivity,

Table 21.1 DSM-5 diagnostic criteria for insomnia disorder

A. A predominant complaint of dissatisfaction with sleep quantity or quality, associated with one (or more) of the following symptoms:
  1. Difficulty initiating sleep
  2. Difficulty maintaining sleep, characterized by frequent awakenings or problems returning to sleep after awakening
  3. Early-morning awakening with inability to return to sleep
B. The sleep disturbance causes clinically significant distress or impairment in social, occupational, education, academic, behavioral or other important areas of functioning
C. The sleep difficulty occurs at least three nights per week
D. The sleep difficulty is present for at least 3 months
E. The sleep difficulty occurs despite adequate opportunity for sleep

Comorbidity specifiers:
  With non-sleep disorder mental comorbidity (including substance use disorders)
  With other medical comorbidity
  With other sleep disorder

Duration specifiers:
  Episodic: symptoms last at least 1 month but less than 3 months
  Persistent: symptoms last 3 months or longer
  Recurrent: two (or more) episodes within the space of 1 year

Source: Adapted from American Psychiatric Association. *Diagnostic and Statistical Manual of Mental Disorders, (DSM-5).* Washington, DC: American Psychiatric Association, 2013.

rates of medical consultations and greater sick leave, there are substantial healthcare costs that are associated with untreated insomnia among cancer patients.[33]

Insomnia also places an enormous toll on an individual's mood. Individuals with insomnia are at a greater risk of developing a psychiatric disorder.[11,34] Indeed, patients with insomnia report more psychological distress, including depression,

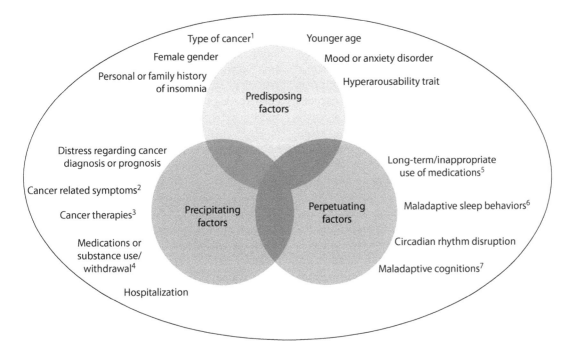

Figure 21.1 Etiological factors of cancer-related insomnia. 1, Insomnia is more prevalent in breast, lung or gynecologic cancer; 2, E.g., cancer pain, fatigue, menopausal symptoms (night sweats or hot flashes) or gastrointestinal and genitourinary disturbances; 3, Including radiotherapy, chemotherapy, hormone therapy or bone marrow transplantation; 4, E.g., some antidepressants, steroids, alcohol, caffeine or nicotine; 5, E.g., Dependence or several years of hypnotic use beyond advised limit; 6 E.g., sleep-interfering activities in the bedroom, excessive amount of time spent in bed or irregular sleep–wake schedule; 7, E.g., faulty attitudes and beliefs about sleep, misconceptions about the cause of insomnia or inaccurate appraisal about sleep difficulty. (Modified from Savard J, Savard M-H. *Sleep Med Clin* 2013; 8(3): 373–387; Savard J, Morin CM. *J Clin Oncol* 2001; 19(3): 895–908; National Cancer Institute. *Sleep Disorders*. http://www.cancer.gov/about-cancer/treatment/side-effects/sleep-disorders-hp-pdq#section/all. Accessed April 24, 2015; Howell D et al. *Support Care Cancer* 2013; 21(10): 2695–2706; Yue HJ, Dimsdale JE. Sleep and cancer. In: Holland JC, Breitbart W, Jacobsen PB, Marguerite SL, Loscalzo MJ, McCorkle R (Eds). *Psycho-oncology*. New York, NY: Oxford University Press, 2010; pp. 258–269; Koul R, Dubey A. *Internet J Pain Symptom Control Palliat Care* 209; 7(1): 1–8.)

anxiety, and substance-related problems, as well as impairments in daytime mood and performance.[11,12,35,36] For example, Davidson et al.[17] found that patients with insomnia were less able to cope with stress and complete their daily activities. Furthermore, Fortner et al.[37] showed that cancer patients with insomnia report less energy, more pain and greater emotional difficulties. Notably, some researchers have criticized these results and the lack of prospective research in the field.[28] There is indeed a bidirectional relationship in this area, and while insomnia can contribute to triggering or worsening conditions such as pain, fatigue, depression, and anxiety, these symptoms can also predispose or worsen insomnia.[28] Nevertheless, overall it is evident that insomnia is associated with a cluster of symptoms in cancer patients that negatively impacts physical and emotional well-being.[12,17,29,38] Clearly, managing insomnia symptoms in cancer patients is important in order to relieve suffering and improve quality of life.[21]

Despite the prevalence and consequences of insomnia in cancer patients, this disorder is often under-reported, under-recognized, and under-treated.[28,39] In a survey of 150 patients, 44% reported having sleep problems during the previous month; however, only 16.6% of patients who reported a sleep problem had told their healthcare providers of this.[40] More recently, in a study of 78 patients with serious insomnia complaints, only four patients (5%) received a recommendation for follow-up or reassessment of their sleep problem.[41] This is a multilayered problem. Cancer patients may not realize that insomnia is avoidable and treatable. Patients may also assume that insomnia is a *temporary* response to their cancer diagnosis and treatments.[24] Some cancer patients may also not discuss their symptoms with their physician because they do not want to detract from their physician's focus of treating cancer or because they do not want to be prescribed a sleeping medication. Indeed, other physical symptoms (e.g., pain or nausea) are discussed more frequently with physicians, oncologists, or oncology nurses. On the other hand, many oncologists and physicians who are caring for cancer patients may not inquire about sleep problems, and when a patient mentions a sleep complaint, they may underestimate the importance of this issue, assume that the insomnia is only temporary and instead focus on other "more urgent problems."[28] Partly as a result of insomnia being underrecognized, treatment for insomnia is inaccessible and is "currently highly limited in oncology settings."[8] Nevertheless, it is well recognized that when insomnia causes personal distress or significant impairment, then treatment is required.[42]

This chapter will provide an overview of the assessment methods and non-pharmacological interventions that can be used for cancer-related insomnia. We will begin by discussing the screening tools that are often used, as well as how and when to assess for insomnia in cancer patients. We will then proceed to discuss a number of non-pharmacological interventions that have been developed, adapted and used to treat insomnia, including cognitive behavioral therapy (CBT), mindfulness-based stress reduction (MBSR), exercise, and yoga.[14] These interventions will be described and the evidence of their efficacy will be reviewed. We will then conclude this chapter by discussing research gaps, as well as future directions in order to improve patients' access to effective treatments for cancer-related insomnia.

SCREENING AND ASSESSMENT OF INSOMNIA IN CANCER PATIENTS

Many researchers and clinicians have advocated for all oncology patients to be regularly screened for insomnia and offered appropriate treatment that is dependent on their underlying clinical presentation.[44] Screening is the initial step in identifying and managing insomnia in cancer patients.[13] It is crucial to screen all cancer patients for insomnia given the high prevalence of insomnia in this group and the detrimental impact of insomnia on quality of life.[15] Furthermore, because sleep disturbances have been shown to change throughout the cancer trajectory, *continual* screening of sleep symptoms is also important.[13] Insomnia screening should be regularly conducted in order to facilitate better triage and referral for further assessment when appropriate.[29] However, problematically, insomnia is seldom included in routine screening of cancer patients.[14,24,45]

In Canada, the revised Edmonton Symptom Assessment System (ESAS) and the Canadian Problem Checklist (CPC) have been largely implemented as routine screening instruments of psychological distress in cancer patients.[46–48] However, the ESAS version that is generally used does not contain any sleep items, but only assesses drowsiness, a symptom that is not specific to insomnia. Patients can report their sleep difficulties by checking "yes" on the CPC sleep item or by identifying sleep difficulties using the "other problem" item of the ESAS. A specific sleep item, rated on a scale from 0 ("best sleep") to 10 ("worst sleep imaginable"), was added to the ESAS in a few studies conducted in patients with advanced cancer.[49] A score of three or higher on this item was found to be associated with a sensitivity of 74% and a specificity of 73% when compared to a clinical score ($\geq$5) on the Pittsburgh Sleep Quality Index (PSQI).[49]

Some guidelines recommend that healthcare providers follow up an initial screening of sleep disturbances by asking: (i) "Do you have problems with your sleep or sleep disturbance on average for three or more nights a week?"; and if yes, (ii) "Does the problem with your sleep negatively affect your daytime functioning?"[24,50] If these questions are answered affirmatively, then a comprehensive assessment of insomnia should follow.[24,50]

Two self-reported tools—the Insomnia Severity Index (ISI) and the PSQI—have also been recommended as screening tools for insomnia in oncological clinical practice and research.[50,51] The ISI is comprised of seven questions that are rated cumulatively using a five-point Likert scale.[52,53] The ISI is designed to evaluate the severity of insomnia on the basis of difficulties falling asleep, nighttime awakenings, early-morning awakenings, dissatisfaction with sleep, noticeability of impairments, distress or worry caused by sleep difficulties, and impaired functioning.[53–55] The ISI has been validated specifically in the context of cancer and found to have equivalent psychometric properties.[51,55,56] A score of eight or greater is used to indicate the presence of clinical levels of insomnia.[51] The PSQI provides a measure of general sleep disturbances. It contains 19 items and yields a global score of 0–21, with higher scores indicating poorer sleep.[44,57] The 19 items generate seven scores regarding subjective sleep quality, sleep latency, sleep duration, habitual sleep efficiency, sleep disturbances, use of sleeping medication, and daytime dysfunction.[50] A total score of five or greater is generally used to screen for sleep disturbances.[50] The PSQI has also demonstrated high test–retest reliability (0.87) and good construct validity in cancer patients with insomnia.[58,59]

A sleep diary is another subjective sleep measure that can help evaluate the severity of insomnia and inform its treatment by highlighting key problem areas in a detailed way (e.g., use of hypnotic medications, bedtime hour, time of falling asleep, number of awakenings, and time spent in bed).[53,60] In addition, clinical interviews, such as the Insomnia Interview Schedule (IIS) and the Duke Structured Interview for Sleep Disorders, contain semi-structured questions that evaluate multiple aspects of sleep difficulties and provide diagnostic guidelines for insomnia.[61,62]

Assessment for insomnia can include multiple sources, including the patient's self-reported sleep difficulty, objective information on the patient's behavioral and physiological manifestations of sleep disturbances, and reports from the patient's significant others vis-à-vis the patient's quality of sleep. A comprehensive sleep assessment should also ideally be accompanied by medical evaluation, evaluation of emotional status, and general sleep difficulties (i.e., Epworth Sleepiness Scale to detect daytime sleepiness), as well as the gathering of information regarding exercise and activity levels, diet and use of medications or other substances so as to rule out the presence of other sleep disorders

and comorbid problems.[13,63] However, when investigating insomnia symptoms in patients with cancer, it is important for clinicians to balance the utility of a comprehensive assessment with the risks and benefits of cumbersome procedures and treatments, as well as cancer patients' willingness and capacity to undergo lengthy evaluations.[21] Indeed, a comprehensive sleep assessment is not realistic in many settings.[21]

CRITICAL REVIEW OF NON-PHARMACOLOGICAL INTERVENTIONS FOR INSOMNIA IN CANCER PATIENTS

Given the side effects of the medications used to treat cancer and/or insomnia (e.g., residual daytime sleepiness) and the risks associated with a long-term usage of hypnotic medications (e.g., dependence),[57,64] there are many advantages and great interest in using non-pharmacological interventions in the treatment of insomnia in cancer patients.[65–67] Moreover, certain behavioral therapies maintain superior therapeutic gains over time compared to pharmacotherapy.[10,68] This chapter will discuss non-pharmacological treatments with an emphasis on evidence-based studies specifically targeting insomnia or sleep disturbance symptoms among cancer populations.

CBT for insomnia

CBT for insomnia (CBT-I) is the most studied non-pharmacological intervention and is considered a safe, effective, and the gold standard treatment for chronic insomnia.[10,66,68] The core objective of CBT-I is to address patients' maladaptive sleep behaviors (e.g., spending too much time in bed) and erroneous cognitive distortions (e.g., unhelpful and faulty beliefs about sleep; see Table 21.2). The format of CBT-I generally varies from between six and eight sessions and combines stimulus control and sleep restriction strategies, cognitive restructuring, and sleep hygiene (see an example of a CBT-I protocol in Table 21.3).

Over the last three decades, there has been a significant volume of research, notably meta-analyses and Cochrane reviews, including randomized controlled trials (RCTs), supporting the efficacy CBT-I in the general population.[69–73]

Table 21.2 Common maladaptive sleep habits and erroneous beliefs specific to insomnia comorbid with cancer

Maladaptive sleep habits

- Excessive time spent resting or in bed
- Daytime napping, often encouraged by treatment team and family to "get rest and sleep"
- Engaging in non-sleep-related activities in bed
- Irregular sleep–wake schedule

Erroneous beliefs

- "If I don't sleep well, my cancer will come back" → increased arousal and performance anxiety → "I really need to sleep (well and more) tonight"
- "I need to sleep more to recuperate" → daytime napping → irregular sleep schedules

Sources: Adapted from Savard J, Savard M-H. *Sleep Med Clin* 2013; 8(3): 373–387; Savard J, Morin CM. *J Clin Oncol* 2001; 19(3): 895–908; Espie CA. *Annu Rev Psychol* 2002; 53: 215–243; Howell D et al. *Support Care Cancer* 2013; 21(10): 2695–2706.

Clinically and statistically significant improvements have been consistently reported in subjective and objective sleep measures such as time to sleep onset, time awake after sleep onset, and duration of awakenings at night. Additionally, participants also report clinically and statistically significant benefits on secondary outcomes, including decreased psychological distress and improved quality of life, with benefits lasting for at least 6 months post-treatment.[72,74–76]

Since 2000, there has been a growing literature of insomnia studies among patients with a cancer diagnosis.[8,77] The common faulty and dysfunctional beliefs about sleep, as well as the maladaptive sleep behaviors of cancer patients, are generally quite similar to those reported among individuals with insomnia with no comorbid conditions. One notable distinction is that many cancer patients are often encouraged by health professionals and their support system to "get rest and sleep" in order to recuperate post-cancer treatments. In the long term, excessive daytime napping and spending too much time in bed can significantly alter sleep and sleep–wake cycles. Also specific to the cancer population are maladaptive, erroneous

beliefs, such as "If I don't sleep well, my cancer will come back," which can increase arousal and performance anxiety at bedtime (e.g., "I really need to sleep tonight"). Some cancer patients also state that the "whys" (e.g., Why me? Why now? Why are other women in remission and not me?) are often prevalent nighttime thoughts that make it difficult to "shut off" and fall asleep. Other concerns and deeply entrenched fears about survival and death include "how will this end?", "how much time do I have left?" or "how will my family and friends cope without me?"[78,79] Importantly, as time passes and the patient realizes that death is less of an imminent threat or that they are surviving their cancer, cancer patients express wanting to return to a regular sleep routine, but have "forgotten how to sleep."

The efficacy of CBT-I in the context of cancer has been reviewed and shows similar success and efficacy as in the general population.[8,77] One of the first quasi-experimental studies found subjective improvements in the number of nighttime awakenings, sleep efficiency, and global sleep quality among cancer survivors following a 6-week treatment program of CBT-I.[80] Two additional uncontrolled CBT-I studies among breast or mixed cancer type patients similarly reported significant improvements in subjective and objective measures such as sleep efficiency, total wake time, and sleep-onset latency as measured using sleep diaries and polysomnography or actigraphy.[81,82] These promising findings encouraged researchers to conduct RCTs. In a recent review,[77] eight RCTs of CBT-I were described and examined.[77,83–89] Five studies' populations consisted of breast cancer patients, and three studies were of men and women with mixed cancer diagnoses. Seven studies included men and women who had completed their cancer treatments. The review reported that CBT-I treatment varied in format and modality: treatment length ranged from 5 to 9 weeks; four treatments were face-to-face, group-based formats, while four were individual-based interventions, with one being internet-based. Control groups also varied, with four studies comparing to waitlist controls, two studies comparing to treatment as usual, and two studies comparing to active treatment groups.

In the largest RCT, comparing standard CBT-I with usual care, Espie et al.[88] found significant improvements in insomnia and nighttime wakefulness, with results persisting for at least 6 months

Table 21.3 An example of a cognitive behavioral therapy for insomnia protocol

Week	Treatment content
1	Presentation of basic information about sleep and insomnia.
2	Introduction of stimulus control therapy and sleep restriction strategies. Stimulus control therapy aims to re-associate the bed and the bed environment with sleep and to establish a regular sleep–wake rhythm by following a specific set of guidelines such as "Go to bed only when you feel sleepy" and "When unable to fall asleep within 20–30 minutes, get out of bed, leave the bedroom and return to bed only when sleepy." Sleep restriction aims to curtail the time in bed to the actual sleep time, thereby resulting in more consolidated and efficient sleep. Time in bed is progressively increased as sleep efficiency improves during treatment.
3	Introduction of cognitive restructuring strategies. Explanation of the cognitive model of emotions and of the five-column technique for restructuring negative automatic thoughts. Patients are guided in identifying their automatic thoughts, challenging the validity of these thoughts and reframing them into more adaptive substitutes.
4	Identification and revision of dysfunctional beliefs about sleep, sleep difficulties and the impact of these sleep difficulties on patients' daily functioning (e.g., "I need 8 hours of sleep to recuperate from my treatments"; "If I do not sleep well, my cancer will come back").
5	Information about sleep hygiene (i.e., effect of environmental factors and health behaviors on sleep) and information about the appropriate usage of hypnotic medications.
6	Evaluation and maintenance of treatment gains and relapse prevention strategies.

Source: Adapted from Garland SN et al. Sleeping well with cancer: A systematic review of cognitive behavioral therapy for insomnia in cancer patients. *Neuropsychiatric Dis Treat* 2014; 10: 1113–1124; Savard J, Morin CM. Insomnia in the context of cancer: A review of a neglected problem. *J Clin Oncol* 2001; 19(3): 895–908.

among 150 mixed diagnosis cancer patients. CBT-I was also associated with increased quality of life and reduced daytime fatigue. Similar successful outcomes were reported in the seven RCTs.[77] In a recent comparative efficacy study of 96 cancer survivors, armodafinil (a psychostimulant) did not improve the reported efficacy (i.e., reductions in insomnia severity and improvements on the PSQI) when it was given in combination with CBT-I.[90] In addition, the efficacy of armodafinil alone was not very different from that shown by placebo alone. These results indicate that CBT-I is a mainstay and that there appears to be no unique or added benefit associated with this medication for treating cancer-related insomnia.

Mindfulness-based stress reduction

Other non-pharmacological interventions that have been studied for the treatment of insomnia include relaxation treatments and relaxation training (e.g., progressive muscle relaxation or autogenic training). Although relaxation training is sometimes a component of CBT-I and other interventions, CBT-I demonstrates larger effects compared to relaxation training as a standalone treatment.[77,93] One treatment that has garnered much attention is MBSR because of its efficacy for improving symptoms of stress, mood disturbance, and high levels of psychological distress.[94,95]

MBSR is one of the most studied and practiced forms of mindfulness. MBSR is a structured treatment modality that is utilized to decrease stress by means of psycho-education, mindfulness mediation, body scan sensory awareness experience, and gentle hatha yoga.[96] Typically, the treatment lasts 8 weeks (2.5-hour sessions) and often includes one full-day "silent" retreat.[97] Despite the primary goal of stress reduction, MBSR has been shown to be efficacious and beneficial for the reduction of numerous symptoms among individuals with many medical conditions, psychiatric and physical, including cancer (for reviews and meta-analyses, see[94,98–101]).

Examining the relationship between the global construct of mindfulness and symptoms of insomnia was motivated by Lundh's[102,103] earlier conceptualization of insomnia, which incorporates aspects of mindfulness. The core tenant of mindfulness is that thoughts and experiences are brought into the present moment in a nonjudgmental fashion.[98,104] Mindfulness can be applied more generally as a way of "being in the world" and more specifically as a discrete, pragmatic practice. As such, mindfulness practice can be applied in either formal or informal meditation practice, such as simply remembering to be present, and as a nonjudgmental attitude as one goes about their daily tasks. Lundh[102,103] explains that there is a theoretical relationship between mindfulness and insomnia, whereby mindfulness enables cognitive deactivation and in so doing decreases cognitive rumination and arousal, which in turn allows for improved sleep quality (i.e., decreasing the likelihood of insomnia symptoms). Moreover, the level of mindfulness would theoretically influence the impact that stress has on an individual. For example, if one were to have an accepting and nonjudgmental disposition, then one would have a decreased psychological and physiological response to stress.

In the context of cancer, the hypothesis that mindfulness is related to insomnia was first indirectly evaluated.[105,106] Earlier research showed that greater mindfulness is associated with decreased stress and improved sleep quality among mixed cancer populations (see Smith et al.[107] for a review).[105–111] Long-term follow-up studies examining physiological biomarkers demonstrated continuing decreases in overall cortisol levels following MBSR programs, primarily due to decreases in evening cortisol levels, which tend to support better sleep–wake patterns.[112]

Some of the earlier uncontrolled studies showed promise for MBSR as a sleep-enhancing intervention. Carlson et al.[105] evaluated a MBSR intervention among cancer patients (primarily breast cancer) and found significant improvements in overall quality of life and sleep quality. At the start of treatment, 40% of their sample reported poor sleep quality, whereas only 20% reported poor sleep quality at the end of treatment.[105] Similarly, Carlson and Garland[106] evaluated an 8-week MBSR intervention among a heterogeneous sample of cancer patients. They found

significant improvements in sleep quality (PSQI), stress (Symptoms of Stress Inventory), mood (the profile of mood states [POMS]), and fatigue (POMS subscale).[106] Given the lack of control groups in these studies, the improvements in reported sleep quality could be attributable to other factors of the MBSR intervention (e.g., psycho-education) or regression to the mean. Moreover, an earlier MBSR study that utilized a control group design did not find significant differences in sleep functioning among breast cancer patients in the MBSR group relative to a placebo group (whereby participants chose stress management techniques such as talking to a friend, exercise, or taking a warm bath).[113] Nevertheless, both groups produced significant improvements on daily diary sleep parameters, and the authors explain that the "control condition in retrospect had many active treatment features."[113] Furthermore, the authors found that even though being assigned to MBSR did not predict significant improvements, *practicing* mindfulness techniques was associated with feeling more rested (as reported through sleep diaries).

In a recent RCT, 317 adults with a non-metastatic cancer diagnosis who had completed chemotherapy and radiation treatments for at least 1 month and met the diagnostic criteria of clinical insomnia compared the efficacy of mindfulness-based cancer recovery (MBCR; an adaptation of MBSR for cancer patients) and CBT-I for improving sleep.[84,114] The study used a non-inferiority design to test whether the novel treatment for sleep (MBCR) was as good as the gold standard (CBT-I). The results indicated that both treatments had clinical efficacy on a variety of sleep outcome measures. MBCR was initially significantly inferior to CBT-I for improving insomnia severity immediately after the program terminated, but MBCR demonstrated non-inferiority at the 3-month follow-up. At follow-up, sleep-onset latency was reduced by 22 minutes in the CBT-I group compared to 14 minutes in the MBCR group. Total sleep time increased by 0.60 hours for CBT-I and 0.75 hours for MBCR. Both treatment groups reported reduced stress and mood disturbance, whereas only CBT-I participants experienced improvements in their sleep quality (on the PSQI) and dysfunctional sleep beliefs (on the Dysfunctional Beliefs and Attitudes about Sleep Scale). Importantly, the authors concluded that CBT-I was superior, as this treatment modality resolved sleep difficulties faster than MBCR.[84,114]

Physical exercise and yoga

Moderate-intensity aerobic exercise (e.g., walking or some forms of yoga) is another type of non-pharmacological treatment that can reduce insomnia symptoms in cancer patients.[115] Exercise may promote sleep and decrease insomnia symptoms by triggering a decline in body temperature in the evening (i.e., the thermogenic effect), impacting circadian rhythms, improving inflammatory immune responses or through the release of serotonin and the resulting amelioration of anxiety or depression.[116,117] In the general population, a handful of studies have suggested that walking and yoga can reduce insomnia.[116,118–120] In cancer patients, the risk of inactivity is greater as cancer treatments often lead to a reduction of physical activity, which results in physical deconditioning and the potential for impaired sleep.[117]

A systematic review by Howell et al.[24] examined best practices in the management of sleep disturbances in adults with cancer and identified a few studies that evaluated the efficacy of exercise therapy in cancer patients.[121–124] Three RCTs have examined home-based aerobic exercise interventions.[121–123] Payne et al.[121] evaluated a home-based moderate-intensity walking intervention (20 minutes a day/four times a week) in breast cancer patients who were receiving hormone therapy. They found that, compared to waitlisted controls, participants in the walking intervention had improved sleep on the PSQI, as well as shorter wake times and less movement as demonstrated by actigraphy.[121] Similarly, Tang et al.[122] found that, compared to usual care controls, a home-based walking intervention (30 minutes a day/three times a week) significantly improved sleep quality (on a translated PSQI) in a mixed group of cancer patients. Lastly, Sprod et al.[123] evaluated a home-based exercise program that included progressive-intensity walking and resistance training components for breast and prostate cancer patients during radiation therapy. They found better self-reported subjective sleep quality (on the PSQI), reduced sleep-onset latency, and increased sleep efficiency post-intervention in the exercise group versus the control group.[123]

Components of exercise have been incorporated into other non-pharmacological interventions. Most notably, gentle yoga is a predominant aspect of MBSR and is considered one of many "mindfulness modes of exercise."[105,125,126] Nevertheless, there has also been a number of evaluations of yoga as a standalone treatment.[125] For example, Cohen et al.[124] conducted a RCT on a Tibetan yoga intervention (of 7 weeks) in patients with lymphoma who were either receiving chemotherapy or had received chemotherapy within the past 12 months. They found that patients in the intervention group reported significantly lower sleep disturbance scores, including better subjective sleep quality (on the PSQI), shorter sleep latency, longer sleep duration, and less use of sleep medications compared to waitlisted controls.[124] Conversely, Bower et al.[127] evaluated the effects of yoga on sleep problems. Although they found that fatigue significantly declined as a result of yoga, they did not find significantly greater effects on subjectively assessed sleep (on the PSQI) in the yoga intervention compared to the health education control group.[127]

Recently, Mustian et al.[128] published a multi-center RCT that examined the effects of yoga on insomnia in mostly female cancer survivors suffering from a sleep disturbance at between 2 and 24 months after surgery, chemotherapy, and/or radiation therapy. The Yoga for Cancer Survivors (YOCAS) program consisted of two 75-minute weekly sessions of gentle hatha and restorative yoga for 4 weeks.[128] Individuals who participated in the YOCAS program (n = 206) showed a significantly greater reduction of PSQI scores at post-intervention compared with individuals receiving standard care (n = 204).[128] Yoga participants also significantly decreased their use of sleep medications by 21%, while control participants increased their use by 5%.[128] Moreover, greater improvements in wake after sleep onset and sleep efficiency, as measured objectively with actigraphy, were also found in yoga participants.[128]

An increasing number of clinical trials have found that cancer patients' sleep improved as a result of moderate-intensity aerobic exercise. However, few trials report statistically significant results.[24] Although the potential for yoga and walking interventions to assist with insomnia in cancer patients is very promising,[129] future research is required in larger and more diverse samples. Indeed, Howell et al.[24] conclude that "[w]hile exercise may have beneficial effects on sleep quality in individuals with cancer, the evidence is inconclusive due to quality biases and the lack of identification of clinically significant levels of insomnia at baseline." Hence, it is still unknown whether these interventions can treat clinical insomnia.

FUTURE DIRECTIONS FOR RESEARCH AND CLINICAL PRACTICE

One of the notable challenges for cancer patients is accessibility to effective insomnia treatments, as they may not want to remain at the cancer center beyond their medical treatment or they may be too weak to leave their homes. The number of treatment sessions can also be difficult for cancer patients who are receiving ongoing cancer treatments. Additionally, embedding CBT-I within a cancer hospital or center—where there is often a lack of mental health professionals trained in CBT-I and associated costs—makes implementing routine CBT-I problematic. These difficulties point to the opportunity for self-administered interventions.[8]

Earlier small-scale studies found a video- and an Internet-based CBT-I to be feasible and to lead to promising outcomes.[86,130] More recently, a RCT conducted in 242 women with breast cancer showed that a video-based CBT-I (VCBT-I) was a highly valuable alternative to a professionally administered CBT-I (PCBT-I), although PCBT-I proved to be the optimal delivery format.[142] More specifically, at post-treatment, both PCBT-I and VCBT-I were associated with significantly greater improvements in most subjective sleep variables and in dysfunctional beliefs about sleep as compared to a no-treatment condition. However, greater reductions of ISI scores, early-morning awakening, depression, fatigue, and dysfunctional beliefs about sleep were observed in PCBT-I patients as compared to VCBT-I patients.[142] Remission rates of insomnia were also larger in PCBT-I than in VCBT-I at that time point.[142]

In another study, Casault et al.[131] tested the efficacy of a self-administered minimal CBT-I (mCBT-I) intervention, offered in bibliotherapy format, combined with three brief phone consultations for treating acute insomnia comorbid with cancer (i.e., duration of less than 6 months). The results indicated that on all sleep measures (i.e., ISI, sleep-onset latency, sleep efficiency, total wake time, wake after sleep onset, and total sleep time) there was a significant improvement from before to after treatment for the mCBT-I group, whereas no significant changes were reported in the control group. There

was also a decrease in the average hypnotic medication usage from pre- to post-treatment. Therapeutic gains were maintained up to 6 months post-intervention. Interestingly, mCBT-I was also associated with a significant decrease of anxiety and depressive symptoms, maladaptive sleep habits and erroneous beliefs about sleep, as well as a significant improvement in subjective cognitive functioning. These results highlight that a minimal intervention delivered early on may suffice to treat acute insomnia in cancer patients. Future studies should examine whether mCBT-I could offer patients a reasonable, comprehensive and affordable intervention by examining not only the efficacy, but also the cost-effectiveness of these therapies in large RCTs.[8] Prevention studies using mCBT-I in high-risk individuals are also warranted.

This chapter has focused on CBT, MBSR, exercise, and yoga. In the future, more controlled studies with large sample sizes are required in order to evaluate the impact of alternative therapies that may improve insomnia in cancer patients, including: bright-light therapy, expressive therapy, expressive writing, healing touch, autogenic training, massage, acupuncture, aromatherapy, music therapy, hypnotherapy, and guided imagery.[24,132–134] A number of low-quality studies have investigated the evidence of these treatments; however, greater evidence is required for establishing these treatments as evidence-based psychotherapies.[134–138]

Future RCTs should be large enough to begin to investigate important differences in cancer stage, cancer type, gender, and other sociodemographics. It would also be important for future research to investigate cancer patients' preferences for how and where they would like to receive psychotherapeutic treatments for insomnia (i.e., the acceptability of interventions at different points of the cancer continuum). Furthermore, it is also important to better understand the acceptability of psychotherapies among oncologists. For example, McCall et al.[139] investigated the knowledge, attitudes, beliefs, and referral practices of oncologists in relation to yoga and found that only a small number of oncologists currently recommend yoga to their patients. Research that surveys patients and oncologists is important for better understanding the barriers to psychotherapeutic intervention for insomnia in cancer patients.

CBT-I and MBSR are multicomponent interventions, and it is therefore important to examine which elements are necessary to improve insomnia and sleep disturbance symptoms among cancer populations, particularly when resources are scarce and highly coveted. For example, it would be important to compare exercise or yoga alone compared to exercise that is incorporated within a full MBSR treatment. Furthermore, exercise and yoga are ubiquitous terms that describe extremely diverse practices.[129] For example, while Mustian et al.[117] identified 11 clinical trials of yoga for insomnia and sleep difficulties in cancer patients, the interventions studied varied tremendously in the number of classes offered, length of sessions, type of yoga postures and poses practiced, intensity of practice, and the population studied. Isolating components and subtypes of interventions that are especially potent will provide patients with the biggest payoff. It would also be useful to determine whether non-pharmacological interventions for insomnia could concurrently address other cancer-related symptoms that are associated with insomnia (e.g., hot flashes, fatigue, and pain) in order to increase treatment efficacy.[8]

Lastly, more clinical studies are needed in order to assess the efficacy of non-pharmacological interventions so as to improve the sleep of patients with advanced cancer. Patients with advanced cancer have high rates of sleep disturbances, which include insomnia and excessive daytime sleepiness.[49,140,141] These patients typically use a number of medications to treat various symptoms (e.g., pain), leading to several side effects. To be able to improve sleep without the recourse to another medication would be extremely advantageous for these patients.

CONCLUSION

This chapter provided an overview of the assessment and treatment of sleep disturbances in cancer patients. It demonstrated that insomnia is highly prevalent in patients with this medical condition. Given the significant toll of insomnia on the patients themselves, as well as on society, systematic screening of this condition in cancer clinics is crucial. Several tools are available, including some very brief ones. When insomnia is detected, it is important to offer an appropriate intervention. While pharmacotherapy is still the most commonly used treatment, CBT-I is considered the treatment of choice for chronic insomnia, and its efficacy in the cancer context is now well established. One

problem with CBT-I is its low accessibility. Self-administered interventions have been developed and have shown positive outcomes. Other non-pharmacological interventions have shown some promise for improving cancer-related sleep disturbances, but comparative studies with CBT-I, for instance, are lacking. In one of the few comparative studies, MBSR was found to produce significant sleep improvements, but these improvements were considered statistically inferior as compared to CBT-I at post-treatment. Some benefits have also been found with physical exercise, but studies have generally included patients who do not fit the criteria of clinical insomnia at baseline. Hence, it is unknown whether physical activity programs are potent enough to treat clinical levels of insomnia in cancer patients. Nonetheless, these alternative treatment options may be considered when patients do not wish to receive CBT-I, do not have access to CBT-I, or have a particular interest in mindfulness-based approaches or exercise/yoga.

REFERENCES

1. International Agency for Research on Cancer. *World Cancer Report 2014*. Geneva: World Health Organization, 2014.
2. Canadian Cancer Society. *Canadian Cancer Statistics 2013*. http://www.cancer.ca/~/media/cancer.ca/CW/cancer%20information/cancer%20101/Canadian%20cancer%20statistics/canadian-cancer-statistics-2013-EN.pdf. Accessed April 24, 2015.
3. World Health Organization. *Cancer*. http://www.cancer.gov/about-cancer/treatment/side-effects/sleep-disorders-hp-pdq#section/all. Accessed April 24, 2015.
4. World Cancer Research Fund International. *Comparing More and Less Developed Countries*. http://www.wcrf.org/int/cancer-facts-figures/comparing-more-less-developed-countries. Accessed April 24, 2015.
5. Canadian Cancer Society. *Cancer Statistics at a Glance*. http://www.cancer.ca/en/cancer-information/cancer-101/cancer-statistics-at-a-glance/. Accessed April 24, 2015.
6. Cancer Research UK. *Cancer Mortality in the UK in 2012*. http://publications.cancerresearchuk.org/downloads/Product/CS_REPORT_MORTALITY.pdf. Accessed April 24, 2014.

7. American Cancer Society. *Cancer Facts and Figures 2014.* http://www.cancer.org/research/cancerfactsstatistics/cancerfactsfigures2014/. Accessed April 24, 2015.

8. Savard J, Savard M-H. Insomnia and cancer: Prevalence, nature, and nonpharmacologic treatment. *Sleep Med Clin* 2013; 8(3): 373–387.

9. Ohayon MM. Epidemiology of insomnia: What we know and what we still need to learn. *Sleep Med Rev* 2002; 6(2): 97–111.

10. Morin CM, LeBlanc M, Daley M, Gregoire JP, Merette C. Epidemiology of insomnia: Prevalence, self-help treatments, consultations, and determinants of help-seeking behaviors. *Sleep Med* 2006; 7(2): 123–130.

11. Palesh OG, Roscoe JA, Mustian KM et al. Prevalence, demographics, and psychological associations of sleep disruption in patients with cancer: University of Rochester Cancer Center–Community Clinical Oncology Program. *J Clin Oncol* 2010; 28(2): 292–298.

12. Savard J, Morin CM. Insomnia in the context of cancer: A review of a neglected problem. *J Clin Oncol* 2001; 19(3): 895–908.

13. National Cancer Institute. *Sleep Disorders.* http://www.cancer.gov/about-cancer/treatment/side-effects/sleep-disorders-hp-pdq#section/all. Accessed April 24, 2015.

14. Davis MP, Goforth HW. Long-term and short-term effects of insomnia in cancer and effective interventions. *Cancer J* 2014; 20(5): 330–344.

15. Savard J, Villa J, Ivers H, Simard S, Morin CM. Prevalence, natural course, and risk factors of insomnia comorbid with cancer over a 2-month period. *J Clin Oncol* 2009; 27(31): 5233–5239.

16. Berger AM. Update on the state of the science: Sleep–wake disturbances in adult patients with cancer. *Oncol Nurs Forum* 2009; 36(4): E165–E177.

17. Davidson JR, MacLean AW, Brundage MD, Schulze K. Sleep disturbance in cancer patients. *Soc Sci Med* 2002; 54(9): 1309–1321.

18. Desai K, Mao JJ, Su I et al. Prevalence and risk factors for insomnia among breast cancer patients on aromatase inhibitors. *Support Care Cancer* 2013; 21(1): 43–51.

19. Trudel-Fitzgerald C, Savard J, Ivers H. Longitudinal changes in clusters of cancer patients over an 18-month period. *Health Psychol* 2014; 33(9): 1012–1022.

20. American Psychiatric Association. *Diagnostic and Statistical Manual of Mental Disorders (DSM-5®).* Washington, DC: American Psychiatric Association, 2013.

21. Dy SM, Apostol CC. Evidence-based approaches to other symptoms in advanced cancer. *Cancer J* 2010; 16(5): 507–513.

22. Sateia M. *The International Classification of Sleep Disorders: Diagnostic & Coding Manual.* Darien, IL: American Academy of Sleep Medicine, 2005.

23. Savard J, Ivers H, Villa J, Caplette-Gingras A, Morin CM. Natural course of insomnia comorbid with cancer: An 18-month longitudinal study. *J Clin Oncol* 2011; 29(26): 3580–3586.

24. Howell D, Oliver TK, Keller-Olaman S et al. Sleep disturbance in adults with cancer: A systematic review of evidence for best practices in assessment and management for clinical practice. *Ann Oncol* 2014; 25(4): 791–800.

25. Savard J, Simard S, Blanchet J, Ivers H, Morin CM. Prevalence, clinical characteristics, and risk factors for insomnia in the context of breast cancer. *Sleep* 2001; 24(5): 583–590.

26. Savard J, Simard S, Hervouet S, Ivers H, Lacombe L, Fradet Y. Insomnia in men treated with radical prostatectomy for prostate cancer. *Psychooncology* 2005; 14(2): 147–156.

27. Otte JL, Carpenter JS, Russell KM, Bigatti S, Champion VL. Prevalence, severity, and correlates of sleep–wake disturbances in long-term breast cancer survivors. *J Pain Symptom Manage* 2010; 39(3): 535–547.

28. Dahiya S, Ahluwalia MS, Walia HK. Sleep disturbances in cancer patients: Underrecognized and undertreated. *Cleve Clin J Med* 2013; 80(11): 722–732.

29. Otte JL, Carpenter JS, Manchanda S et al. Systematic review of sleep disorders in cancer patients: Can the prevalence of sleep disorders be ascertained? *Cancer Med* 2015; 4(2): 183–200.

30. Lee K, Cho M, Miaskowski C, Dodd M. Impaired sleep and rhythms in persons with cancer. *Sleep Med Rev* 2004; 8(3): 199–212.

31. Deimling GT, Kahana B, Bowman KF, Schaefer ML. Cancer survivorship and psychological distress in later life. *Psychooncology* 2002; 11(6): 479–494.

32. Caplette-Gingras A, Savard J, Savard MH, Ivers H. Is insomnia associated with cognitive impairments in breast cancer patients? *Behav Sleep Med* 2013; 11(4): 239–257.

33. Sivertsen B, Overland S, Bjorvatn B, Maeland JG, Mykletun A. Does insomnia predict sick leave? The Hordaland Health Study. *J Psychosom Res* 2009; 66(1): 67–74.

34. Bardwell WA, Profant J, Casden DR et al. The relative importance of specific risk factors for insomnia in women treated for early-stage breast cancer. *Psychooncology* 2008; 17(1): 9–18.

35. Irwin MR. Depression and insomnia in cancer: Prevalence, risk factors, and effects on cancer outcomes. *Curr Psychiatry Rep* 2013; 15(11): 404.

36. Baglioni C, Battagliese G, Feige B et al. Insomnia as a predictor of depression: A meta-analytic evaluation of longitudinal epidemiological studies. *J Affect Disord* 2011; 135(1–3): 10–19.

37. Fortner BV, Stepanski EJ, Wang SC, Kasprowicz S, Durrence HH. Sleep and quality of life in breast cancer patients. *J Pain Symptom Manage* 2002; 24(5): 471–480.

38. Chen ML, Yu CT, Yang CH. Sleep disturbances and quality of life in lung cancer patients undergoing chemotherapy. *Lung Cancer* 2008; 62(3): 391–400.

39. Pachman DR, Barton DL, Swetz KM, Loprinzi CL. Troublesome symptoms in cancer survivors: Fatigue, insomnia, neuropathy, and pain. *J Clin Oncol* 2012; 30(30): 3687–3696.

40. Engstrom CA, Strohl RA, Rose L, Lewandowski L, Stefanek ME. Sleep alterations in cancer patients. *Cancer Nurs* 1999; 22(2): 143–148.

41. Siefert ML, Hong F, Valcarce B, Berry DL. Patient and clinician communication of self-reported insomnia during ambulatory cancer care clinic visits. *Cancer Nurs* 2014; 37(2): E51–E59.

42. Wilson SJ, Nutt DJ, Alford C et al. British Association for Psychopharmacology consensus statement on evidence-based treatment of insomnia, parasomnias and circadian rhythm disorders. *J Psychopharmacol* 2010; 24(11): 1577–1601.

43. Yue HJ, Dimsdale JE. Sleep and cancer. In: Holland JC, Breitbart W, Jacobsen PB, Marguerite SL, Loscalzo MJ, McCorkle R (Eds). *Psycho-oncology*. New York, NY: Oxford University Press, 2010; pp. 258–269.

44. Koul R, Dubey A. Insomnia in oncology; an overview. *Internet J Pain Symptom Control Palliat Care* 2009; 7(1): 1–8.

45. Everitt DE, Avorn J, Baker MW. Clinical decision-making in the evaluation and treatment of insomnia. *Am J Med* 1990; 89(3): 357–362.

46. Bultz B, Groff S, Fitch M. Screening for distress toolkit working group. Canadian problem checklist. In: *The Guide to Implementing Screening for Distress, the 6th Vital Sign: Moving Toward Person-centered Care, Part A: Background, Recommendations, and Implementation.* Toronto, ON: Canadian Partnership against Cancer, 2009, 34.

47. Bruera E, Kuehn N, Miller MJ, Selmser P, Macmillan K. The Edmonton Symptom Assessment System (ESAS): A simple method for the assessment of palliative care patients. *J Palliat Care* 1991; 7(2): 6–9.

48. Nekolaichuk C, Watanabe S, Beaumont C. The Edmonton Symptom Assessment System: A 15-year retrospective review of validation studies (1991–2006). *Palliat Med* 2008; 22(2): 111–122.

49. Delgado-Guay M, Yennurajalingam S, Parsons H, Palmer JL, Bruera E. Association between self-reported sleep disturbance and other symptoms in patients with advanced cancer. *J Pain Symptom Manage* 2011; 41(5): 819–827.

50. Buysse DJ, Reynolds CF 3rd, Monk TH, Berman SR, Kupfer DJ. The Pittsburgh Sleep Quality Index: A new instrument for psychiatric practice and research. *Psychiatry Res* 1989; 28(2): 193–213.

51. Savard MH, Savard J, Simard S, Ivers H. Empirical validation of the Insomnia Severity Index in cancer patients. *Psychooncology* 2005; 14(6): 429–441.

52. Morin CM. *Insomnia: Psychological Assessment and Management.* New York, NY: Guilford Press, 1993.

53. Savard J, Simard S, Ivers H, Morin CM. Randomized study on the efficacy of cognitive-behavioral therapy for insomnia secondary to breast cancer, part I: Sleep and psychological effects. *J Clin Oncol* 2005; 23(25): 6083–6096.

54. Morin CM. *Insomnia Severity Index.* http://www.seinstitute.com/wp-content/uploads/2012/03/Insomnia-Index.pdf. Accessed May 4, 2015.

55. Bastien CH, Vallieres A, Morin CM. Validation of the Insomnia Severity Index as an outcome measure for insomnia research. *Sleep Med* 2001; 2(4): 297–307.

56. Smith S, Trinder J. Detecting insomnia: Comparison of four self-report measures of sleep in a young adult population. *J Sleep Res* 2001; 10(3): 229–235.

57. Derogatis LR, Morrow GR, Fetting J et al. The prevalence of psychiatric disorders among cancer patients. *JAMA* 1983; 249(6): 751–757.

58. Backhaus J, Junghanns K, Broocks A, Riemann D, Hohagen F. Test–retest reliability and validity of the Pittsburgh Sleep Quality Index in primary insomnia. *J Psychosom Res* 2002; 53(3): 737–740.

59. Beck SL, Schwartz AL, Towsley G, Dudley W, Barsevick A. Psychometric evaluation of the Pittsburgh Sleep Quality Index in cancer patients. *J Pain Symptom Manage* 2004; 27(2): 140–148.

60. Carney CE, Buysse DJ, Ancoli-Israel S et al. The consensus sleep diary: Standardizing prospective sleep self-monitoring. *Sleep* 2012; 35(2): 287–302.

61. Morin CM, Barlow DH. *Insomnia: Psychological Assessment and Management.* New York, NY: Guilford Press, 1993.

62. Edinger J, Wyatt J, Olsen M et al. Reliability and validity of the Duke Structured Interview for Sleep Disorders for insomnia screening. *Sleep* 2009; 32: A265–A265.

63. Schutte-Rodin S, Broch L, Buysse D, Dorsey C, Sateia M. Clinical guideline for the evaluation and management of chronic insomnia in adults. *J Clin Sleep Med* 2008; 4(5): 487–504.

64. Nordin K, Berglund G, Glimelius B, Sjoden PO. Predicting anxiety and depression among cancer patients: A clinical model. *Eur J Cancer* 2001; 37(3): 376–384.

65. Casault L, Savard J, Ivers H, Savard MH, Simard S. Utilization of hypnotic medication in the context of cancer: Predictors and frequency of use. *Support Care Cancer* 2012; 20(6): 1203–1210.

66. National Institutes of Health. National Institutes of Health State of the Science Conference statement on Manifestations and Management of Chronic Insomnia in Adults, June 13–15, 2005. *Sleep* 2005; 28(9): 1049–1057.

67. Morgenthaler T, Kramer M, Alessi C et al. Practice parameters for the psychological and behavioral treatment of insomnia: An update. An American Academy of Sleep Medicine report. *Sleep* 2006; 29(11): 1415–1419.

68. Morin CM, Bootzin RR, Buysse DJ, Edinger JD, Espie CA, Lichstein KL. Psychological and behavioral treatment of insomnia: Update of the recent evidence (1998–2004). *Sleep* 2006; 29(11): 1398–1414.

69. Montgomery P, Dennis J. Cognitive behavioural interventions for sleep problems in adults aged 60+. *Cochrane Database Syst Rev* 2003; 1: CD003161.

70. Morin CM, Culbert JP, Schwartz SM. Nonpharmacological interventions for insomnia: A meta-analysis of treatment efficacy. *Am J Psychiatry* 1994; 151(8): 1172–1180.

71. Murtagh DR, Greenwood KM. Identifying effective psychological treatments for insomnia: A meta-analysis. *J Consult Clin Psychol* 1995; 63(1): 79–89.

72. Smith MT, Perlis ML, Park A et al. Comparative meta-analysis of pharmacotherapy and behavior therapy for persistent insomnia. *Am J Psychiatry* 2002; 159(1): 5–11.

73. Irwin MR, Cole JC, Nicassio PM. Comparative meta-analysis of behavioral interventions for insomnia and their efficacy

in middle-aged adults and in older adults 55+ years of age. *Health Psychol* 2006; 25(1): 3–14.

74. Roscoe JA, Kaufman ME, Matteson-Rusby SE et al. Cancer-related fatigue and sleep disorders. *Oncologist* 2007; 12(Suppl 1): 35–42.

75. Edinger JD, Hoelscher TJ, Marsh GR, Lipper S, Ionescu-Pioggia M. A cognitive–behavioral therapy for sleep-maintenance insomnia in older adults. *Psychol Aging* 1992; 7(2): 282–289.

76. Morin CM, Kowatch RA, Barry T, Walton E. Cognitive–behavior therapy for late-life insomnia. *J Consult Clin Psychol* 1993; 61(1): 137–146.

77. Garland SN, Johnson JA, Savard J et al. Sleeping well with cancer: A systematic review of cognitive behavioral therapy for insomnia in cancer patients. *Neuropsychiatric Dis Treat* 2014; 10: 1113–1124.

78. Harvey AG. Pre-sleep cognitive activity: A comparison of sleep-onset insomniacs and good sleepers. *Br J Clin Psychol* 2000; 39(Pt 3): 275–286.

79. Harvey AG, Tang NK. (Mis)perception of sleep in insomnia: A puzzle and a resolution. *Psychol Bull* 2012; 138(1): 77–101.

80. Davidson JR, Waisberg JL, Brundage MD, MacLean AW. Nonpharmacologic group treatment of insomnia: A preliminary study with cancer survivors. *Psychooncology* 2001; 10(5): 389–397.

81. Simeit R, Deck R, Conta-Marx B. Sleep management training for cancer patients with insomnia. *Support Care Cancer* 2004; 12(3): 176–183.

82. Quesnel C, Savard J, Simard S, Ivers H, Morin CM. Efficacy of cognitive–behavioral therapy for insomnia in women treated for nonmetastatic breast cancer. *J Consult Clin Psychol* 2003; 71(1): 189–200.

83. Dirksen SR, Epstein DR. Efficacy of an insomnia intervention on fatigue, mood and quality of life in breast cancer survivors. *J Adv Nurs* 2008; 61(6): 664–675.

84. Garland SN, Carlson LE, Stephens AJ, Antle MC, Samuels C, Campbell TS. Mindfulness-based stress reduction compared with cognitive behavioral therapy for the treatment of insomnia comorbid with cancer: A randomized, partially blinded, noninferiority trial. *J Clin Oncol* 2014; 32(5): 449–457.

85. Fiorentino L, McQuaid JR, Liu L et al. Individual cognitive behavioral therapy for insomnia in breast cancer survivors: A randomized controlled crossover pilot study. *Nat Sci Sleep* 2010; 2: 1–8.

86. Ritterband LM, Bailey ET, Thorndike FP, Lord HR, Farrell-Carnahan L, Baum LD. Initial evaluation of an Internet intervention to improve the sleep of cancer survivors with insomnia. *Psychooncology* 2012; 21(7): 695–705.

87. Berger AM, Kuhn BR, Farr LA et al. Behavioral therapy intervention trial to improve sleep quality and cancer-related fatigue. *Psychooncology* 2009; 18(6): 634–646.

88. Espie CA, Fleming L, Cassidy J et al. Randomized controlled clinical effectiveness trial of cognitive behavior therapy compared with treatment as usual for persistent insomnia in patients with cancer. *J Clin Oncol* 2008; 26(28): 4651–4658.

89. Savard J, Simard S, Ivers H, Morin CM. Randomized study on the efficacy of cognitive–behavioral therapy for insomnia secondary to breast cancer, part II: Immunologic effects. *J Clin Oncol* 2005; 23(25): 6097–6106.

90. Roscoe JA, Garland SN, Heckler CE et al. Randomized placebo-controlled trial of cognitive behavioral therapy and armodafinil for insomnia after cancer treatment. *J Clin Oncol* 2015; 33(2): 165–171.

91. Espie CA. Insomnia: Conceptual issues in the development, persistence, and treatment of sleep disorder in adults. *Annu Rev Psychol* 2002; 53: 215–243.

92. Howell D, Oliver TK, Keller-Olaman S et al. A pan-Canadian practice guideline: Prevention, screening, assessment, and treatment of sleep disturbances in adults with cancer. *Support Care Cancer* 2013; 21(10): 2695–2706.

93. Edinger JD, Wohlgemuth WK, Radtke RA, Marsh GR, Quillian RE. Cognitive behavioral therapy for treatment of

chronic primary insomnia: A randomized controlled trial. *JAMA* 2001; 285(14): 1856–1864.

94. Chiesa A, Serretti A. Mindfulness based cognitive therapy for psychiatric disorders: A systematic review and meta-analysis. *Psychiatry Res* 2011; 187(3): 441–453.

95. Carlson LE. Mindfulness-based interventions for physical conditions: A narrative review evaluating levels of evidence. *ISRN Psychiatry* 2012; 2012: 651583.

96. Kabat-Zinn J, Hanh TN. *Full Catastrophe Living: Using the Wisdom of Your Body and Mind to Face Stress, Pain, and Illness.* New York, NY: Bantam Dell - Random House, Inc.; 2005.

97. Kabat-Zinn J, Massion AO, Kristeller J et al. Effectiveness of a meditation-based stress reduction program in the treatment of anxiety disorders. *Am J Psychiatry* 1992; 149(7): 936–943.

98. Baer RA. Mindfulness training as a clinical intervention: A conceptual and empirical review. *Clin Psychol Sci Pract* 2003; 10(2): 125–143.

99. Ledesma D, Kumano H. Mindfulness-based stress reduction and cancer: A meta-analysis. *Psychooncology* 2009; 18(6): 571–579.

100. Grossman P, Niemann L, Schmidt S, Walach H. Mindfulness-based stress reduction and health benefits. A meta-analysis. *J Psychosom Res* 2004; 57(1): 35–43.

101. Hofmann SG, Sawyer AT, Witt AA, Oh D. The effect of mindfulness-based therapy on anxiety and depression: A meta-analytic review. *J Consult Clin Psychol* 2010; 78(2): 169.

102. Lundh L-G. The role of acceptance and mindfulness in the treatment of insomnia. *J Cogn Psychother* 2005; 19(1): 29–39.

103. Lundh L-G. An integrative model for the analysis and treatment of insomnia. *Scand J Behav Ther* 2000; 29(3–4): 118–126.

104. Hayes SC, Follette VM, Linehan M. *Mindfulness and Acceptance: Expanding the Cognitive-Behavioral Tradition.* New York, NY: Guilford Press, 2004.

105. Carlson LE, Speca M, Patel KD, Goodey E. Mindfulness-based stress reduction in relation to quality of life, mood, symptoms of stress, and immune parameters in breast and prostate cancer outpatients. *Psychosom Med* 2003; 65(4): 571–581.

106. Carlson LE, Garland SN. Impact of mindfulness-based stress reduction (MBSR) on sleep, mood, stress and fatigue symptoms in cancer outpatients. *Int J Behav Med* 2005; 12(4): 278–285.

107. Smith JE, Richardson J, Hoffman C, Pilkington K. Mindfulness-based stress reduction as supportive therapy in cancer care: Systematic review. *J Adv Nurs* 2005; 52(3): 315–327.

108. Lerman R, Jarski R, Rea H, Gellish R, Vicini F. Improving symptoms and quality of life of female cancer survivors: A randomized controlled study. *Ann Surg Oncol* 2012; 19(2): 373–378.

109. Branstrom R, Kvillemo P, Moskowitz JT. A randomized study of the effects of mindfulness training on psychological well-being and symptoms of stress in patients treated for cancer at 6-month follow-up. *Int J Behav Med* 2012; 19(4): 535–542.

110. Henderson VP, Clemow L, Massion AO, Hurley TG, Druker S, Hebert JR. The effects of mindfulness-based stress reduction on psychosocial outcomes and quality of life in early-stage breast cancer patients: A randomized trial. *Breast Cancer Res Treat* 2012; 131(1): 99–109.

111. Hoffman CJ, Ersser SJ, Hopkinson JB, Nicholls PG, Harrington JE, Thomas PW. Effectiveness of mindfulness-based stress reduction in mood, breast- and endocrine-related quality of life, and well-being in stage 0 to III breast cancer: A randomized, controlled trial. *J Clin Oncol* 2012; 30(12): 1335–1342.

112. Carlson LE, Campbell TS, Garland SN, Grossman P. Associations among salivary cortisol, melatonin, catecholamines, sleep quality and stress in women with breast cancer and healthy controls. *J Behav Med* 2007; 30(1): 45–58.

113. Shapiro SL, Bootzin RR, Figueredo AJ, Lopez AM, Schwartz GE. The efficacy of mindfulness-based stress reduction in the treatment of sleep disturbance in women with breast cancer: An exploratory study. *J Psychosom Res* 2003; 54(1): 85–91.

114. Garland SN, Carlson LE, Antle MC, Samuels C, Campbell T. I-CAN SLEEP: Rationale and design of a non-inferiority RCT of mindfulness-based stress reduction and cognitive behavioral therapy for the treatment of insomnia in CANcer survivors. *Contemp Clin Trials* 2011; 32(5): 747–754.

115. Youngstedt SD. Effects of exercise on sleep. *Clin Sports Med* 2005; 24(2): 355–365, xi.

116. Passos GS, Poyares DLR, Santana M, Tufik S, de Mello M. Is exercise an alternative treatment for chronic insomnia? *Clinics* 2012; 67(6): 653–659.

117. Mustian KM, Janelsins M, Peppone LJ, Kamen C. Yoga for the treatment of insomnia among cancer patients: Evidence, mechanisms of action, and clinical recommendations. *Oncol Hematol Rev* 2014; 10(2): 164–168.

118. Passos GS, Poyares D, Santana MG et al. Effects of moderate aerobic exercise training on chronic primary insomnia. *Sleep Med* 2011; 12(10): 1018–1027.

119. King AC, Pruitt LA, Woo S et al. Effects of moderate-intensity exercise on polysomnographic and subjective sleep quality in older adults with mild to moderate sleep complaints. *J Gerontol A Biol Sci Med Sci* 2008; 63(9): 997–1004.

120. Newton KM, Reed SD, Guthrie KA et al. Efficacy of yoga for vasomotor symptoms: A randomized controlled trial. *Menopause* 2014; 21(4): 339–346.

121. Payne JK, Held J, Thorpe J, Shaw H. Effect of exercise on biomarkers, fatigue, sleep disturbances, and depressive symptoms in older women with breast cancer receiving hormonal therapy. *Oncol Nurs Forum* 2008; 35(4): 635–642.

122. Tang MF, Liou TH, Lin CC. Improving sleep quality for cancer patients: Benefits of a home-based exercise intervention. *Support Care Cancer* 2010; 18(10): 1329–1339.

123. Sprod LK, Palesh OG, Janelsins MC et al. Exercise, sleep quality, and mediators of sleep in breast and prostate cancer patients receiving radiation therapy. *Community Oncol* 2010; 7(10): 463–471.

124. Cohen L, Warneke C, Fouladi RT, Rodriguez MA, Chaoul-Reich A. Psychological adjustment and sleep quality in a randomized trial of the effects of a Tibetan yoga intervention in patients with lymphoma. *Cancer* 2004; 100(10): 2253–2260.

125. Mustian KM. Yoga as treatment for insomnia among cancer patients and survivors: A systematic review. *Eur Med J Oncol* 2013; 1: 106–115.

126. Elkins G, Fisher W, Johnson A. Mind–body therapies in integrative oncology. *Curr Treat Options Oncol* 2010; 11(3–4): 128–140.

127. Bower JE, Garet D, Sternlieb B et al. Yoga for persistent fatigue in breast cancer survivors: A randomized controlled trial. *Cancer* 2012; 118(15): 3766–3775.

128. Mustian KM, Sprod LK, Janelsins M et al. Multicenter, randomized controlled trial of yoga for sleep quality among cancer survivors. *J Clin Oncol* 2013; 31(26): 3233–3241.

129. DiStasio SA. Integrating yoga into cancer care. *Clin J Oncol Nurs* 2008; 12(1): 125–130.

130. Savard J, Villa J, Simard S, Ivers H, Morin CM. Feasibility of a self-help treatment for insomnia comorbid with cancer. *Psychooncology* 2011; 20(9): 1013–1019.

131. Casault L, Savard J, Ivers H, Savard MH. A randomized-controlled trial of an early minimal cognitive–behavioural therapy for insomnia comorbid with cancer. *Behav Res Ther* 2015; 67: 45–54.

132. Harsora P, Kessmann J. Nonpharmacologic management of chronic insomnia. *Am Fam Physician* 2009; 79(2): 125–130.

133. Cheuk DK, Yeung WF, Chung KF, Wong V. Acupuncture for insomnia. *Cochrane Database Syst Rev* 2012; 9: CD005472.

134. Farrell-Carnahan L, Ritterband LM, Bailey ET, Thorndike FP, Lord HR, Baum LD. Feasibility and preliminary efficacy of a self-hypnosis intervention available on the web for cancer survivors with insomnia. *E-J Appl Psychol* 2010; 6(2): 10.

135. Kashani F, Kashani P. The effect of massage therapy on the quality of sleep in breast cancer patients. *Iran J Nurs Midwifery Res* 2014; 19(2): 113–118.

136. Cerrone R, Giani L, Galbiati B et al. Efficacy of HT 7 point acupressure stimulation in the treatment of insomnia in cancer patients and in patients suffering from disorders other than cancer. *Minerva Med* 2008; 99(6): 535–537.

137. Jeste N, Liu L, Rissling M et al. Prevention of quality-of-life deterioration with light therapy is associated with changes in fatigue in women with breast cancer undergoing chemotherapy. *Quality Life Res* 2013; 22(6): 1239–1244.

138. Neikrug AB, Rissling M, Trofimenko V et al. Bright light therapy protects women from circadian rhythm desynchronization during chemotherapy for breast cancer. *Behav Sleep Med* 2012; 10(3): 202–216.

139. McCall MC, Ward A, Heneghan C. Yoga in adult cancer: A pilot survey of attitudes and beliefs among oncologists. *Curr Oncol* 2015; 22(1): 13–19.

140. Sela RA, Watanabe S, Nekolaichuk CL. Sleep disturbances in palliative cancer patients attending a pain and symptom control clinic. *Palliat Support Care* 2005; 3(1): 23–31.

141. Akechi T, Okuyama T, Akizuki N et al. Associated and predictive factors of sleep disturbance in advanced cancer patients. *Psychooncology* 2007; 16(10): 888–894.

142. Savard J, Ivers H, Savard M-H, Morin CM. Is a video-based cognitive behavioral therapy for insomnia as efficacious as a professionally administered treatment in breast cancer? Results of a randomized controlled trial. *Sleep* 2014; 37(8): 1305–1314.

Index